WITHDRA

THE
CORROSION AND OXIDATION
OF METALS
SECOND SUPPLEMENTARY VOLUME

ULICK R. EVANS

M.A., Sc.D., F.R.S., C.B.E.

Honorary Fellow, King's College, Cambridge
Emeritus Reader in the Science of Metallic Corrosion,
University of Cambridge

EDWARD ARNOLD

First published 1976 by Edward Arnold (Publishers) Ltd
25 Hill Street, London W1X 8LL

ISBN: 0 7131 2536 5

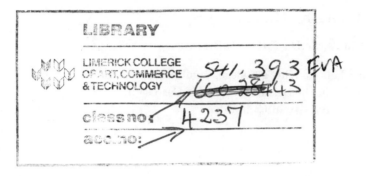
To T. P. Hoar, Sc.D., PhD. and J. E. O. Mayne, D.Sc.,
in appreciation of their scientific achievements and personal friendship.

Printed in Great Britain by
Butler & Tanner Ltd., Frome and London

Preface

The 1106 page book entitled 'The Corrosion and Oxidation of Metals' was published in 1960. By 1968, results of much new work were available and the reader required a new edition or a supplementary volume to bring his knowledge up to date. It was decided that the provision of a supplementary volume would be more beneficial than a new edition (which would inevitably have contained much material from the 1960 book reprinted unchanged). The supplement policy represents economy of money, time and shelf space. The plan of issuing supplements is clearly even more sensible at the present time and has been applauded by reviewers of the 1968 book. However, there is evidence that many readers, and even some librarians, may be unfamiliar with the use of supplementary volumes. The intended mode of usage is made clear in the Publisher's Note (p. vii).

The second supplement consists especially of new information published during the period 1968–75, but a few early researches, which today appear to possess an importance not realized in the year of publication, are recalled in summarized form. The need for such a volume will be obvious to anyone who has read the findings of the Hoar Committee, embodied in their 'Report of the committee on Corrosion and Protection' (H.M.S.O. 1971); this report states that in the UK alone over £310M could be saved annually if the knowledge available to corrosion specialists was put to practical use. Since the Hoar Committee held their last session, the situation has developed in two ways. First, the economic plight of the UK and many other countries has made it highly desirable that every means of economy shall in future be utilized. Secondly, fresh researches have produced a wealth of information of greater potential value than those made available by the researches of 1960–68 but also more inaccessible; some of the most significant findings are recorded in papers printed in 'Conference Volumes'— the very existence of of which is unknown to many of those whom the papers should interest.

In the present (second) supplement, the buried treasure has been disinterred; the papers printed in conference volumes and various specialized journals not read by the ordinary engineer or technologist have been summarized, and the information, systematically arranged, made available to readers.

The tendency for papers published today to carry the names of several authors (six authors to one paper is not uncommon) poses a serious problem to the writer of a new book. If all these authors are mentioned in the text of the book, the reader is presented with a mass of names mostly meaning nothing to him. Yet to omit the names of all those whose careful observations have provided essential material for the book, would be utterly unjust; one argument, sometimes put forward to excuse the omission of references, is that the average reader will not look them up; this is only too true if the absurd plan is adopted by which the reference list is placed at the end of each chapter so that reading must be interrupted and a veritable research undertaken to locate it. In the present book, it is believed that every experimenter's name is quoted in an easily accessible footnote reference—as is obviously his due, since without the experimental work the book could not have been written at all. In cases where there are more than two authors to a given paper, I have generally quoted in the text only the person whom I believe (perhaps mistakenly) to have been the guiding spirit; I have tried to bring in names which will be known to the reader, or are likely to become known. Where it has seemed best to mention no names of persons, I have quoted the laboratory or the country of research. To anyone who feels that these plans do less than perfect justice to the experimenters, I offer my sincere apologies.

If at all possible, the corresponding chapters in the 1960 and 1968 books should be read; it may be recollected that the chapter titles of the 1960 book were retained in the 1968 supplement. The same titles have been used in the second supplement, except for small changes in the case of chapters IX and XIX, and the disappearance of the mathematical chapters XX, XXI and XXII; in presenting a research where the author has provided mathematical treatment for his results, it has seemed best to place this in the chapter devoted to the particular type of corrosion studied.

The new chapter XIX differs from all others in containing unpublished experimental work on interference colours; it is designed to explore certain anomalies brought out by the late Dr Vernon, which may have puzzled those who would be inclined to use interference colours as a rough guide to the thickening of film; it also provides a simple explanation of the fact that some lack of perfection in the film (imperfect transparency, imperfect reflection or imperfect smoothness) is needed for the attainment of bright colours. Discussion with friends well qualified to express an opinion led to the conclusion that the information is more likely to reach potentially interested persons if published in the book rather than if presented in a journal.

Every effort has been made to include references to recent work, but in the case of the latest papers it has only been possible to quote the title and reference.

I desire to thank all those who have rendered valuable help by (1) supplying information, (2) granting permission for the use of diagrams based on those in their papers and (3) reading and commenting on sections of the new volume submitted in draft; their cooperation has, it is believed, contributed very greatly towards achieving what it is hoped will be a result acceptable to the reader. A list of some who have helped in these three ways are given in the acknowledgement list on p. vi; it is feared that it may be incomplete and I would ask those whose names are omitted, to forgive me and to accept my thanks. Very special acknowledgement is due to Dr T. Hurlen, Dr E. Heitz, Dr M. J. Pryor and Dr I. D. G. Berwick, who have given permission for work to be quoted in detail from their important papers. Finally, I would acknowledge the help of Mrs E. E. Johnson and Miss B. M. Gorse in connection with the typing.

U. R. E.

University Department of Metallurgy
Pembroke Street
Cambridge
1975

Acknowledgements

Approval for the reproduction of diagrams has kindly been given by M. J. Blackburn, R. Bruno, M. Cohen, M. G. Fontana, C. S. Giggins, J. F. Henriksen, S. Hettiarchchi, T. P. Hoar, H. V. Hyatt, G. Okamoto, R. N. Parkins, F. S. Pettit, I. M. Ritchie, S. Schuldiner, J. C. Scully, M. O. Speidel, R. W. Staehle, Z. Szklarska-Smialowska, G. C. Wood.

The following friends have kindly read chapters or sections or have supplied valuable information:

Miss D. M. Brasher, S. C. Britton, K. A. Chandler, J. B. Cotton, T. E. Evans, H. Grubitsch, T. P. Hoar, L. Horner, B. B. Hundy, J. Kruger, C. Maitland, N. Maruko, J. E. O. Mayne, M. Pourbaix, N. Sato, J. F. Stanners, C. A. J. Taylor, J. E. Truman, R. B. Waterhouse, A. C. Wilson.

Publisher's Note

The three volumes of 'The Corrosion and Oxidation of Metals' (1960, 1968 and 1976) should be kept together on the shelf and read together. If a reader desires information on some branch of corrosion he will consult the table of contents and select the appropriate chapter. If he is interested in Atmospheric Corrosion this will be chapter XIII and the 1960 book should be opened at that chapter, with the first and second supplements, also open at chapter XIII, placed to the left and right. This will give the reader an up-to-date picture of the subject, not merely empirical but based on scientific principles.

It is recognized, however, that many readers who wish to become acquainted with recent work will possess knowledge based on other texts, which may not in all cases give prominence to the basic principles necessary for an understanding of material in this second supplement. To help them, a preamble has been provided for each chapter, summarizing the state of knowledge in 1968 and providing the principles assumed to be familiar to anyone who wishes to understand the chapter in question. The reader whose time is limited may choose to read all the preambles and then read the text of those chapters which he judges to be pertinent to his own problems.

Contents

A*

Conference Volumes Quoted in this Book

with Abbreviations adopted in Footnotes

'First International Congress on Metallic Corrosion' held in London in 1961. Report published by Butterworths (London) in 1962 [Abbreviated as *'London Cong.'*]

'Fourth International Congress on Metallic Corrosion' held in Amsterdam in 1969. Report published by National Association of Corrosion Engineers (Houston, Texas) in 1972. [Abbreviated as *'Amsterdam Cong.'*]

'Third European Symposium on Corrosion Inhibitors' held at Ferrara in 1970. Report published by Ferrara University in 1971. [Abbreviated as *'Ferrara Symp.'*]

Conference on Localized Corrosion, held at Williamsburg, Virginia, Dec. 6–10, 1971. Volume published by National Association of Corrosion Engineers in 1972 (NACE 3). Editors R. W. Staehle, B. F. Brown, J. Kruger and A. Agrawal. [Abbreviated as *'Williamsburg Conf.'*]

'Fifth International Congress on Metallic Corrosion', held in Tokyo in 1972. Extended Abstracts. [Abbreviated as *'Tokyo Cong.'* *Ext. Abs.*]

CHAPTER I

The Approaches to Corrosion

Preamble

Picture presented in the 1960 and 1968 volumes. The first chapter of the 1960 book opens with an explanation of such terms as *Corrosion, Erosion* and *Oxidation*. It is shown that the subject of Corrosion in its widest sense possesses potential interest to several classes of persons, including Pure Scientists, Engineers, Architects, Economists, and Health Authorities. Attention is then turned to the various manners of regarding the subject, which presents a different appearance according to the angle from which it is viewed. Different ways of regarding Corrosion will appeal to different types of mind, but there is benefit to be gained by each from an endeavour to understand alternative aspects of the subject and thus gain sympathy with those whose approach is different.

Perhaps the most remarkable feature of the subject is this. At the start of the century, the facts of Corrosion—so far as they were known at all—appeared to be a confused jumble, without rhyme or reason. Scientific research has, however, shown that these apparently discordant events really obey simple laws, capable of being interpreted on rational principles and susceptible of representation, in many cases, by simple mathematical expressions. The discovery of law and order in what had previously appeared to be chaos makes an appeal to the intellect, and Corrosion would seem to have a strong claim for inclusion in a scientific curriculum, even if there were no utilitarian reason for its study. In fact, however, a strong argument for teaching Corrosion can be based on economic grounds; Vernon's classical calculation of the annual cost of preventive measures and metal losses in the UK alone as being of the order of 600 million pounds is today regarded as an underestimate. Why then is it that, whilst reactions representing combination of different non-metals are seriously studied by the Organic Chemist, and the alloying of different metals receives detailed attention from the Metallurgist, little consideration is given to the combination of metals with nonmetals? Perhaps the unfortunate name ('Corrosion') attached to the subject may have something to do with its neglect, but the main reason is probably a lack of time for its study; in all branches of education a large number of fresh subjects are clamouring for a share in the timetable. It is a matter of regret rather than of surprise, that, although it is widely agreed that anticorrosion measures ought to start at the drawing board, the engineering designer of today is often someone who has received no instruction in the principles of Corrosion. For this he can hardly be blamed, but those responsible for drawing up an engineering curriculum should surely give more attention to the claims of Corrosion. If corrosive influences are not taken into account by designers, accurate calculations based on purely mechanical principles can lead to most dangerous situations.

Developments during the period 1968–1975 and Plan of the new Chapter. The main event of the period under review has been the publication of the Hoar report, which has dealt in detail with most of the matters discussed briefly in the first chapter of the 1960 and 1968 books. The subjects discussed in the report include the education problem—the importance of bringing existing knowledge of Corrosion to the notice of Engineers and others; the situation at present is far from satisfactory. The report also emphasizes economic possibilities—the savings which could be achieved if existing knowledge were put to use.

There has been another development during this six year period. Previously, Corrosion possessed potential interest only to a few classes of people; the Chemist, the Engineer, the Architect, the Economist and the Health Authority have been noticed above. Lately Corrosion has become a matter which concerns every owner of a motor car, and in 'developed' countries that means a large fraction of the population.

The present chapter starts with a section on Education in Corrosion based largely on the Hoar report, then proceeds to discuss the interest of Corrosion to different classes of persons and ends with a section on the Corrosion of cars, largely based on the authoritative views of Hundy.

Education in Corrosion
The Hoar Committee. The Committee appointed to report on corrosion and protection released its findings in 1971.[1] After some introductory remarks, an estimate is made of the national annual cost of corrosion and protective measures. This is considered to be £1365M, of which £250M is connected with building and construction, £280M with marine problems and £350M with transport. The potential savings envisaged are estimated at £310M, of which £100M would be derived from changes of exhaust system materials in cars and improved awareness at the design stage.

The Committee then proceeds to the changes in education which would seem necessary if the situation is to be improved. They write: 'The prevalence of costly failures caused by corrosion suggests that insufficient attention is paid to corrosion and protection in the training of engineers, designers and architects.' This view accords with that previously expressed by another (parliamentary) committee which had investigated the failure of steel bolts in nuclear power stations. In their report they had recommended that nuclear engineering teams should be strengthened by the inclusion of men interested in corrosion problems and possessing technical competency to handle them. Examining the curricula at different institutions, the Hoar Committee states that the courses which place most emphasis on Corrosion and Protection are those offered in departments of Metallurgy and Materials Science, although the time devoted to the subject varies from six hours at one university to over

1. 'Report of the Committee on Corrosion and Protection. A Survey of Corrosion and Protection in the U.K.' Chairman, T. P. Hoar, H.M.S.O., London, 1971.

a hundred hours at certain other universities and polytechnics; they consider the overall position in the departments of Metallurgy and Materials Science to be satisfactory. The position in departments of Engineering is not so satisfactory. There are many institutions in which no instruction at all is provided in Corrosion and Protection. As regards departments of Chemistry, corrosion study is now included in some degree courses, and the Committee recommends that this should be encouraged.

As regards the general situation, the Committee feels that a focal point for all interests connected with Corrosion and Protection in the country should be provided, and that this can best be done by the establishment of a National Corrosion and Protection Centre, with an initial staff of not less than ten well qualified people and appropriate assistance. The Centre, it is thought, should be situated adjacent to an existing scientific establishment.

The Committee also recommends that more effort should be made to ensure that engineers, draughtsmen, designers and architects receive tuition in corrosion and protection during their undergraduate and professional training. This could be achieved if the appropriate syllabuses contained specific references to the subject. In addition, short courses of a specialized character should be provided to cater for those already engaged in industry.

Regarding the use made of the large amount of information already available in the country, they consider that the means of exchange between corrosion technologists is satisfactory. However, they find that only a few industries, notably the oil, chemical, aircraft and nuclear power industries, pay serious attention to corrosion problems at the design stage; these particular industries are either compelled to control corrosion if a process is to work at all, or are vitally concerned with the avoidance of accidents arising from corrosion damage. Other industries exhibit a wide range of corrosion awareness, extending from excellent to deplorable; the less corrosion conscious concerns have little idea where to obtain information about Corrosion, even when it has become a pressing problem.

As regards the means of possible saving, they make fairly detailed enquiry into cases in two selected areas of industry. One concerns transport, more particularly car exhaust systems; this matter is summarized on p. 7.

The report has aroused much interest, and two editorial comments[1] discuss the foundation of a Professorship in Corrosion Science at UMIST (University of Manchester Institute of Science and Technology) in the light of the recommendations of the Hoar Committee. This Research Centre under Professor Graham Wood, the first holder of the Chair mentioned, already exceeds the size recommended in the Hoar Report, and the high standard of research already published gives great hope for the future. One of the aims of the Centre has been defined by Ellinger;[2] it is 'to establish fruitful cooperation between the researchers and industry, and to make sure that discoveries are put to practical use in reducing the cost of the nation's corrosion bill'. An outline of of the teaching and research programme has been provided by Procter.[3]

1.   *Corr. Prev. Control* 1972, **19**, (3) 5. *Brit. Corr. J.* 1972, **7**, 56.
2.   M. L. Ellinger, *Paint Manufacture* 1972, **42**, (10)42.
3.   R. P. M. Procter, *Brit. Corr. J.* 1973, **8**, 194.

A two day conference organized by the Institution of Mechanical Engineers for the discussion of the Hoar report led to some interesting remarks, well summarized by Gilbert.[1] It seemed to have been thought by some of those present that figures representing corrosion losses have little meaning, but that the penalties of neglecting precautions can be very serious; the most severe losses are those caused by the unexpected breakdown of an important installation. One sentence in Gilbert's comments deserves quoting, since it applies to many other corrosion meetings: 'The people attending the Conference were . . . almost certainly more representative of those who provide corrosion advice than of those who need to receive it.'

Role of Academic Institutions in Spreading Corrosion Knowledge. A useful review of the corrosion courses provided in Academic Institutions in the UK has been provided by Shreir.[2] The education of corrosion technologists and technicians is also discussed by West.[3] Scully[4] emphasizes the need for cooperation between separate disciplines in tackling corrosion problems.

A scheme of collaboration between Professor Pourbaix, working for 'Cebelcor' at Brussels, and several American authorities, particularly in Florida, will be noted with satisfaction.[5] Relations between various organizations in the UK are discussed by Campbell.[6]

Interest of Corrosion to different Classes of Persons

General. In the 1968 book separate paragraphs were devoted to the concern of corrosion to five different classes, namely (1) the Pure Scientist, (2) the Engineer, (3) the Architect, (4) the Economist, (5) the Health Authority. These five classes will be briefly considered in the present chapter, and then a longer discussion will be provided of an entirely new class of interested person, the Motorist.

The Pure Scientist. During the period 1968–75, much work has been carried out on the subject of this book by pure scientists mainly interested in learning how Nature works, without particular utilitarian motives or desire for financial enrichment. They have, however, rarely described their work as 'Corrosion'—possibly because the term is unattractive to those who appreciate 'purity'. Much of the work has consisted in the study of the growth of oxide-films, and is rightly described as 'oxidation'; where the film is produced by electrochemical methods, the terms 'polarization' and 'passivation' have been appropriately used.

However described, much of the work has been of high quality, yielding results that are full of interest. Several modern techniques (largely developed for other purposes) have been found valuable. Low energy electron diffraction

1. P. T. Gilbert, *Brit. Corr. J.* 1972, **7**, 97.
2. L. L. Shreir, *Brit. Corr. J.* 1970, **5**, 11.
3. J. M. West, *Brit. Corr. J.* 1968, **3**, 213.
4. J. C. Scully, *Ericeira Conf.* 1971, p. 3.
5. M. Pourbaix, *Cebelcor Rapp. tech.* **159** (1969); F. N. Rhines and M. Pourbaix, *Cebelcor Rapp. tech.* **192** (1971).
6. H. S. Campbell, *Brit. Corr. J.* 1971, **6**, 187.

(LEED) deserves special mention since it provides information about the outermost atomic layers of a crystal surface. A more recent technique is Auger electron spectroscopy which yields accurate chemical analyses of surface layers as thin as 5 to 15 Å. It has been found possible to convert commercial LEED instruments, which have been on the market for some time, into Auger spectrometers of high resolving power by means of electronic additions.

The principles underlying Auger spectrometry have been conveniently presented by Southworth.[1] When a solid surface is bombarded with electrons of energy between 1 and 3 keV (which is sufficiently strong to ionize some of the atoms) a few electrons, originally situated at the K energy level, are knocked out, leaving behind holes into which other electrons, previously at the (higher) L level, may fall. Whenever this happens, energy equal to $E_K - E_L$ is released, and may be transferred to another electron at L level which may then be ejected from the solid; such an electron is known as an Auger electron, and its energy is $E_K - 2E_L$. Since both E_K and E_L are characteristic of the atomic species, any given substance will emit Auger electrons when bombarded with an electron beam of a certain energy value characteristic of an atom present in the substance, but not, to any great extent, if the beam has an energy value slightly higher or slightly lower. It would thus appear possible, by plotting $N(E)$ (the number of electrons possessing energy E) against E, to identify the atoms present in the layer close to the surface (owing to the low energy of the Auger electrons, only those liberated within 2 or 3 interatomic distances from the surface will escape unchanged; those originating at deeper levels will lose energy in collisions and will not emerge as such). However, the number of Auger electrons is small compared to electrons produced in other ways, such as back scattered primary electrons and low energy secondary electrons. Thus the curve obtained on plotting $N(E)$ against E will show only a small irregularity at the value of E corresponding to the atomic species (or one of the atomic species if we are dealing with a compound); such a curve fails to provide precise information about the composition. If, however, $dN(E)/dE$ is plotted against E, very marked peaks are obtained at the characteristic values of E, and the Auger spectrum thus obtained provides accurate information regarding the composition of the surface layers. The differentiation can be performed very accurately by an electronic method.

Some of the most hopeful developments are connected with Ellipsometry; the efforts of Winterbottom and of Kruger (this volume, p. 393) have produced apparatus, embodying early principles originated by Tronstad, which allow the rapid film-growth taking place on an originally film-free surface to be followed, along with simultaneous study of electrochemical happenings. Radioactive tracer techniques have thrown much light on film-growth. These and other experimental methods will receive notice in the appropriate chapters of the present volume.

1.   H. Southworth, *New Scientist* 1973, **60**, 342.

The Engineer and Architect.   The somewhat unsatisfactory training of engineers and architects in Corrosion has been brought out in the report of the Hoar Committee, mentioned above. Some further information regarding the engineering aspect of the subject may, however, be added. As regards metal collapse, an analysis of metal failures recorded by a large American firm, quoted by Fontana,[1] deserves attention. Of 313 cases, 56·9% were due to Corrosion, the rest being mechanical. Of the corrosion failures, 31·6% were described as 'general'; the next most frequent type was stress corrosion cracking, which accounted for 21·6%; pitting contributed 15·7%, inter-granular attack 10·2%, but corrosion fatigue only 1·8%. This low figure is surprising. It would be interesting to know whether some of the 'mechanical' failures would have been included in the 'corrosion fatigue' class, had those who made the diagnosis been more familiar with the symptoms of that type of failure.

An article on the importance of design in structures, emphasizing the importance of avoiding crevices as far as possible, has been contributed by Chandler.[2]

The Economist.   The economic importance of Corrosion has already been emphasized in connection with the Hoar report, but some additional information is available. The economic aspect of selection of materials and protective methods in chemical industries has been discussed by Koelbel and Schulze.[3] Attention should also be given to an article by Turner[4] entitled 'Corrosion can be fought and fought economically'.

A book by Meadows[5] entitled 'The Limits to Growth' may arouse interest. A review by Gregg states that it is 'required reading for all who wish to follow the coming debate on the predicament of mankind. Meanwhile corrosion scientists must surely feel that they have a special role to play in minimizing the wastage of irreplaceable metals through corrosion.' The economics of protection schemes, with special reference to motor cars, is discussed by C. L. Wilson.[6] Figures for the costs of replacements of pipes, tanks, taps and fittings are provided by Treadaway,[7] who states that 13·5% of the annual maintenance bill is directed to repair and replacement of plumbing systems, representing an annual national bill of £61M. Estimates for the cost of protection by painting and galvanizing are published by Brace and Porter.[8]

Various publications have appeared which suggest how corrosion troubles can be minimized by suitable choice of materials. Ross[9] has discussed the

  1.   M. G. Fontana, *Corrosion (Houston)* 1971, **27**, 129; esp. Table I, p. 131.
  2.   K. A. Chandler, *Brit. Corr. J.*, Supplementary Issue 1968, p. 42.
  3.   H. Koelbel and J. Schulze, *Amsterdam Cong.* 1969, p. 26.
  4.   T. H. Turner, *Corr. Sci.* 1969, **9**, 793.
  5.   D. L. Meadows, 'The Limits to Growth', 1972 (Earth Island, London). Review by S. J. Gregg, *Brit. Corr. J.* 1972, **7**, 145.
  6.   C. L. Wilson, *Anticorrosion* 1971, **18**, (8) 4.
  7.   K. J. Treadaway, Paper contributed to Symposium on 'Metals in Domestic Services', May 6, 1970.
  8.   A. W. Brace and F. C. Porter, *Met. Mat.* 1968, **2**, 169.
  9.   T. K. Ross, *Chem. Engineer* 1971, No. 247, p. 95.

selection of materials for chemical plant. Chapnick[1] has considered the use of polymers to replace metals. A disturbing feature, however, is that some publications dealing with the choice of materials hardly mention Corrosion.

The Health Authority. The dangers of plumbosolvent waters have aroused concern in recent years. In 1972, nine patients were treated in hospitals in northern Scotland for poisoning by lead derived from soft, acid, moorland water which had been stored in lead-lined tanks. This aroused concern in Glasgow, which has a soft water supply. Tests were carried out on water samples drawn early in the morning, and these were found to contain lead at 18 times the concentration regarded as safe by the World Health Organization ($100 \mu g/l$); the contamination was worst in houses fitted with lead-lined storage tanks or considerable lengths of lead piping. A report to the Corporation suggested reduction in the acidity of the public supply as a cheaper method than replacement of lead pipes by those of another metal.[2] The subject of plumbosolvency has been discussed by Campbell,[3] who confirms the generally accepted view that it occurs in waters containing organic acids or much free carbonic acid, but that it is often associated with presence of nitrates; a protective film can often be formed if $Ca(HCO_3)$ is present, but chlorides tend to interfere with the development of such films.

Work continues to be carried out on the corrosion of metal implants in the body. An inter-disciplinary meeting held on the subject[4] seemed to reach the conclusion that titanium is the most promising material for the purpose, being highly resistant to the fluids concerned. Praise was given to a Co–Cr alloy, but it was stated that stainless steel implants, when removed, frequently showed signs of corrosion, and an example was presented of corrosion around a screw-hole in a vitallium plate. Further information regarding the various materials is provided by Revie and Greene.[5] They pay special attention to sterilizing methods, stating that sterilization by steam and dry heat decreases the subsequent corrosion rate of stainless steel; an increase of the time or temperature of sterilization can depress the corrosion rate still further. The performance of titanium is affected to a lesser extent by sterilization, whilst that of vitallium is hardly affected at all. The effect of surface preparation is discussed in a second paper; an electropolished surface is least corroded, whilst sand blasted finishes show less resistance than others.

The Corrosion of the Motor Car

Extension of Interest in Corrosion. The high proportion of motor cars which terminate their useful life largely as a result of corrosion, coupled with the great increase in the number of families owning one or more cars, has led to a sudden increase in the number of people claiming an 'interest' in corrosion. The Corrosion Scientist, who is probably himself a car owner, must

1. P. Chapnick, *The Sciences* (N.Y. Acad. Sci.), Sept. 1972, p. 12.
2. *The Times*, May 26, 1972.
3. H. S. Campbell, *Water Treatment and Examination*, 1971, **20**, 11; esp. p. 21.
4. *Met. Mat.* 1970, **4**, 179. See esp. papers by I. R. Scholes, F. W. Bultitude and D. F. Williams.
5. R. W. Revie and N. D. Greene, *Corr. Sci.* 1969, **9**, 755, 763 (2 papers).

take account of the changed situation. It may affect him directly if he should be asked to use his specialist knowledge in improving protective schemes applied to cars by the manufacturer; in other cases, he may have to answer questions put to him in private conversation by indignant owners of corroding vehicles.

It should be noticed that the interest, although widely diffused through the population, does not extend to all branches of the subject. The success achieved in preventing the perforation of condenser tubes or heating systems —matters which some years ago were causing acute anxiety but now produce only occasional trouble—excites no enthusiasm; for the average man today, corrosion means the corrosion of cars—particularly *one* car. At the same time, the views expressed should not be written off as uninformed. The Automobile Association has exhorted its members to 'know about corrosion', and has produced a book[1] to help them.

The remark is heard today that, for the car manufacturer, Corrosion is something that is welcomed, since it increases his sales. Printed comments are usually more restrained, but certain manufacturers have come under censure for failing to apply protective measures believed to be available. In 1970, a journal[2] made this complaint against the British car industry: 'American manufacturers are using galvanized steel for underbodies in view of the use of salt on roads in cold weather. British manufacturers object, urging welding difficulties, unsuitability for the new electrophoretic painting process, and expense.' On the other hand, a speaker from a British industrial concern who took part in a two day Symposium on Corrosion and Prevention in motor vehicles, held in 1968,[3] had expressed a desire to produce a corrosion resisting car, provided that the public were prepared to pay the appropriate price for it.

The reader may care to study the papers presented at this Symposium, which cover most aspects of the problem. He will probably feel that the expressed desire to provide a corrosion resisting car at an appropriate price was sincere, even though he may have his misgivings as to whether the methods in which speakers placed confidence will really suffice to give the desirable reliability.

Much depends upon the addition to the price needed. This book is not the right place for a sermon on the ethics of price fixation; but the realist will probably agree that if a corrosion resisting car would command a price which would even slightly increase the profits of the industry, that car would appear on the market; conversely, if the highest price obtainable would involve a reduction in total annual profits, it would not appear. The determining factor is the addition to the price which would leave annual profits unchanged. This may be greater than would at first sight seem necessary. If, as assumed in the passage quoted above, galvanizing is the complete solution of the

1.   'Know about Corrosion' (Automobile Association), 1972.
2.   *Drive* 1970, Autumn Number, p. 26.
3.   'Corrosion Problems in Motor Vehicles', Instn mech. Engrs and Brit. Corr. Group joint Symposium, March 1968. *Instn mech. Engrs Proc.* 1967–68, **182**, Pt 3J.

problem (and not all authorities take that view), the cost of galvanizing must clearly be added; incidental costs, arising from the greater difficulty of welding galvanized steel, will be another legitimate charge. Those two items in costing are not likely to be challenged, but a third reason for a price increase may have escaped the notice of some who discuss the question. If each car on an average lasts longer, a greater profit margin per car is needed to keep unchanged the industry's total annual profit.

Willingness to pay the appropriate extra price will vary with the outlook of the purchaser. If the car is to be a status symbol, he may actually prefer a relatively cheap vehicle which will have a short life—thus enabling him always to possess a fairly recent model. In contrast, the man whose car represents an essential factor in his livelihood will almost certainly be willing to pay a considerably enhanced sum, if that is necessary for reliability. For instance, someone living at a distance from his place of work, in a village which has become deprived of public transport owing to the fact that his neighbours, like himself, have become car owners, will be faced with a major crisis if his car fails without warning. In general, however, sudden unexpected break-downs occur from mechanical rather than from chemical causes.

No car last for ever. If it is true that the proportion of cars which end their lives through corrosion instead of through mechanical failure is greater today than some years ago, this could be ascribed to one of two causes. It may be that the mechanical features of the design have been so much improved that the mechanical life is prolonged until it exceeds the chemical life; or it may be that the manufacturer has been remiss in regard to the measures provided for protection against corrosion. The first explanation would leave the owner without a grievance; being human, he generally adopts the second.

Hundy[1] has collected some useful information. He writes: 'In recent years the question of vehicle corrosion has been receiving increasing attention. As the car owning public has increased, more individuals have come into per-sonal contact with cases where vehicles have had to be scrapped due to corrosion, while the engine and other components were still good for several more years.

'There is a public feeling that cars today have built-in obsolescence, that they corrode far more quickly and have to be scrapped earlier than vehicles built in the past. Nothing could be further from the truth . . . Data from the UK and Sweden . . . indicates that the average age at which a car is scrapped is about 12 years. More detailed information [from Sweden] shows that since 1964/5 there has been a steady improvement in the service life expectancy from 9·5 to 11·8 years . . . examination of available data for cars in the UK for 1031 30 suggest a service life expectancy of only 8 years. It is apparent that the car manufacturers have achieved an improvement in vehicle life over the last 30 to 40 years, despite worsening environmental conditions.'

Since these figures refer only to two countries, they cannot be said to settle the question on a world-wide basis. But the evidence, so far as it is available, seems to suggest that the first explanation of the increase in corrosion

1. B. B. Hundy, *The Metallurgist* 1973, **5**, 119.

failures may be the right one. But that makes it all the more desirable to improve the protective measures adopted.

Corrosion due to Salt on Roads. It is essential to use de-icing chemicals in wintry weather to maintain safe driving conditions; on account of its cheapness, salt is likely to remain the chemical employed. The plan of mixing calcium chloride with sodium chloride has been tried, but Hundy states that 'all the evidence shows that this material only aggravates the corrosion problem. More expensive materials such as urea have been used instead of salt, to minimize corrosion in expensive structures such as the Severn Bridge, but economics preclude the wider use of such materials.'

Many trials have been carried out with inhibitors added to the salt, but it is considered that to reduce corrosion in this way would involve an expenditure of about £50M in the UK; although the cost of vehicle deterioration produced by salt exceeds that value, Hundy considers it unlikely that such expense would be borne by the public authorities.

It should be noted that, although the chipping of the paint by grit and the uptake of salt at the bare points thus produced occurs in winter, the actual corrosion occurs mainly in summer on days of high humidity. This was demonstrated definitely by work in Finland,[1] but most authorities consider that in countries like the UK corrosion is a summer phenomenon. Unless the humidity is very low, the salt present at the bare points will take up moisture until the concentration of the solution produced is in equilibrium with the moisture of the gas phase. This means that in very humid weather the salt concentration will be lowest. It is likely that the corrosion rate will be highest in humid weather. Measurements by Bishop[2] show that, under conditions of intermittent spraying by salt solution, the corrosion rate increases with salt concentration up to about $0\cdot1\%$, then becomes fairly constant up to $1\cdot0\%$, afterwards diminishing. The drop in the corrosion rate at high salt concentrations is probably due to diminished oxygen solubility. Early work[3] on corrosion of steel by *potassium* chloride solution under partly immersed conditions shows that over a considerable range of concentration the corrosion rate is proportional to the oxygen solubility; the same would probably be true with sodium chloride. A reasonable deduction from these measurements is that corrosion will normally be fast under humid conditions when the salt solution is relatively dilute.

Results of a three-year study of Corrosion in different districts, carried out by the Automobile Association,[4] show that the corrosion of cars is slowest in the industrialized region around Swansea and Cardiff, where rainfall is high, and most rapid in East Anglia and Kent, where rainfall is low—despite the fact that the salt usage is low in the latter two districts. This is almost certainly due to the fact that heavy rain washes away mud and grit, which would otherwise keep the salt in position on the metal.

1.   P. Asanti, *Instn mech. Engrs Proc.* 1967–68, **182**, Pt 3J, 73.
2.   R. B. Bishop, TRRL Report (Department of the Environment) LR 489, esp. Fig. 1.
3.   U. R. Evans and T. P. Hoar, *Proc. Roy. Soc.* (*A*) 1932, **137**, 343; esp. Fig. 9, p. 357.   4.   *The Times*, Feb. 6, 1974.

Precautions against Salt Corrosion. Two main methods are used to combat troubles due to salt. At the Symposium of 1968, hope seemed to be felt that the use of zinc-coated steel for the car body would solve the problem. It was admitted that this involves welding difficulties; but it was believed that the use of a special galvanized steel carrying a thick coat on one side and a thin one on the other would supply the answer to welding problems. Today this system is used on some cars—apparently with satisfactory results from the corrosion standpoint. On the other hand, some manufacturers are contemplating the use of another type of galvanized iron in which the zinc layer has been converted to an iron–zinc alloy. Such a covering will have different electrochemical properties from unalloyed zinc, and it is not certain whether it will provide equally good protection against rusting. The information hitherto published is insufficient to make a judgement on that matter.

There is little doubt that the presence of zinc on the steel does retard rusting. The mechanism of its action has not, it would seem, been clearly established; the suggestion which follows is speculative, but may provisionally be accepted until, as is highly desirable, the matter has been subjected to scientific study.

Where the paint has been chipped away by a stone particle carrying salt, this would, of course, set up fairly rapid corrosion—leading ultimately to perforation—if no zinc were present. If there is a zinc layer covering the steel, the corrosion will extend not inwards but sideways, since zinc is anodic to steel. At the outset the protection is 'sacrificial'; the zinc is destroyed and the steel, being the cathode, remains unattacked. Clearly, when the corrosion has spread sideways for a certain distance, destroying the zinc layer initially present between paint and steel, protection will cease; but it is likely that by that time a solid zinc compound will have been deposited on the steel by the cathodic reaction, since the cathodic reduction of oxygen will produce OH^- ions which will combine with Zn^{2+} ions from the anodic reaction to give zinc hydroxide or perhaps a basic chloride. Even if this is later converted to iron hydroxide, the rust formed in physical contact with the steel should be of a fairly protective character, and corrosion is likely to be slow.

Hundy is inclined to doubt whether galvanizing will be the answer to the corrosion problem, except in special cases. Apart from the welding difficulty already mentioned, the drawing properties of galvanized steel are held to be inadequate, and some types of pressing cannot be carried out at all. Electropainting is likely—he thinks—to be regarded as the prime defence against corrosion. In this process, which was developed independently on both sides of the Atlantic, the car body is passed fully immersed through a tank of water-based paint, being made the anode, with cathodes situated along the sides of the tank; the EMF used may vary between 150 and 500 V according to the system. The principle of the process is described elsewhere (1968 book, p. 24; this volume, p. 283). The resin and pigment constituents of the paint are anionic and plate out as a uniform film over the surface of the car body, whilst the cationic constituent is liberated at the cathodes. Hundy provides pictures showing the improvement obtained by electropainting as compared with the protection provided with the older painting methods; he also provides pictures showing that the electropainting method has been improved over the years. Nevertheless, his information has some disturbing features, and suggests that the designers of this process may still have some way to go. He writes: 'Despite good effective cleaning and phosphating, batches of steel sheet from different steelmaking plants, and from within the same plant, show variable corrosion resistance after painting in the standard manner. In some cases this could lead to corrosion resistance lower than acceptable. The reason for this phenomenon still remains uncertain, though it is probably associated with variations in finishing techniques in the steel mills.' It would seem that a really satisfactory painting procedure for steel (a material which does not claim to resist corrosion in the unprotected condition) should be able to prevent corrosion—whatever the finishing techniques at the steel mills may have been. It is understood that active steps are being taken to provide what is needed. It is likely that for some purposes a combination of two methods may be used—namely a suitable zinc coating covered with a superior paint. Hundy mentions the design of a new omnibus in which a zinc coat is followed by powder epoxy coatings to ensure maximum freedom from corrosion in the chassis regions.

Humphreys[1] has discussed the spread of corrosion from places where damage from chips has occurred, arguing that if 'chipping back to metal does not occur', corrosion will not set in. Chipping to metal can be avoided (1) by *increasing* the adhesion of the electroprimer to the metal or (2) by *reducing* the adhesion of the subsequent coats to the electroprimer, thus causing any chip failure to occur at another interface. Of these, the first method is preferable but the second should not be discounted. He emphasizes the importance of using a compatible system of phosphate, primers and top coats.

Corrosion of Exhaust Systems. Corrosion by salt is only one of the car owner's troubles; H_2SO_4 and HNO_3 in the exhaust gases can shorten the life of a steel exhaust system. The report of the Hoar Committee already mentioned considers the possibility of savings by the use of superior material. The exhaust system usually supplied at the time of the report was made of relatively thin unprotected mild steel; the replacement cost was between £5·50 and £34·50 (this refers to British cars manufactured on the assembly line). A weighted-mean replacement cost, allowing for a preponderance of lower priced vehicles, was considered to be about £14. The average life of an exhaust system is about two years, and with 12·25 million cars in use, the annual replacement bill is about £86M. The production of an exhaust system with a life expectancy of approximately six years is technically possible, and with a material such as aluminized steel should only involve a small extra expense. Allowing, however, as much as £1 per exhaust system, the annual replacement cost with an improved material would be only £31M; the net national saving obtainable from the use of an improved material would seem to be about £55M.

Indian trials[2] made on a vehicle run for 4000 km show that Al-sprayed

1.   J. Humphreys, *Trans. Inst. Met. Finishing* 1973, **51**, 204.
2.   R. Singh and B. Sanyal, *Labdev. J. Sci. Tech. (Kanpur)* 1970, **8A**, 30.

steel performs better as a material for exhaust systems than galvanized, zinc-sprayed or epoxy coated steel. The acids which produce serious attack on unprotected steel include H_2SO_4 and H_2SO_3 derived from S in the fuel, but also HCl and HBr; the latter is derived from $C_2H_4Br_2$, present as an additive.

Corrosion of Interiors. An experienced coach builder, quoted by A. C. Wilson,[1] has stated that the doors of most cars start to corrode on the *insides*—not on the surfaces exposed to the road; apparently this is due to water, probably largely condensation, running down the windows and penetrating inner surfaces of the doors, reaching places where the flange of the door panel has been clinched over, forming a sort of 'hem'. Corrosion almost invariably starts at these hems, and the trouble, once started, is almost impossible to arrest. Wilson observes 'this seems to me a serious weakness in design', but Hundy remarks 'Electropainting can help . . . as does zinc-rich weld-through primer applied before assembly of the door.' The same hem construction (also known as 'clinched seam construction') is used at other points in the car; Hundy does not regard it as a serious starting point for corrosion except in the case of doors. It is interesting to note that, except at the hems and the exhaust system, some cars seem to suffer little corrosion, provided that the underside has been coated from the first with the substance known as 'underseal' which prevents abrasion from stones etc. Hundy, however, states that 'even today cars still run into trouble with mud pack areas around the inside of the wings and sometimes in the sill sections'. Wilson quotes the case of a vehicle (described as coming into 'the lower price-range of quality cars') which ran 80 000 miles in six years and suffered only minimal rusting except for the hems and the exhaust system; the latter had not been treated with underseal and required five replacements during the first four years; during the last two years no replacement was necessary, suggesting that the manufacturers had adopted a protective system—perhaps that suggested in the Hoar report. However, the apparent improvement may have been due to the fact that mild winters have led to less than the usual amount of salt being applied to the roads. Hundy suggests that some of the early failures were due to fatigue and that improved design has now avoided this.

Corrosion of Brake-fluid Lines. Linder[2] has described Swedish recommendations regarding materials for brake-fluid lines. Copper-brazed steel tubes are insufficiently resistant. Phosphorus-deoxidized copper and stainless steel are considered to be the most suitable materials, but the low fatigue strength of copper, if used, calls for care in design. As regards cost, copper held an advantage in 1972. It is stated that both copper and stainless steel are liable to corrode at fittings and fastening points.

Electrostatic Coating. An important development is the application of powder, generally epoxy powder, by an electrostatic procedure. These

1. A. C. Wilson, Report, June 1973; with comments by B. B. Hundy, Sept. 1973.
2. B. Linder, *Werkstoffe u. Korr.* 1972, **23**, 187.

have been discussed by Meyer[1] with special relation to the possibilities in the protection of automobiles. He mentions specially use on rims, bumpers, headlights, horns, coolers, springs and brakes.

Testing Methods. Various methods of testing cars or car materials for corrosion resistance, both in the laboratory and in the field, were described in the 1968 Symposium (this volume, p. 8). In the laboratory, reliance still seems to be placed on salt spray tests in different forms. Some of the outdoor tests include full-scale salt-splash and stone-throwing contrivances; if, however, the actual corrosion arising from salt on roads takes place on humid days in summer, this is the stage which requires most study.

Further References

Books. Wranglén[2] has provided an excellent account of the subject of corrosion, based on an introductory course for engineers which he has been giving for many years at Stockholm. Pourbaix[3] has contributed a book based on lectures delivered at Brussels; his authoritative approach is mainly thermodynamic, but the kinetic aspects of corrosion are not neglected. A series of volumes entitled 'Advances in Corrosion Science and Technology' has been ably edited by Fontana and Staehle,[4] aided by a Board of acknowledged authorities; the first three volumes contain a long series of review papers, in which 18 experienced authors have played their parts. Wormwell[5] has provided a survey of the distinguished work carried out at Teddington over a period of 44 years. Fontana[6] has lectured on the 'Perspectives on Corrosion of Materials'.

Numerous other full-scale volumes dealing with special aspects of corrosion have appeared; they will be mentioned in the appropriate chapters of this supplementary volume.

Papers. 'Korrosion und korrosionsschutz im Automobilbau'.[7]
'Bridge building between laboratory corrosion data and process plant'.[8]
'Applications of Electrochemistry in Corrosion Science and in Practice'.[9]
'Cars made to last'.[10]

1. B. D. Meyer, *Verfkroniek* 1973, **46**, 130.
2. G. Wranglén, 'An Introduction to Corrosion and Protection of Metals' (Institut för Metallskydd), 1972.
3. M. Pourbaix, 'Lectures on Electrochemical Corrosion' (Plenum Press), 1973.
4. M. G. Fontana and R. W. Staehle (Editors), 'Advances in Corrosion Science and Technology' (Plenum Press), Vol. I (1970), II (1972), III (1973) and IV (1974).
5. F. Wormwell, 'Corrosion of Metals Research: 1924–1968' (H.M.S.O., London), 1973.
6. M. G. Fontana, *Corrosion* (*Houston*) 1971, **27**, 129.
7. G. Buss, *Galvanotechnik* 1973, **61**, 377.
8. M. Pourbaix, *Corr. Sci.* 1974, **14**, 25.
9. C. Edeleanu, *Brit. Corr. J.* 1973, **8**, 200.
10. Discussion arranged for March 27, 1974, Inst. Met. Finishing.

'Science and Technology in the Service of Society'.[1]
'Changing priorities in corrosion work'.[2]
'Metric Units in Corrosion Technology'.[3]
'The passive film on iron; an application of Augur electron spectroscopy'.[4]

The (British) Department of Industry has set up a Committee under the chairmanship of J. B. Cotton with the aim of demonstrating the benefit to industry of applying proved techniques of corrosion control.[4]

1.   N. Hackerman, *Corrosion (Houston)* 1973, **29,** 85.
2.   S. C. Britton, *Brit. Corr. J.* 1974, **9,** 130.
3.   A. de S. Brasunas, *Mat. Performance* 1974, **13,** (5) 30.
4.   R. W. Revie, B. G. Boker and J. O'M Bockris, *J. electrochem. Soc.* 1975, **122,** 1460.
5.   *The Metallurgist* 1975, **7,** 94.

CHAPTER II

Simple Oxidation of Single Metals

Preamble

Picture presented in the 1960 and 1968 volumes. In 1923, Pilling and Bedworth divided the metals into two classes, according as the oxide formed would (if uncompressed) occupy a smaller or larger volume than the amount of metal destroyed in producing it. In the first case (represented by ultra-light metals), the oxide-film formed should, they argued, be porous and non-protective; the oxidation should therefore obey a rectilinear growth law. In the second case, the film should be protective, and the oxidation rate should be proportional to the reciprocal of the film thickness—leading to the well-known parabolic growth law. They appeared to obtain confirmation by a study (1) of calcium, an ultra-light metal, which they found to oxidize at a constant rate (until a moment when the oxidation rate suddenly started to increase rapidly, probably owing to rise in temperature), and (2) of copper, a dense metal, which was found to obey the parabolic law. Later work confirmed their findings for copper, and established parabolic growth for a number of other dense metals; but their results on calcium were not confirmed when experiments came to be carried out on what was probably purer material.

Pilling and Bedworth assumed that oxidation proceeded by oxygen diffusing inwards through the film. Later, however, it was shown by Pfeil that film thickening was largely produced by metal diffusing outwards to meet the oxygen, and this is indeed to be expected for the heavier metals, where the size of the metal cation is usually smaller than that of the oxygen anion. Pfeil's work was conducted on iron where the presence of layers of three different oxides complicated the situation; but subsequent work on other metals has supported the idea that outward movement of metal is a common mechanism; sometimes outward movement of metal and inward movement of oxygen can take place simultaneously.

Perhaps the main defect in some of the early reasoning was the assumption that compressional stresses are always beneficial (in the sense that they oppose oxidation). This is far from being the case; a small compressional stress, tending to press together the sides of any cracks present may indeed be beneficial, but larger stresses may have the opposite effect—as shown by the behaviour of metals like zirconium. Cases are known where a metal exposed to oxygen at a slightly elevated temperature will develop an oxide-film at a rate which at first declines as the film thickens, according to the parabolic law (or perhaps by the cubic law, which causes an even more rapid decline of oxidation rate). Then suddenly, after a certain thickness has been reached, 'breakaway' occurs and thereafter the film thickens rapidly by an approximately rectilinear law. This should occasion no surprise. A film containing internal stress (whatever the origin) carries strain energy, and if there is either local rising as a blister or more extensive detachment, the energy content of the system will be reduced. But spontaneous blistering or peeling will only occur if the accompanying drop of strain energy exceeds the work needed for detachment over the area involved; this will not be the case for very thin films. If the strain energy is due to discrepancy of volume between metal destroyed and oxide produced, the strain energy per unit area will be proportional to thickness, and there is some critical thickness above which detachment will not conflict with the laws of energy. Another factor, however, must be taken into account in deciding the thickness at which spontaneous breakaway is possible. If thickening is taking place by the outward movement of metal (or by the inward movement of cation vacancies), a number of vacant sites will be produced at the metal–oxide interface, and these could join together to form cavities; such a development will certainly favour detachment, since no energy of separation need be provided for areas where a cavity between metal and oxide already exists.

In 1933, a great advance was made when Wagner produced an equation which expressed the parabolic rate constant in terms of electrochemical quantities, notably the EMF of the cell metal/oxygen. The movement of material through the film is pictured as taking place partly as ions (generally cations) migrating under the electrical field due to that EMF and partly as particles diffusing under a concentration gradient; for the outer surface of the film (considered to be in equilibrium with the oxygen of the gas phase) contains less metal than the inner surface (in equilibrium with the metallic phase). Since, however, a relation (due to Einstein) exists between migration under an electric potential gradient and diffusion under a concentration gradient, it becomes possible to express the situation solely in terms of the electrical values (or, as an alternative, solely in terms of the diffusional values). Wagner found that the values of the parabolic rate constants calculated for the oxidation of copper at 1000°C (in oxygen at different pressures) agreed remarkably well with the experimentally determined values. In 1951, he obtained similar agreement for the formation of iodide films on copper, and for sulphide and bromide films on silver. The agreement (see Table XXVIII on p. 849 of the 1960 book) is impressive. It provides good confirmation of the idea that the so-called 'simple oxidation' is itself an electrochemical phenomenon, in the sense that there is anodic attack upon the metal and cathodic reduction of oxygen at the inner and outer surfaces of the film respectively.

It is, however, well established that the parabolic equation is only obeyed in a certain range of temperatures; other equations are obeyed outside this range. About 1955–59 three attempts were made to arrive at a general equation which would represent film-growth at any fixed temperature, based solely on the assumption that an ion will only move from one site to the next when it possesses sufficient energy to surmount the intervening energy hump. The three papers (two from Cambridge and one from Oslo) represented independent thinking. Essentially the same conclusion was reached in all three places, but the Norwegian paper carried the argument further than the others. By permission of the Author, T. Hurlen, it is reproduced as an Appendix to this chapter.

The general equation expresses the rate of thickening by the difference between two exponentials representing respectively movement aided or opposed by the electric field (1968 book, pp. 298–303). At low temperatures, where the film will be very thin and the field consequently very strong, movement against the field can be neglected, so that one exponential can now be omitted. In these circumstances, the equation reduces by an approximate transformation to the inverse logarithmic law. At high temperatures, movement in both directions must be taken into account but the later terms of the expansion of e^x or e^{-x} can be neglected. In certain circumstances the equation will then reduce to the parabolic law. This accords with the experimentally established facts. Experiments by Mayne and Gilroy on iron exposed to dry air for a year at ambient temperature showed that the invisible film formed thickened by the inverse logarithmic law (not by the direct logarithmic law as had sometimes been imagined; the two laws are difficult to distinguish in short experiments). Over the higher temperature range, obedience to the parabolic law had been established by earlier investigators.

It should be emphasized that the derivations of the parabolic and inverse logarithmic laws as being the two limiting cases of the general equation involve approximations. (For details, see 1960 book, pp. 823–7.) It also involves certain assumptions which may not always be justified—for instance, contact at all points between film and metal. If the growth depends on the movement of metal outwards (or cation vacancies inwards), vacancies will be created at the metal–oxide interface which may coalesce to form definite cavities between metal and oxide-film, reducing the area in contact. Such cavities have been observed experimentally, but they are not always formed; if, for instance, the oxide-film at the temperature in question is sufficiently plastic to subside into the cavities, the reduction of contact area may not be serious. Or again the vacancies may move into the metal. If, however, the vacancies remain at the interface, the oxidation rate will clearly fall off more quickly than the parabolic law would predict on the assumption of constant contact area. Simple calculation (1960 book, pp. 836–7) shows that the law to be expected is now logarithmic—either the direct logarithmic law which is well established for many metals over a temperature range lower than that in which the parabolic law is obeyed, or a new logarithmic law found for copper under circumstances where the direct logarithmic law fails. Here again, therefore, there is good agreement between experimental facts and predictions from theory made without *ad hoc* assumptions.

Thus, without any assumption other than the generally accepted fact that a particle can only move from one low energy site to a neighbouring site separated by a high energy region if it possesses sufficient energy, we can arrive at three equations—all well established experimentally:

Parabolic Equation $\qquad\qquad\qquad\qquad y^2 = K_1 t + K_2$

Inverse Logarithmic Equation $1/y - 1/y_0 = K_3 \log (K_4 t + 1)$

Direct Logarithmic Equation $\qquad y - y_0 = K_5 \log (K_6 t + 1)$

Here y is the film thickness at time t, (y_0 being the thickness when $t = 0$), whilst the K's are constant at constant temperature.

However, logarithmic growth may arise in other ways. If, for instance, the film is porous, but the pores sometimes become blocked, probability considerations lead to two equations, according as the pores are taken to be self-blocking or mutually blocking (the substance formed in one pore exerting lateral pressure tending to close up neighbouring pores). The first assumption leads to an asymptotic equation and the second to a logarithmic equation.

The argument used in derivation of the general equation also assumes equilibrium between oxide and gas phase at the outer interface and between oxide and metal at the inner interface. In the opening stages of film-growth, where the movement through the very thin film could be rapid, it may well be that the surface reactions will not proceed sufficiently quickly to preserve equilibrium; thus at the outset, neither equation is likely to be obeyed.

Finally, if, instead of assuming that growth occurs uniformly over the surface producing a film of uniform thickness, it is assumed that formation of a three-dimensional film (as opposed to a two-dimensional film of oxide or adsorbed oxygen) starts at certain points favourable to nucleation and spreads sideways until the various oxide-covered areas join up, sigmoid growth curves must be expected, and are in fact obtained.

Other equations, such as the cubic relationship ($y^3 = K_7 t + K_8$), have been put forward, and also alternative derivations of the three equations already mentioned.

It is not claimed that the simple derivations suggested in this preamble represent agreed opinion. Moreover, from time to time, an experimenter announces that he has obtained results which cannot be represented by any of the equations put forward by 'the Armchair Scientists'. This is not surprising; indeed it is rather remarkable that the production of results inconsistent with the equations mentioned above is not more frequent. Certainly, in the admittedly rare cases where it is possible to compare experimental values for a rate constant with those calculated from theory, the agreement (notably in the work of Wagner already mentioned) has been most satisfactory. The situation, soberly considered, suggests that the Armchair Scientists have not been far wrong.

Developments during the period 1968–1975 and Plan of the new Chapter.

In 1968 a fair understanding existed of the theoretical basis of equations obeyed during the period when the oxidation rate is controlled by passage of material through a reasonably uniform film; earlier and later periods present different problems, and these are now being studied, theoretically and experimentally. In the very early period when the film is extremely thin, the rate at which material could pass through it would be extremely rapid, and it is unlikely that the reactions at the two interfaces (or perhaps even the supply of oxygen) would keep pace with this; thus different processes assume control of the rate. During the later periods, also, new factors are introduced. It might, at first sight, appear that if the corrosion rate falls off with film thickness, it is only necessary to wait until a sufficient thickness is attained to reduce the corrosion rate to any desired extent. Unfortunately, when the film becomes thick, it is likely to break down, either because cavities are formed at the base (which happens if metal is moving outwards rather than oxygen inwards), or because strains arise in the film substance (most likely if oxygen is moving inwards).

In the new chapter, some recent papers of a general character are first reviewed—first those representing theoretical thinking, and then those describing new experimental work. There follows a summary of papers which describe the behaviour of individual metals. The presentation starts with iron, about which more information is available than for most other metals. The other metals of the iron group are then considered followed by the B-groups of the periodic system, and finally the A-groups. This arrangement may appear arbitrary, but it seems to present the matter in a form more likely to interest the reader than the adoption of some strictly logical sequence. Even so, it cannot be claimed that the various papers link up to form a continuous argument or to provide a coherent picture of the situation; the assemblage of papers is more like a patchwork quilt. In the present stage of knowledge this is probably inevitable. In a few years time, additional work may fill in the gaps, and possibly some reader who feels unhappy about the presentation in the present chapter, may himself take up the pen and provide something more intellectually satisfying. At the end of the chapter, Hurlen's presentation of the general equation, with slight modification and addition, is presented as an Appendix.

The oxidation of a 'pure' metal in 'pure' oxygen is a phenomenon rarely met with in industrial or engineering experience; the arguments and experimental results summarized in the present chapter will therefore interest the 'pure scientist' rather than the 'practical man', who will find more information helpful to his problems in the chapters that follow.

Theoretical Discussion

Limits of application of Wagner's oxidation theory. An assessment of Wagner's treatment of oxidation by Lacombe and his colleagues[1] deserves study. Up to the present time, few oxides have been used to provide a quantitative experimental test of the theory. The best examples are Cu_2O and CoO, which are considered by many authorities to obey the principles satisfactorily. Why, asks Lacombe, is it that the kinetics of many other metals seem to fail in their compliance?

In several cases an answer can be provided. Cr is held to fail because of the formation of a second oxide, CrO_3, which is volatile. In the case of NiO, lack of plasticity may cause a porous layer to be formed, introducing a factor not taken into account in Wagner's argument. Experiments with gas mixtures such as $CO-CO_2$ or H_2O-H_2 introduce another disturbing factor—namely the rate of gas exchange. Where there are several layers composed of different oxides, new complications come in. As regards the oxidation of Co, Lacombe considers that here the agreement between theory and observation is good;

1.   F. Morin, G. Beranger and P. Lacombe, *Oxidation Met.* 1972, **4**, 51.

but as regards Cu, a comparison of the data available presents conflicting results; a good correlation—he thinks—still remains to be established.

Recent views on the mechanism of film-growth. An approach to growth laws which has more in common with Wagner's original paper of 1933 than with his later treatments of the subject, has been adopted by Ritchie,[1] who makes some sound new points. He observes that whilst uncharged atoms can move under a concentration gradient, charged particles move partly under a potential gradient; the total current (due to both gradients) is

$$j = -D \, dC/dy - \mu z q \, dV/dy$$

where C is the concentration, q the electronic charge and z the particle charge number; the diffusivity D and mobility μ are connected by the Nernst–Einstein relation $D/\mu = kT/zq$, and if V, the EMF involved, is known, the current and hence the oxidation rate can be evaluated. In general, however, V is not exactly obtainable. The most tractable approximation, in Ritchie's opinion, is 'to neglect all space-charge effects and to assume that the electric field is homogeneous and due to the uncompensated charges at the oxide interface'. The real situation may be more complicated. If no simplifying assumptions, such as quasi-neutrality, are introduced, an exact solution would necessitate the solution of a set of five simultaneous equations.

There is also the question of the influence exerted by the reactions occurring at one or both interfaces. It has generally been assumed that, if the reaction proceeding at the oxide/O_2 interface is rate-controlling, the growth law *must* be rectilinear. Ritchie and Hunt[2] suggest that this is not necessarily the case. If the electron concentration is uniform, and if electrons are assumed to be taken up by the sequence

$$O_2 \text{ (gas)} \xrightarrow{1} O_2 \text{ (adsorbed)} \xrightarrow{2} O \xrightarrow{3} O^- \xrightarrow{4} O^{2-}$$
$$\xrightarrow{5} O_2^- \xrightarrow{6} O^{2-}$$

it is clear that, whichever step is rate-determining, the velocity must be constant, leading to a rectilinear growth law. If step 1 is rate-determining, the growth rate will be proportional to the oxygen pressure in the gas phase, but that is not the case if steps 5 or 6 are rate determining; under some conditions, the rate established may be almost independent of pressure. If the electron concentration is not uniform, control by a boundary reaction does not necessarily lead to a rectilinear law. If, for instance, a Boltzmann distribution is assumed to exist, then a wide variety of growth equations become possible, with the rectilinear, parabolic, cubic and direct logarithmic laws as special cases. The argument should be studied in the original paper.

The complexity of the situation is not, however, entirely due to events at or near the surface. In a lengthy mathematical discussion, Wagner[3] has himself shown that, even if control by transport processes is assumed (the effect of

1. I. M. Ritchie, Chap. 10 of the book, 'Chemisorption and Reactions of Metallic Films' (ed. J. R. Anderson), Academic Press, 1971.
2. I. M. Ritchie and G. L. Hunt, *Surface Sci.* 1969, **15**, 524.
3. C. Wagner, *Corr. Sci.* 1973, **13**, 23.

experimenter may draw the wrong conclusion from his accurate measurements. The whole reputation of the subject is in jeopardy. A mathematician who is accustomed to thinking in terms of dimensions may automatically apply tests to every equation on which his eye alights. Such a man, looking through a journal and chancing to notice in a paper on Corrosion an equation which fails to pass his tests, may be excused if he thinks, 'I know nothing of this subject "Corrosion", but clearly it cannot be regarded seriously as a branch of Science.' He may be unaware that the paper in question contains conscientious work and accurate measurements. The effect of his censure could affect the reputation of others working on the subject, and indeed prejudice the prospect of financial support for work which they are planning.

The matter deserves attention from those concerned with the basic educational requirements for research students. If the accurate measurements made during a research are to lead to accurate conclusions, it is necessary that the experimenter shall have an elementary knowledge of certain not very formidable branches of Mathematics; one of these is Dimensional Analysis; another is Curve-fitting—especially the Principle of Least Squares.

Recent Experimental Work

Effect of an applied Electric Field on Oxidation Rate. An important research by Ritchie[1] on the oxidation of nickel shows that an applied electric field will accelerate or decrease the oxidation rate according to its polarity;

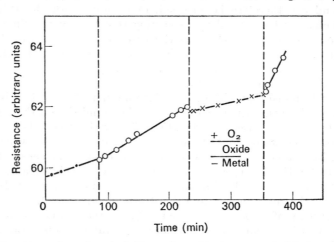

Fig. 1   The effect of an electric field on the oxidation rate of nickel at 371°C. No field applied ● ; O_2/oxide interface negative with respect to metal ○ ; O_2/oxide interface positive × (I. M. Ritchie, G. H. Scott and P. J. Fensham).

the effect reverses itself according as the experiment is conducted above or below 350°C. The effect of applying the field in different directions is shown schematically in Fig. 1. The results confirm the theoretical views held by Ritchie, but are not inconsistent with certain other theories.

1.   I. M. Ritchie, G. H. Scott and P. J. Fensham, *Surface Sci.* 1970, **19**, 230.

A study of the effect of a field on the oxidation of copper, carried out in Sircar's laboratory,[1] has shown acceleration or retardation according to polarity; the results appear to support the Cabrera-Mott theory.

Leach and Nehru[2] have shown that the oxidation of uranium in moist oxygen or moist argon can be retarded by the application of a field.

The effect of Pressure on Cavity Production. Research in Cohen's laboratory[3] shows that an effect of increased gas pressure is to prevent the formation of voids between oxide and metal, and thus allow oxidation to continue without the abatement which the voids would introduce. The oxidation of annealed iron is slow at 10 torr because the plastic deformation of the oxide is insufficient for maintenance of adequate contact between oxide and metal. The voids formed by condensation of cation vacancies soon start to isolate the oxide from the retreating metal, so that oxidation is to some extent stifled. Thus the oxidation curve flattens out, rising again when the separated oxide layer fails, and flattening again when contact between oxide and metal is once more lost. At high pressures, oxide and metal are squeezed together, so that the curves are continuous. Annealed iron which develops voids oxidizes more slowly than cold-worked iron, which provides vacancy sinks in the metal, thus suppressing void formation; it is also believed that there is more rapid initial diffusion through the oxide formed on cold-worked iron, either because there are leakage paths or a steeper cation-vacancy gradient.

Stresses in Films. Francis and Hodgson[4] point out that few oxides grow through the movement of the cation only or that of the anion only. Given a high oxide/metal volume ratio, a small contribution from anion diffusion may produce strain, even though the oxidation process in question is mainly due to cathode movement. Thus with a Pilling–Bedworth ratio of 2·0, a 1% contribution by anion diffusion may be equivalent to a 1% strain in the oxide. This, they suggest, is particularly relevant to the complex mixed oxide phase formed on alloys designed for high temperature use under conditions of thermal cycling.

There are, however, many different ways in which stresses in oxide-films can originate. The subject has been discussed by Appleby and Tylecote[5]. During the oxidation of nickel the movement of vacancies into the metal is considered to be the only mechanism producing stress; a similar movement is a significant source of stress during the oxidation of copper, but here the 10·5% volume decrease which accompanies the formation of Cu_2O from $CuO + Cu$ is also an important cause. The wedge effect which occurs when oxide penetrates into metal along grain-boundaries contributes to the stress

1. P. K. Krishnamoorthy, S. C. Sircar (with S. K. Ray), *Acta. Met.* 1968, **16**, 1461; 1969, **17**, 1009; 1970, **18**, 519.
2. J. S. L. Leach and A. Y. Nehru, *Corr. Sci.* 1969, **9**, 447.
3. D. Caplan, M. J. Graham, M. Cohen, G. I. Sproule and R. J. Hussey, *Corr. Sci.* 1970, **10**, 1, 9 (2 papers).
4. J. M. Francis and K. E. Hodgson, *Amsterdam Cong.* 1969, p. 64.
5. W. K. Appleby and R. F. Tylecote, *Corr. Sci.* 1970, **10**, 325.

found in copper oxides around 940°C. In oxides growing predominantly through inward diffusion of anions, such as ZrO_2, the stress is initiated at the metal–oxide interface as a result of the high oxide–metal volume ratio.

Stringer[1] has provided a useful review of stress generation and relief in growing oxide-films. He states that, whilst it is easy to demonstrate the existence of stress, it is less easy to establish its origin. Simple arguments based on the Pilling–Bedworth ratio are considered to be inadequate by themselves, but the qualititive agreement with experiment suggests that the volume-conservation principle must be involved. Magnesium is apparently the only metal in which tensile stresses have been observed; its oxidation deserves to be studied in greater detail. Additional sources of stress include epitaxial constraints, for which there is some evidence in the case of Cu and Ta. Another possible cause is a vacancy gradient; vacancy injection into the metal has been demonstrated in a number of cases, but, in Stringer's opinion, is not a general cause.

The stress generation in films formed on zirconium during thermal oxidation at between 400° and 700°C, or during anodic oxidation at lower temperatures, has been studied in Leach's laboratory.[2] Optical methods and scanning electron microscopy have been used. The stresses present in oxide grown on thin foil can be relieved more easily than those present in oxide grown on thick foil. The stresses appear to arise from the fact that the films grow predominantly through the inward migration of oxygen; the volume ratio of the oxide is high, and the stresses are largely compressive. They can cause mechanical failure of the film, with consequent change of the kinetic situation.

Work in Leach's laboratory[3] has also thrown light on stresses present in zirconia films formed by anodic action. The stress seems to become lower as the growth rate increases; microscopic examination of the surface of anodized zirconium specimens shows that with increasing current density there is a marked decrease in the occurrence of features in the film which might be attributed to the presence of compressive stresses.

Oxidation of Special Types of Surface

Oxidation of Cleaved Faces. The behaviour of freshly cleaved tellurium shows some interesting features.[4] If the crystals are cleaved carefully at room temperature, the faces exposed are mainly prismatic, and these suffer oxidation, giving a fine-grained smooth film. Crystals cleaved by impact at a low temperature (e.g. in liquid nitrogen) tend to expose areas of coarse surface with non-prismatic faces; on these the oxide grows more extensively and with a nodular structure.

Oxidation of Liquid Metals. The behaviour of molten metals has been reviewed by Drouzy and Mascré,[5] who state that knowledge regarding the

1.   J. Stringer, *Corr. Sci.* 1970, **10**, 513.
2.   J. S. L. Leach and F. H. Hammad, *Tokyo Cong. Ext. Abs.* 1972, p. 309.
3.   P. A. Brook, V. R. Howes, J. S. L. Leach and A. Y. Nehru, *Amsterdam Cong.* 1969, p. 272.
4.   G. D. Hopkins, I. M. Ritchie and J. V. Sanders, *Oxidation Met.* 1969, **1**, 209.
5.   M. Drouzy and L. Mascré, *Met. Rev.* 1969, **14**, 25.

kinetics is still deficient. Parabolic as well as rectilinear growth is known; logarithmic or inverse logarithmic growth is rare. Magnesium at low oxygen pressures oxidizes by the burning of the vapour. Small amounts of a second metal present greatly reduce the oxidation rate. Thus the addition of even 0·004% Be to aluminium containing 10% Mg will greatly slow down the oxidation owing to the formation of a BeO film. Small amounts of zinc added to molten cadmium will reduce the oxidation rate.

Experiments on evaporated metal. Ritchie[1] points out that whilst, for the study of the formation of thick oxide-films at high temperatures, foil specimens possess advantages. information about thin films formed in the early stages of oxidation is best obtained from experiments on evaporated metal, since it is possible to start with material that is free from oxide. His experiments suggest that there is adsorption first as O_2 and then as O, which can be incorporated either as a solid solution or as a suboxide; this is in due course followed by formation of true oxide. At the adsorption stage, the MO pairs possess a dipole moment; the energy content of the 'system can be reduced if half of the pairs reverse their polarity, and this leads to the start of the incorporation and oxide phase production. Much of the information provided was obtained by low energy electron diffraction (LEED).

Oxidation of Individual Metals.

Iron. Wingrove[2] has discussed the iron-oxygen equilibrium diagram, comparing the versions put forward by different investigators. She points out that Bénard considers that the low oxygen boundary of the wüstite area is practically coincident with the composition FeO, but that most other authorities think that the wüstite area does *not* include FeO.

Since three layers are often present on oxidized iron, knowledge of the oxidation involves (1) the conversion of Fe to wüstite, (2) the conversion of wüstite to magnetite and (3) the conversion of the latter to haematite. The last of these is perhaps the subject regarding which, there was, some years ago, considerable doubt. However, important information regarding the oxidation of magnetite to Fe_2O_3 has been obtained by Feitknecht and his colleagues.[3] Magnetite crystals smaller than 3000 Å always yield $\gamma\text{-}Fe_2O_3$, whilst larger particles are first oxidized to a substance of cubic structure with a composition intermediate between Fe_3O_4 and Fe_2O_3, which then undergoes disproportionation into Fe_3O_4 and $\alpha\text{-}Fe_2O_3$ in the final stages of the reaction. Below 220°C, $\gamma\text{-}Fe_2O_3$, although metastable, remains unchanged indefinitely; at higher temperatures it is transformed to the stable $\alpha\text{-}Fe_2O_3$. Certain impurities, especially Al, stabilize the γ-form.

Several investigators have reached the conclusion that whilst, of the three layers formed during the oxidation of iron, the wüstite and Fe_3O_4 phases grow

1. I. M. Ritchie, Chap. 10 of 'Chemisorption and Reactions of Metallic Films' (ed. J. R. Anderson), Academic Press, 1971.
2. J. Wingrove, *J. Iron Steel Inst.* 1970, **208**, 258.
3. K. J. Gallagher, W. Feitknecht and U. Mannweiler, *Nature* 1968, **217**, 1118.

by the outward movement of cations, the Fe_2O_3 layer grows by oxygen moving inwards. This view has been tested by Bruckman and Simkovich,[1] by means of marker experiments; the markers consist of platinum wire 0·08 mm diameter. The first experiments seemed to indicate that there is outward movement through the Fe_2O_3 layer which is reduced to Fe_3O_4 at its inner interface; owing to the movement of Fe ions outwards, the thickness of the Fe_2O_3 layer increases with time, in spite of the reduction. In order to decide the question under conditions uncomplicated by the existence of more than one layer, work was conducted on films consisting entirely of Fe_3O_4 (without metal substrate); this was converted to Fe_2O_3 during the experiment. It is calculated that, when transformation to Fe_2O_3 is complete, the marker position should be 1/18 of the total thickness of the Fe_2O_3 layer below the interface with the gas phase, assuming that growth is connected with the movement of cations. It is evident that the difference between the cases corresponding to the anion diffusion and cation diffusion mechanisms respectively is only a small one, but photo-micrographs seem to indicate that the situation attained supports the idea of the movement of cations. The experimenters are probably right in their conclusions, but in view of the small differences between the expected situations corresponding to the two mechanisms, further investigation would be welcomed.

The cracking of oxide-scales on iron has been studied by Bruce and Hancock,[2] using a novel method. A freely supported rod is exposed to vibration in an oxidizing atmosphere. Normally there is a gradual change of frequency as the oxide-film thickens, but if it cracks, there may be a sudden change. Such sudden changes are noted on iron, whereas on nickel the change is continuous, showing that the scale is adherent and free from cracks; this goes far to explain the superior protection given by films on nickel. (The reader will notice, however, that the experiments on the two metals were carried out at different temperature ranges.)

Hancock's calculations,[3] based on Young's modulus and plasticity, serve to explain the friability of surface scales on iron formed in the range 570° to 800°C, and the better adhesive properties of oxide-scales formed on nickel in the range 800° to 1000°C. He remarks: 'For a thorough understanding of oxidation behaviour, the mechanical properties of the surface oxides are of equal importance to the diffusion kinetics of the oxidation process'.

The breakdown of scales formed on pure iron with only 0·003% of carbon has been studied by Boggs and Kachick.[4] Blistering occurs when oxidation takes place at 500°C at 10 torr oxygen, with alternating periods of high and low oxidation rate—a phenomenon attributed to the cation vacancies arriving more quickly than sinks are provided; this leads to cavities. Later Fe_3O_4 is oxidized to Fe_2O_3, with production of strains, cracking and oxygen leakage, which in turn leads to the formation of a fresh layer of oxide. At 700 torr there are no blisters, and a smooth, nearly parabolic curve is obtained on

1.   A. Bruckman and G. Simkovich, *Corr. Sci.* 1972, **12**, 595.
2.   D. Bruce and P. Hancock, *J. Inst. Met.* 1969, **97**, 140, 148.
3.   P. Hancock, *Werkstoffe u. Korr.* 1972, **21**, 1002.
4.   W. E. Boggs and R. H. Kachick, *J. electrochem. Soc.* 1969, **116**, 424.

plotting oxidation against time. The presence of cementite in the substrate also tends to prevent blistering, probably by providing additional sinks.

The effect of pressure on the oxidation of iron in the range 350° to 400°C has been studied by Graham and Cohen.[1] They find that the initial rate increases with increasing pressure, but that after 180 minutes the weight gain is independent of pressure. During the early (rapid) stage, there is cation diffusion outwards through the magnetite film and this is rate-controlling. Later, α-Fe_2O_3 is probably formed and the rate becomes slow, and it is thought that the inward diffusion of oxygen is now rate-controlling. There are certain apparent discrepancies with the work of Boggs, quoted above, but the experimental conditions were not the same.

The behaviour of iron above 800°C in an inert gas containing small amounts of O_2 has been studied by Rahmel.[2] He considers that the rate-controlling step is the transport of oxygen from the gas phase to the surface of the scale layer, which consists exclusively of FeO. Under conditions of laminar flow, the rate constant is proportional to the O_2 content of the gas mixture as well as to the square root of the gas-flow velocity; it is almost independent of the temperature.

Cobalt. The initial oxidation rate of cobalt in the temperature range 320° to 520°C has been found to follow the direct logarithmic law;[3] and to be independent of the oxygen pressure.

Interesting *moiré* patterns have been observed on thin films of cobalt oxide, obtained by annealing metallic cobalt in H_2 and then in argon, oxidizing at 320° to 520°C in pure dry O_2, and finally dissolving away the metal with a solution of I_2 in 10% KI. The patterns are attributed to the overlapping of two similar lattices with a slight relative disorientation.

Nickel. An investigation in Cohen's laboratory[4] shows that cold-worked nickel becomes oxidized more quickly than annealed nickel; on the former the apparent parabolic constant shows a high initial value but tends to decrease with time. On annealed nickel the value of the parabolic constant is lower, the grain-size larger, and both of these change little with time. The results can be interpreted on the view that the grain-boundaries provide paths of easy diffusion through the NiO film. The rate constant is 1000 times smaller when the NiO layer forms a single crystal, so that easy diffusion paths are absent. These results confirm previous work on iron, where also it had been found that the cold-worked material was oxidized more quickly; but the existence of several layers here complicates interpretation.

If the interpretation offered for nickel is correct, it may be necessary to revise some earlier views. The new work shows that two grades of pure nickel similarly treated yielded different oxidation rates by producing different

1.   M. J. Graham and M. Cohen, *J. electrochem. Soc.* 1969, **116**, 1430.
2.   A. Rahmel, *Werkstoffe u. Korr.* 1972, **23**, 96.
3.   B. Chattopadhyay and K. J. C. Measor, *J. Mat. Sci.* 1969, **4**, 457; *Nature* 1968, **220**, 1319.
4.   D. Caplan, M. J. Graham and M. Cohen, *J. electrochem. Soc.* 1972, **119**, 1205. D. Caplan and M. Cohen, *Tokyo Cong. Ext. Abs.* 1972, p. 293.

grain-sizes in the oxide layer. It is known that certain alloying conditions produce changes in the oxidation rate, and these differences have often been explained by assuming a Wagner–Hauffe valency effect. It now appears at least possible that the effect has really been due to the influence of the alloying constituent on grain-size.

Nickel has proved a suitable metal for the study of the opening stages of oxidation. The important researches of Germer[1] deserve study. Oxygen is initially adsorbed upon a {110} surface with unit sticking probability. At $\frac{1}{3}$ coverage there are two different phases. Between $\frac{1}{2}$ and $\frac{3}{4}$ coverage at 300°C there is a set of superstructures of gradually increasing O-content, made up of units uniformly distributed. During removal of O by heating in H_2 at 300°C, sharp spots representing the superstructures are seen in reverse order until the surface is clean.

This work reveals the formation of several successive structures before ever the first monolayer is complete. At each step there is a pattern of O and Ni atoms on the surface, and at any particular step there will be more O and less Ni than at the preceding step. With increasing O-content for the surface layer, the probability of oxygen sticking becomes progressively lower. The sixth step consists of the formation of true NiO, showing an epitaxial relationship with the metallic nickel below.

There had been a lack of reproducibility in the earlier work, and this was due to the presence of CO in the residual gas, which affects the sequence of the various patterns. Recent work in Cohen's laboratory[2] has demonstrated that cavities in NiO scales contain CO_2 at appreciable pressures and that the amount of cavity produced is related to the carbon content of the metallic nickel.

Another research in Cohen's laboratory[3] has shown that the oxidation rate of zone-refined nickel varies greatly according to the pre-treatment. Two surface conditions were studied:

(1) *Hot-bare conditions* produced when the oxide-film remaining after electropolishing in H_2SO_4 has been removed by reduction in hydrogen and the specimen is brought up to the experimental temperature before oxygen is admitted.

(2) *Furnace-raised conditions* produced by raising an electropolished or etched specimen to the experimental temperature in the presence of oxygen.

The hot-bare specimens suffered oxidation at 600°C very much more quickly than the furnace-raised specimens. The values of the parabolic rate constant could vary by a factor of 10^4 according to the character of the oxide present at the start of the experiment. The results can be explained if it is assumed that the activation energy for diffusion through the lattice is 52 kcal/mole or more, whereas for movement along leakage paths such as the boundaries of

1.   L. H. Germer, J. W. May and R. J. Szostak, *Bull. Amer. phys. Soc.* 1967, **12**, 549.
2.   M. J. Graham and D. Caplan, *J. electrochem. Soc.* 1973, **120**, 769.
3.   M. J. Graham, G. I. Sproule, D. Caplan and M. Cohen, *J. electrochem. Soc.* 1972, **119**, 883.

grains or subgrains it is only 37 kcal/mole. On the hot-bare material at 700°C the initial oxidation is extremely rapid and the film contains many mismatch boundaries which act as paths for easy diffusion. This rapid oxidation rate leads to varying degrees of separation between oxide and metal and hence there is poor reproducibility at 700°C. At 500° and 600°C there is no separation, and the parabolic law is obeyed faithfully, with little change of the rate constant with time. The oxide separation at 700°C is due to the formation of voids at the metal–oxide interface, probably as a result of cation vacancies moving inwards. For the first hour of oxidation at 700°C insufficient voids are, it seems, nucleated to reduce the oxidation rate to any great extent; but when the oxide thickness has reached about 2μm, more complete interfacial detachment occurs. The originally formed oxide still continues to thicken, but a new layer of oxide is produced at the base of the voids, apparently through the presence of traces of CO, which acts as an oxygen-carrier, transporting oxygen between the oxide-film forming the roof of a void and that starting to grow on the metal base. Thus a duplex scale develops, as revealed by electron microscope photographs.

The concept of easy paths was later[1] used to explain the different oxidation rates of Ni single crystals. Ni was electropolished in H_2SO_4 and the oxidation rates determined by a manometric method; the specimens were examined by electron diffraction, X-ray emission and electron microscopy both before and after analysis. Where a single orientation persists during the oxide-film formation, the oxidation rate is slow; this is the case for {112} faces oxidized at 600°C. Where twin-boundaries appear, affording easy paths, as on {111} faces, oxidation is more rapid. On {100} it is more rapid still, since here two sets of twin-boundaries provide 'a greater density of incoherent boundaries and hence more rapid Ni transport'.

The oxidation of nickel at 250°C has been discussed by Wagner,[2] basing his arguments in part on the tracer research of Boreskov.[3] The dissociation of molecular O_2 is assumed to be the rate-determining step. The velocity is inversely proportional to the square of the thickness of the NiO film. This is explained on the assumption that the rate of dissociation of molecular O_2 at the NiO–gas interface is proportional to the local concentration of excess electrons, which increases with decreasing distance from the Ni–NiO interface.

Iridium. The only solid oxide of Ir known is IrO_2, which possesses a rutile structure. It decomposes at 1124°C in the presence of 1 atm. O_2. Nevertheless, although IrO_2 is stable at room temperatures, no oxidation occurs on metallic Ir even on long exposure. The matter has been studied by Fortes and Ralph.[4] They found that a layer of adsorbed gas is formed, which can be removed by raising the applied field. Specimens placed in a furnace at 700°C in air took 80 s to reach the furnace temperature. Treatments shorter than 70 s produced

1. M. J. Graham, R. J. Hussey and M. Cohen, *J. electrochem. Soc.* 1973, **120**, 1523.
2. C. Wagner, *Corr. Sci.* 1970, **10**, 641.
3. G. K. Boreskov, *Disc. Faraday Soc.* 1966, **41**, 263.
4. M. A. Fortes and B. Ralph, *Proc. Roy. Soc. (A)* 1968, **307**, 431.

no oxide; those longer than 90 s gave complete coverage with oxide. In order to obtain surfaces incompletely oxidized, times between 75 and 80 s were used. In such cases {111} and {110} regions remained free from oxide, whilst {001}, {113} and {102} planes became covered.

Copper. The conductivity of a CuO semiconductor has been studied in Indian work.[1] The addition of Li increases the number of positive holes and improves conductivity; Ga decreases the number of positive holes and diminishes conductivity.

Sub-microscopic faceting has been studied by Swanson and Uhlig.[2] It occurred when a Cu surface was heated in H_2, the {111} faces being favoured, or in N_2, in which the {100} faces were favoured; it is possible that this was due to traces of O_2 in the nitrogen. The subsequent formation of a thin film when the surface was heated in O_2 at 200°C was determined by the rearranged structure rather than by the original orientation of the single crystal. This may explain the marked effect of gaseous pre-treatment of copper either in the monocrystalline or polycrystalline state, and also suggests why there has been lack of agreement between results previously published.

Zinc. Lucas[3] has studied the oxidation of zinc at 300° to 400°C. The oxidation of single crystals varies slightly with their orientation, the rate for polycrystalline zinc being intermediate. The growth law can be expressed as $W = Kt^n$. Here n varies at 300°C between 0·29 and 0·46, but at 390°C the variation is smaller, namely between 0·43 and 0·46. The surfaces develop whiskers, and high values of n represent a copious whisker growth. The addition of 1% Pb reduces oxidation rate and whisker formation.

Lead. The thermal oxidation of lead powder is found to obey the parabolic law;[4] pseudo-tetragonal PbO is formed. Values found for the activation energy suggest that oxidation occurs by the diffusion of O through interstitial tetrahedral sites in the PbO.

Aluminium. The defect structure of thin oxide-films formed on aluminium has been studied by Pryor.[5] Films less than 50 Å thick contain a higher Al/O ratio than thicker films; apparently they contain loosely bound electrons in surface traps, for they eject electrons when irradiated in blue light; as the oxide-film becomes thicker, the power to emit electrons diminishes. The metal excess could be explained in three ways:

(1) deficiency of O-ions with presence of trapped electrons (F-centre structure),
(2) excess of Al-ions in normally vacant sites, and
(3) excess Al-ions in interstitial sites.

1. D. P. Bhattacharyya and P. N. Mukherjee, *J. appl. Chem.* 1972, **22**, 889, esp. Fig. 9, p. 895.
2. A. W. Swanson and H. H. Uhlig, *J. electrochem. Soc.* 1971, **118**, 1325.
3. R. W. Lucas, *J. Inst. Met.* 1971, **99**, 69.
4. N. E. Bagshaw, J. R. Feeney and M. R. Harris, *Brit. Corr. J.* 1969, **4**, 301.
5. M. J. Pryor, *Oxidation Met.* 1971, **3**, 523.

Pryor favours the second explanation. γ-Al_2O_3 has a pseudo-spinel structure. On a true spinel there are 24 cation sites: in γ-Al_2O_3 only $21\frac{1}{3}$ of the 24 sites can be occupied if it is to possess stoichiometric composition. It is considered that crystalline oxide is essentially stoichiometric γ-Al_2O_3, but that the so-called 'amorphous' material present in thin films (which shows strong pseudo-spinel reflection but has poorly developed long-range structure) contains a lower number of cation vacancies. This explains the higher a.c. resistivity of the thin films; an amorphous film (128Å thick) has a resistivity of 1×10^{10} ohm-cm, whereas a crystalline film (217 Å thick) gives only $4\cdot6 \times 10^7$ ohm-cm.

The oxidation of aluminium around 525°C has been studied by Gray and Pryor.[1] Two layers are formed. There is an inner formation of crystalline γ-Al_2O_3 which occurs as a series of expanding cylinders spreading laterally from the points where they originate, and obeying the laws for expanding circles (see 1960 book, p. 939). There is an outer film of amorphous Al_2O_3 thickening according to the parabolic law in a manner completely independent of the development of the inner crystalline material. The outer film is the result of the outward diffusion of Al^{3+} ions, and there is a sharp break in the value of the parabolic rate constant at 525°C. The inner crystalline film is due to inward movement of oxygen and there is no break at 525°C; the inner crystalline material contributes nothing to film-resistance.

Zirconium. Japanese work[2] suggests that the breakaway noted in Zr and its alloys may be connected with a polymorphic transformation. High temperature X-ray research shows that the black layer formed before the breakaway contains a considerable amount of metastable tetragonal oxide. The colour change from black to white occurs simultaneously with the breakaway and with a polymorphic transformation to monoclinic oxide.

The defect structures of ZrO_2, and also those of Nb_2O_5, and Ta_2O_5, have been studied by Kofstad.[3] He considers that oxygen vacancies are the most important factor in causing variation of composition, except perhaps in the part of the range most deficient in oxygen, where interstitial cations may be the main cause. At high pressures of gaseous oxygen, there is small deviation from stoichiometry, and the effect of impurities becomes important.

Niobium. Sheasby[4] has studied the oxidation of Nb between 450° and 720°C. During the initial oxidation there is approximate obedience to the parabolic law and the scale appears compact; later blisters and cracks appear and there is a transition to linear kinetics. The Nb_2O_5 layer contains lenticular fissures parallel to the surface, giving the scale a laminated structure.

An interesting study of the oxidation of $Nb_{12}O_2$; to Nb_2O_5 is provided by Browne and Anderson.[5] The structure of $Nb_{12}O_{29}$ is an open one, with channels

1. T. J. Gray and M. J. Pryor, *Mat. Sci. Res.* 1969, Chap. 26, p. 446 (Plenum Press).
2. T. Nakayama and T. Koizumi: see *Corr. Abs. (Houston)* 1969, p. 154.
3. P. Kofstad, *Corrosion (Houston)* 1968, **24**, 379.
4. J. S. Sheasby, *J. electrochem. Soc.* 1968, **115**, 695.
5. J. M. Browne and J. S. Anderson, *Proc. Roy. Soc. (A)* 1974, **339**, 463.

which permit an easy solid state reaction. As a result, the oxidation can proceed at temperatures as low as 110°C; above 440°C, a second mechanism, involving reaction and rearrangment at the crystal surface, becomes competitive. Both end products have a structure similar to that of the starting material, being derived from it by crystallographic shear.

Tantalum. Stringer and Dooley[1] have studied the oxidation of Ta and Ta-base alloys in the range 800° to 1050°C. They find that on Ta the main product is Ta_2O_5, which grows, adhering to the metal substrate, under considerable compressional stress, but ruptures when a critical thickness is reached. This sequence of alternate growth and failure is repeated, leading to a laminated scale, which grows at a rate approximately rectilinear on a macroscopic time-scale. Above 800°C the growth seems to be a 'repeated parabolic' process, the thickening during the short adherent stages between the ruptures being parabolic. As a consequence the rectilinear rate constant depends on the lamination thickness and thus on the crystal orientation.

Chromium. The production of facets on certain planes of chromium has been studied by Lee and his colleagues.[2] These planes are stable in ultra-high vacuo, but develop facets when heated at very low oxygen pressures. The faceting is reversible and the smooth surface may be regenerated by heating in the absence of oxygen. The activation energy for this transformation is high. The {100} surface of Cr is stable in O_2, but the {110} surface develops {100} facets. The analogous changes have been studied on Mo and W; here the {100} surfaces develop facets of the {110} and {211} types, the facets being preceded by the formation of various ordered structures of chemisorbed oxygen.

Molybdenum. Studies of the initial stages of oxide-film formation by Lee[3] have revealed three stages involving (1) a chemisorbed oxygen layer, (2) a thin, even layer of oxide and (3) small nuclei of crystalline oxide showing one or more epitaxial relationships with the substrate. It is believed that a contact potential difference exists across the growing film which influences the dissolution rate of substrate cations in the oxide. At first, this dissolution could proceed too fast for the rate of arrival of oxygen to support; thus the oxidation rate is determined by processes occurring at the oxide–gas interface. Later, the field-assisted ion-solution becomes rate-determining. At the point where the kinetics of oxide growth change, substrate facet planes with low work function support an enhanced oxidation rate, leading to the growth of local centres of oxide, which may be identified with epitaxial nuclei.

Tungsten. The formation of isolated oxide particles preceding the growth of a uniform film has been observed on several metals and may be a general phenomenon. The changes have been studied on tungsten by Brenner.[4]

1. J. Stringer and R. B. Dooley, *Amsterdam Cong.* 1969, p. 346; *Corr. Sci.* 1970, **10**, 265.
2. H. M. Kennett, A. E. Lee and J. M. Wilson, *Proc. Roy. Soc.* (*A*) 1972, **331**, 429.
3. A. E. Lee (partly with H. M. Kennett), unpublished work.
4. S. S. Brenner and W. J. McVeagh, *J. electrochem. Soc.* 1968, **115**, 1247.

Here the particles appear to be of irregular shape. It is thought that oxygen penetrates into the metal lattice and forms a supersaturated solution, from which nuclei of oxide are precipitated.

Lee[1] has studied the processes which occur between the initial chemisorption stage up to the thicknesses suitable for examination by the gravimetric and similar techniques commonly used today; he used low pressures (e.g. 10^{-8} torr O_2). In previous work it had usually been thought that the oxide–metal interface remains parallel to the original clean metal surface. It is now shown, however, that faceting occurs on the {100} faces, which is difficult to ascribe to a decrease in interfacial energy, since the surface energies for {100} and {111} faces are much the same. The facet formation is ascribed to a difference between the evaporation rates of W (as WO_3) from different planes. The facets formed on a {100} face consist of {110} types, themselves made up of a hill-and-valley structure involving {211} planes. The {110} planes do not develop facets.

The chemisorption of oxygen on W ribbons between 1415 and 1650 K has also been studied by Lee[2]. On {100} faces there are three states which appear before the first monolayer is completed (at 10^{15} atoms atoms/cm^2). Formation of a true oxide phase is not apparent until the coverage reaches one monolayer. Before this is achieved, the observations show that:

(1) Up to half a monolayer there is decreasing sticking probability, low dipole moment, low activation energy and low polarizability.
(2) Between 5 and 8·5 $\times$ 10^{14} atoms/cm^2, a constant value for the sticking probability is met with, along with high dipole moment and moderate activation energy.
(3) Above that range, the dipole moment becomes zero.

Avery,[3] using the LEED method, has studied the oxidation of a {110} surface of a tungsten crystal in 2×10^{-3} torr O_2 at 1000 K for 80 s. The patterns obtained are thought to result from three pairs of twin-related WO_3 {11$\bar{1}$} nuclei each faceted to pyramids exposing three well developed {110} type surfaces; the nuclei exhibit the same monoclinic symmetry as exists in bulk WO_3.

Further References

Review Articles. An interesting survey of oxidation research over a period of 50 years, commencing with the work of Tammann, which was earlier than that of Pilling and Bedworth, is contributed by Wagner.[4] A seminar on the Oxidation of Metals and Alloys,[5] organized by the American Society of Metals in 1970, included papers by J. Bénard, N. F. Mott, J. S. Kirkaldy, W. W. Smeltzer, C. E. Birchenall, G. C. Wood and others. Atten-

1. A. E. Lee and K. E. Singer, *Proc. Roy. Soc.* (A) 1971, **323**, 513.
2. A. E. Lee and D. A. Pethica, *Proc. Roy. Soc.* (A) 1969, **309**, 141.
3. N. R. Avery, *Surface Sci.* 1972, **33**, 107.
4. C. Wagner, *Werkstoffe u. Korr.* 1970, **21**, 886.
5. 'Oxidation of Metals and Alloys', A.S.M. Seminar Books, 1972.

tion may be drawn to a discussion of the properties and reactions of clean metal surfaces by Gasser.[1]

Hayfield[2] has contributed an interesting review of the development of ellipsometry, with a tribute to the early work of Tronstad, and its continuation by Winterbottom after Tronstad's untimely death. Much information regarding adsorbed layers, largely based on ellipsometry, was provided at a Faraday Society Symposium.[3]

Papers. 'Ellipsometric apparatus for surface investigation, especially film thickness'.[4]

'Point defect diffusion and oxidation kinetics'.[5]

'A kinetic study of the initial oxidation of Ta (110) using oxygen K, X-ray emission'.[6]

'Stress generation and adhesion in growing oxide-scales.'[7]

'Some factors influencing the adherence of oxides on metals.'[8]

'The mechanical properties and breakdown of surafce oxide films at elevated temperatures.'[9]

'Corrosion in aqueous solution and corrosion in gases at elevated temperatures—Analogies and discrepancies.'[10]

1. R. P. H. Gasser, *Quarterly Rev. (Chem. Soc.)* 1971, **25**, 223.
2. P. C. S. Hayfield, 'Ellipsometry in Corrosion Technology', Chap. 2 in Vol. II of 'Advances in Corrosion Science and Technology', editors M. G. Fontana and R. W. Staehle (Plenum Press), 1972.
3. *Symp. Faraday Soc.* 1970, **4**. See esp. paper by J. O'M. Bockris, M. A. Genshaw and V. Brusic (p. 177) and introduction by P. C. S. Hayfield (p. 7).
4. F. Gorn and K. J. Vetter, *Z. phys. Chem. (neue Folge)* 1972, **77**, 317.
5. F. Morin, *Oxidation Met.* 1973, **6**, 65, 79 (2 papers).
6. P. B. Sewell, D. F. Mitchell and M. Cohen, *Surface Sci.* 1972, **29**, 173.
7. J. Stringer, *Werkstoffe u. Korr.* 1972, **23**, 747.
8. R. F. Tylecote and W. K. Appleby, *Werkstoffe u. Korr.* 1972, **23**, 855.
9. P. Hancock and R. C. Horst, Chap. 1 of 'Advances in Corrosion Science and Technology', edited by M. G. Fontana and R. W. Staehle (Plenum Press), 1974.
10. C. Wagner, *Werkstoffe u. Korr.* 1974, **25**, 327.

APPENDIX TO CHAPTER II

The General Oxidation Equation

T. Hurlen[1]

Introduction. On the basis of the theory of absolute reaction rates (Eyring[2]), it is possible to derive a general oxidation equation for cases of control by rate of transport of material in the oxide-film. In two limiting cases, the general equation may be replaced by simpler forms known as the parabolic and inverse logarithmic equations. Under certain conditions it may satisfactorily be replaced also by a cubic type of equation.

The general equation and its importance to metal oxidation studies seem first to have been pointed out by Evans,[3] who derives a simplified, formal expression for this equation and also showed how it could be reduced to the more conventional oxidation equations. On this basis, Hoar[4] later calculated the limiting conditions of temperature and oxide-film thickness under which the general equation for some metals (Fe, Zn, Cu) may be satisfactorily represented either by the inverse logarithmic equation or by the parabolic equation.

Gulbransen[5] seems to have been the first to apply the theory of absolute reaction rates to metal oxidation. He thereby derived an absolute expression for the parabolic rate constant.

To our knowledge, nobody has so far applied the absolute rate theory in deriving the general oxidation equation.

From the absolute rate theory, it is possible to obtain an expression for the *net rate* of transport of a species in an ideal solution under the action of a field established by an electrical potential gradient, with accompanying concentration gradient. Expression (3) of p. 553 of the Glasstone–Laidler–Eyring book[2] can be written

$$v = \lambda \frac{kT}{h} \, e^{-\Delta G_0^\ddagger/kT} \left[C \, e^{-\alpha \lambda z F \mathrm{d}\phi/kT\mathrm{d}x} - \left(C + \lambda \frac{\mathrm{d}\ln c}{\mathrm{d}x} \right) e^{-(1-\alpha)\lambda z F \mathrm{d}\phi/kT\mathrm{d}x} \right]$$

where v represents the *net* rate of transport, R is the gas constant, k the Boltzmann constant, h the Planck constant, T the Kelvin temperature, F the Faraday, $\Delta G_0^\ddagger$ the standard chemical energy of activation, C the concentration of the species in question, ϕ the inner potential, x a distance in the direction of transport, z the charge carried by a single migrating entity,

1. T. Hurlen, *Acta chem. Scand.* 1959, **13**, 695; some changes and additions were made in 1973.
2. S. Glasstone, K. J. Laidler and H. Eyring, 'Theory of Rate Processes' (McGraw-Hill, New York), 1941.
3. U. R. Evans, *Rev. pure and appl. Chem., Australia* 1955, **5**, 1.
4. T. P. Hoar, *J. chim. Phys.* 1956, **53**, 826.
5. E. A. Gulbransen, *Ann. N.Y. Acad. Sci.* 1954, **58**, 830.

λ the distance between equilibrium positions for these entities, and α the so-called symmetry factor.

Here the expression outside the bracket represents the specific rate which would occur at unit concentration in absence of a field, but which would lead to no *net* transfer, since transfer would be taking place at equal rates in both directions. The two terms inside the bracket introduce the effect of concentration and of the field; the first term brings in the accelerating effect of the field on movement in one direction, and the second term the retarding effect on movement in the opposite direction.

The equation can conveniently be used in the form

$$v = \lambda C \frac{kT}{h} \mathrm{e}^{-\Delta G_0 \ddagger} \, {}^T \left[\mathrm{e}^{-\alpha \lambda z F \mathrm{d}\phi / RT \mathrm{d}x} - \left(1 + \lambda \frac{\mathrm{d} \ln C}{\mathrm{d}x} \right) \mathrm{e}^{(1-a)\lambda z F \mathrm{d}\phi / RT \mathrm{d}x} \right] \quad (1)$$

According to the theories of Frenkel[1] and Schottky,[2] ionic crystals usually contain reversible defects (ions in interstitial positions and ion vacancies in connection with an equivalent number of quasi-free electrons and electron holes). It seems to be well recognized that the transport of material in ionic crystals may be described as a transport of such defects. As a normal crystal may, with good approximation, be regarded as an ideal solution of reversible defects in a perfect crystal, the rate of transport of any one type of defect in the crystal should be represented by equation (1) above.

The thickening of an oxide-film on a metal certainly requires transport of material across the film. The defects, which may take part in this transport, generally are:

$$\left. \begin{array}{l} \mathrm{O}_{\mathrm{v}}^{\cdot\cdot} \text{ (oxide ion vacancies)} \\ \mathrm{O}_{\mathrm{i}}^{-} \text{ (oxide ion interstitials)} \\ \mathrm{M}_{\mathrm{v}}^{z/} 1 \text{ metal ion vacancies)} \\ \mathrm{M}_{\mathrm{i}}^{z+} \text{ (metal ion interstitials)} \end{array} \right\}$$

where the symbols $\cdot$ and $/$ represent a positive and a negative surplus charge.

During oxidation, oxide–ion vacancies and metal–ion interstitials will move outwards and oxide–ion interstitials and metal–ion vacancies inwards through the film.

If we now choose the outward direction (from the metal to the gas) as our single reference direction or x-direction (instead of having to consider two transport directions), the total net rate of transport of defects through a plane at a distance x from the metal/oxide interface is given by:

$$v_x \text{ (total)} = \sum_i \pm v_{i,x} \quad (2)$$

$v_{i,x}$ means here the rate of transport of the defect species i in the x-direction and is given by equation (1) when $(\mathrm{d}\phi/\mathrm{d}x)$ and $(\mathrm{d}C/\mathrm{d}x)$ are considered gradients also in this direction (not in the actual transport directions). The signs $+$ and $-$ apply to positively and negatively charged species, respectively. In this way, all the rate terms in (2) become positive.

1.   J. Frenkel, *Physik* 1926, **35**, 652.
2.   W. Schottky, *Z. phys. Chem.* (B) 1930, **11**, 163.

When + and − are put together as in the above equation, it will in this paper always mean that the upper sign applies to positive species and the lower to negative ones.

As both the concentration and the concentration gradient of a defect species as well as the electric potential gradient at any given time may vary from point to point within the film, it is obvious that v_x (total) also may vary It is a question, therefore, how the momentary rate of oxidation is connected to the various momentary rates of transport in the oxide-film.

What is actually measured, however, by gravimetric and volumetric methods of investigation of oxidation of metals, is the rate of entrance of oxygen from the gas phase into the oxide phase. The other rate processes in the system (such as the transfer of metal ions from the metal phase into the oxide phase and the migration of ions in the oxide phase) are not directly susceptible to measurements of this kind, and information on these processes must be obtained either by theoretical deductions from the oxidation measurements or by other types of measurements

The measured oxidation rate can be assumed to be equal to the rate of entrance of oxygen into the oxide, provided the concentration of chemisorbed oxygen on the oxide surface does not change during oxidation. On this basis, the oxidation rate, $v(ox)$, may be represented by the following formula:

$$v(ox) = \lim_{x \to y} v_x \text{ (total)} \tag{3}$$

where y is the oxide-film thickness, and where the various rates must be expressed in equivalent units (number of equivalents per unit area per unit time).

Equation (3) combined with (1) and (2) may thus be regarded as a general oxidation equation for cases of control by rate of transport of material in the oxide-film. In practice, one also has to add the requirement that the area undergoing oxidation must be constant with time as experimental oxidation rates usually are determined on the basis of the apparent surface area of the specimens at the start of the experiment.

The general equation will be somewhat easier to handle if we introduce the following approximations:

$$\left.\begin{array}{l} v(ox) = \dfrac{1}{V}\mathrm{d}y/\mathrm{d}t \\[4pt] \mathrm{d}\phi/\mathrm{d}x = \Delta\phi/y \\[2pt] \mathrm{d}C/\mathrm{d}x = \Delta C/y \\[2pt] F/\Delta\phi = \Delta G \\[2pt] \alpha = \tfrac{1}{2} \end{array}\right\} \tag{4}$$

and the following substitution:

$$V \lambda kT/h \cdot e^{-\Delta G_0 \ddagger /RT} = K \tag{5}$$

where V is the equivalent volume of the oxide, $\Delta\phi$ the total potential increase in the film from metal to gas, ΔC the total concentration increase (for species i) in the film from metal to gas, and ΔG the energy of formation of one gram equivalent oxide at the actual oxidation conditions.

On this basis (equations (1) to (5) above) the general oxidation equation may be written

$$\frac{dy}{dt} = \sum_i \pm K_i C_{i,y}\left[e^{-\lambda_i z_i \Delta G/2RTy} - \left(1 + \lambda_i \Delta C_i / y C_{i,y}\right)e^{\lambda_i z_i \Delta G/2RTy}\right] \quad (6)$$

where $C_{i,y}$ means the limiting value of C_i for $x = y$ (i.e. when approaching the oxide–gas interface).

Deductions from the General Equation. The general oxidation equation (6) is rather clumsy. Under special conditions, however, it may be replaced by equations of simpler forms. This greatly increases its applicability and facilitates the testing of its validity.

Following previous treatments by Evans[1] and Hoar,[2] we shall in the following discuss how the more classical parabolic (Wagner) and inverse logarithmic (Mott–Cabrera[3]) equations may be regarded as limiting cases of the general oxidation equation. It shall also be shown how the general equation under certain conditions may be approximately, though satisfactorily, replaced even by a cubic type of equation.

The Parabolic Equation. At sufficiently high temperatures and large film thicknesses, the general equation (6) may be reduced to the simpler form:

$$\frac{dy}{dt} = \frac{1}{y}\sum_i \mp K_i \lambda_i C_{i,y}\left(\frac{z_i \Delta G}{RT} + \frac{\Delta C_i}{C_{i,y}}\right) \quad (7)$$

which is the parabolic oxidation equation ($y^2 = kt$) in its differentiated form.

This simplification is easily performed by extending the exponential functions in (6) and omitting the higher powers of the exponents. This is very nearly the same as omitting p^2 in the expression $6 + p^2$, where in our case p is given by:

$$p = \mp \lambda z \Delta G/2RTy \quad (8)$$

The parabolic equation should thus be expected to represent the oxidation rate within about 10% accuracy when $p^2 < 6/10$, which means:

$$y > \mp 0 \cdot 65 \lambda z \Delta G/RT \quad (9)$$

A similar requirement has previously been pointed out by Hoar.[4]

The Inverse Logarithmic Equation. At sufficiently low temperatures and small film thicknesses, one of the exponential terms in (6) vanishes as compared to the other one, and it is easily seen that the general equation under these conditions may be reduced to:

$$\frac{dy}{dt} = \sum_i K_i C_{i,y}\, e^{\mp \lambda_i z_i \Delta G/2RTy} \quad (10)$$

1. U. R. Evans, *Rev. pure and appl. Chem., Australia* 1955, **5,** 1.
2. T. P. Hoar, *J. chim. Phys.* 1956, **53,** 826.
3. N. Cabrera and N. F. Mott, *Rep. Progress Phys.* 1949, **12,** 163.
4. T. P. Hoar, *J. chim. Phys.* 1956, **53,** 826.

Through integration and some simplification we obtain:

$$\frac{1}{y} = \sum_i \pm 2 \cdot 303 \frac{2RT}{\lambda_i z_i \, \Delta G} \left[\log \frac{K_i C_{i,y}}{y} + \log t \right] \tag{11}$$

assuming $y = 0$ when $t = 0$.

For small values of y, $\log (1/y)$ may be considered constant as compared to $1/y$, whereby (11) represents a so-called inverse logarithmic oxidation equation $1/y = A - B \log t$. It also gives an interpretation of the constants (A and B) in this equation.

In integrating (10) to obtain (11), one has to make the requirement that $e^{\pm \lambda z \Delta G / 2RT y} \ll 1$ and in reducing (6) to obtain (10), the necessary requirement is that $e^{\pm \lambda z \Delta G / 2RT y} \ll e^{\mp \lambda z \Delta G / 2RT y}$. Of these requirements, the former seems to be strongest. As this is the most difficult one to evaluate, however, we shall here assume that the other requirement is sufficient.

The inverse logarithmic equation should thus represent the general equation within about 10% accuracy when

$$y < \mp \lambda z \, \Delta G / 2 \cdot 32 RT \tag{12}$$

This requirement also has previously been pointed out by Hoar.[1]

The Cubic Equation. Under conditions at which $e^p - e^{-p}$ with sufficient accuracy might be replaced by ap^2, where a is a constant and p as given by equation (8), the general equation reduces to:

$$\frac{dy}{dt} = \frac{a}{y^2} \sum_i K_i C_{i,y} \left(\frac{\lambda_i z_i \Delta G}{2RT} \right)^2 \tag{13}$$

when $\Delta C / C_y$ is negligible compared to 1.

This is the cubic oxidation equation $y^3 = k_c t + \text{const.}$) in its differentiated form, and it gives an interpretation of the cubic rate constant (k_c).

In accordance with the requirements suggested there the following film thickness limits may be given for the validity of the cubic approximation:

$$\mp \lambda z \, \Delta G / 1 \cdot 72 RT > y > \mp \lambda z \, \Delta G / 2 \cdot 32 RT \tag{14}$$

To our knowledge, this is the first explanation of the cubic oxidation equation not implying any assumptions as to the semiconductor properties of the oxide. The cubic equation derived by Cabrera and Mott[2] and the one derived by Engell, Hauffe and Ilschner[3] are only valid for p-type conducting oxides.

Other Equations. The general oxidation equation, as derived above, is only applicable to the growth of films which maintain uniform thickness, remain unbroken and are in contact at all points with the metallic basis. It would cease to be accurate if the film develops whiskers, although these add to the weight of oxide less than might at first sight be expected. It will not be true if periodically the film breaks away, blisters or peels, and will therefore be unhelpful for materials like zirconium and its alloys where such phenomena

1.   T. P. Hoar, *J. chim. Phys.* 1956, **53**, 826.
2.   N. Cabrera and N. F. Mott, *Rep. Progress Phys.* 1949, **12**, 163.
3.   H. J. Engell, K. Hauffe and B. Ilschner, *Z. Elektrochem.* 1954, **58**, 478.

are an essential feature of oxidation. If breakdown occurs whenever a film has reached such a thickness that the compressional strain arising from volume difference between metal and oxide reaches a value capable of providing the energy needed for detachment, it may be expected that, in the absence of other complications, the breakdown will occur at regular time intervals, in that case the curve relating weight increase to time will, viewed on a macroscopic scale, be roughly straight and the oxidation over long periods will obey a roughly rectilinear law, although during the short periods between the breakdowns there may be obedience to a parabolic, or even a cubic, law. If, through accumulation of vacancies at the metal–oxide interface and their union to form cavities, the area of contact between metal and oxide decreases, there may be conformity either to the classical logarithmic law or a new logarithmic law; derivation of these laws will be found elsewhere (1960 book, pp. 836–7). The classical logarithmic law is commonly met with at relatively low temperature where the oxide is insufficiently plastic to subside into cavities which are being established and thus maintain contact everywhere between metal and oxide; however the same relationship can be established in other ways—such as blocking of pores (pp. 834–6) and control by electron movement, instead of ionic movement (pp. 829–33). The new logarithmic law has been established for copper under conditions where the classical logarithmic law is not obeyed.

Effect of Other Elements on Oxidation and Film-growth

Preamble

Picture presented in the 1960 and 1968 volumes. Chapter II was concerned with the oxidation of essentially unalloyed metals, although the sections dealing with iron included discussion of mild steels containing substantial quantities of carbon and manganese. In chapter III, the effect of alloying additions, often introduced with a view to improving oxidation resistance, receive consideration. These may enter the oxide-film when oxidation takes place, increasing or decreasing the number of lattice defects; if the valency of the element added differs from that of the main metal, it should be possible, in the absence of complicating factors, to predict whether the concentration of vacant sites in the latter will be increased or diminished, and whether the oxidation rate will be made greater or less. The rules governing such a prediction are due to Hauffe, and the fact that they do not always point to the situation established by experiment should not obscure the fact that Hauffe's work (combined with the principles expounded at an earlier date by Wagner has greatly enhanced our understanding of the oxidation process.

In other cases, the alloying element does not enter the main oxide phase, but accumulates in the metallic phase, or appears as a separate oxide phase at the interface between the metal and the main oxide-film, thus reducing the area of contact between metal and the main oxide. Mathematically, this case is similar to that where cavities appear between metal and oxide, leading to the logarithmic law. Portevin found that the oxidation of steel free from aluminium obeys the parabolic law at 900°C; when aluminium is present, the experimental curves relating oxidation with time depart from that law and become closer to curves which would represent logarithmic growth. Chromium is commonly added to steel to increase the oxidation resistance. It has a greater affinity for oxygen than has iron, and a Cr_2O_3 film is quickly formed at the base of the main scale, but will only protect if it remains uncracked. Wood has pointed out that iron alloys containing only a little Cr oxidize more rapidly at the outset than pure iron, although later they may sometimes oxidize more slowly, since, starting from Cr_2O_3 nuclei, lateral spread occurs until an obstructive film extends over the whole interface. The Cr_2O_3 layer probably owes its protective character to a low ionic conductivity, thickening slowly by movement of cations outwards. Local failure, if it occurs, is due to lifting and cracking. Subsequent behaviour depends on the concentration of Cr remaining at the alloy–scale interface; if this is low, non-protective, stratified scale can develop.

Another matter discussed in chapter III of both books is the effect of a second constituent in the external gas phase. Sulphur compounds usually promote attack. The oxidation of steel at 1000°C by combustion products of fuel (whether coke, coal, oil, producer-gas or coal-gas) is greatly increased by the presence of small amounts of SO_2. Vernon showed that air containing small amounts of H_2S produces on copper interference tints at ordinary temperatures—similar to those formed at slightly elevated temperature on exposure to pure air; the parabolic growth law is obeyed at ordinary temperature in air containing H_2S and in pure air at higher temperatures—although in pure air at ordinary temperature thickening falls off much more quickly than the parabolic law would predict. The films causing interference tints at ordinary temperature are not pure oxide, but oxide containing a certain amount of sulphur ($10\cdot0$ to $15\cdot7\%$ of the oxygen), which presumably increases the number of defects, and thus the rate of film thickening.

A particularly serious type of oxidation occurs above a certain temperature when the scale formed on steel contains either V or Mo; the V is usually derived from oil fuel, whilst the Mo is a constituent of the steel. This is generally attributed to the formation of liquid in the scale or to gaseous oxidation products. Most authorities attribute the damage caused when oil fuel contains V to the formation of a scale that is partly liquid; V_2O_5 is rarely formed as such in the ash, but if Na_2SO_4 is present, a number of complex vanadates can be formed; mixtures of V_2O_5 and Na_2SO_4 are liquid at temperatures well below the melting-point of pure V_2O_5. In general, serious trouble seems to occur only when a liquid phase is present.

Developments during the period 1968–1975 and Plan of the new Chapter. Engineering progress has led to a demand for materials capable of withstanding oxidation at very high temperatures. Much research has been devoted to oxidation resistant alloys, generally based on iron, cobalt or nickel with chromium or aluminium (sometimes with silicon added) to provide a protective layer; interest is shown in the possibility of improving oxidation resistance by adding small amounts of rare-earth metals such as yttrium. Since the films produced by chromium and aluminium are often remarkably protective so long as they remain unbroken, much attention has been devoted to the causes of 'breakaway', but it cannot be claimed that there is universal agreement on that subject. Some useful presentations of the facts as established experimentally have been published; a comprehensive review by Wood deserves special attention.

The first main section of the chapter is devoted to a classification of the various oxidation systems met with in different alloys, mainly based on Wood's ideas. This is followed by a discussion of various alloy systems, starting with ferrous alloys, for reasons explained at the start of chapter II. After these sections, oxidation in the presence of carbon and/or sulphur compounds receives attention, including the formation of sulphide films. The chapter ends with a discussion of catastrophic oxidation in the presence of vanadium compounds.

General

Classification of Systems. A review of the high temperature oxidation of alloys has been provided by Wood.[1] He points out that there is not, and

1. G. C. Wood, *Oxidation Met.* 1970, **2**, 11.

never can be, a single or comprehensive theory of alloy oxidation; rather, there is a sequence of special cases. Even in a single alloy system, several classifications of behaviour are possible, depending on the composition of the oxidizing atmosphere, the temperature, pressure and time; also alloys of a single system may present different behaviours according to the composition of the member under study. For instance, in the case of an alloy AB, in which A is the more noble and B the less noble metal, the following ranges can be distinguished:

(1) A relatively narrow range of composition containing little B, where the oxide produced is almost exclusively AO, at least in the external scale.
(2) An intermediate range where both AO and BO are produced.
(3) A relatively wide composition range near pure B, where BO is produced exclusively.

The classification of alloy systems preferred by Wood (which is similar to, but not identical with, that of Bénard) is shown below, along with examples of the different cases:

Class I

Only *one* of the elements (B) oxidizes, giving BO.
(a) The element B is the *minor* constituent and oxidizes
   (1) *internally*, giving BO particles in a matrix of A; for instance, Ag–Si alloys containing little Si produce SiO_2 particles in a silver matrix.
   (2) *externally*, giving a single layer of BO above an alloy matrix depleted in B; Ag–Si alloys richer in Si produce an external SiO_2 layer.
(b) The element B is the *major* constituent and oxidizes exclusively
   (1) leaving the non-oxidizable metal A dispersed in BO; Cu Au alloys rich in copper provide examples.
   (2) leaving the non-oxidizable metal A in an A-enriched zone behind the BO scale; Ni–Pt alloys come into this category, as do Fe–Cr alloys which are rich in Cr (those poorer in Cr belong to category I(a)(2)).

Class II

In this class *both* A and B oxidize to give AO and BO, the oxygen pressure in the atmosphere being greater than the equilibrium dissociation pressure of either oxide.

(a) AO and BO combine to give
   (1) a *single solid solution*; Ni–Co alloys behave in this way.
   (2) a *double oxide*, often a spinel, which may produce a complete surface layer of variable composition; this happens in certain Fe–Cr alloys. There may be particles incorporated in a matrix of AO, as in the case of certain Ni–Cr alloys.
(b) AO and BO are virtually *insoluble* in one another.
   (1) The less noble metal B is the *minor* component, and an internal oxide of BO lies beneath a mixed layer of AO and BO; this occurs in certain Cu–Ni, Cu–Zn, Cu–Al alloys, and indeed in many other systems.
   (2) the less noble metal B is the *major* component, so that no internal

oxidation is now observed. In practice, the second phase may not be present in the outer regions of scale, because AO may grow rapidly to produce outer regions exclusively of this oxide. Sometimes the outer regions are oxidized to higher oxides; for instance on Cu–Si alloys a CuO layer is found outside the Cu_2O layer.

The various systems are shown schematically in Fig. 2.

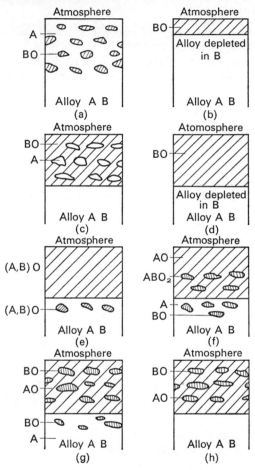

Fig. 2   Schematic representation of modes of oxidation of alloy AB, where B is the less noble metal. (a) Minor metal B only oxidizes, giving internal oxide BO in matrix of A. (b) Minor metal B only oxidizes, giving external oxide BO above alloy depleted in B. (c) Major element B only oxidizes, giving particles of A in a matrix of BO. (d) Major element B only oxidizes, giving external oxide BO above alloy depleted in B. (e) A and B oxidize to give solid solution of compound of variable composition. (A,B)O. (f) A and B oxidize to give compound ABO_2 dispersed in a matrix of AO. (g) A and minor component B oxidize to give insoluble oxides, with BO in a matrix of AO. (h) A and major component B oxidize to give insoluble oxides, with AO in a matrix of BO. (G. C. Wood).

Wood's comments regarding the Wagner–Hauffe rules should receive study. He points out that the over-optimistic application of the rules has led to disappointment at their lack of generality, but closer scrutiny often shows that universal obedience could not reasonably be expected in real situations. In predicting the situation from the rules, an understanding of the true defect structure of the parent oxide is necessary; many oxides really contain several types of defects, possibly in association; the rules as generally set out assume that the foreign cations are entering normal cation positions in the parent oxide at their anticipated valency, and that a parabolic growth relationship of the Wagner type exists, rather than a mechanism involving, say, grain-boundary diffusion. Ions often have only limited solubility in the parent oxide, so that, when the addition of the minor constituent is high, the rules will not necessarily be valid. The ionic radius is a pertinent parameter in such considerations. In ideal cases it is assumed that the dopant is homogeneously distributed through the scale, that the parent and dopant oxidize at the same rates, and that the rates of diffusion of the two cations in the scale are equal. In practice probably this occurs rarely, if ever.

Perhaps some readers may feel that it is less surprising that the rules sometimes fail than that they are ever obeyed at all! Any disappointment felt today at the discovery of cases where the rules fail should not blind us to the great services which both Wagner and Hauffe have conferred in indicating the principles of oxidation; there is some analogy between the attitude sometimes adopted towards Wagner and Hauffe and that adopted towards Pilling and Bedworth, who still deserve our gratitude, even though the situation is known today to be less simple than appeared from their paper of 1923.

Causes of Breakaway. Wood[1] provides a short discussion of breakaway, which deserves study. He considers that in the case of the Fe–Cr–base alloys, the evidence supports a mechanism in which adhesion between the initial doped Cr_2O_3 scale and the underlying alloy is lost through stress developed in the system, or through coalescence of vacancies into cavities at the interface between alloy and oxide; this produces ballooning, often above grain-boundaries and at the corners of specimens. Such changes are possible even when oxidation is carried out at a constant temperature; but in general they occur more readily during cooling, owing to the differential contraction between alloy and oxide. If the lifted oxide cracks sufficiently to admit the atmosphere, or if it is removed completely, the Cr-depleted alloy becomes exposed to the oxidizing gas at a high temperature—producing a much more serious situation than had existed during the first heating of the material. Clearly the development of a healing layer has now become more difficult.

Similar breakaway phenomena can occur in the corresponding Ni–Cr alloys. Potentially the situation would appear to be worse because the Cr-depletion at the interface between alloy and oxide is greater and more lasting than for Fe–Cr alloys. In practice, however, isothermal breakaway is rarer because scale adhesion is better owing to a more irregular and

1. G. C. Wood, *Oxidation Met.* 1970, **2**, 42.

interlocked alloy–oxide interface, partly caused by the lower alloy inter-diffusion coefficient.

Composition of an Alloy needed for Protection—as influenced by Spalling. An important paper by Whittle[1] presents the theoretical conditions under which a protective layer may be expected to be formed on an alloy AB, in which the constituent A is selectively oxidized,* giving an oxide-film AO_V, which will protect the metal because diffusion in it is slow. Clearly the mole-fraction of A in the alloy at the interface with the oxide must not fall below that level corresponding to the thermodynamic stability of AO_V; as to whether this condition can be fulfilled, will depend, of course, on the mole-fraction in the bulk of the alloy, $N_A b$, as well as on the inter-diffusivity coefficient, D, in the alloy phase, and the parabolic rate constant, k_p; it is the latter which decides whether the protective film can thicken quickly and thus reduce the rate of consumption of A at the intercept. If there is no spalling, the situation starts with a concentration at the interface equal to $N_A b$. If, however, the film, having successfully provided protection during the opening stage, suddenly breaks, exposing an alloy surface seriously depleted, the prospect of re-establishing protection is far less good. Thus the value of $N_A b$ needed to smother oxidation is higher for an alloy which suffers spalling than for one which develops a film capable of remaining unbroken. Whittle develops a set of inequalities defining the situations, using the symbol ϕ to represent $(\pi k/2D)^{1/2}$.

(1) If $N_A b < \phi$, there is insufficient A to produce a protective oxide at all.
(2) If $\phi < N_A b < 1 - (1 - \phi)^2$, there is sufficient A to form a protective film at the outset, but not enough to re-establish protection when once the initial film has broken.
(3) If $1 - (1 - \phi)^2 < N_A b < 1 - (1 - \phi)^3$, there is sufficient A to heal the scale at a single crack, but not to cope with rapid spalling.
(4) If $N_A b > 1 - (1 - \phi)^3$, there is sufficient A to cope with rapid spalling.

Oxidation of various Alloy Systems

Iron–Aluminium Alloys. Boggs[2] has studied alloys containing different amounts of Al. Below 570°C alloys containing less than 0·09% Al develop two phases (Fe_3O_4 and α-Fe_2O_3) and become oxidized more quickly than iron containing no Al. Above 0·09% a third phase, $FeO.Al_2O_3$, appears between the metal and the lower (Fe_3O_4) layer. This interferes with oxidation, so that the rate is now slower than that of unalloyed iron. Above 570°C, γ-Al_2O_3 starts to appear; between 800° and 850°C the film is almost pure alumina, and oxidation is very slow. As oxygen pressure is increased, there is a slight increase of oxidation rate up to 50 torr, then a sharp increase between 50 and 160 torr, with only a slight additional increase up to 700 torr. Below 50 torr the initial oxide-film tends to separate from the metal,

1. D. P. Whittle, *Oxidation Met.* 1972, **4**, 171.
2. W. E. Boggs, *J. electrochem. Soc.* 1971, **118**, 906.

* Whittle's convention (A being less noble than B) is the opposite of that adopted by Wood.

and cavities are formed, in which Al is preferentially oxidized so as to form a barrier. Above 160 torr, film separation ceases—perhaps for reasons suggested on p. 25.

Localized oxidation at the higher temperatures studied leads to corn-like nodules of iron oxide. The amount of nodular oxide increases with temperature and with the pressure of water vapour. Nodules formed in absence of moisture carry an outer shell of Fe_2O_3, whilst the interior, consisting of Fe_3O_4, FeO and some Al_2O_3, is porous in character.

Rolls and Bateman[1] also bring out the fact that the protective scale formed on these alloys can be non-adherent. Hot-bend tests carried out in the furnace show that alloys with 0·1 to 0·2% aluminium tend to develop breakaway. The weak adhesion is most marked in the thicker scales containing a high proportion of structural defects.

An oxidation resistant steel containing 2% to 3% Si and 0·5 to 1% Al was developed at the Fulmer Institute several years ago at a time of nickel shortage. Today it is again arousing interest,[2] since at temperatures up to 900°C its resistance properties appear to be comparable to those of 18/8 Cr/Ni stainless steel.

The invisible air-formed film present on alloyed iron may be different from that on unalloyed iron. Work in Hoar's laboratory,[3] based on X-ray photoelectron spectroscopy, has shown that the air-formed film on pure iron is Fe_2O_3, whilst that on iron containing 1·5% Al (by weight) is largely Fe_3O_4; the difference is doubtless due to the reducing property of the Al.

Iron–Chromium Alloys. The long term oxidation rate of an Fe–Cr alloy is largely determined by the power of the film to resist breakaway, since an unbroken film is remarkably protective. The breakaway has been studied at Teddington,[4] where the oxidation of an alloy containing 10% Cr has been kept under observation at various temperatures between 20°C and 600°C. Perhaps the most instructive observations were those made at 600°C, although the results were somewhat irreproducible. Curves connecting weight gain and time were obtained for specimens in different surface conditions. Most of the specimens showed a small weight gain in the first few minutes, followed by a quiescent period—sometimes prolonged. The thickness increase seldom exceeded 1600 Å during the first 70 minutes; after that most specimens showed little further change for some hours, and many of them for a few days. Breakaway, when at last it did occur, was usually sharp. It was followed by nearly rectilinear thickening at about 3 μm/day, with no tendency to protective kinetics up to thicknesses of almost 40 μm. Some specimens showed temporary accelerations to much greater rates.

The onset of breakdown appears to be associated with the formation of localized thickenings which occur preferentially at sites of poor fit in the

1. R. Rolls and G. J. Bateman, *Amsterdam Cong.* 1969, p. 336. *Brit. Corr. J.* 1970, **5**, 122.
2. *Met. Mat.* 1970, **4**, 10.
3. T. P. Hoar, M. Talerman and P. M. A. Sherwood, *Nature Phys. Sci.* 1972, **240**, (101) 116.
4. A. Fursey, B. Kent and G. O. Lloyd, *Amsterdam Cong.* 1969, p. 354.

polycrystalline layer of hexagonal oxide. These nodules may start to be formed early in the life of a protective scale, but at first grow comparatively slowly, perhaps for some days, before the overall weight gain accelerates sharply. Coarse whiskers are often formed in association with these nodules, although their distribution is irregular. Breakaway leads not only to recti-linear kinetics but to the formation of a scale consisting almost entirely of a sesquioxide of varying Cr-content, with no detectable quantity of either spinel or wüstite. It is thought that breakaway results from the development of a porous structure in which active growth takes place by reaction of molecular O_2 close to the scale–alloy interface, so that the solid state diffusion processes normally operating in protective scales are by-passed. The Ted-dington authors remark that breakaway phenomena have been discussed for many years in terms of the cracking or ballooning of a chromium enriched scale to expose a layer of impoverished alloy, with rapid oxidation ensuing and producing a large nodule of stratified scale. Such a picture would involve the formation of a relatively thick layer of almost pure Cr_2O_3, with a chromium depleted zone some μm thick in the underlying alloy. In the Teddington work, however, the Cr-content of the protective scales, although somewhat variable, is comparatively low and the protective scales are ex-tremely thin; moreover, no depleted zone has been observed. The authors find it difficult to explain the breakaway in the manner commonly favoured, for even if a thin depleted zone were to be formed during the protective phase, protective kinetics should quickly be re-established after breakaway. The structure of the thin layer present at the alloy–scale interface suggests the possibility of a self-perpetuating micro-cracking mechanism. This would accord with the observation that a few specimens showed temporary ac-celerations to rates very much faster than those normally encountered after breakaway, so that the linear rates may perhaps be interpreted in terms of a rapid oxidation growing over a very small fraction of the total surface at any given moment. The authors recognize that a detailed mechanism ac-counting for the observations must await the results of more detailed examination of the structure of nodules.

The apparent discrepancy between the Teddington results and those ob-tained by earlier workers may be due to the fact that the earlier work was generally carried out on alloys richer in chromium and at a higher tem-perature.[1]

The transport properties and defect structure of the $(Fe,Cr)_2O_3$ formed on Fe–Cr alloys has been studied at Leatherhead.[2] It is found that Cr_2O_3 is a cation deficient oxide, with p-type conductivity and predominatingly cation transport; the presence of iron will reduce the concentration of cation vacancies and tend to suppress the p-type conductivity. In contrast, Fe_2O_3 has interstitial iron and oxygen vacancies and shows n-type conductivity. Solid solutions possess a lower concentration of defects and lower transport rates than the pure oxides, and it would seem that the change-over between

1.   G. C. Wood, Private Comm. Oct. 1, 1973.
2.   P. K. Footner, D. R. Holmes and D. Mortimer, *Nature* 1967, **216**, 54; K. A. Hay, F. G. Hicks and D. R. Holmes, *Werkstoffe u. Korr.* 1970, **21**, 917.

the two types ought to occur when the alloy contains about 20% Cr; and that there should be a low oxidation rate at about this composition; actually, experiments show a minimal value of the parabolic growth constant at about 18% Cr.

In studying the composition of films formed on these alloys, conventional electron probe micro-analysis may lead to confusion between the film and underlying alloy; this can be avoided by examining the films after stripping from the metallic basis—a plan adopted at Teddington by Fursey and others.[1] They find that the oxide produced on exposure of iron carrying 10% Cr to air at 600°C has initially a higher Cr-content than the alloy, but that its Fe-content increases with time; in contrast, the Cr-content of the oxide grown on iron containing 20% Cr increases with time from 5 min to 5 days; the Cr is not uniformly distributed, being concentrated in thick nodule networks.

The early stages of oxidation of Fe–Cr and Ni–Cr alloys (5 to 30% Cr) at 600°C in oxygen at 1 atm. pressure has been studied by Chattopadhyay and Wood.[2] Stripped films studied by transmission electron microscopy show that substantial amounts of Fe and Ni oxides are produced before steady-state, healing layers containing Cr are developed. On dilute Fe–Cr and very dilute Ni–Cr (dilute in Cr), the grain- and subgrain-boundaries are covered with oxide thicker than the rest, probably due to their efficiency as cation-vacancy sinks or to more frequent local diffusion paths to the oxide above. For the alloys richer in Cr, healing is most rapid in these locations, owing to a more rapid diffusion of Cr to the alloy–oxide interface so that the oxide is specially thin above regions of defect structure.

A Czech study[3] of Cr - and Cr–Ni steels in the range 1000° to 1100°C indicates that the results of short term and long term tests on heat resistance do not always agree; long term exposure often produces much more attack than relatively short term tests would predict. This may be due to surface depletion in Cr, which falls off from 24% to 7% near the metallic surface. In some cases there may be internal oxidation. Cyclic heating (alternate heating and cooling) reduces attack on a steel containing 24% Cr, but increases that on a steel with 24% Cr and 20% Ni.

An interesting comparison of Fe–Cr, Ni–Cr and Co–Cr alloys, carried out in Wood's laboratory,[4] deserves study. In an atmosphere containing water vapour and argon, both Ni–Cr and Co–Cr alloys show a marked minimum for the parabolic rate constant at 20% Cr. There is a very adherent surface-scale at this composition—perhaps for reasons rather similar to those advanced above in connection with the Leatherhead work. In contrast, an iron alloy containing 20% Cr shows breakaway after initial formation of Cr_2O_3.

Tedmon[5] has pointed out that the Fe–Cr alloys suffer nitridation as well

1. A. Fursey, B. Kent and S. R. J. Saunders, *J. Microscopy* 1973, **99**, 147.
2. B. Chattopadhyay and G. C. Wood, *J. electrochem. Soc.* 1970, **117**, 1163.
3. M. Vyklický and M. Meřička, *Brit. Corr. J.* 1970, **5**, 162.
4. G. C. Wood, I. G. Wright, T. Hodgkiess and D. P. Whittle, *Werkstoffe u. Korr.* 1970, **21**, 900. F. H. Stott, G. C. Wood and M. G. Hobby, *Oxidation Met.* 1971, **3**, 103; F. H. Stott and G. C. Wood, *Corr. Sci.* 1971, **11**, 799.
5. C. S. Tedmon, *Amsterdam Cong.* 1969, p. 212.

as oxidation when heated in air; this is very localized; metallographic examination does not show a continuous surface layer of nitride, but large, acicular particles of Cr_2N dispersed along grain-boundaries and to a smaller extent within the grains. The weight gain measured in any given time increases with the Cr-content.

The oxidation of steel specimens which have been coated with a Cr or Cr–Al alloy by the chromizing process has been studied in the laboratory of the (British) Gas Council.[1]

Iron–Chromium–Nickel Alloys. The role of nickel in Fe–Cr–Ni alloys which are largely austenitic has been discussed by Hobby and Wood.[2] A paper from Poland[3] discusses the possibility of reducing the nickel content and yet obtaining resistance to oxidation. It is found that addition of Si improves oxidation resistance but adversely affects creep resistance at high temperatures. Of the alloys tested, that with 25% Cr, 20% Ni and 2% Si was found to have the highest oxidation resistance.

Sharp[4] has discussed the oxidation of alloys of different compositions. Ferritic alloys containing 0% to 35% Ni were found to form Cr_2O_3 scales, but an increase in the Ni-content was found to increase the oxidation rate. Austenitic alloys tend to form two layers of scale, with $(Fe,Cr,Ni)_2O_3$ outside and $(Cr,Fe,Ni)_3O_4$ below it. The outer layer is thickest above the centre of the grains and may be absent over the grain-boundaries of the alloy, since here a rapid renewal of Cr permits the formation of a more protective form of oxide.

An analysis of the oxide-scale formed on stainless steel oxidizing in steam at 800°C has been provided by Ericsson.[5]

Other Iron Alloys. The advent of natural gas with a very low sulphur content has made it important to decide on materials most suitable to resist oxidation by combustion products in absence of sulphur. Information furnished by the (British) Gas Council will be welcomed.[6] Good performance by iron containing 1% Al and 1% Si, also iron containing 2% Si and 2% Cr, should be noted.

It has been found that the presence of B and P in the scale formed on iron at 900–1100°C renders it more porous; apparently these elements hinder the absorption of cation vacancies into the metal.[7]

The behaviour of iron alloy drops during a fall through an oxidizing atmosphere may interest some readers.[8]

1. N. R. Chapman, R. Micklethwaite and G. A. Pickup, *Amsterdam Cong.* 1969, p. 760; also *Anticorrosion* 1970, **17**, (12) 10.
2. M. G. Hobby and G. C. Wood, *Oxidation Met.* 1969, **1**, 23.
3. E. Raliszewski, St. Mrowec and H. Serwicki, *Werkstoffe u. Korr.* 1968, **19**, 211.
4. W. B. A. Sharp, *Amsterdam Cong.* 1969, p. 291; also *Corr. Sci.* 1968, **8**, 717.
5. T. Eriksson, *Amsterdam Cong.* 1969, p. 306.
6. J. A. von Fraunhofer and G. A. Pickup, *Amsterdam Cong.* 1969, p. 235.
7. F. Vadopivec and L. Kosec, *Archiv Eisenhüttenwesen* 1969, **40**, 425.
8. J. B. See and N. A. Warner, *J. Iron Steel Inst.* 1973, **211**, 44.

Cobalt-base Alloys. Kofstad[1] has studied the oxidation of the Co–Cr system and finds three different mechanisms. With unalloyed Co, there is accurate obedience to the parabolic law; the controlling process is diffusion of Co through a CoO film. With alloys containing Cr up to 25%, the controlling process is the same, but the situation is modified by porosity and a dispersion of spinel particles. Above 25%, there is diffusion of Cr through a Cr_2O_3 scale.

Further information about the oxidation of super-alloys containing Co and/or Ni is provided by Queneau.[2]

The effect of yttrium on the high temperature oxidation of cobalt containing 30% Cr has been studied by Beltran.[3] He states that 1% Y has generally been used, but that a smaller quantity suffices to improve resistance; his research is concerned with the effect of 0·1% Y. The alloy containing no Y develops a fairly pore-free Cr_2O_3 film at 900°C, but at 1000–1200°C nodules occur consisting of a thin Cr_2O_3 film covered with a very porous spinel layer and a layer of large grained CoO interspersed with a few large pores. The presence of Y inhibits the formation of spinel and CoO, both at 900°C and 1000°C, and only small amounts are formed at 1100–1200°C. The benefit conferred by Y is connected with the prevention of nodule formation, and also with the build-up of a sub-surface grain-boundary network in advance of the oxide–metal interface. This serves as a sink for vacancies, accounting for improved scale adhesion.

Nickel-base Alloys. The oxidation of the binary alloys Ni–Cr, Ni–Al, Ni Si, Ni–Co and Ni–Mn has been studied in Wood's laboratory.[4] Experiments on Ni–Cr, Ni–Al and Ni–Si alloys in O_2 at 600°C show that some NiO is formed before an oxide layer containing the less noble metal appears at the base of the scale and retards attack; this occurs most readily near the grain-boundaries. On Ni–Co alloys it is principally (Ni,Co)O, whilst on Ni–Mn alloys it is principally (Ni,Mn)O; some higher oxide is present in each case. Under conditions where healing layers are not developed, the grain-boundaries and sub-structural defects become covered with thicker oxide.

The formation and breakdown of Al_2O_3 scales on Ni–Al alloys heated at high temperatures has been studied by Wood and Stott.[5] Alloys of this type are much used in gas turbines, since the oxides formed on them (mixtures of Al_2O_3, with Cr_2O_3 if Cr is present) are not excessively volatile and are capable of restricting mass transport. The formation of an initially complete and protective layer of Al_2O_3 becomes easier as the Al-content is increased and also when the temperature is raised, but after its development it may fail mechanically in certain regions. Under breakaway conditions, Ni containing 7% Al oxidizes rapidly, as NiO nodules are formed on the exposed

1. P. Kofstad and A. Z. Hed, *Amsterdam Cong.* 1969, p. 196.
2. P. E. Queneau, Thesis, Delft, 1971, pp. 53–63.
3. A. M. Beltran, *Amsterdam Cong.* 1969, p. 186.
4. G. C. Wood and B. Chattopadhyay, *Corr. Sci.* 1970, **10**, 471; *Oxidation Met.* 1970, **2**, 373.
5. G. C. Wood and F. H. Stott, *Brit. Corr. J.* 1971, **6**, 247; *Corr. Sci.* 1971, **11**, 799.

surfaces, although the rate declines as a healing Al_2O_3 layer is formed below them. On Ni containing 12·5% Al, the re-formation of an Al_2O_3 layer is rapid, so that breakaway produces no serious effect; even if there is local lifting and cracking, fresh Al_2O_3 develops rapidly without serious kinetic or morphological changes.

The adhesion of the oxide-films formed on 80/20 Ni–Cr alloys has been studied by Japanese investigators.[1] Earlier work had shown that 0·1% Be is effective in preventing spalling in tests providing for alternate heating and cooling periods. Al and Si have now been studied as additives, but they seem less satisfactory. There are two types of scale which restrain spalling. One is a dense uniform Cr_2O_3 layer with a subscale penetrating into the substrate; this type is obtained with 0·1% Be. The other type has an outer layer of NiO and $NiO.Cr_2O_3$ with a dense Cr_2O_3 layer below it; such a structure appears on the pure binary alloy after the second oxidation period.

Russian work[2] on the effect of Ca, Zr and La suggests that the presence of NiO in the scale is a sign of poor oxidation resistance, under conditions of either continuous or cyclic heating. During oxidation periods of 100 hours, Ca is found to reduce the formation of NiO at 1100–1200°C, whilst La and Zr prevent it completely.

A study of nickel-base alloys has been carried out by Fleetwood and Whittle.[3] They found that a protective scale was formed quickly, but that subsequent spalling sometimes occurred; this was reduced by small additions of Si, Ce or Al.

The oxidation of Ni–Cr–Al alloys between 1000° and 1200°C has been studied by Giggins and Pettit.[4] Materials based on the Ni–Cr–Al system are extensively used at high temperatures where mechanical strength combined with oxidation resistance is required. Experiments in 0·1 atm. O_2 for 20 hours or more show that, when once steady state conditions have been established, the alloys fall into three distinct classes according to their compositions:

I   Alloys with low Cr- and Al-contents form a continuous, external NiO scale and a discontinuous subscale of Cr_2O_3 and Al_2O_3.

II   Those with higher Cr- and Al-contents form a continuous, external Cr_2O_3 scale but Al is still internally oxidized.

III Those with still larger Al-contents form a continuous, external Al_2O_3 scale with only a small amount of Cr in solution.

The situation at the different stages of the oxidation process is indicated in Fig. 3, which also shows that external NiO and $Ni(Cr,Al)_2O^2$ can be formed during the transient oxidation period prior to the time when steady state conditions are established. Compositional limits of alloys falling into the three classes are presented in Fig. 4.

The addition of Si, Y and other elements to these alloys has been found

1.   A. Takei and K. Nii, *Tokyo Cong., Ext. Abs.* 1972, p. 295.
2.   L. L. Zhukov and I. M. Plemyannikova, *Corr. Control Abs.* 1970, No. 3, p. 10.
3.   M. J. Fleetwood and J. E. Whittle, *Brit. Corr. J.* 1970, **5**, 131.
4.   C. S. Giggins and F. S. Pettit, *J. electrochem. Soc.* 1971, **118**, 1782.

effective in lessening oxidation. The causes have been studied by Wood and his colleagues,[1] using scanning electron microscopy. Experiments at 1000° to 1300°C in O_2 at 1 atm. show that a virtually complete film of SiO_2 is revealed on the alloys containing Si. For Y alloys the evidence is less clear, but the facts suggest a blocking layer.

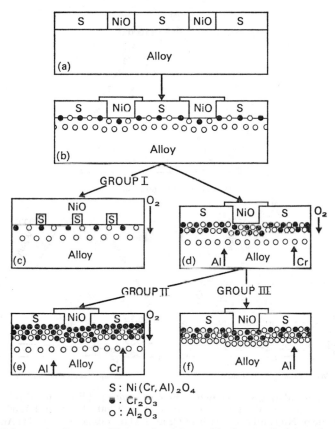

Fig. 3 Schematic diagram illustrating the oxidation mechanism for Ni–Cr–Al alloys. Group I represents alloys with small contents of Cr and Al. Group II is relatively rich in Cr, whilst Group III is rich in Al and Cr (C. S. Giggins and F. S. Pettit).

Another paper by Wood and Boustead[2] concerns the effect of yttrium and gadolinium. It was found that Y decreases the oxidation rate, but does not prevent 'typical breakaway' on dilute Cr alloys at 1200°C. Gd also decreases the oxidation rate and tends to increase the time which elapses before the occurrence of breakaway, which is limited to isolated regions. A pure Fe–Cr

1. G. C. Wood, J. A. Richardson, M. G. Hobby and J. Boustead, *Corr. Sci.* 1969, 9, 659.
2. G. C. Wood and J. Boustead, *Corr. Sci.* 1968, 8, 719.

c

alloy shows only slight departure from the parabolic law (in a direction indicating increased protection), but for the ternary alloys containing either Y or Gd, the oxidation–time curves become almost asymptotic. It is believed that a layer of yttrium oxide retards oxidation

Weigel and Track[1] find that the improvement in scale adhesion produced

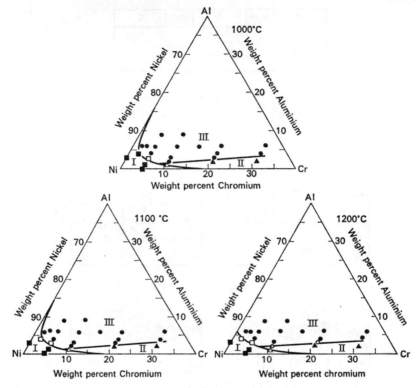

Fig. 4   Isothermal diagrams showing the compositional limits for three oxidation mechanisms of Ni–Cr–Al alloys in 0·1 atm. O_2 at 1000°, 1100° and 1200°C. (I) An external scale of NiO and a subscale of Cr and Al oxides is formed. (II) An external layer of Cr_2O_3 and a subscale of Al_2O_3 is formed. (III) Only an external layer of Al_2O_3 is formed (C. S. Giggins and F. S. Pettit).

by the addition of Ca or Ce to strong nickel-base alloys is obtained only in a certain temperature range. Above a limiting temperature, which varies with the alloy composition but generally lies between 1040° and 1100°C, the presence of Ca or Ce confers no advantage and sometimes makes behaviour worse.

Copper-base Alloys.   The influence of various additions to copper has been studied in Scully's laboratory.[2] Selective oxidation was carried out

1.   K. Weigel and W. Track, *Werkstoffe u. Korr.* 1972, **23**, 1.
2.   M. D. Sanderson and J. C. Scully, *Corr. Sci.* 1970, **10**, 55, 165 (2 papers).

under the conditions used by Price and Thomas (1960 book, p. 76). Of the additions studied, beryllium was the most effective. In air, copper containing 2% Be oxidizes in the classical Wagner manner. Be and Cu oxides are nucleated simultaneously, and the Cu oxide-film grows faster, forming the outer layer; eventually a complete inner layer of BeO is formed which prevents further oxidation.

Addition of Al in small amounts provides less complete protection; but with sufficient addition, tarnish resistance can be obtained. Si additions improve the situation; the alloy containing 7·5% Al and 2% Si is recommended for resisting oxidation in air at 800°C; in the case of damage, the oxide-film is self-healing.

The oxidation of a copper alloy containing 3% Al and 2% Si has been studied in Pryor's laboratory.[1] The parabolic growth of γ-Al_2O_3 is found to have a much lower Arrhenius constant on the alloy than on pure Al. It is thought that copper ions are introduced into the γ-Al_2O_3 structure, thus diminishing the number of diffusion paths, especially at high temperatures; this reduces the temperature sensitivity. When the copper ions occupy cation vacancies, additional electrons are needed to preserve electrical neutrality, and this enhances the n-type structure and increases the electronic conductivity; at the same time, the concentration of cation vacancies falls, so that the ionic conductivity is reduced. At low temperatures the increase of electronic conductivity prevails over the decrease of ionic conductivity; at high temperatures the reverse is the case.

The oxidation of Cu–Mn alloys is discussed in an Indian paper.[2] All the alloys show parabolic growth, with the rate constant increasing as the Mn-content is increased up to 10%, and then diminishing; at about 20% Mn it is similar to that of unalloyed copper and at 40% very much lower.

The internal oxidation of Cu–Pd and Cu–Pt alloys has been studied by Wagner.[3] Cu_2O is formed as an external scale and is also produced within the alloy owing to the inward diffusion of oxygen. Conditions under which this internal oxidation occurs are indicated, but it is considered that a further theoretical analysis would be desirable.

The kinetics of the reaction of Cu and Ag in liquid sulphur have been studied by Richert.[4]

Aluminium–copper Alloys. Brock and Pryor[5] have compared the behaviour of Al–Cu alloys with that of unalloyed Al at different temperatures. Unalloyed Al oxidized up to 300–350°C probably obeys the inverse logarithmic law; from 350–425°C thickening is parabolic and the film is amorphous γ-Al_2O_3. About 425°C in addition to the amorphous oxide, cylinders of crystalline γ-Al_2O_3 are formed, starting from nuclei situated at

1. S. R. J. Saunders and M. J. Pryor, *J. electrochem. Soc.* 1968, **115**, 1037.
2. A. K. Lahiri, P. K. Panda and T. Banerjee, *Tokyo Cong. Ext. Abs.* 1972, p. 291.
3. C. Wagner, *Corr. Sci.* 1968, **8**, 889.
4. A. Richert, *Z. phys. Chem. (neue Folge)* 1960, **23**, 355.
5. A. J. Brock and M. J. Pryor, *Corr. Sci.* 1973, **13**, 199.

the interface between metal and amorphous oxide and spreading out according to the law of expanding circles (1960 book, p. 939).

At 525°C and above, alloys containing Cu obey the same growth law for the crystalline phase as pure Al, namely

$$W_c = \rho\delta(1 - e^{-\pi v^2 \Omega t^2})$$

where δ is the terminal crystal thickness, ρ the density, v the radial growth velocity, assumed to be constant, Ω the crystal frequency and t the time. However at 475° and 500°C, this expression requires modification, since nucleation is not immediate. Also it seems that amorphous oxide is not formed so quickly over the growing cylinders of crystalline material as on the regions between them; circular depressions suggest that the amorphous oxide is locally thinner—presumably over the sites of the cylinders.

The growth of the crystalline cylinders into the metal is ascribed to inward diffusion of oxygen, but the amorphous film is thought to grow by movement of Al ions outwards (more slowly on the areas covered by the crystalline cylinders than on the parts between those areas).

Increase of Cu-content in the range $0 \cdot 1$–4% causes a decrease in the crystal nucleation density and in the depth of intrusion of the crystals into the underlying alloy; it increases the rate of formation of amorphous γ-Al_2O_3 between the crystals, and also the rate of radial growth of the crystals.

Oxidation in presence of Carbon and/or Sulphur Compounds

Linear Growth Rate in gas mixtures containing Carbon Dioxide. Smeltzer[1] has attributed the rectilinear growth of wüstite on iron and iron–nickel alloys exposed to CO_2–CO mixtures at 1000°C to control of the change by a boundary reaction such as

$$CO_{2\ gas} = O^{2-}_{\ ads} + CO_{gas} + 2\oplus$$

This would of course be independent of film thickness.

A research at Didcot[2] establishing the conditions under which non-protective scale can be formed on mild or low-alloy steel deserves notice; in $CO_2/CO/H_2O$ mixtures at 500°C the factors causing the scale to be non-protective and to thicken by the rectilinear law are the presence of water vapour and a high proportion of CO_2. In non-protective oxide, the mode of transport is by the diffusion of gaseous O_2 along open pores. When the scale is protective, transport occurs by cations moving outwards, but this leaves vacancies which can condense to produce cavities; oxygen can, in effect, cross cavities by the dissociation of iron oxides at one side and re-formation of oxide on the other; both CO and H_2 act as oxygen carriers, forming CO_2 and H_2O respectively.

The influence of pressure on the long-period oxidation of mild and low-alloy steels in CO_2 containing small quantities of CO and H_2O has received much study.[3] Early results showed that oxidation was usually protective in

1. W. W. Smeltzer, L. A. Morris and R. C. Logan, *Amsterdam Cong.* 1969, p. 221.
2. J. E. Antill, K. A. Peakall and J. B. Warburton, *Corr. Sci.* 1968, **8**, 689.
3. D. Goodison, R. J. Harris and P. Goldenbaum, *Brit. Corr. J.* 1969, **4**, 293.

character, with obedience to a parabolic or cubic law, but that in some cases irregularities in the curve occur, indicating breakaway. Later results show that as gas pressure is increased, the time needed for breakaway is reduced, and the rate of development of excrescences is increased on killed mild steel, but not always on low-alloy steels. However, at 380°C and above, the rectilinear rate established after breakaway on mild and low-alloy steels is essentially dependent on the temperature and H_2O pressure, and may be unaffected by total gas pressure and partial pressure of CO.

Corrosion of boiler steels by fuel gases in municipal incinerators has received study;[1] the accumulated deposits contain S, Cl, Zn, Al and K, occasionally also Pb and Cu.

Iron in presence of Sulphur Compounds. Rahmel[2] has studied the behaviour of iron in various gas mixtures. When the CO_2/CO ratio is low, so that the formation of oxide is impossible, pure sulphide layers will be formed if H_2S is present. If reaction with CO_2 is possible, an intimate mixture of FeO and FeS appears, the structure of which varies with conditions. In mixtures of O_2–N_2–SO_2, iron oxides are thermodynamically stable, whilst in CO–CO_2–COS, FeS is to be expected on thermodynamic grounds. In practice, the 'stable' compound is formed if either diffusion through the scale or phase boundary reaction at the metal–scale interface is rate determining. If diffusion in the gas phase or the reaction at the scale–gas interface is rate determining, then simultaneous formation of FeO and FeS takes place.

The behaviour of iron in pure argon containing small amounts of SO_2 has been studied by Birks.[3] Between 500° and 900°C, there is an initial period of rectilinear growth during which the constant rate is controlled by the diffusion of SO_2 through a boundary layer in the gas to the scale surface. SO_2 cannot produce a simple FeS scale, but gives rise to a lamellar FeO FeS structure. When the scale reaches such a thickness that the ionic movement through it would be comparable to the rate of SO_2 supply, only those lamellar blocks which are suitably aligned (e.g. those standing perpendicular to the metal surface) continue to grow. This produces a very rough surface. Eventually, under some conditions, a layer of oxide free from sulphide begins to form on the outside. At 550°C, after an initial slow growth of sulphide-free oxide, very rapid reaction occurs, resulting in the formation of bursts of oxide–sulphide agglomerate all over the surface.

Iron–Chromium Alloys in presence of Sulphur Compounds. A second paper by Birks[4] extends the work to Fe–Cr alloys, with results somewhat comparable to those obtained on unalloyed iron.

The corrosion of unalloyed iron and also Fe–Cr alloys in H_2S has been

1. P. D. Miller and H. K. Krause, *Corrosion (Houston)* 1971, **27**, 31.
2. A. Rahmel and J. A. Gonzalez, *Werkstoffe u. Korr.* 1971, **22**, 283. A. Rahmel, *Tokyo Cong. Ext. Abs.* 1972, p. 301.
3. T. Flatley and N. Birks, *J. Iron Steel Inst.* 1971, **209**, 523.
4. R. P. Salisbury and N. Birks, *J. Iron Steel Inst.* 1971, **209**, 534.

studied in Japanese work.[1] A plot between log W and log t, where W is the weight gain in time t, shows three steps when the experiment is carried out at 400°C. The value of a in the equation $W = bt^a$ is about 0·5 to 0·6 in step 1, approximately 1·0 in step 2, and between 0·6 and 0·8 in step 3. There are always two layers, and a platinum marker is found located at the outer side of the interface separating the two layers.

Sulphide-tarnishing on tin. Two papers from Sheffield[2] on the tarnishing of tin in polysulphide solutions deserve attention, since they introduce factors not met with in ordinary oxidation. The film grows in two stages; in the first stage only a single layer is observed, which in the second stage is partially superseded by a film consisting of two layers; the kinetics are logarithmic for the first stage and rectilinear for the second stage. It seems likely that there are two processes at work from the start. At the outset the second process makes only a negligible effect, so that the growth law is apparently logarithmic at first, becoming rectilinear later. Simultaneous dissolution of the film is observed; stirring, which has a negligible effect on film growth, has a marked effect on dissolution; polysulphide concentration affects film formation and also dissolution.

Catastrophic Oxidation in presence of Vanadium Compounds.

Oil-ash Corrosion. Webster[3] states that sodium vanadyl vanadate melts below 900°C and is formed when the ratio V/Na in the ash is 85/15. The molten phase can then slag off an oxide-film which would otherwise be protective. A relatively successful approach to the problems of petroleum heaters has been the development of Ni–Cr alloys; these are particularly useful for the hangers supporting heaters in combustion chambers. Rare earths (0·5% Y or Ce) improve resistance.

Hoch[4] states that, although the addition of Cr to Ni increases resistance to melts containing V, all other likely constituents of Ni alloys promote corrosion—notably Mo, which has the greatest detrimental effect, especially when present in excess of 5%; W is also detrimental at high temperatures. Sulphates in the ash increase corrosion by lowering the melting-point and by increasing the O_2 solubility.

Indian work[5] shows that V_2O_5 and some of its compounds accelerate oxidation when in liquid form, whilst small amounts of Na_2SO_4 further stimulate oxidation by altering the fluidity of the molten slag at lower temperatures; however it acts as a diluent at higher concentrations and at high temperatures. The low oxidation rate of nickel-base alloys is attributed to the formation of $Ni_3(VO_4)$.

The behaviour of Ni–Cr alloys in presence of oil-ash is discussed by Horda

1. K. Yamamoto, M. Izumiyama and K. Nishino, *Tokyo Cong. Ext. Abs.* 1972, p. 303.
2. S. Wolynec and D. R. Gabe, *Brit. Corr. J.* 1972, **7**, 126, 134 (2 papers).
3. T. M. Webster, *Met. Mat.* 1969, **3**, 197.
4. P. Hoch, *Werkstoffe u. Korr.* 1970, **21**, 630.
5. A. K. Lahiri, H. R. Thilakan and T. Banerjee, *Amsterdam Cong.* 1969, p. 264.

and Swales,[1] who provide triangular diagrams for Ni–Cr–Fe systems with lines of equal corrosion for 700°, 800°, 900° and 1050°C.

In 1962, an objection was raised to the idea that the catastrophic oxidation is due to the low melting-point of mixtures containing V_2O_5, on the grounds that in one set of experiments there was no sudden increase of oxidation rate at the melting-point of the mixture used (1968 book, p. 40). This objection can be answered by a study of measurements carried out many years previously by Brasunas,[2] who had produced equilibrium diagrams for several systems and showed that there are mixtures melting at a temperature lower than the MP of either component. In the case of the V_2O_5–Fe_2O_3 system, a thermal-arrest diagram indicates that a steel carrying an iron oxide film would have that film dissolved off by contact with a mixture containing V_2O_5, even at temperatures below the MP of the mixture free from iron. This would seem to remove the apparent objection.

Corrosion due to Sodium Salts. Seybolt[3] states that the sulphidation–oxidation found in gas turbines is generally ascribed today to a liquid film of Na_2SO_4 formed by interaction between the sulphur of the fuel and the NaCl of sea air. Bergman[4] states that the corrosion problems associated with molten Na_2SO_4 or V-compounds in gas turbines used for marine service (helicopters and surface vessels) have generally been alleviated by empirical methods connected with coatings and fuel treatments; no panacea has been found to permit of unrestricted use of gas turbines in highly corrosive environments. Tests on the use of rare-earth additions in Ni-base alloys have been carried out by Elliott and Ross.[5] The alloys were tested at 870° to 1200°C with S supplied by molten Na_2SO_4 or by argon containing a low concentration (e.g. 0·2%) of SO_2. Rare earths in super-alloys lead to better scale properties and so delay the catastrophic hot-corrosion attack. Traces of lead, often found in gas-turbine fuels, seriously reduce the resistance of nickel-base alloys at 900°C to corrosion induced by Na_2SO_2.[6]

Further References

Books. 'Oxidation of Metals and Alloys' (Amer. Sci. Metals), 1971.
'Techniques of Metals Research' (Editors, R. A. Rapp and R. F. Bunahah), Vol. IV, Part 2 (Wiley). Especially Chap. 10A dealing with 'Oxidation of Metals', by G. C. Wood.

Review Article. 'Fundamental factors determining the mode of scaling of heat resistant alloys'.[7]

1. W. Horda and G. L. Swales, *Werkstoffe u. Korr.* 1968, **19**, 679.
2. A. de S. Brasunas, Thesis, Massachussetts Inst. of Tech. 1952; also private letter dated June 8, 1973.
3. A. U. Seybolt, *J. electrochem. Soc.* 1969, **116**, 287C.
4. P. A. Bergman, *J. electrochem. Soc.* 1969, **116**, 288C.
5. P. Elliott and T. K. Ross, *Werkstoffe u. Korr.* 1971, **22**, 531.
6. D. Chatterji, D. W. McKee, G. Romeo and H. C. Spacil, *J. electrochem. Soc.* 1975, **112**, 941.
7. G. C. Wood, *Werkstoffe u. Korr.* 1971, **22**, 491.

Papers. 'Interaction between vanadium in gas turbine fuels and liquidation attack'.[1]

'Oxidation of Ni–7·5% Cr and Ni–7·5 Cr–2·5 Sm_2O_3 at 900° and 1100°C'.[2]

'High temperature oxidation of Ni/20% Cr alloys containing dispersed oxide phases'.[3]

'Review of the diffusion-path concept and its application to the high temperature oxidation of binary alloys'.[4]

'Long term oxidation of super-alloys'.[5]

'Zum Mechanismus der Hochtemperatur-oxidation von Kupfer–Aluminium Legierungen'.[6]

'Some aspects of the hot corrosion of cobalt-base alloys'.[7]

'Effect of heat treatment variations on hot-salt corrosion resistance of nickel-base super-alloys'.[8]

'The use of isotope reactions for the study of oxygen transfer from H_2O and from CO_2 to metals and oxides'.[9]

'Oxidation of nimonic 80A at 800°C in low oxygen pressures'.[10]

1. N. S. Bornstein, M. A. Decresciente and H. A. Roth, *Corrosion (Houston)* 1972, **28**, 264.
2. J. Stringer and A. Z. Hed, *Oxidation Met.* 1971, **3**, 571.
3. J. Stringer, B. A. Wilcox and R. I. Jaffee, *Oxidation Met.* 1972, **5**, 11 (two further papers on pp. 49, 59).
4. A. D. Dalvi and D. E. Coates, *Oxidation Met.* 1972, **5**, 113.
5. O. L. Angerman, *Oxidation Met.* 1972, **5**, 149.
6. K. Hauffe and F. Ofulue, *Werkstoffe u. Korr.* 1972, **23**, 351.
7. M. N. Richards and J. Stringer, *Brit. Corr. J.* 1973, **8**, 167.
8. J. Billingham, J. Lauriosen and R. E. Lawn, *Corr. Sci.* 1973, **13**, 623.
9. H. J. Crabe, *Ann. N.Y. Acad. Sci.* 1973, **213**, 110.
10. R. K. Wild, *Corr. Sci.* 1973, **13**, 105.

CHAPTER IV

Electrochemical Corrosion

Preamble

Picture presented in the 1960 and 1968 volumes. The dry oxidation discussed in chapters II and III produces a film which becomes increasingly protective as it thickens. At relatively high temperatures the rate of thickening may be inversely proportional to the thickness already attained. At lower temperatures the thickening rate falls off more quickly. At ambient temperature, in reasonably dry air, the thickening practically ceases at a stage when the film is still insufficiently thick to be visible; thus, in absence of liquid water, 'direct oxidation' can be said to produce 'no damage'.

In presence of liquid water, the position may be very different. It is possible now for the oxygen to be taken up at one place and for the metal to pass into the liquid as ions at another. Thus a protective film may not be formed even in those cases where the ultimate product is a solid oxide or hydroxide. For instance, in the case of iron partly immersed in a sodium chloride solution, oxygen may be reduced to OH^- at the water-line, where oxygen can be readily replenished; this cathodic change requires a supply of electrons. Simultaneously iron atoms may pass into the ionic condition at places below the water-line; this anodic change furnishes electrons, which pass upwards to supply the needs of the cathodic reaction at the water-line. The cathodic and anodic products (sodium hydroxide and ferrous chloride) are both freely soluble. Where they meet, generally at points well away from the metal surface, ferrous hydroxide is precipitated and quickly oxidized to a ferric or ferroso-ferric compound; if oxygen is present in sufficient amount, a ferric or ferroso-ferric compound may be the first substance to be precipitated. Whether the precipitate is ferrous hydroxide, ferric hydroxide ('rust'), magnetite or a ferroso-ferric hydroxide, hardly affects the result. Most of the precipitation takes place *out of contact with the metallic surface*, so that *no protective film* is formed, except near the water-line, where any iron ions escaping from the metal find themselves in an alkaline environment; here a film may appear, sufficiently thick to show interference tints.

This mechanism has been established without doubt. The alkali and the ferrous salts can be recognized by simple chemical tests, the precipitation of rust or other compound is visible to the eye, whilst the current flowing has been measured and found equivalent, in the sense of Faraday's law, to the corrosion rate as measured by loss of weight. This direct verification of the electrochemical mechanism is only possible where the anodic and cathodic areas are fairly large and well separated. Where these areas are very small, or where a single point alternately acts as anode and cathode, more sophisticated methods are needed to correlate electrochemical measurements with corrosion rate. In such cases the argument is less direct, but the agreement between the corrosion as measured and that calculated on the assumption of an electrochemical mechanism provides good support for the idea that the corrosion is really an electrochemical phenomenon.

Whilst it is customary to distinguish between 'direct oxidation' (e.g. that produced by exposure to hot air) and 'electrochemical corrosion' (e.g. that produced by partial immersion in a salt solution), it should be pointed out that even in the first case electric currents are probably flowing. When an oxide-film is formed on metal heated in air, an anodic reaction is proceeding at the metal/oxide interface and a cathodic reaction at the oxide/gas interface; ions are passing through the film, and also electrons. However, the electrons here are moving in a direction roughly normal to the surface, whereas in the wet corrosion of half-immersed metal they are moving roughly parallel to the surface. The result is that a film which is to some extent protective is formed in the first case, but not in the second.

There are many other ways in which a current connected with corrosion can be set up. Strong currents often arise when two dissimilar metals are in contact. Such contacts produce relatively rapid attack on the base metal forming the anode, whereas the noble metal forming the cathode may be prevented from corroding in a liquid which would otherwise attack it. This method of preventing corrosion was used as early as 1824, but nearly a century was to elapse before it was generally appreciated that corrosion currents could be generated in many other ways besides the junction of dissimilar metals. One example has already been provided by the 'differential aeration' currents flowing between a well aerated cathodic zone at the waterline and an anodic zone lower down. Several others could be cited. If, on copper immersed in a salt solution, the liquid is rapidly moving over one part (only) of the surface, so that copper ions are here washed away, the area of rapid liquid movement is anodic and the relatively stagnant region cathodic. If the metal is partly in contact with dilute salt solution and partly with concentrated salt solution (as may happen on metal piles near the mouth of a river), the part exposed to the concentrated salt solution may suffer anodic attack, owing to the formation of complex ions.

Although the electrochemical mechanism may often allow corrosion to continue unabated for long periods, the results are generally only serious if the attack is concentrated on a small area. This will happen if the anodic area is very small compared to the cathodic area. In such a case, the current is often determined by the amount of oxygen reduced on the large cathode, but the effect will be concentrated on the anodic area, and if this is small, the intensity of corrosion (i.e. the corrosion per unit area of the part affected) can become very high. A practically important example is the so-called deposit attack produced in brass condenser tubes at places where mud or other debris has settled, screening a small patch from oxygen, and causing an anode of limited area. In extreme cases, intense localized attack may produce small depressions or pits sometimes bounded by flat crystal faces, but often hemispherical. Pitting is most commonly met with on materials which normally cover themselves with a protective film; such materials include aluminium and stainless steel. Under special conditions this protective film may break down locally, giving rise to pits. The total destruction of metal may be small, but owing to the localization the result can be serious.

Developments during the period 1968–1975 and Plan of the new Chapter. The outstanding feature of research carried out between 1968 and 1973 has been the detailed study of localized corrosion—especially pitting. In the 1960 book a section discussing 'Spreading, Healing or Pitting' occupied about a quarter of the fourth chapter. In the 1968 Supplement the corresponding chapter contained a section devoted simply to 'Pitting', and occupying nearly half the length. In the present chapter the various sections dealing with 'Pitting Conditions' occupy more than half.

The new supplementary chapter is divided into three main parts. First some general researches carried out under pitting conditions are discussed—including early work today almost forgotten but still possessing importance for those who desire to understand pitting. On the basis of these results—old and new—an attempt is made to provide a theory of the pitting process, bringing out points common to all metals. It cannot be claimed that there exists today complete agreement regarding the mechanism of pitting; but it is hoped that recent results, partly unpublished, may provide, when brought together, an acceptable picture of the subject. Next, researches on individual metals, mostly carried out under pitting conditions, receive attention. Then, important researches carried out under non-pitting conditions are surveyed; one section concerns the important question as to whether 'pure chemical corrosion', as opposed to 'electrochemical corrosion', is possible; this is largely based on the work of Heitz. The comparison between results obtained by various methods of measuring corrosion receives attention; the good agreement itself provides support for the electrochemical mechanisms. The chapter ends with a discussion of the effect of flow-rate, also largely based on the work of Heitz.

General Researches carried out under Pitting Conditions

Procedures. Two main types of procedure have been used in studying pitting. In the first, the specimen is held by means of a potentiostat at a potential within the passive range, in a solution free from chloride. When passivity has been established, a small measured amount of chloride is added; after a certain induction period, pits appear, visible under the microscope; the current needed to maintain the potential at the chosen value increases as the pits grow. Since this method involves anodic conditions, detailed consideration is deferred to chapter VII. The method provides quantitative information, and has been useful to investigators of the mechanism. It could be criticized on the grounds that the specimen is being subjected to an anodic current from an external source—which is not generally the case in ordinary service experience.

Accordingly, many investigators have preferred a procedure in which the metal is exposed to a solution chosen to produce pitting in the absence of an applied anodic current. The appropriate liquid varies with the material to be studied. Sometimes ferric chloride solution is used, but this also could be criticized on the grounds that it does not represent service experience. In the study of iron, a solution containing sodium chloride with an inhibitor such as sodium carbonate, hydroxide or chromate in an amount designed to prevent corrosion over the main surface whilst allowing it to proceed at specially sensitive points, has been used. This plan also has been criticized on the ground that it is not realistic; actually, however, such conditions do really occur in service on occasions when an attempt has been made to prevent corrosion by adding an inhibitor, but the amount added has been insufficient to cope with the chloride present. Under these borderline conditions, the dangerous combination of small anode and large cathode leads to intense attack, but it should be noted that the attack, although localized and penetrating, does not always attain the regularity and symmetry of the hemispherical or polyhedral pits produced by the electrochemical method.

In the method based upon incomplete inhibition, the geometrical placing of the specimen greatly influences the result. In almost any solution, corrosion will start locally at isolated points, but if the corrosion product is allowed to spread over the surface, the corrosion will cease to be localized. For instance, on a vertical iron specimen partly immersed in a chloride solution, attack will start at isolated points, but will spread downwards and outwards, producing arch-shaped areas undergoing corrosion. Even if sodium carbonate or sodium chromate is present, the attack may cease to be localized if the solid corrosion product spreads over the metallic surface, protecting it from the renewal of inhibitor and/or oxygen. If the specimen is placed horizontally, being supported above the bottom of the containing vessel, and if the corrosion of the lower surface is studied, there is a better prospect of the corrosion remaining localized, because the heavy corrosion product will sink away from the metal. This method, based on the use of the lower surface of a horizontal specimen, was much employed in researches conducted in the thirties; but its possibilities seem to be neglected today, and some of the earlier work must be recalled.

Inclusions as the Points of Origin of Pitting.   Early work by Homer[1] at Birmingham on the behaviour of steel in a solution containing NaCl, with small amounts of Na_2CO_3 to localize the attack, showed that frequently attack took place at the sites of sulphide inclusions. Not all the sulphide inclusions acted as starting points of corrosion. The proportion of inclusions which set up attack was sometimes as low as 1%, although often higher. Several types of iron and steel were tested; in many of them the sulphur mainly occurred as manganese sulphide. A rimming steel used, however, contained iron sulphide inclusions as well as manganese sulphide, and these were found to behave in the same way. In ingot iron the inclusions were nearly all iron sulphide, but here only a small proportion acted as corrosion centres, and the attack was slower. Silicate inclusions existed in several of the materials; these were never found to act as centres of attack; alumina also failed to set up corrosion.

Homer generally presented the lower surface of a horizontal plate to the

1.   C. E. Homer, Iron Steel Inst. Corr. Comm. 2nd report (1934), p. 225.

liquid, so as to avoid the spreading of attack; he was able, however, to obtain localization on an upper surface if the ratio of carbonate to chloride was carefully adjusted. Tronstad[1] employed small amounts of chromate to restrict attack and obtained evidence that attack on iron was associated with sulphide inclusions. In the case of stainless steel, sulphide inclusions have been shown to be the centres of pitting in researches carried out in France,[2] Poland[3] and Sweden.[4] The Polish investigators point out that sulphide inclusions generally occur along with oxide particles, since the latter act as nuclei for sulphide crystallization during the solidification of steel; the sulphide often forms a shell surrounding an oxide centre. During corrosion the sulphide dissolves, leaving narrow crevices around the oxide particles, and these crevices are favourable to corrosion. The oxide remains unattacked, and consequently micro-probe analysis reveals the presence of oxide, and not of sulphide. There is today ample evidence of the importance of sulphide in setting up pitting—a matter about which there was at one time some scepticism.

Wranglén[5] has pointed out that in soil and water, the sulphur content of iron has little effect on general corrosion, but increases local corrosion at welds, crevice corrosion and pitting. In acid, the fact that FeS is a good electronic conductor and MnS a poor one becomes important. The addition of sulphur to pure iron may increase corrosion—perhaps 100 times, since (1) the couple Fe/FeS causes pitting, (2) there is adsorption on the iron surface of sulphide ions, which catalyse both anodic attack and also cathodic production of hydrogen. The presence of Mn reduces (1), but does not prevent (2) since MnS is more soluble in acids than FeS. In contrast, Cu counteracts (2), since Cu_2S is relatively insoluble in acids.

Wranglén explains the fact that some MnS inclusions act as starting points for pitting, whilst others do not, as follows: 'Both active and inactive sulphides are usually normal MnS inclusions, containing about 90% MnS. However, the active sulphides seem to be surrounded by a sub-microscopically divided MnS precipitate which makes the matrix closest to the active sulphides particularly reactive. This seems to be a consequence of rapid cooling of the steel and too brief annealing prior to rolling, insufficient time being given for coalescence of the sulphide phase.' Such an explanation would seem to suggest a practical method of minimizing the objectionable effect of sulphur in steel.

French work[6] with stainless steel produced from extremely pure material confirms the importance of sulphide inclusions in relation to pitting. It is found that, whereas commercial 18/10 stainless steel held at $+0.80$ V_0 (sat.

1. L. Tronstad and J. Sejersted, *J. Iron Steel Inst.* 1933, **127**, 425; cf. G. D. Bengough, p. 437.

2. A. Portevin and L. Guitton, *Comptes rend.* 1937, **204**, 125.

3. M. Smialowski, Z. Szlarska-Smialowska, M. Janik-Czachor, M. Rychcik and A. Szummer, *Amsterdam Cong.* 1969, p. 651; *Corr. Sci.* 1969, **9**, 123.

4. G. Wranglén, *Williamsburg Conf.* 1971, p. 462.

5. G. Wranglén, *Corr. Sci.* 1969, **9**, 585.

6. B. Rondot, M. da C. Belo and J. Montuelle, *Comptes rend.* 1972 (Serie C), **274**, 1028.

calomel scale) in N/10 NaHCO₃ loses its passivity when only 3·3 g/l NaCl is added, a specially prepared alloy (of composition 18/14) free from sulphide inclusions requires as much as 180 g/l NaCl. Measurements of the potential at which loss of passivity occurs suggest, however, that the situation is complicated, and that the presence of Mn greatly influences the results. Assuming that the rupture potential is a criterion of the danger of loss of passivity, it would seem that, in absence of Mn, steel with less than 0·005% sulphur is superior to that with a higher sulphur content, but that the reverse is true if the material contains 1·25% Mn. If S is completely absent, the presence of Mn does not influence the rupture potential.

On the other hand, the importance of sulphide or other inclusions must not be exaggerated. Wood's belief (this volume, p. 77) that pitting generally originates at flaws in the oxide film is probably correct; only a few of these flaws are associated with inclusions.

Difficulties in pitting theory connected with Resistance. About 1970, the rapid attack at a pit was generally explained on the principle of the cathode/anode ratio. The main portion of the surface, covered with a protective film, was thought to serve as cathode and the small break in the film as anode (Fig. 5). If the current flowing is determined by the reduction of oxygen at the large cathodic area, the corrosion, concentrated at the small anode, may be expected to be intense.

However, several authorities find difficulty in accepting this explanation.

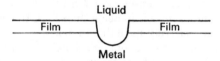

Fig. 5   Schematic diagram to suggest views held regarding the mechanisms of pitting. The film-covered part of the metallic surface serves as a large cathode on which considerable reduction of O_2 to OH^- takes place. The corresponding anodic attack is concentrated on the small gap in the film, and is thus intense, explaining the phenomenon of pitting.

They point out that the potential to be expected at a film-covered (passive) area will be very much higher than that prevailing on the active region where corrosion is taking place. The value of iR will be too small to account for such a difference in potential between anode and cathode—at least in the case of an open pit not covered by a membranous roof. In order to obtain compliance with Ohm's law, they have invoked the presence of something which would introduce a high resistance, and thus raise the value of iR. Vetter[1] considers that a continuous film of chloride may be present on the surface within the pit and has developed a procedure for detecting such a film. Frankenthal[2] ascribes the high resistance to hydrogen bubbles, whilst

    1.   K. J. Vetter and H. H. Strehblow, *Ber. Bunsen Ges.* 1970 **74**, 449, 1024; 1971, **75**, 822 (with A. Willgallis); also *Williamsburg Conf.* 1971, pp. 32, 111, 116, 240.
    2.   H. W. Pickering and R. P. Frankenthal, *J. electrochem. Soc.* 1972, **119**, 1297, 1304 (2 papers); *Williamsburg Conf.* 1971, p. 261.

Sato[1] considers that the smooth surface of the pits is covered with an oxide-film of the type involved in electropolishing. There is little doubt that all these things do occur in certain cases. The hydrogen bubbles can often be seen at low power magnification. Evidence of a layer having 'the properties of a metal chloride with a given solubility product' has been obtained by Isaacs,[2] who found that an increased concentration of Cl⁻ in the solution diminishes the current flowing. The presence of a 'polishing film' on the surface of hemispherical pits has been confirmed by Yahalom,[3] using an electron microscope. He studied open pits on nickel previously passivated in sulphate or phosphate solution, to which Cl⁻ was afterwards introduced, the potential being kept high enough to cause breakdown; in every case a solid film was found on the surface of the pit; the thickness, inferred from the interference colours, was about 400 Å—in contrast with a thickness of about 25 Å on passive nickel. Yahalom accepts Sato's views with certain reservations.

It could be remarked here that it is probably the presence of the polishing film which produces the smooth, hemispherical form in certain cases, for reasons explained on p. 157; in the absence of such a film, flat faces corresponding to crystal planes would be expected, as explained below. Where the pits are bounded by flat crystal faces, the high resistance can be explained without assuming a film or gas bubble. To produce a smooth crystal face starting from a surface which is not smooth on an atomic scale involves the anodic removal of all small projections or clumps of disorganized atoms. These will be removed preferentially, since here each atom is in contact with a fewer number of neighbours than would be the case on a truly flat and perfect surface, so that less energy is needed for escape into the liquid. During this process, the attack will be concentrated on very small areas where the projections are situated. When once the disorganized material has been removed, the flat crystal faces are maintained by removing atoms one layer at a time. During any stage in this removal, an atom situated at a kink site such as A (Fig. 6) will be removed more easily than a typical atom in the plane below (such as X), having fewer neighbours. The removal of A will leave atom B in a favourable situation for removal; the removal of C will follow, and so on. Attack on the lower plane will not generally start until all the atoms in the upper plane have been removed. At any one moment, the attack is concentrated on a point of atomic dimensions. Now the resistance of the approach to a circular area of radius r' is approximately

$$\int_{\infty}^{r'} \frac{1}{k \cdot 2\pi r^2}\, dr \quad \text{or} \quad \frac{1}{2\pi r' k}$$

where k is the specific conductivity of the liquid. If r' is of atomic dimensions, this resistance can be very great even though the liquid is one possessing high specific conductivity. Thus the difficulty disappears; there is no need to postulate a film, bubble or other resistance-raising factor in cases where evidence is lacking for the presence of such a factor.

1. N. Sato, T. Nakagawa, K. Kudo and M. Sakashita, *Trans. Japan Inst. Met.* 1972, **13**, 103.
2. H. S. Isaacs, *J. electrochem. Soc.* 1973, **120**, 1456.
3. J. Yahalom, *Tokyo Cong. Ext. Abs.* 1972, p. 441.

An article by Hoar[1] on 'bright pitting' deserves study. It mainly concerns the pits formed by breakdown of oxide-film at positive potentials; such pits are crystallographically not well defined and tend to be roughly hemispherical; they show a degree of electrobrightening, and have been described as

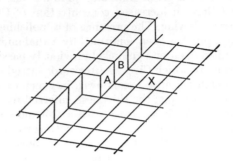

Fig. 6   Schematic diagram explaining why etching produces flat faces along crystal planes. The atom A at the kink site on an incomplete layer is more likely to be removed than a typical atom X on the complete layer below—having fewer neighbours to exert attractive force and oppose removal. When A has been removed, the new kink site atom B becomes specially easy to remove. This explains why attack on an incomplete layer is likely to continue in preference to the starting of attack on the complete layer below.

'polished pits'. (Crystallographic features are more common on pits formed at low potentials.) In the pits, anodic current is passing through the polishing film by anodic transport, the current density being usually between 10^{-6} and 10^{-11} amps/cm^2. Hoar calculates that the iR drop involved is often only about 10 mV, although it may be higher in special cases. The manner in which the film produces brightness is discussed on p. 157 of this volume.

Generalized Theory of Pitting.   If a metallic specimen carrying an oxide-film is plunged into a corrosive liquid (e.g. a salt solution), anodic attack will generally start at isolated points, representing points where the film is absent or faulty. There is good reason to think (1960 book, p. 176) that the film on metal containing internal stresses is continually cracking, but that in presence of air, the cracks generally heal themselves by the formation of fresh oxide. If the dry metal specimen is suddenly immersed in liquid there may be points where the film has chanced to crack just before the moment of immersion; anodic attack may sometimes, but not always, develop at these points. These may be the places where the film is consistently imperfect, as, for instance, around the edges of certain types of inclusions; these could also provide points of anodic attack. It is likely that the decision as to whether corrosion will or will not continue at a break in the film is determined immediately, and that where pitting is set up it really commences at the moment of immersion, even though some time may have to elapse before it can be detected by the optical or electrical instruments

1.   T. P. Hoar, *Williamsburg Conf.* 1971, p. 112.

commonly available. However, other pits may be initiated considerably later, namely at moments where a fresh crack occurs; such cracking may be connected with the fact that strain energy in the metal can be relieved if the film cracks or blisters, but it can probably arise in other ways. Sato[1] has made the interesting suggestion that the high field (10^6 V/cm), set up at places where a current (connected with corrosion) is passing through the film, could cause cracking by electrostriction.

Whatever the cause of the fresh cracking, it is likely that only a few of the cracks develop into pits, the proportion varying greatly with the conditions; the potential, the chloride content and the temperature will be important factors. At a given potential, the probability of a pit forming at a point suddenly laid bare will increase with the chloride content, but it is unlikely that there is a critical chloride concentration below which the probability is zero and above which it is finite; similarly, at a given chloride content, the probability will rise with potential, but it is doubtful whether careful study of the situation would establish a critical potential below which he probability suddenly becomes zero.

The probability that a fresh crack will become a place of fresh corrosion will be greatly influenced by the existence of other points of corrosion in the neighbourhood; if there are many such corroding points around the site of the crack, the potential at that site will be raised and the chance of new attack reduced. Evidence of this was provided by experimental work on 'crack–heal' carried out in 1946 (see 1960 book, p. 177; Fig. 40).

If the conditions are favourable to continued attack, this will only remain ocalized, forming a pit, if they are also unfavourable to the lateral extension of that attack. It is most likely to remain localized if an inhibitor is present in a concentration sufficient to confine the attack to the most susceptible points. Even here the geometrical situation is all important, since, if corrosion product is able to spread over the surface, it may act as a screen, preventing the access of inhibitor or of oxygen, which, if present, would produce OH^-—itself an inhibitor.

The specific effect of Cl^- and other halide ions is probably connected with the fact that when sites in an oxide-lattice which would normally be occupied by O^{2-} become occupied by Cl^- ions, the number of vacant cation sites in the neighbourhood must increase, if neutrality is to be maintained. Consider, for instance, a place where imperfection of the lattice structure—perhaps connected with stresses which locally increase the interatomic distance—facilitates detachment of ions from the metallic surface. At such a place, a slow movement of ions through the film will start. This will in most cases involve movement of both anions and cations in opposite directions. In absence of Cl^-, the movement of, say, Al^{3+} and O^{2-} should not affect the concentration of vacant sites. But if Cl^- is present in the liquid, a proportion of the anions moving inwards through the film will now be Cl^-, and if neutrality is to be preserved, Al^{3+} must move out from the film into the liquid at a rate greater than would be the case if O^{2-} was the only anion involved, leaving an increased number of vacant cation sites. When once,

1. N. Sato, *Electrochim. Acta* 1971, **16**, 1683.

at the point in question, the concentration of vacant cation sites in the film has become abnormally high, movement of ions through the film will become much faster than elsewhere, and the pit will propagate with increasing velocity. The formation of acid by hydrolysis of the chloride will probably be a very important accelerating influence in the later stages. The surprising low value of the pH at points of localized anodic attack has been measured accurately in the case of stress corrosion cracking by the freezing method of Floyd Brown (this volume, p. 327); but it has been established for pitting also, notably by Butler.[1]

Most of what has been stated above is applicable to all, or most, metals. The marked contrast between the behaviour of different metals seems generally to be connected with the seat of the cathodic reaction. On aluminium, where the pure oxide is a good insulator, it is probable that the oxide-film plays no part in the cathodic reaction; the cathodic area could be particles of deposited metallic copper, if there are traces of copper in the water or, as an impurity, in the metal. Possibly, in alloys where a second metal present in solid solution in the oxide raises the electronic conductivity, the oxide-film may itself act as cathode; again there may be a second phase which provides cathodic particles. For some of the heavier metals, where the pure oxide possesses electronic conductivity, the film itself probably provides the cathodic area; the intense localized corrosion met with at breaks in mill-scale on steel is well known. On copper tubes carrying a carbon film, pitting set up at breaks in the carbon layer has been established.

Researches on Individual Metals, mostly under Pitting Conditions

Iron. An important research has been carried out at Teddington by Butler and his colleagues[1] on the pitting of pure iron, steel and various alloys containing chromium and copper exposed to various solutions at different temperatures. In general, pitting was found to be a phenomenon characteristic of dilute salt solutions or pure water; more concentrated solutions tended to give general corrosion or etching. Thus steel exposed to a boiling solution of a sodium, calcium or magnesium salt became pitted in very dilute solutions but not in concentrated ones. On pure iron the pits formed were often straight-sided polygons, sometimes square, rectangular or triangular, and occasionally hexagonal. These were bounded by crystal planes, but circular pits sometimes appeared—apparently where mechanical working had disorganized the crystal structure. After an induction period (during which there was no visible change) pitting started at various points. Subsequent attack was governed by the flow of corrosion product from the *primary* pits; for instance corrosion product sinking under gravity would set up *secondary* pits at points below (the result, it may be suggested, of oxygen screening). In this research neither inclusions nor debris on the surface was observed to create sites for attack. The growth rate in the early stages was difficult to estimate, but seemed to correspond to an anodic current of about 0·5 to 1·0 amps/cm^2. This is comparable to figures quoted by other investigators for

1. G. Butler, H. C. K. Ison, A. D. Mercer, P. Stretton and J. G. Beynon, *Brit. Corr. J.* 1972, **7**, 168; *Corr. Sci.* 1967, **7**, 385.

current flowing from electrochemically generated pits in *acid* chloride solutions.

By means of very small Sb/Sb oxide electrodes (5–25 μm in diameter), it was found possible to study the variation of pH in and around the pits; at the centre of a pit, the value might be as low as 2·0, but it gradually rose on passing outwards, and reached about 8·0 at remote points.

The pitting of iron in a solution containing potassium chloride along with sufficient potassium hydroxide to restrict the attack to isolated points, has been studied at Brno by Toušek.[1] He found that the number of pits per unit area increased with rising KCl concentration (at constant KOH) and decreased with rising KOH concentration (at constant KCl). The number of pits was found to be a rectilinear function of the time, t, and the rate of pitting a rectilinear function of t^3. In his later papers Toušek develops equations enabling him to calculate, on the basis of measured values of the current density (CD) and the number of pits, the dimensions of the active pits and the CD at the places where corrosion is occurring. He finds that the aggressive action of SO_4^{2-} is considerably weaker than that of Cl^-.

The work of Hoar, Leach, Kruger and others on pitting under anodic conditions will be discussed in chapter VII.

A discussion of the prediction of pitting rates in pipes took place at a conference in 1968 devoted to 'Corrosion and Protection of pipes and pipelines'. It has been reviewed by Potter.[2] One author seemed to regard the incorporation of Cl^- ions into a hydrated oxide-film formed on iron as the condition likely to lead to pitting. Pitting was felt to be a particularly vexing type of attack on pipes, because an initial period of immunity is followed by an erratic series of failures; these are determined by the progress of the deepest pits. The views of Collins and Fuller[3] may be studied; they consider that, for making predictions, a fitted gamma-distribution is superior to other statistical methods.

Iron-chromium alloys, including Stainless Steel. The Teddington work mentioned above was extended to cover alloys containing Cu and Cr. In particular, pitting initiation was studied on alloys made from high purity iron with various Cr contents between 1·0% and 11·3%. The introduction of even 1% Cr altered the shape of the pits, which, instead of being polygonal, became approximately circular and were located at grain-boundaries—especially at the junction of such boundaries. As the Cr content was increased further, the pits tended to become elongated in form, with the longer diameter located in the grain-boundary. At the highest Cr concentrations the attack was predominantly intergranular; secondary pits within individual grains were polygonal in shape. In contrast, Scully,[4] working on austenitic stainless steel, reports that pitting penetrates downwards more quickly than sideways.

1.   J. Toušek, *Werkstoffe u. Korr.* 1970, **21**, 21; *Corr. Sci.* 1972, **12**, 1, 15 (2 papers).
2.   E. C. Potter, *Amsterdam Cong.* 1969, p. 39.
3.   H. H. Collins and A. G. Fuller, *Brit. Corr. J.* 1971, **6**, 97; see discussion by W. K. Johnson and D. J. Fuller, p. 98.
4.   I. S. McCulloch and J. C. Scully, *Corr. Sci.* 1969, **9**, 707.

The pitting and etching of austenitic stainless steel has been studied at Milan under Bianchi.[1] A solution containing ferric chloride, aluminium chloride and lithium chloride was used, generally, without applied anodic current. When the Fe^{3+} concentration was low, there were many small round pits. When a high Fe^{3+} concentration was used, there were fewer pits, but they were large and generally hexagonal. Square pits, sometimes met with at lower concentrations, ceased to be formed at high concentrations. The best method of obtaining square pits was to use potentiostatic control, keeping the potential at 520 mV relative to the Ag/AgCl electrode.

The important result of this Italian work was to show that the pitting depended greatly on the state of the film present on the stainless steel before its immersion in the pitting liquid. If the specimen had been heated at 150°C in dry air for 140 hours the pitting susceptibility was small, whereas a specimen heated at 300°C showed a high pitting susceptibility. The difference was not due to the greater thickness of the film, since if a specimen which had been heated at 300°C was then treated at 150°C, it assumed the same pitting characteristics as a specimen originally treated at 150°C. It is believed that the oxide formed at 150°C displays n-type conductivity, whilst that formed at 300°C is a p-type conductor. The current falls off with time if the film has been formed at 150°C, but increases with time if it has been formed at 300°C.

The pitting of stainless steel has also been studied by Isaacs and Kissel,[2] using a solution containing ferric chloride and hydrochloric acid. They consider their work to confirm that of Bianchi. Slight divergences of results may be due to the fact that the liquid was different. On electropolished specimens the number of active pits declined with time of immersion in the solution, but this decline was much slower if the material had been heated in air after electropolishing but before immersion in the liquid. The decline was slower if the heating had been carried out at 165°C than at 110°C, and slower still after heating at 240°C. If after electropolishing the specimen was cathodically treated in the polishing solution and not heated, the rate of decline was greatly increased. This research shows how greatly the state of the film on the surface affects pitting behaviour.

Japanese work by Sato and his colleagues,[3] using a stainless steel electrode rotating in dilute sulphuric acid to which chloride had been added, provides important quantitative results. One series of experiments, based on 0·2M sodium chloride and 0·1M sodium sulphate with sulphuric acid added to give various pH values, shows that, for a given number of pits, the corrosion current is constant in time. When, however, an extra pit appears, the current suddenly increases. By using a high rotation rate (2550 rpm) they obtain information about a single pit, since no second pit then arrived within the experimental period. For a single pit, the current passing at time t was $k_i(t + a)$, where k_i and a are constants. Pit generation on the average

1.   G. Bianchi, A. Cerquetti, F. Mazza and S. Torchio, *Amsterdam Cong.* 1969, p. 614; *Corr. Sci.* 1970, **10**, 19; 1972, **12**, 493.

2.   H. S. Isaacs and G. Kissel, *J. electrochem. Soc.* 1972, **119**, 1628.

3.   N. Sato, T. Nakagawa, K. Kudo and M. Sakashita, *Williamsburg Conf.* 1971, p. 447; *Trans. Japan Inst. Met.* 1972, **13**, 103.

proceeded linearly with time, although there was some scatter. The equation was

$$n = k_p(\tau_n - \tau_0)$$

where n is the number of pits and τ_0 is the 'incubation time', probably associated with the initial adsorption of Cl^-. This differs from the 'induction time', t, for the opening of the first pit. The constant k_p increases nearly linearly with the potential maintained by the potentiostat. The current from a single pit is proportional to the area of the pit-mouth, giving a CD of 8 amps/cm^2, irrespective of potential. The dissolution rate is similar to that obtained in electropolishing work. The shape of the pits is approximately hemispherical, but the ratio of depth to radius increases as the pit grows.

The current passing during pitting under the electrochemical procedure has been studied in Poland by Szklarska-Smialowska,[1] using iron containing 13% or 16% Cr. The anodic current increases with time, being proportional to t^b. In cases where b is 2, all the pits are found to nucleate simultaneously. Where b exceeds 2, the number increases with time; b can be as high as 6. In this research, the pits were hemispherical and bright, and the currents were similar to those accompanying an electropolishing process—in agreement with the work of Sato just mentioned.

The pitting of austenitic stainless steels has also been studied by Forchhammer and Engell.[2] They found pitting to be promoted by increase of chloride content and temperature: also by decreasing homogeneity. Cold-work produced only slight effect.

The pitting of iron–chromium alloys containing 5·6% and 11·6% Cr has been studied by Stolica.[3] He found that curves produced under galvanostatic conditions showed an induction period depending on the Cl^- concentration, and a reaction period characterized by oscillations; a minimum concentration of Cl^- ions was necessary for attack. Russian work reported by Tomashov[4] was devoted to the effect of introducing various elements (Si, V, Mo or Re) into 18/14 Cr/Ni stainless steel immersed in ferric chloride solution. Three stages were noted in some cases: (1) surface destruction at the site of local impurities, (2) formation of pits capable of re-passivation, and (3) development of steadily growing pits. When the addition amounted to 5%, change stopped at the first stage. When it was only 2·5%, change generally stopped at the second stage. When there were no additions, all three stages occurred.

The film found covering the pits in stainless steel apparently represents the original passive film, which has been undermined. In work on thin foil[5] the film formed on the face where the pits start to form appears to give a similar electron diffraction pattern to that found where the pit has penetrated to the back surface.

1.   Z. Szklarska-Smialowska and M. Janik-Czachor, *Brit. Corr. J.* 1969, **4**, 138.
2.   P. Forchhammer and H. J. Engell, *Werkstoffe u. Korr.* 1969, **20**, 1.
3.   N. Stolica, *Corr. Sci.* 1969, **9**, 205.
4.   N. D. Tomashov, G. P. Chernova and O. N. Markova, *Williamsburg Conf.* 1971, p. 363.
5.   J. Yahalom, L. K. Ives and J. Kruger, *J. electrochem. Soc.* 1973, **120**, 384.

Nickel. Polish work[1] on nickel in contact with solutions containing both Cl^- and SO_4^{2-} shows that there is the greatest development of pits in respect of width and depth when the ratio of Cl^- to SO_4^{2-} is 0·75. At higher values of this ratio development is slower. At the high Cl^- concentrations, besides pitting, there is a uniform attack on the main surface.

Aluminium. In general, aluminium suffers pitting only in waters containing chloride and traces of copper. The question has been asked whether the presence of copper is absolutely necessary for pitting. Johnson[2] has confirmed the occurrence of pitting in copper-free waters containing Cl^- and/or HCO_3^-, but states that in practice copper is almost invariably found where pitting has occurred. It might be suggested that in cases where pitting has developed in waters free from copper, there was copper either as an impurity or as an alloying constituent in the metal. Wranglén[3] has provided pictures of pitting under such conditions; the cathodic points appear to be places where copper has been reprecipitated as metal on the surface, or where a particle of Al_3Fe occurs, nearly penetrating through the surface film of alumina.

The manner in which Cl^- promotes pitting has been studied in detail by Pryor.[4] He believes that it is due to a defect structure set up where Cl^- ions exchange with O^{2-} ions on the Al_2O_3 lattice, electrical neutrality being maintained by the passage of the appropriate number of Al^{3+} ions from the oxide into solution. This leads to the production of vacancies at cation sites, and thus facilitates the passsage of Al^{3+} ions outwards through the film. Although in the absence of oxygen rapid pitting only occurs above a certain critical potential, he found it at a much lower range of potential if oxygen is present. In fact, he considers that under service conditions all pitting of aluminium and its alloys is produced in a region well below the so-called critical potential—the significance of which may have been exaggerated. In contrast, Böhni and Uhlig[5] stress the importance of a critical pitting potential which is depressed by the presence of Cl^- but elevated by additions of nitrate, chromate, acetate, benzoate or sulphate. They report a relationship between the amount of inhibitor needed to produce inhibition and the Cl^- activity; $\log [Cl^-]$ plotted against $\log$ (inhibitor activity) gives a straight line. The mechanism is interpreted on the basis of a competitive adsorption of Cl^- and oxygen at suitable sites on the metallic surface.

The influence of Cl^- concentration on pitting has been studied by Bogar and Foley,[6] who plotted the logarithm of the inverse induction time against $\log [Cl^-]$. The points often fall close to two straight lines intersecting one another. The gradient can be as high as 3·0 to 4·8 at the higher Cl^- concentrations. This may indicate that the reaction is

$$Al + 4Cl^- \longrightarrow AlCl_4^-$$

1. Z. Szklarska-Smialowska, *Corr. Sci.* 1972, **12**, 527.
2. W. K. Johnson, *Brit. Corr. J.* 1971, **6**, 200.
3. G. Wranglén, *Metallers och Ytskydd (Stockholm)* 1967, p. 108.
4. M. J. Pryor, *Williamsburg Conf.* 1971, p. 2.
5. H. Böhni and H. H. Uhlig, *J. electrochem. Soc.* 1969, **116**, 906.
6. F. D. Bogar and R. T. Foley, *J. electrochem. Soc.* 1972, **119**, 462.

An investigation by Metzger and others[1] on the pitting of high purity Al in 16% HCl has shown that pitting occurs only at sites where the film has suffered mechanical damage during cooling. When aluminium is cooled from above 400°C, small voids appear to be formed below the oxide-film, by condensation of some of the excess lattice vacancies, except near the boundaries of grains and subgrains which act as sinks for vacancies. Attack starts at the voids and spreads over the whole surface except for unattacked strips at the grain-boundaries which are left standing up in relief. Water-quenched and pre-polished quenched specimens behave differently; the latter show special attack at grain-boundaries, due to film-cracking connected with localized shear.

Although nitrate normally acts as an inhibitor to Al, the presence of NO_3^- in a Cl^- solution produces more rapid attack than a Cl^- solution free from NO_3^-. In addition to the cathodic reduction of oxygen, alternative cathodic reactions become possible, such as the reduction of NO_3^- to ammonia, which, in the case of alloys containing copper, will then dissolve out the copper. The subject is discussed by Foley and his colleagues.[2]

An important study of the pitting of Al has been carried out at Manchester by Richardson and Wood.[3] They used a scanning microscope to study the surface and reached the conclusion that the pits arise from flaws in an otherwise inert oxide-film, and then proceed to undermine that film. The electron micrographs show cracking, peeling and undermining of the film—which optical micrographs had failed to reveal. In some cases hydrogen bubbles seem to help in dislodging the film covering the surface; but this is not a necessary condition for pitting, since some experiments were carried out at a potential where hydrogen evolution would be impossible. The micrographs are considered to 'provide unambiguous evidence of the inert role played by the surface oxide-film during pitting propagation', and one of them 'provides very strong circumstantial evidence for pitting initiation at the bases of flaws in the surface film'. The surface oxide-film is considered to function as an insulating layer. 'No Cl^- contamination of the film, either by an exchange process or by a field-induced entrance mechanism, is envisaged as essential, nor is the solubility of the film considered a crucial factor. This is not to say that these processes' [which constitute essential features of certain theories] 'do not perhaps occur . . . merely that they are not important in pitting. . . . Far from being an autocatalytic process from the outset, as has been almost universally assumed, the present work would support the view that pitting may well occur at the fastest rate during the early stages of immersion, where the cathode/anode area [ratio] for the electrode as a whole is at its greatest. This is not to say that chemical changes within the anolyte [e.g. increased acidity] cannot accelerate the pitting rate during the later stages of immersion. But the latter phenomenon must be recognised as the result rather than the cause of pitting.'

1. O. P. Arora, J. A. Isasi and M. Metzger, *Corrosion (Houston)* 1969, **25**, 445.
2. R. T. Foley (with A. M. McKissich, A. A. Adams and K. E. Eagle), *J. electrochem. Soc.* 1970, **117**, 1459; 1972, **119**, 1692.
3. J. A. Richardson and G. C. Wood, *Corr. Sci.* 1970, **10**, 313.

Further experimental data is available at Manchester, accumulated through work, still partly unpublished, carried out under Wood; it is summarized in a paper with four colleagues,[1] dealing mainly with aluminium. The materials studied include high purity Al specimens carrying air-formed films, others carying thick films obtained by anodization and also alloys containing Fe and Cu. Electron microscopy shows that pitting generally starts at flaws in films; the flaws develop into pits by metal dissolution, followed by the under-mining of the surface film. The authors draw a clear distinction between 'residual flaws' such as copper-rich segregates, and 'mechanical flaws' produced during the relief of internal stresses. The former often act as cathodic points, whilst the latter provide gaps in the film where anodic attack starts, leading in due course to a detectable pit. There is, however, an intermediate type of 'mechano-residual flaw'. For instance, iron-rich segre-gates, probably $FeAl_3$, although apparently permitting the growth of anodic films upon their surfaces, nevertheless cause cracking of the film at the periphery, providing a place where pitting can be set up.

On Wood's theory, pit initiation is not viewed as a separate process from pit propagation. As the flaws develop into pits, the residual flaws become less important in their contribution to the cathodic reaction—much of which now occurs at places within the pits. Thus although the total corrosion current increases, the true current density at the anodic points diminishes. It is recognized, however, that a decrease of pH in the pits may cause an accelera-tion in the pitting rate, which must be superimposed on the other effects described.

The flaw population is shown by the electron micrographs to be orienta-tion defined; certain grains contain a high flaw density and others a low one. Pitting has been studied under conditions of simple immersion (without applied current), but also under the influence of applied polarization from an external source. As the potential of the applied anodic pulses is raised, the nature of the attack becomes progressively less crystallographic; the visible pits become smaller, brighter, more uniform in size and more regularly hemispherical in shape. In experiments with external polarization, there is first an 'induction period' followed by a sharp rise in the current. It has often been supposed that pitting only starts at the onset of this current rise. On Wood's theory, flaws are developing into pits from the moment of immersion; sensitive electron optical methods show that pits are already present before the current rise; moreover a slight current rise during the induction period can be detected with a sensitive meter. The so-called 'induction period' is considered to be merely the period during which the pits are too small to produce an easily measurable change in the current or to be visible by con-ventional methods. Wood and his colleagues discuss critically the concept of a 'breakdown potential' supposed to be associated with the first visible signs of pitting; they recognize that the term may be useful in connection with prac-tical problems, but consider that it has no real fundamental significance. The 'breakdown potential' represents that step in the pitting process at

1. G. C. Wood, W. N. Sutton, J. A. Richardson, T. N. K. Riley and A. G. Malherbe, *Williamsburg Conf.* 1971, p. 526.

which the combined effects of increasing area on which anodic attack is proceeding, and decreasing ability of local cathodes to provide the current corresponding to the increasing anodic attack, require that the potentiostat is called upon to supply a larger proportion of the required current.

In a more recent paper,[1] the effect of different solutions upon Al carrying faulty oxide-films is discussed, as a result of experiments based on measuring impedance changes. It is found that, in chromate and dichromate solutions, mechanical flaws originally present in air-formed films tend to be healed; films formed by anodizing are thinned down to the thickness of a passive film. Little attack is detected in sulphate solutions until after long periods. In chloride solutions, and to a lesser extent in bromide and iodide solutions, pitting commences rapidly by preferential dissolution of metal transiently exposed at flaw-bases in air-formed anodic films; undermining of the relatively inert surface layer then occurs. In fluoride solutions, the oxide-film is removed, apparently by undermining, and is replaced by a complex basic fluoride. It is pointed out that the mechanical flaws do not represent places where the metal is persistently bare, but are produced rather by crack–heal processes on the immersion of the specimen carrying the film. Healing is believed to occur by the passivation of the bare metal and by partial plugging of the mechanical flaws with alumina. Eventually, as the stress is relieved, crack–heal events become less frequent. The impure oxides at the base of residual flaws, although healed by contact with chromate or dichromate solution, remain largely semiconducting in character and initially act as cathodes in any electrochemical reactions, although they can in time become sites for corrosion.

Titanium.   T. R. Beck[2] has studied the formation of pits on titanium in bromide solution. The results depend on the macroscopic geometry of the specimen. In one research pits were obtained at the edges of thin foil specimens and also on the circumference of circular holes cut in specimens. The titanium was held at a constant potential, and the anodic current passing generally became fairly constant with time. As the level of the potential chosen was increased, the current passing increased also. No pitting was obtained below a certain threshold potential, this steady state pitting potential decreased in rectilinear manner with log $[Br^-]$ irrespective of pH value; (the same curve was obtained in HBr as in KBr).

Authoritative information about titanium pitting is provided by papers delivered at Williamsburg.[3]

Copper.   The pitting of copper pipes used in hot-water systems has been a very serious practical problem, although the proportion of pipes giving trouble is relatively small. Behaviour varies greatly with the methods of producing and surface treating the pipes, and also with different supply waters. Waters drawn from certain wells are found to contain an unidentified

1.   J. A. Richardson and G. C. Wood, *J. electrochem. Soc.* 1973, **120**, 193.
2.   T. R. Beck, *J. electrochem. Soc.* 1973, **120**, 1310, 1317.
3.   *Williamsburg Conf.* 1971. Papers by T. R. Beck (p. 644), M. F. Ard Rabboh and P. J. Boden (p. 653) and J. B. Cotton (p. 676).

inhibitor which prevents attack on pipes where other purer waters cause trouble; this 'beneficial' substance is probably derived from pollution with sewage!

An excellent summary of the position has been provided by Campbell,[1] who points out that there are two types of pitting. The first occurs when the copper pipes carry films of carbon as a result of manufacturing conditions; this type is met with in hard waters, being more common in the cold part of the system. A simple explanation, not universally accepted, is that the carbon film acts as a large cathode, and the copper exposed at breaks in it as small anodic points. Most manufacturers now clean their pipes before delivery to remove the carbon films. The second type of pitting occurs in soft hot water and produces deep pits of small cross section found to contain hard Cu_2O beneath small mounds also consisting mainly of Cu_2O but often containing basic copper sulphate. This type proceeds slowly, 6 years frequently being needed for perforation. MnO_2, a good cathodic depolarizer, will stimulate the pitting, if present; but pitting has been found in the absence of Mn, especially where an adherent oxide-film is present. Cathodic protection by aluminium rods is found to be effective where a water is known to cause pitting.

Campbell[2] has also discussed different theories of copper pitting, presenting the views of Lucey[3] at length, and those of Mattsson[4] and Pourbaix[5] briefly; the papers of the two latter, based on conditions prevailing in Sweden and Belgium respectively, could with advantage be studied at source, but the salient features may be mentioned here.

In Sweden hard-drawn tubes are commonly used, and the carbon films are destroyed in the final drawing procedure; this simplifies the problem. Mattsson suggests that in hot-water tubes, pitting need not be feared if the pH value exceeds 7·4 and the HCO_3^-/SO_4^{2-} ration stands above unity. SO_4^{2-} decreases the proportion of the current carried by CO_3^{2-}, and thus decreases the chance of obtaining a protective film of basic carbonate in close union with the copper surface; basic sulphate is more soluble than basic carbonate.

Pourbaix states that the Brussels water supply does not produce pitting on tubes of electrolytic copper which are free from carbon films, but may lead to pitting if such films are present. Tests carried out over periods reaching three years show that whenever the potential has exceeded a certain value (generally +350 to +420 mV, on the hydrogen scale), pitting has occurred. The presence of cuprous chloride in pits can be detected by exposing corroding specimens to light; a potential drop indicates that the pits contain CuCl. Pourbaix's picture of the pitting process is an extension of that put forward in early work by May.[6] At first the Cu corrodes at about +220 to +240 mV, with the formation of CuCl, but this hydrolyses to give solid Cu_2O and hydro-

1. H. S. Campbell, *Water Treatment and Examination* 1971, **20**, 11.
2. H. S. Campbell, *Williamsburg Conf.* 1971, p. 625.
3. V. F. Lucey, *Brit. Corr. J.* 1972, **7**, 36.
4. E. Mattsson and A. M. Fredriksson, *Brit. Corr. J.* 1968, **3**, 246.
5. M. Pourbaix, *Cebelcor Rapp. tech.* 157 (1969); **167** (1969); **186** (1970); **201** (1971) (with J. van Muylder).
6. R. May, *Trans. Inst. marine Engrs* 1937–38, **49**, 171; 1950–51, **62**, 291 (with P. T. Gilbert).

gen chloride which passes out into the water, and thus causes no damage. Later the Cu_2O becomes covered at some points with a malachite film, which hinders the removal of HCl, and leads to the formation of cavities filled with acid liquid. The situation is presented in a modified Pourbaix diagram, and it is shown that, to avoid trouble, the potential at all points must be kept below $+350$ to $+420$ mV.

A paper by von Franqué[1] presents practical experience and laboratory tests based on Pourbaix's work, designed to assess the possibility of predicting the propensity of a water to produce pitting.

Zinc.   Little work has been carried out on pitting of zinc during the years following 1968, but a *résumé* of knowledge based on early work may be welcomed, since zinc is almost the only metal liable to be pitted in pure water. The contrast with aluminium is striking; the film on aluminium prevents attack by pure water, but pitting occurs if traces of chloride are present, especially if copper is deposited to provide points for the cathodic reaction. Zinc, however, is pitted by pure water, whereas the addition of chloride at first increases the number of pits per unit area, but in larger amounts replaces pitting by relatively innocuous general corrosion. It is unlikely that here the pitting in 'pure' water is due to unintended traces of chloride or other impurity. The careful work of Bengough[2] showed in 1927 that specially prepared water produced pits on the purest zinc then obtainable. His even earlier work with O. F. Hudson had shown that the pits produced on a vertical surface tend to fall below one another on vertical lines. This phenomenon was later studied by Eurof Davies,[3] who found that the corrosion product produced at one pit would slide down the surface, and produce a secondary pit at the place where it lodged. This was ascribed to oxygen screening, producing small anodes at the points screened. Davies produced trenching by tying or pressing a fibre against a zinc surface so as to produce a line screened from oxygen. He demonstrated the flow of electric current between the cathodic area accessible to oxygen and the screened anode. These and other observations explain the behaviour of zinc in solutions of different concentrations. The corrosion of plates suspended 6 cm below the surface of distilled water is increased by the addition of small quantities of sodium (or potassium) chloride (or sulphate); but even where the total corrosion is increased, the intensity of corrosion may be diminished, owing to the attack being distributed over a larger area; higher concentrations of chloride depress the total amount of corrosion produced in a given time.

It should be noted that fairly intense localized corrosion can be produced in relatively concentrated salt solutions if the oxygen screening due to the corrosion product is avoided. Specimens of zinc sheet immersed horizontally in $N/12$ K_2SO_4 solution for seven weeks developed isolated hemispherical humps of gelatinous zinc hydroxide on the lower surface, and conspicuous

1.   O. von Franqué, *Werkstoffe u. Korr.* 1968, **19**, 377.
2.   G. D. Bengough and O. F. Hudson, *J. Inst. Met.* 1919, **21**, 59. G. D. Bengough, J. M. Stuart and A. R. Lee, *Proc. Roy. Soc. (A)* 1927, **116**, 425; 1928, **121**, 88.
3.   U. R. Evans and D. Eurof Davies, *J. chem. Soc.* 1951, p. 2607.

local corrosion was found below these humps. There was no general corrosion or blanketing, since any zinc hydroxide which did not adhere to the metal sank by gravity into the liquid.[1]

The fact that zinc needs no chloride to produce pitting might seem to indicate that it is essentially different from all other metals. It is noteworthy, however, that, if the electrochemical method is used, in which zinc is subjected to anodic polarization so as to produce passivity, pitting only occurs when chloride has been added. Eurof Davies and Lotlikar[2] found that a higher concentration of chloride is needed at pH 11 than at pH 9·2. This seems to bring zinc more into line with other metals. Under service conditions, an external source of current will rarely be provided.

General Researches not concerned with Pitting Conditions

Corrosion Products of Iron. Interest has been shown in the corrosion products of iron, and particularly in those formed by the oxidation of ferrous hydroxide; this interest may have been aroused by the variety of colours exhibited, but much work has been carried out on the crystal structure. Certain compounds are associated with rapid corrosion, but whether this is the result of their structure or of their composition is uncertain. For instance, the presence of β-FeOOH seems to be generally regarded as a danger signal, but it usually contains Cl, presumably replacing OH to a small extent on the lattice; Keller[3] states that β-FeOOH is only formed if Cl^- or F^- is present. It is stable at low pH values, but if the pH is raised, there is a change to α-FeOOH or haematite (Fe_2O_3). In contact with metallic iron, it accelerates the corrosion rate, probably owing to continuous liberation of Cl^- ions. Indian research[4] shows that β-FeOOH produced by the action of NaCl on iron, is changed to lepidocrocite (γ-FeOOH) on washing, with some cubic oxide, detected by X-ray diffraction.

The possible corrosion of the minor phases present in steel deserves more attention than it has generally received; it is often assumed that they merely act as cathodes to the iron forming the main phase, stimulating attack upon it. However, they may be attacked themselves. Staehle[5] has studied the attack on iron carbide; under certain conditions of pH and potentials, nitrates may promote a rapid attack on the carbide particles.

The question of 'pure chemical corrosion'. In the preamble to this chapter, attention was given to cases where the cathodic and anodic reactions take place on distinct areas; but there was also mention of cases where at a single point anodic and cathodic changes might occur at different times. As an example, let us consider a liquid amalgam (for instance, mercury containing in solution a small amount of some reactive metal like zinc)—a case studied by Wagner and Traud.[1] If this is exposed to a slightly acid solution

1.   U. R. Evans, 'Corrosion of Metals' 1926, pp. 85, 245 (Edward Arnold).
2.   D. Eurof Davies and M. M. Lotlikar, *Brit. Corr. J.* 1966, **1**, 149.
3.   P. Keller, *Werkstoffe u. Korr.* 1969, **20**, 102.
4.   G. P. Sharma, G. F. Singhania and S. N. Pandey, *Labdev. J. Sci. Tech.* (*Kanpur*) 1968, **6A**, 1.
5.   R. W. Staehle, *Cebelcor Rapp. tech.* **177** (1970).

(containing an anion which forms a salt of the reactive metal freely soluble in the acid liquid) with oxygen uniformly distributed throughout the liquid, there is no reason why any part of the amalgam surface should have a greater tendency to be anodic (or cathodic) than any other; integrated over time, the anodic (and cathodic) reaction will be uniformly distributed over the whole surface. Even in such a case, it is customary to divide the reaction into anodic and cathodic parts (the passage of the reactive metal into the liquid, and the reduction of oxygen). This allows the potential (known as a 'mixed potential') to be calculated by studying the values of the current i flowing when the potential V is raised or lowered to different levels: log i is plotted against V, and the 'Tafel slopes', obtained in the parts where the curves become straight lines (see, for instance, 1968 book, p. 61) are extrapolated back; the intersection point provides the potential. It might, however, be argued that in situations where there are no areas specially consecrated to anodic or cathodic reactions, we are dealing with a 'simple chemical change', and that the division into anodic and cathodic reactions is merely one of convenience; if that standpoint is accepted, the application of the labels 'electrochemical' or 'chemical' depends on the mentality of the person discussing the matter, and further argument would be without profit.

However, some clear thinking by Heitz[2] has shown that a 'pure chemical reaction' could be envisaged, which would not be capable of separation into anodic and cathodic parts. In the cases hitherto discussed, the anodic and cathodic reactions, whether proceeding at different points at the same time or at different times at the same point, would involve the formation of *two* 'activated complexes'. There might, however, be a direct interaction (between the metal, acid and oxygen, in the case mentioned above) giving rise to only *one* activated complex. It should be possible to distinguish experimentally the situations where there are two complexes from those where there are only one; if two mechanisms can be distinguished experimentally, they cannot be regarded as two different ways of expressing the same thing.

A study of the literature up to 1968 led Heitz to believe that no 'pure chemical reaction' had been clearly established, in the case of metals; the only established case concerned the dissolution of semiconductors in bromine solution—a matter investigated by Gerischer[3] and later the case of aluminium corrosion in anhydrous phenol by Neufeld.[4] Preliminary considerations showed that certain reactions must clearly have an electrochemical mechanism; these include all cases where the corrosion rate is dependent on potential—as had been shown in the Wagner–Traud paper quoted above. In other cases, geometrical considerations would make difficult the formation of a single activated complex as a transition state (a prerequisite of a 'pure chemical reaction'); it would be prevented by steric hindrance.[5]

1. C. Wagner and W. Traud, Z. Elektrochem. 1938, **44**, 391.
2. E. Heitz, Habilitationsschrift, Frankfürt a/M, 1968.
3. H. Gerischer and I. Wallem-Mattes, Z. phys. Chem. (neue Folge) 1969, **64**, 187.
4. P. Neufeld and A. K. Chakrabarty, Corr. Sci. 1972, **12**, 517.
5. E. Heitz, Corrosion of Metals in Organic Solvents, in Adv. in Corr. Sci. and Technology, Vol. 4, p. 149.

Heitz decided to study situations where a pure chemical reaction might reasonably be expected. He selected as suitable to his purpose a study of the corrosion of metals by formic, acetic, propionic or butyric acids in a solution containing their lithium salts; oxygen was present in the main experiments. He found that the change proceeded as a reaction of the first order with respect to oxygen. In the absence of oxygen, copper and nickel were hardly attacked at all; zinc and iron were corroded with evolution of hydrogen in approximately the amount to be expected if the sole cathodic reaction was the reduction of H^+ to H_2; alternative reactions such as the reduction of the acid to the corresponding alcohol were looked for, but they seemed in fact to take place only to a very minor extent.

In presence of oxygen, it was possible to calculate the corrosion velocity on the assumption that the mechanism was 'electrochemical' with two (not one) activated complexes. The current i flowing was measured when the

TABLE I

| Acid | Calculated values of iron corrosion (mA/cm^2) obtained by | | Observed value obtained by chemical analysis |
	Tafel method at high current value	Resistance polarization method at low current value	
Formic	0·31	0·34	0·35
Acetic	0·063	0·078	0·067
Propionic	0·028	0·033	0·024
Butyric	0·0008	0·0009	0·0005

potential V was raised above or depressed below that prevailing in the current-less state, and the relation between i and V was obtained experimentally. Two methods then became available for the calculation (1968 book, pp. 320–4); one method depends on the value of di/dV at low current values where a rectilinear relationship exists between i and V, and the other on the situation at high current where such a relationship exists between $\log i$ and V. The two calculated values were compared with the value of the corrosion as obtained by chemical analysis, and the values, reproduced in Table I, show satisfactory agreement. It was concluded that the mechanism was 'electrochemical' in the sense of the word used above; no evidence was obtained for a 'pure chemical reaction' with only a single activated complex.

Heitz also carried out an extensive study of corrosion by monocarboxylic acids.[1] An important conclusion drawn from this work is that the mechanism does not change on passing from aqueous to non-aqueous conditions.

Electrochemical methods of measuring Corrosion. The reasonable agreement between the corrosion as obtained by two electrochemical methods

1. E. Heitz, *Ber. Bunsen Ges.* 1969, **73**, 1085; *Werkstoffe u. Korr.* 1970, **21**, 360, 457 (with M. Huković and K. H. Maier). Also Private Comm., March 7, 1973.

with that obtained by chemical analysis (shown in Table I) provides strong support for an electrochemical mechanism. Heitz[1] points out that electrochemical methods of measurement have the advantage that one and the same method can be used for corrosion velocities differing by several orders of magnitude. The method based on the determination of the polarization resistance, R_p at low values of the polarization, followed by application of the Stern–Geary equation

$$I_{corr} = \frac{1}{2 \cdot 3} \cdot \frac{b_a \cdot b_c}{b_a + b_c} \cdot \frac{1}{R_p}$$

requires a knowledge of the Tafel slopes b_a and b_c. It is sometimes thought that they must be obtained by separate experiments, since the Tafel slope involves studies at high values of the polarization; this has, indeed, been regarded as a disadvantage of the method. However Mansfeld[2] states that both the polarization resistance and the Tafel slopes can be obtained with precision in a single set of experiments by studying polarization over the range ± 25 mV, and has demonstrated this for the case of iron in N H_2SO_4. A computer is helpful in calculating b_a and b_c, but satisfactory results have been obtained by the use of hand calculators.

Another comparison between corrosion rates as measured (1) by the intersection of the prolongation of Tafel lines, (2) by the volume of hydrogen evolved, and (3) by weight loss, appears in a paper by Aramaki.[3] These refer to corrosion of iron in acid with and without an imine inhibitor. Agreement, although not perfect, is reasonably satisfactory.

Pražák[4] has found that the relationship of polarization resistance R_p (as obtained by the use of a.c.) and corrosion velocity K is $K = c/(R_p)^n$, where the value of the constants ($c = 1 \cdot 24$ and $K = 1 \cdot 09$) remain much the same for all the four metals studied (Fe, Cu, Zn and Ag) and for different types of control (activation control in the attack of acids on Fe and the transport of oxygen or oxidizing agent in some other cases studied). The slight departure from unity in the value of n is connected with the exchange current which flows in each direction even when there is no corrosion; if this were negligibly small compared to the corrosion current, n would be unity.

The relation between the corrosion rate and polarization resistance has been discussed by Hoar.[5] He considers that the Stern–Geary analysis is of somewhat limited quantitative application. The question of the polarization resistance method of measuring corrosion rate has been discussed from several standpoints at a Cebelcor Congress;[6] the contribution by Epelboin deserves special consideration.

Japanese work[7] has been directed to establishing correct conditions to be observed in using the polarization resistance method for predicting cor-

1. E. Fot and E. Heitz, *Werkstoffe u. Korr.* 1967, **18**, 529; 1968, **19**, 763.
2. F. Mansfeld, *Corrosion (Houston)* 1972, **28**, 468.
3. K. Aramaki, *J. Electrochem. Soc.* 1971, **118**, 1553.
4. M. Pražák, *Werkstoffe u. Korr.* 1968, **19**, 845.
5. T. P. Hoar, *Corr. Sci.* 1967, **7**, 455.
6. *Cebelcor Rapp. tech.* 209 (1972).
7. M. Nagayama, K. Goto and Y. Otake, *Tokyo Cong. Ext. Abs.* 1972, p. 413.

rosion rate. Difficulties are connected with the fact that the anodic polarization does not attain a steady value until a certain time has elapsed; this is due to the need to charge the electrical double layer and also to the time-variation of the surface condition. It is now recommened that measurements should be made 10 s after the start of the experiment. This is considered to be just long enough to charge the electric double layer but short enough to avoid disturbance due to other causes.

An instrument embodying the Stern–Geary method for measuring corrosion has been described in a paper from Portsmouth.[1] It has been used both in the laboratory and on board ship, and is available in both manually operated and automatic forms.

Other work related to the measurement of corrosion rate. LaQue[2] has discussed the effect of specimen size on the results of corrosion testing.

In measuring corrosion in the early stages, sensitive reagents are welcome. Shome[3] states that N-benzyl-phenylhydroxylamine, a new analytical reagent, is 'much more effective than cupferon for analytical work'.

Yahalom and his colleagues[4] have discussed the use of radio-tracer methods which will detect small amounts of corrosion likely to be missed by microbalance methods or by spectroscopic analysis.

The removal of corrosion product from test specimens without attack on the metal is a necessary condition for the accurate measurements of corrosion velocity. Sanyal and his colleagues[5] have provided a review covering eleven different metals.

Some useful information regarding accurate values of reference electrodes is provided by Pourbaix and Van Muylder[6], who also discuss the precision obtained by the Debye–Hückel formulae—with or without correction, Verink[7] has provided a new procedure for obtaining Pourbaix diagrams, based on the direct use of the Nernst equation. He states that the classical method of obtaining the diagrams presents difficulty to some students, who have found the new method easier. The final results are of course identical to those presented in the Pourbaix atlas.

Influence of Flow-rate on Corrosion in moving liquids. An Important report by Heitz[8] brings out the point that corrosion may be either independent of the flow-rate or strongly influenced by it, according to the value of the velocity constant. If oxygen is needed for the attack, the rate of its supply will be given by the equation

$$V_1 = k_1(O_{2\text{solution}} - O_{2\text{surface}})$$

1.   J. C. Rowlands and M. N. Bentley, *Brit. Corr. J.* 1972, **7**, 42.
2.   F. L. LaQue, *Tokyo Cong. Ext. Abs.* 1972, p. 3.
3.   S. C. Shome, Private Comm. Dec. 24, 1972.
4.   A. Aladjem, A. S. Roy and J. Yahalom, *Amsterdam Cong.* 1969, p. 810.
5.   T. K. Grover, D. P. Sinha and B. Sanyal, *Labdev. J. Sci. Tech.* 1968, **6A**, 167.
6.   M. Pourbaix and J. van Muylder, *Cebelcor Rapp. tech.* **158** (1969).
7.   E. D. Verink, *Corrosion (Houston)* 1967, **23**, 371.
8.   E. Heitz, Europäische Gemeinschaft für Kohle u. Stahl, 4th Congress, 1968.

where k_1 is a physical constant, whilst its consumption in the corrosion reaction will be given by $V_2 = k_2(O_{2\text{surface}})$ where k_2 is a chemical (electro-chemical) constant. In the steady state $V_1 = V_2$, and either of these provides a measure of the corrosion velocity (V). It follows that

$$V = \frac{\text{oxygen solubility}}{(1/k_1 + 1/k_2)}$$

Curves relating to corrosion of iron pipes by air-saturated media show that when

$$k_2(O_2) < 5 \text{ g/m}^2, \text{ days,}$$

TABLE II

System	Flow-type	Dependence on			Critical value of Reynold's number
		n	r	C	
Rotating disc	L	$n^{0.5}$	independent	C	wr^2/v
	T	n	$r^{0.9}$	C	$= 10^4$ to 10^5
Concentrically rotating cylinder	L	n	r	C	Ud/v
	T	n	r	C	$= 4\cdot6 \times 10^3$
Freely rotating cylinder	T	$n^{0.7} \to r$	$r^{0.7} \to r$	C	
Passage through tube	L	$u^{0.33}$	$r^{0.66}x^{0.66}$	C	ud/v
	T	u	$r^{-0.1}$	C	$= 2\cdot3 \times 10^3$
Passage over plate	L	$u^{0.5}$	$x^{0.5}$	C	
	T	$u^{0.9}$	$x^{0.9}$	C	
Natural convection on vertical plate	L	—	$h^{-0.23}$	$C^{1.25}$	

the corrosion is practically independent of the flow-rate, whereas when

$$k_2(O_2) > 30 \text{ g/m}^2, \text{ days,}$$

it is strongly dependent on the flow-rate.

However, in applying these equations, various factors must be taken into account. Where a protective film, the maintenance of which requires the rapid replenishment of oxygen, is responsible for good performance, this may fail to be achieved at very low flow velocities; also at very high velocities, corrosion may set in again, because the film is removed by erosion. At both high and low flow-rates, the attack may be localized and intense, owing to the combination of small anode and large cathode. In such cases the presence of chlorides may become important. Heitz reproduces curves due to Butler and Stroud for waters passing through steel tubes; if much chloride is present, the corrosion rate increases steadily with the flow-rate; in the absence of chloride, it first increases with increasing flow-rate and then falls off again.

D

Thus, whereas at very low flow-rates, the chloride concentration has a relatively small effect on the velocity, it may have an enormous effect at high flow-rates. In the case of stainless steels (of both the 13% Cr and 18/8 Cr/Ni types), the value of k_2 will normally be small, and it would be expected that the corrosion rate would be little influenced by the flow-rate. This is indeed found to be true, provided that the flow-rate does not fall below a minimal value needed to maintain passivity.

Table II, due to Heitz, shows for various systems (rotating discs or cylinders, flow through tubes etc.), the power of the rotation speed (n), radius (r) and concentration (C) which determines the transport of material under laminar (L) and turbulent (T) conditions; also the critical value of Reynold's number at which laminar flow gives place to turbulent flow. Here v is the kinematic viscosity (cm^2/s). For the rotating disc, w represents $2\pi n$ times the angular velocity ($1/s$). For flow through tubes, U represents the initial velocity (cm/s) and u the mean velocity, x the length (cm) and d the diameter (cm). For natural convection, h is the plate-height (cm).

Further References

Conference Volumes. Several conferences have been held at which valuable papers were presented. These include the 3rd International Conference on the Passivity of Metals held at Cambridge in 1970, the Williamsburg Conference on Localized Corrosion held in 1971 and the 5th International Congress on Corrosion, held at Tokyo in 1972.

At the Tokyo Congress, one of the plenary lectures, by I. L. Rosenfeld, deals with pitting and other forms of localized attack, and should be studied. In his view the term localized corrosion covers pitting, crevice corrosion, stress corrosion cracking, intergranular attack, contact corrosion, filiform attack, tuberculation and deposit attack. He distinguishes causes related to the metallic phase from those related to the electrolyte. The former include local activation of the surface owing to displacement of oxygen, structural heterogeneity of alloys, micro-defects, separation of new phases, stresses in micro-volumes and local variation of chemical composition. The latter include local changes in the composition of the liquid, sometimes arising from hydrolysis or anodic polarization, as well as from an insufficient supply of electrolyte to individual sections of the surface, and removal of anodic reaction products.

Papers. 'Review of literature on pitting published since 1960'.[1]

'Critical potential for growth of localized corrosion of stainless steel in chloride media'.[2]

'Characteristics of localized corrosion of steel in chloride solutions'.[3]

'Pitting of molybdenum-bearing austenitic stainless steel'.[4]

'Corrosion of an iron rotating disc'.[5]

1.   Z. Sklarska-Smialowska, *Corrosion (Houston)* 1971, **27**, 223.
2.   T. Suzuki and Y. Kitamura, *Corrosion (Houston)* 1972, **28**, 1.
3.   A. Pourbaix, *Corrosion (Houston)* 1971, **27**, 449.
4.   R. J. Brigham, *Corrosion (Houston)* 1972, **28**, 177.
5.   N. Vardhat and J. Newman, *J. electrochem. Soc.* 1973, **120**, 1682.

'Localized Corrosion of Stainless Steel during food processing'.[1]

'Differential aeration corrosion of a passivating metal under a moist film of locally variable thickness'.[2]

'The Problem of Localized Corrosion'.[3]

'Etch Pitting—Theory and Observation'.[4]

'The Pitting of Fe–Cr–Ni alloys'.[5]

'Potential Scanning of Stainless Steel during Pitting Corrosion'.[6]

'Pitting and Sulphide Inclusions in Steel'.[7]

'Contribution to the study of localized corrosion of aluminium and its alloys'.[8]

'Initiation of Pitting at Sulphide Inclusions in Stainless Steel'.[9]

1. J. A. Richardson and A. W. Godwin, *Brit. Corr. J.* 1973, **8**, 258.
2. R. Alkire and G. Nicolaides, *J. electrochem. Soc.* 1973, **121**, 183.
3. F. L. LaQue, *Williamsburg Conf.* 1971; introductory section, p. 147.
4. M. B. Ives, *Williamsburg Conf.* 1971, p. 78.
5. S. Szklarska-Smialowska, *Williamsburg Conf.* 1971, p. 312.
6. H. S. Isaacs, *Williamsburg Conf.* 1971, p. 158.
7. G. Wranglén, *Corr. Sci.* 1974, **14**, 331.
8. P. L. Bonora, G. P. Ponzano and V. Lorenzelli, *Brit. Corr. J.* 1974, **9**, 108, 112 (2 papers).
9. G. S. Eklund, *J. electrochem. Soc.* 1974, **121**, 467.

CHAPTER V

Soluble Inhibitors

Preamble

Picture presented in the 1960 and 1968 volumes. The electrochmicale corrosion discussed in chapter IV proceeds quickly in cases where both anodic and cathodic products are freely soluble. If one of these is sparingly soluble, it may form an obstructive film and inhibit the reaction to a greater or lesser degree. Cathodic inhibitors are the less effective, retarding the corrosion rather than preventing it. Anodic inhibitors can be very effective if added in sufficient quantity, but the concentration needed increases if anions like Cl⁻ are present. If too little inhibitor is added, corrosion may still occur, restricted to the more sensitive parts of the surface. In certain cases the total corrosion is determined by the cathodic reaction (e.g. reduction of oxygen); it will be greatest when the cathodic area is large, and if the effect is then concentrated on a small anodic area, the intensity of corrosion may be greater than if no inhibitor had been added at all. Thus anodic inhibitors, although effective if rightly used, can be highly dangerous if the amount needed has been miscalculated.

Although many cases of inhibition can be attributed to the formation of an obstructive film of some known compound either as an anodic or as a cathodic product, there are cases where it would seem that molecules of the inhibitor are adsorbed at places where corrosion would otherwise commence, and hinder the process without the formation of any definite compound. Such cases include the use of various organic compounds to retard the action of acids on metals.

Some of the substances present in supply waters have a mild inhibitive action. Thus hard waters containing calcium bicarbonate are less corrosive to steel pipes than soft waters. The rise of pH connected with the cathodic reaction will often precipitate calcium carbonate in close contact with a steel surface, and the attack will be slowed down. Small quantities of certain organic matters in the waters alter the character of the calcium carbonate layer, producing the so-called 'egg-shell scale', which is very effective in protecting the metal.

Developments during the period 1968–1975 and Plan of new Chapter. Features of the period under review include (1) interest in unsaturated organic compounds as inhibitors, (2) the important work in Horner's laboratory on inhibition by onium compounds, (3) the demonstration that the Hammett constant, which has long been used by organic chemists in discussing the acceleration of purely chemical reactions, can be usefully applied to the retardation of corrosion reactions; (4) advances in knowledge based on the use of radioactive tracers.

The first main section of the new supplementary chapter is devoted to inhibition in acid media, with work on unsaturated organic compounds, organic compounds containing sulphur and onium compounds. The next main section is devoted to inhibition in near-neutral media, including studies of the action of azelates and benzoates based on the use of radioactive tracers. A discussion of inhibition by chromates follows. There is then a section dealing with inhibitors for non-ferrous metals. This is followed by a discussion of inhibition under various service conditions; the influence of substances already present in natural waters which can be considered as inhibitors must here be taken into consideration. Practical cases discussed include inhibition in water supply pipes and central heating systems and the prevention of corrosion by coolants in motor vehicles.

The influence of substituent groups in organic inhibitors is then discussed; the concept of the Hammett constant is here found to be helpful. Synergistic effects obtained with a combination of inhibitors receive attention.

An attempt is then made to gather together the conclusions, and thus provide a picture of the mechanism of inhibition. Some repetition of facts and arguments presented earlier in the chapter is inevitable; if the picture is acceptable, this may, perhaps, be forgiven. The chapter closes with some remarks on measuring methods.

In the present chapter the word 'inhibition' will be used to cover cases where the addition of an 'inhibitor' to a solution which would otherwise attack metal readily, greatly reduces the speed of attack. In most cases the retardation only continues so long as the inhibitor is present; if the specimen is moved to a similar solution containing no inhibitor, the rate of attack returns to a high value. In cases where there is a marked change of behaviour, continuing after the retarding substance has been removed, the word 'passivity' would appear more appropriate. This was the original use of the word 'passivity' in the days of Faraday and Schönbein. Since that time, the word has been used in other senses, and that is largely responsible for apparent differences of opinion regarding the causes of passivity—a matter discussed elsewhere.[1] It is, however, difficult to distinguish sharply between inhibition and passivity. Cases are known where, after treatment in a solution containing an inhibitor, it becomes possible to move the specimen to a solution containing no inhibitor, or, in some cases, a smaller concentration of inhibitor than had been needed at the outset, and still to obtain a considerable reduction in the rate of attack. There are also cases where the addition of an inhibitive chemical produces a reduction in the corrosion rate which, although well worth obtaining, is far from complete. The addition of certain substances to the acid used in pickling for the removal of mill-scale from steel provides an example; in such a case it is appropriate to use the word 'restrainer' rather than 'inhibitor'.

It has become customary to indicate the degree of retardation by means of the 'percentage inhibition', defined as $(1 - W/W_0)100$, where W_0 is the corrosion produced in absence of inhibitor, whilst W is the rate in its pre-

1. U. R. Evans, *London Cong.* 1961, p. 1; esp. p. 8.

sence; naturally the value of W depends on the concentration of inhibitor present.

Inhibitors for Iron in Acids

Unsaturated Organic Compounds. Putilova[1] has continued her valuable work on the inhibition of attack on steel by hydrochloric acid. Increase in the acid concentration, which causes marked increase of corrosion in the absence of inhibitor, is found to have much less effect when an inhibitor is present. A rise of temperature produces an increase in protective action, which suggests that inhibition cannot be solely caused by physical adsorption, which would fall off with temperature. Small additions (10^{-5} M) sometimes stimulate dissolution, in cases where a high concentration provides protection.

Other Russian work[2] has been concerned with the use of acetylene derivatives and compounds containing nitrogen. The combined use of both types together produces a synergistic (mutually helpful) effect, which is enhanced by rise of temperature. Even better results are obtained by triple mixtures containing KCl in addition to the two organic compounds. The addition of Sn^{2+} and Cr^{3+} cations improves the protective action of acetylene compounds; apparently this is associated with the catalytic influence of these cations on the polymerization of the unsaturated bodies to produce resinous films. Other examples of synergistic effects are presented later.

Italian work by Trabanelli and his colleagues[3] on inhibition by unsaturated hydrocarbons and alcohols shows that 1-Decyne at very low concentrations (10^{-4} M) produces 99% inhibition of attack on N and 5N H_2SO_4. The mechanism has been studied by means of iV curves.

Amines and Imines. Inhibition by polymethylene–imines has been investigated by Hackerman.[4] The general formula can be written $(CH_2)_n\ NH$: inhibition is good when n is 10 or more, but poor at 5. There is some doubt about the mechanism, but importance is attached to the effect of strain in the ring, which is higher at $n = 10$ than at $n = 6$. Inhibition may be connected with the formation of polymer molecules by ring-opening reactions which take place most readily when the strain is high. These polymers, probably including straight chain and large ring compounds, provide good protection.

The polymethylene diamines have also been studied by Hackerman. The effect of increasing the length of the chain $NH_2—(CH_2)_n—NH_2$ between $n = 2$ and $n = 12$ was investigated. The lowest members were effective inhibitors. No improvement resulted on increasing n from 3 to 8. Beyond that point there was an apparent improvement, attributed to decrease of solubility. Hetero alkylated amine inhibitors are discussed by Rosenfeld.[5]

1.   I. N. Putilova, E. N. Chislova and A. M. Lolua, *Corr. Control Abs.* 1970, (3) 39.
2.   V. V. Vasil'yev, *Corr. Control. Abs.* 1970, (4) 23.
3.   F. Zucchi, G. L. Zucchini and G. Trabanelli, *Ferrara Symp.* 1970, p. 121.
4.   K. Aramaki, E. McCafferty and N. Hackerman, *J. electrochem. Soc.* 1968, **115**, 1007; 1972, **119**, 146.
5.   I. L. Rosenfeld, *Brit. Corr. J.* 1975, **10**, 3.

Other amines have been found to be useful inhibitors; hexamethylene-tetramine hydroiodide has shown promising results in connection with the pickling of steel in HCl.[1] Long chain n-alkyl isoquinolinium compounds are excellent inhibitors for steel against H_2SO_4, HCl and HBr, providing 99% protection in the best cases.[2]

Compounds containing Sulphur. The effect of thiourea on the corrosion rate of steel in H_2SO_4 and HCl has been studied by Federov.[3] It appears that adsorbed molecules prevent corrosion, but that the decomposition products promote it, with the formation of sulphides. Thus small concentrations (10^{-5} mole/l) reduce corrosion, but at high concentrations the adsorbed film breaks up, and the corrosion rate increases. Thiourea is an effective inhibitor for the corrosion of mild steel in nitric acid, as shown by Indian work[4] based on observation of the temperature rise which occurs if corrosion takes place; good correlation was obtained between the results of the temperature rise method and those furnished by weight loss. Thiourea was found to be capable of arresting the corrosion rate even when it was added after the reaction had been proceeding for some time.

Indian work[5] shows thiobenzoic acid to be an excellent inhibitor in protecting Al against 0·5M HCl, although hydroxy-benzoic acid has poor inhibitive properties. Evidently –OH and –COOH groups on a benzene ring are ineffective in preventing corrosion; –SH has the desired effect, greatly increasing the cathodic polarization.

Inhibition by sulphoxides was studied in 1965 by Horner,[6] who found dibenzyl sulphoxide to be very effective in preventing the corrosion of iron, zinc and aluminium in dilute hydrochloric acid free from oxygen. The actual inhibition was, however, attributed mainly to a secondary product, namely dibenzyl sulphide.

The matter was further elucidated by Italian work.[7] Radioactive tracer studies, based on ^{35}S, has shown the sulphide to be more strongly adsorbed than the sulphoxide, and also to be a better inhibitor. However, the sulphoxides themselves possess some inhibitive power, which depends on the nature of the substituent groups present. Six sulphoxides have been compared, carrying different groups (dimethyl, dibutyl, etc.). Dimethyl-sulphoxide actually stimulates corrosion, apparently by forming a soluble complex; the other five are inhibitors.

Onium Compounds. Important studies on phosphonium and arsonium compounds (conveniently grouped as 'onium compounds') have continued in

1. E. D. Mor, V. Scotto and C. Wruhl, *Brit. Corr. J.* 1972, **7**, 276.
2. R. J. Meakins, *Brit. Corr. J.* 1973, **8**, 230.
3. Y. V. Fedorov, M. V. Uzlyuk and V. M. Zelenin, *Corr. Control Abs.* 1970 (10) 29.
4. B. Sanyal and K. Srivastava, *Brit. Corr. J.* 1973, **8**, 28.
5. S. C. Makwana, N. K. Patel and J. C. Vora, *Werkstoffe u. Korr.* 1973, **24**, 1036.
6. H. Ertel and L. Horner, *Ferrara Symp.* 1965, p. 71.
7. G. Trabanelli, F. Zucchi, G. Gullini, V. Carassiti and G. L. Zucchini, *Brit. Corr. J.* 1969, **4**, 212, 267 (2 papers).

Horner's laboratory.[1] He has reached the conclusion that the onium compounds are not the true inhibitors, but are merely compounds providing cations which, by virtue of their charge, move to areas where the cathodic reaction would normally take place; there the uptake of electrons leads to the formation of secondary products as two-dimensional layers. Some of the inhibitors retard the cathodic and others the anodic reaction; a combination of cathodic and anodic inhibitors produces a result which is at least additive; sometimes, indeed, the effect of the combination is even better than would be expected if the two inhibitors acted separately. For instance, 2-mercapto-benzthiazole (an anodic inhibitor) along with a phosphonium salt (a cathodic inhibitor) produces unexpectedly good inhibition. The protection may be enhanced by pre-treatment of the iron with a solution containing the salt of a noble metal; copper is the most effective. An interesting feature of these inhibitors is that, in some cases, protection continues even when the acid employed has ceased to contain an inhibitor. For instance, the percentage inhibition with acid containing an inhibitor is often 90% or more; if the specimen is washed and placed in fresh acid containing no inhibitor, there may still be about 70% inhibition.

The identity of the secondary product which is mainly responsible for the inhibition has been established in several cases. Triphenyl-benzyl-phosphonium chloride produces triphenylphosphine as a layer only one or two molecules thick. The identification of the secondary product can often be carried out by normal analytical procedure, but the elegant nucleation method developed by Karagounis[2] is extremely useful. Here a quantity of the substance suspected to be the secondary product is melted, and cooled to a point just below the melting-point where no spontaneous crystallization occurs. The iron powder, previously coated with the inhibitor, is introduced; if the suspicion is correct, this will start crystallization. Fluorescence spectroscopy, as well as ultra-violet light adsorption methods, have also been employed in the identification of secondary inhibitors.

Inhibitors for Iron in Near-neutral Media

Azelates. Important studies of the inhibitive action of azelates (salts of $COOH(CH_2)_7COOH$) have been carried out by Mayne and Page.[3] A radio-active tracer method was used to detect the distribution of the product (probably ferric azelate, perhaps basic) on the surface. The azelate solution used contained, in addition to ordinary sodium azelate, a small amount of azelate in which the carbon was present as the isotope ^{14}C. Steel immersed in 0·05N azelate at pH 6 is found to take up a quantity of azelate ions which would provide roughly a monolayer of ferric azelate if it were assumed that the distribution was uniform. The actual distribution was ascertained by

1. L. Horner, F. Röttger, H. Ertel, H. Hinrichs, G. Strube and H. J. Dörges, *Werkstoffe u. Korr.* 1964, **15**, 123, 228; 1965, **16**, 35; 1971, **22**, 833, 864, 867, 924, 930; 1972, **23**, 6, 9; 1973, **24**, 860.
2. G. Karagounis and H. Reis, *Z. Elektrochem.* 1958, **62**, 865.
3. J. E. O. Mayne and C. L. Page, *Brit. Corr. J.* 1970, **5**, 95; 1972, **7**, 111, 115 (2 papers); 1975, **10**, 99; *Corr. Sci.* 1972, **12**, 679.

bringing the specimen after treatment into contact with a fine-grained photographic plate, which underwent darkening through radiation from the radioactive isotope. The darkening was found to be far from uniform, and the distribution of the dark patches provided clear evidence that the formation of ferric azelate tended to be localized, being presumably greatest at defects in the oxide-film. At any point the thickening would continue until it was sufficient to prevent further movement through the film; thus a specially dark point would seem to indicate a local defect. Inspection suggested that there was more azelate at grain-boundaries than elsewhere; statistical calculations indicated that this was indeed the case. The enhanced reactivity of grain-boundary regions is independent of impurity segregation and probably depends on discontinuity in the epitaxial growth of oxide at grain boundaries in the substrate metal. Evidence supporting non-uniform distribution is important for reasons which will be explained later.

Immersion of iron in a deaerated solution of sodium azelate results in breakdown of the original oxide-film after some hours,[1] probably through reductive dissolution (anodic attack on the iron being compensated by cathodic reduction of the ferric oxide film, producing ferrous ions which pass into solution). Aerated solutions of moderate concentration inhibit attack at pH values as low as 4·8.

A drop of azelate solution placed on a horizontal iron surface produces a fine deposit of particles detected by the radioactive method, the largest being a few μm in size. The abundance of these particles is greatest in the central regions of the drop-site, the place where anodic attack would occur if there were no inhibitor; evidently at such points a greater thickness will be needed to block attack than elsewhere. The deposits are not removed by washing, which suggests that they are firmly bound to the surface in insoluble form as a ferric azelate (perhaps basic). The uptake is greater at pH 6 than at pH 5— which suggests that the doubly ionized species, $-COO(CH_2)_7COO-$, is more important than the single species, $NaCOO(CH_2)_7COO-$. The smallness of the quantity taken up—corresponding to less than a monolayer if averaged over the whole surface—is ascribed to the fact that the uptake is localized at the points susceptible to attack (such as grain-boundaries, where the discontinuity of atomic arrangement normally makes it easier for an atom to escape into the liquid); here the film must continue to thicken until it reaches a thickness sufficient to arrest further escape, but the mean thickness (averaged over the whole surface) is much smaller.

In Mayne's earlier work[2] it was shown that lead azelate was a more efficient inhibitor than the corresponding sodium or calcium salt; for a given cation the inhibition improved with the length of the chain up to 8 or 9 carbon atoms. It was suspected that the effect of the cation was due to the deposition of metallic lead on the iron surface, but the experimental methods first tried failed to detect lead on the surface. Lately[3] Auger spectroscopy—a very

1. J. E. O. Mayne, *Tribune du Cebedeau*, 1968, No. 300.
2. J. E. O. Mayne and E. H. Ramshaw, *J. appl. Chem.* 1960, **10**, 419; A. J. Appleby and J. E. O. Mayne, *J. Oil Colour Chem. Assoc.* 1967, **50**, 897.
3. J. E. O. Mayne, S. Turgoose and J. M. Wilson, *Brit. Corr. J.* 1973, **8**, 236.

sensitive method—has revealed the presence of lead, probably in metallic form, on the surface of cold-rolled Swedish iron, which had previously been immersed in 10^{-14}M lead azelate for 24 hours. It is believed that the lead is deposited by the cathodic reaction of the local cells set up, and that the cathodic reduction of oxygen proceeds more smoothly on the lead particles than on iron. This means that the production of OH^- by the cathodic reaction and also the oxidation of Fe^{2+} to Fe^{3+} by the anodic reaction will proceed more readily. As a result a smaller concentration of azelate is needed to ensure that conditions are such that the solubility product of the ferric basic azelate (which is assumed to be the blocking substance) is exceeded, so that its precipitation can occur, with prevention of corrosion. Thus lead azelate should be a more effective inhibitor than sodium azelate—as is found to be the case.

Benzoates. Studies of benzoates at Swansea and Cambridge show reasonable agreement on facts but differences in interpretation. Slaiman and Eurof Davies[1] found that sodium benzoate was inhibitive only in the presence of oxygen. A specimen of iron immersed in deaerated 0·1N benzoate showed a potential which remained in the active range, becoming constant at about −0·75 V (saturated calomel electrode); if the solution was aerated, the potential rose into the passive range, becoming nearly constant at about −0·15 V. They found, however, that when the concentration of the benzoate exceeded a critical value (C_B), there was no inhibition; the concentration above which corrosion set in depended on the pH value according to the relationship

$$pH = A \log_{10} C_B + C$$

They attributed the corrosion to the fact that oxygen solubility in a salt solution decreases with the salt concentration, and that the rate of supply of oxygen to the metal surface might, at high benzoate concentrations, be insufficient to establish passivity.

The measurements of Page and Mayne[2] gave results in reasonable agreement. They found that mild steel, carrying an air-formed oxide-film of known thickness and immersed in a sodium benzoate solution at pH 5·0, suffered rapid corrosion in 0·1N and 0·0001N solutions but less corrosion at 0·01N; there was complete inhibition at 0·001N; thus the inhibitor functioned satisfactorily only over an intermediate range of concentration. They pointed out that, if a series of basic ferric benzoates exists, the one which is thrown down will depend on the relationship between the concentrations of the benzoate ion and OH^-; in the intermediate (inhibitive) range of concentration, the compound deposited possessed some specific protective property, which was lacking in that deposited in the higher (corrosive) range. They even considered the possibility that the protective body might be, not a basic ferric benzoate, but γ-Fe_2O_3 which, they remark, 'is generally associated with anodic passivation and inhibition processes on iron', adding, pertinently,

1.   Q. J. M. Slaiman and D. Eurof Davies, *Ferrara Symp.* 1970, p. 739; *Corr. Sci.* 1971, **11**, 671, 683 (2 papers).
2.   C. L. Page and J. E. O. Mayne, *Corr. Sci.* 1972, **12**, 679.

that, 'If this is the case, then the question of the inhibitive role of the benzoate ion remains to be answered.'

Page's tracer experiments based on benzoate containing ^{14}C atoms certainly suggest that the blocking substance is a basic ferric benzoate, not an oxide. The results indeed show analogy with his experiments on azelate, and suggest that if the oxygen supply is sufficient, the rapid cathodic reduction of oxygen sets up a high anodic current density; thus Fe^{2+} is largely oxidized to Fe^{3+}, and basic ferric benzoate is formed at weak points in the film, blocking the reaction. 'It was demonstrated', they write 'by means of high-resolution autoradiography that the surface-bound inhibitor was contained in a deposit of very small particles, the largest of which were a few microns in size. It is thought that these particles were composed of some form of "ferric benzoate".'

A detailed study of the action of benzoate has been carried out by Kelly[1] using oxygen-free solutions. He was interested in the active range and in hydrogen evolution; his results may not be directly applicable to the practical use of benzoate as an inhibitor, which requires the presence of oxygen.

Chromates. Before discussing recent work at Teddington under Miss Brasher, and that at Cambridge under Mayne, it may be helpful to recall some early observations.

It has long been known that iron placed in a solution of K_2CrO_4 remains uncorroded and unchanged in appearance; in a solution containing chloride as well as chromate, corrosion may occur; if the amount of chromate present is sufficient to protect the main surface but not specially sensitive points, the large cathode–anode area ratio leads to intensified attack. Even though the total corrosion, as measured by weight loss, is much smaller than that which would occur in a chloride solution of the same concentration but free from chromate, the chromate-containing chloride will cause perforation of a sheet specimen in a period during which chromate-free solution of the same chloride concentration produces only slight thinning, the attack being well spread out. This has been established more than once.[2] The intensity of attack produced by the chloride–chromate mixture is due to the fact that chromate is essentially an anodic inhibitor as has been shown by electrochemical experiments carried out in a cell divided by a porous partition into anodic and cathodic compartments.[3] Addition of chromate to the anodic compartment greatly diminishes the current, whereas addition to the cathodic compartment produces a change which is small but variable if the solution is nearly neutral (in an alkaline solution there is definite reduction of the current, and in an acid solution a large increase). The inhibitive action at the anode is probably due to the fact that any iron entering the liquid as Fe^{2+} ions formed by anodic attack immediately interacts with CrO_4^{2-} to give solid matter containing Fe and Cr; if this is formed in physical contact with the metal surface, it is likely to be protective; the precipitation

1. E. J. Kelly, *J. electrochem. Soc.* 1968, **115**, 1111.
2. U. R. Evans, *J. Soc. chem. Ind.* 1925, **44**, 163T; 1927, **46**, 347T.
3. E. Chyżewski and U. R. Evans, *Trans. Electrochem. Soc.* 1939, **76**, 215.

produced by mixing solutions of $FeSO_4$ and K_2CrO_4 can be demonstrated in a test-tube. Such an explanation of the inhibition is confirmed by the fact that the film formed on metallic iron immersed in a chromate solution does in fact contain a certain amount of chromium—as was detected by a chemical method in early work.[1]

Later, the quantity of chromium present in the film was estimated at Teddington by a radioactive tracer method,[2] and measurements made at Cambridge[3] gave rather similar numbers for the chromium content of the surface layer. Regarding the iron content, there was at first less agreement, but it is now established that both metals are present. The Teddington measurements show that at pH 2·4 chromium oxide represents only about 37% of the whole. The film formed at pH 4 has been examined at Cambridge. In the main research, the Mössbauer technique was used, but the results obtained by X-ray and electron diffraction also provided important information. A comparison between the film formed in chromate solution with that formed by air exposure showed differences in composition, structure and properties. Films were stripped by the bromine method, but in order to remove ferrous bromide, it was necessary to heat to about 300°C, which changed the crystal structure from cubic to rhombohedral; it is unlikely that the ratio Cr/Fe was affected. X-ray examination by the powder method was carried out to ascertain the composition, using the values of Wretblad,[4] who had studied the system Fe_2O_3–Cr_2O_3 by the X-ray powder method; he found the rhombohedral oxides to form an unbroken series of solid solutions, with Fe_2O_3 and Cr_2O_3 as end-members. Wretblad measured the lattice parameters, which vary in a nearly rectilinear manner with the composition; his figures permit anyone to ascertain the composition of a solid solution $(Fe,Cr)_2O_3$ by merely measuring the parameter. Application of this procedure showed that a solid solution of composition $Fe_2O_3 + 3Cr_2O_3$ was present in the films examined at Cambridge; but, since lines of the magnetic oxides of iron were also observed, the actual chromium content of the film as a whole was probably considerably less than the 75% (molar) suggested by that formula.

Recent ellipsometric work by Szklarska-Smialowska and Staehle[5] receives a rather different interpretation. The film formed on iron on anodic treatment in chromate is an iron oxide with negligible amounts of Cr, which thickens according to the logarithmic law. However, at open circuit potential a film of lower refractive index is formed which thickens according to a cubic law; regarding its composition, the authors say, 'we will not speculate . . . because more studies are necessary. But such a low index of refraction shows that the

1.   T. P. Hoar and U. R. Evans, *J. chem. Soc.* 1932, p. 2476.
2.   D. M. Brasher and A. D. Mercer, *Trans. Faraday Soc.* 1965, **61**, 803; D. M. Brasher, J. G. Beynon, K. S. Rajagopalan and J. G. N. Thomas, *Brit. Corr. J.* 1970, **5**, 264.
3.   G. M. Bancroft, J. E. O. Mayne and P. Ridgeway, *Brit. Corr. J.* 1971, **6**, 119.
4.   E. P. Wretblad, *Z. anorg. Chem.* 1930, **189**, 329.
5.   Z. Szklarska-Smialowska and R. W. Staehle, *J. electrochem. Soc.* 1974, **121**, 1146.

film cannot be composed of γ-Fe_2O_3, Fe_3O_4 and Cr_2O_3. We presume that this film is some combination of iron hydroxide and chromium hydroxide.'

Metaphosphates as Inhibitors. A substance largely used in the treatment of natural water is the metaphosphate glass prepared by the thermal dehydration of NaH_2PO_4, and sometimes known as *sodium hexametaphosphate*. It is one of the few inhibitors suitable for the treatment of drinking water, or for use in other situations where toxicity would impose danger. Behaviour varies with temperature, the flow-rate of water (if moving) and the concentration of other divalent cations (if present). Butler[1] finds that, in absence of divalent cations, it can either stimulate or retard attack, according to conditions. In motionless water, corrosion is actually increased by a rise in metaphosphate concentration, but falls off again with larger additions, so that considerable inhibition is achieved, provided that Cl^- is absent; if it is present, there is stimulation. The efficacy of metaphosphate is greatly improved, however, if Ca^{2+} is present, especially under conditions of flow, where the simultaneous presence of Ca^{2+} and metaphosphate can reduce the corrosion rate to an acceptable figure—similar to (or better than) that obtained under quiescent conditions. In one test lasting 84 days with exposure to water containing 2500 ppm NaCl, 60 ppm Ca^{2+} and 100 ppm metaphosphate, the deposit had the composition $(Na,H)FeCa(PO_3)_5.8H_2O$; this may be a definite compound, but the possibility that it is a mixture cannot be excluded.

There is some disagreement about the mechanism of the inhibitor. Butler believes that in absence of divalent cations metaphosphate is primarily an anodic inhibitor, but in their presence a cathodic one. Polish investigators believe that it is not the real inhibitor, and that only after $HPO_4{}^{2-}$ has been formed by hydrolysis does inhibition occur. The tendency of a metaphosphate solution to suffer hydrolysis is generally recognized, but whether this is essential to the protection of the metal is a point on which opinion seems to be divided.

Inhibitors for Non-Ferrous Metals

Magnesium. Important observations by Vermilyea[2] and his colleagues have thrown light on the mechanism of attack on magnesium and have led to the discovery of inhibitors. Experiments were performed in $10^{-4}M$ $MgSO_4$ treated with sufficient KOH to produce pH 10. Magnesium placed in this solution suffers little weight change up to 10^5 seconds, and then there is a weight gain, due to formation of $Mg(OH)_2$ on the surface. The interpretation offered is that there is first a dissolution of a primary film, probably MgO, producing a solution supersaturated with respect to $Mg(OH)_2$, and after a time this compound is precipitated—explaining the sudden weight gain; the film, studied under the electron microscope by the carbon replica method, is found to consist of platelets; when $NaIO_4$ is present, the surface becomes much smoother. If the mechanism is that suggested, the dissolution rate of

1. G. Butler, *Ferrara Symp.* 1970, p. 753; see discussion on p. 774.
2. D. A. Vermilyea, *J. electrochem. Soc.* 1969, **116**, 1179, 1487 (with C. F. Kirk); also *Trans. Faraday Soc.* 1969, **65**, 561 (with W. Vedder).

MgO becomes a matter of interest; substances which retard its dissolution should restrain attack on metallic Mg. It was found that periodate, germanate, vanadate and tellurate retard the dissolution of MgO in alkaline solutions, whilst sodium dodecyl sulphonate retards it in acid solution. At pH 9 to 10 the dissolution rate of metallic Mg was found to be diminished in solutions containing $NaIO_4$ by a factor of 10.

Aluminium. Important work on inhibitors effective in reducing the corrosion of aluminium and its alloys has been carried out at Ahmedabad by Desai.[1] Encouraging results have been obtained with diethylamine, which in N HCl produces almost 100% inhibition at 43·5 ml/l concentration; this result was obtained on three different aluminium alloys. Even at higher HCl concentration, efficiencies varying between 97% and 99·5% were obtained. This inhibitor can be usefully employed in connection with cathodic protection; it greatly reduces the CD needed for the prevention of attack. Other inhibitors worth considering are diethylamine triamine and triethylamine tetramine, also certain aldehydes and substituted anilines.[2] Aniline itself accelerates corrosion at low concentrations but retards it when present in sufficient amount. Methylaniline produces no acceleration and, in short tests, the inhibition increases with increase of acid concentration; in N HCl the efficiency is relatively low (67%), but increases to 93% with time, whereas in 3N HCl it decreases from an initial 93% to 64%.

Inhibition in alkaline solutions is a more difficult problem, possibly because of the specific adsorption of OH^- on the surface. Sodium alginate has been found to produce some inhibition, but it is not complete, and at high alkali concentrations the film formed is found to detach itself.[3] Hydroxy-compounds (e.g. aminophenols) and complexing compounds, are sometimes efficient inhibitors in dilute alkali, but may actually stimulate corrosion in strongly alkaline solution.[4]

The use of surface-active substances as inhibitors for aluminium has been studied by Vermilyea.[5] Monododecyl phosphate prevents attack almost completely at pH 1·3, but ceases to be effective at pH 8·7. It is believed that the surface-active agents attach themselves to the Al_2O_3 surface by the inorganic group (phosphate being especially effective), and produce a hydrophobic surface, so that the access of water is prevented, and the dissolution of the Al_2O_3 film averted.

Copper and Brass. The use of benzthiazole (BTA) for the prevention of copper corrosion has given most satisfactory results. Much useful infor-

1.   M. N. Desai, S. M. Desai, C. B. Shah, Y. B. Desai and M. H. Gandhi, *Brit. Corr. J.* 1971, **6**, 269.
2.   M. N. Desai, G. J. Shah, D. K. Shah and M. H. Ghandhi, *Anticorrosion* 1972, **19** (3) 16.
3.   M. N. Desai and S. M. Desai, *Corr. Sci.* 1970, **10**, 233.
4.   K. Schwabe and C. Michels, *Corr. Sci.* 1968, **8**, 285.
5.   D. A. Vermilyea, J. F. Brown and D. R. Ochar, *J. electrochem. Soc.* 1970, **117**, 783.

mation has been provided by Bolognesi,[1] regarding inhibition in solutions containing NaCl, Na_2SO_4 and $NaNO_3$. In flowing sulphuric acid, Ross and Berry[2] have found that BTA greatly reduces attack. Under stagnant conditions and at slow flow-rates, inhibition was maximal at 0·01M concentration—irrespective of whether the acid was aerated or not; under aerated conditions, higher concentrations of BTA reduced inhibitor efficiency and a concentration exceeding 0·04M exerted a stimulating effect. It would appear that BTA is a more effective inhibitor for the hydrogen evolution reaction than for the oxygen reduction reaction. The film formed has been identified by means of infra-red light, and is found to be a polymer of the cuprous salt of BTA; the composition is nearly stoichiometric. Altura and Nobe[3] have studied the hydrogen evolution rate, which is found to fall as the BTA concentration rises; the reaction order and the Tafel slope remain essentially unchanged, indicating that the mechanism of the hydrogen evolution reaction is the same, whether inhibitor is present or absent. Poling[4] has studied the film produced by means of infra-red light, and states that the polymer layer is 'inert to a host of organic solvents' and acts as a physical barrier.

Inhibitors for brass in different solutions have been studied by Desai.[5] In the case of 63/37 Cu/Zn brass exposed to nitric acid, the attack by 4N HNO_3 was reduced to below that caused by 3N HNO_3 by a sufficient concentration of inhibitor.[6] O-anisidine is a mixed inhibitor, but o-phenetidene and o-toluidine act at local cathodic areas. A study of the attack on 60/40 brass has revealed the cause of the inhibition.[7] The rapid attack on copper alloys by HNO_3 is connected with an autocatalytic cycle involving the formation of HNO_2 (1960 book, p. 326). Aromatic anilines react with HNO_2 to give diazo compounds, and thus interrupt the catalytic chain. Even at 0·007 mol/l nitro- and carboxylic-substituted anilines protect 60/40 brass almost completely in 2N and 3N HNO_3.

Inhibition of the attack of HCl on 63/37 brass has also been studied by Desai,[8] who found that in aniline substituents at the *ortho* position affect efficiency; ethoxy and methyl groups act favourably, but methoxy unfavourably.

Potassium persulphate solutions are extremely corrosive to 63/37 brass in absence of inhibitor, but Desai[9] found that 2·17 ml/l of phenylhydrazine produces 99% inhibition, whilst 2% of thiourea produces 95% inhibition during a one hour test; this is a mixed inhibitor, retarding the cathodic and, to a lesser extent, the anodic reaction.

1. P. L. Bonora, G. P. Bolognesi, P. A. Borea, G. L. Zucchini and G. Brunoro, *Ferrara Symp.* 1970, p. 685.
2. T. K. Ross and M. R. Berry, *Corr. Sci.* 1971, **11**, 273.
3. D. Altura and K. Nobe, *Corrosion (Houston)* 1972, **28**, 345.
4. G. W. Poling, *Corr. Sci.* 1970, **10**, 359.
5. M. N. Desai, *Werkstoffe u. Korr.* 1973, **24**, 707.
6. M. N. Desai, Y. C. Shah and B. K. Punjani, *Brit. Corr. J.* 1969, **4**, 309.
7. M. N. Desai and G. H. Thanki, *Ferrara Symp.* 1970, p. 175.
8. M. N. Desai, S. S. Rana and B. K. Punjani, *Ferrara Symp.* 1970, p. 159.
9. M. N. Desai and Y. C. Shah, *Corrosion (Paris)* 1970, **18**, 25.

Considerable success has been achieved in preventing the attack upon 63/37 brass by NaOH; 2% of glucose gives satisfactory retardation.

Zinc. Acrylonitrile has been found to inhibit the corrosion of zinc by sea-water;[1] its efficiency is greater than in NaCl or Na_2SO_4. The difference is ascribed to 'the nature of the corrosion products, which are of smaller particle size in sea-water and so increase the number of active sites for the adsorption of acrylonitrile. The presence of Mg is also favourable as it increases the ease of adsorption of unsaturated organic compounds.' Infra-red spectroscopy shows that the nitrile is chemi-adsorbed and gives rise to polymerization products.

Inhibition under various Service Conditions

Inhibitive action of substances present in natural waters. It has been mentioned that substances normally present in some waters affect corrosion behaviour; traces of certain unidentified organic impurities may prevent the pitting of copper; these seem to be the same substances which produce a protective 'egg-shell') type of scale on steel (assuming $Ca(HCO_3)_2$ also to be present); the substances responsible probably originate from sewage pollution. The natural inhibitors present in many surface waters are considered by Storace[2] to be compounds of the δ-lactone type.

However, constituents present in larger amounts may greatly affect behaviour, and particularly influence the action of recognized inhibitors; their effect may be favourable (as in the case of Ca^{2+} and Mg^{2+}) or unfavourable (Cl^-). Sea-water corrodes less quickly than NaCl solution owing to the formation of a clinging cathodic deposit containing Ca and/or Mg, later converted by Fe^{2+} to produce a relatively protective rust; in one 128–day test, sea-water caused only about 1/3 of the weight loss produced by N/10 NaCl (1960 book, p. 165).

Water-Supply Systems. Campbell[3] has discussed the formation of protective and non-protective $CaCO_3$ as the cathodic product in various supply waters. He points out that the Langelier index (1960 book, p. 164) has long been regarded as the criterion for predicting whether a water will or will not produce a protective scale. However, the character of the scale laid down when the index is positive is extremely important. If an adherent and coherent layer is formed over the whole of the metal surface, this will stop attack. Many waters, however, deposit rough nodular scales, which have little protective value. The flocculation treatment often given to waters for removal of organic matter may affect the results, by removing the substance which had been responsible for the desired type of scale. From the corrosion standpoint, slow filtration through sand is preferable to a flocculation method. Campbell quotes experiments by McCauley, who has shown that $CaCO_3$ deposited on cleaned steel is protective if microcrystalline, but not if present

1.   E. D. Mor and A. M. Beccaria, *Brit. Corr. J.* 1973, **8**, 25.
2.   G. Storace, see *Corr. Control Abs.* 1970, (2) 21.
3.   H. S. Campbell, *Water Treatment and Examination* 1971, **20**, 11; cf. H. S. Hall, p. 27.

as nodules; a small amount of polyphosphate is favourable to the desired microcrystalline character. The addition of polyphosphate is generally ineffective unless sufficient calcium is present. This is hard to reconcile with the theory of Hall, who ascribes the protection to a particular type of iron hydroxide rather than to $CaCO_3$, but Campbell agrees that Hall's suggestion cannot be ruled out. The paper should be studied in the original.

Domestic Central-Heating Systems. The choice of inhibitor for water in a central-heating system naturally depends on the metals involved. Where copper is the metal in question, BTA has proved effective. A suggestion by Pyx[1] that inhibited ethylene glycol (as used in motor vehicles), suitably diluted with water, should be applied to central-heating systems has been criticized by von Fraunhofer. The latter recommends a triple mixture of benzoate, nitrite and molybdophosphate for self-priming automatic systems. Brown states that there are three types of proprietary preparations on the market intended primarily for motor vehicles. One is recommended for general application, a second for systems with aluminium predominating, and a third for systems with cast iron predominating. Should it be decided to use one of these in a central-heating system, it will be all important to select the right type. Butler and Mercer state that in a properly designed and maintained central-heating system, corrosion problems are infrequent. In cases where trouble arises, inhibitors are recommended. Glycol, they point out, is not non-toxic, and care would have to be taken that there is no possibility of the central-heating water coming into contact with the water feeding the household taps. The risk of inadvertent mixing is greatest with the self-priming type of installation. They are doubtful whether the dilution proposed by Pyx would provide the required degree of protection.

Drane[2] states that the benzoate–nitrite mixture introduced by Vernon and Wormwell for motor vehicle systems (1960 book, p. 171) is equally effective for central-heating systems. He states that 'the decrease of nitrite content [with time] does not appear to cause any problems. In fact, once the protective film on the cast iron has been formed, subsequent protection can be left to the benzoate, provided that the benzoate level does not drop unduly.' It appears that a biocide is necessary to control growth, von Fraunhofer[3] states that the benzoate–nitrite mixture provides 'an excellent food for certain fungi'.

Motor-Cooling Systems. A useful Symposium was organized by the Institution of Mechanical Engineers in 1968 (this volume, p. 8), in which the merits of various inhibitive mixtures for cooling water were brought out. In one well known car, the coolant used contains sodium mercaptobenzthiazole and triethanolamine in ethylene glycol, this is changed every twelve months. However the conditions here may be rather exceptional; in this engine no copper alloy has contact with the coolant, whilst the aluminium

1. Pyx, *Met. Mat.* 1971, **5**, 160. See comments by R. J. Brown, p. 221, J. A. von Fraunhofer, p. 222, G. Butler and A. D. Mercer, p. 222.
2. C. W. Drane, *Brit. Corr. J.* 1971, **6**, 39.
3. J. A. von Fraunhofer, *Brit. Corr. J.* 1971, **6**, 23.

alloy employed contains silicon. Dulat[1] states that in another type of car, where there are brass, aluminium alloy and cast iron surfaces to be protected, excellent inhibition has been obtained with a mixture containing borax, mercaptobenzthiazole, silicate and lime; after five years of service, no noticeable corrosion of the aluminium or brass parts has occurred. Indian tests[2] have shown that several inhibitive mixtures give satisfactory results. These include the mixture of sodium benzoate and nitrite, already mentioned. The benzoate mainly controls the anodic reaction, but to a lesser extent the cathodic reaction; the nitrite controls the anodic reaction. Other mixtures recommended involve nitrite, phosphate, chromate and borax. Here the fluids tested were mixtures of glycerine and water, which are much used for anti-freeze purposes in internal combustion engines and marine hydraulic systems. If borax is used NaOH must be added to render the liquid alkaline, since although an aqueous solution of borax may have a pH of about 9, the value in the presence of glycerine is about 4·8.

Factors influencing Efficiency of Inhibitors.

Influence of Substituent Groups. Many of the organic compounds which inhibit corrosion contain an S or a N atom which is generally a member of a ring; the other members of the ring generally consist of C atoms, usually accompanied by one or more hydrogen atoms attached to each; it is believed that the compound attaches itself to the metal through the S or N atom. It has been noticed that often, if one of the H atoms attached to a C atom is replaced by some 'substituent group', such as $-NH_2$, $-NO_2$, $-CHO$ or $-COOH$, the inhibitive power is altered—frequently improved. The cause of this improvement is naturally a matter of great interest.

It is well known that if two hydrogen atoms on a carbon ring are replaced by some substituent, these may be attached to adjacent carbon atoms, or to carbon atoms at one remove, or to carbon atoms at two removes; the products formed in the three cases are known as the *ortho-*, *meta-* and *para-*compounds. It has been found that, whereas the *ortho-* and *para-*compounds have usually similar chemical properties, the *meta-*compounds generally behave differently. Even more important is the nature of the substituent. An electrophilic group like $(NH_3)^+$, which tends to acquire an electric charge, behaves differently from a nucleophilic group, which tends to dispose of charge, like $(COO)^-$; the first can be described as an electron sink and the second as an electron source. It should not, however, be thought that a whole electron moves from the benzene ring in the first case or into the ring in the second; the charge which moves is generally less than one electronic charge, and it is not taken up exclusively by the carbon atom to which the substituent group is attached, but is distributed unequally over the ring as a whole; if the ring is attached through the S or N atom to a metal surface, the electron density in the metal at the point of attachment may well be changed. Since cathodic reactions consume electrons and anodic reactions furnish them, it is reason-

1. J. Dulat, *Brit. Corr. J.* 1968, **3**, 190.
2. K. R. S. Nigam and B. Sanyal, *Labdev. J. Sci. Tech. (Kanpur)* 1970, **8A**, 120.

able to expect that the nature of a substituent will sometimes retard one or other of these reactions, and therefore retard corrosion.

Organic chemists, studying reactions which have no connection with metallic corrosion, have placed the subject on a quantitative footing. Largely owing to the work of Hammett,[1] the relation between the rate constant of a reaction (k_0) when no substituent is present and that measured (k) when the substituent is present is expressed by the equation $\log (k/k_0) = \rho\sigma$, where ρ is a constant depending on the type of reaction but independent of the substituent, whilst σ (known as the Hammett constant) depends on the substituent but not on the type of reaction. Thus in hydrolysis, ρ will have a different value from that observed in bromination, but will have the same value whether the substituent is $(NH_3)^+$ or $(COOH)^-$; σ will be different for $(NH_3)^+$ or $(COOH)^-$ but each of these groups will keep its own value whether the reaction is bromination or hydrolysis. It should be noted that this is only true of *meta*- and *para*-compounds. In the case of *ortho*-compounds, no constant values of ρ and σ can be assigned. Indeed, Hammett's principle does not apply to *ortho*-compounds. Hammett's book, when published in 1940, contained fairly accurate values of σ for over 40 groups, and fairly accurate values of ρ for over 40 different types of reactions or equilibrium systems (which obey a similar principle, but need not here be considered); since 1940, fresh work has added greatly to the data available.

In such cases a straight line can be obtained when $\log k$ is plotted against σ for different substituents; different reactions will have different gradients, since ρ will be different. A point which must interest the corrosion specialist is that the lines pass without change of gradient from the region representing acceleration of a reaction (σ being positive and $k > k_0$) to that representing a retardation (σ being negative and $k < k_0$). If the principle applies to retardation, it may reasonably be expected that the influence of the substituent on the electron density at the attachment point might serve to explain the alteration of the inhibitive power. If it is found that inhibition shows a rectilinear relationship with the Hammett constant—as is indeed the case—that is a most satisfactory situation, for the values of the Hammett constant have been established by chemists who are completely disinterested in corrosion reactions, in order to elucidate their own problems. It cannot be objected that the explanation is based on some *ad hoc* assumption, made by corrosion investigators to solve their own difficulties; an *ad hoc* assumption lays the person making it open to the charge of answering one question only by asking another.

The demonstration that values for inhibition efficiency can be related to the electron density is largely due to Russian, Japanese and Polish workers. A few examples will be given. Altybeeva, Levin and Dorokhov[2] have studied the relation between the chemical structure of various compounds containing nitrogen and their inhibitive powers, plotting the quantity $\log (z/(100 - z))$,

1.   L. P. Hammett, 'Physical Organic Chemistry' (McGraw-Hill), 1940, Chap. VII. See also C. D. Johnson, *Chem. Ind.* 1973, p. 119 and references quoted.

2.   A. L. Altybeeva, S. Z. Levin and A. P. Dorokhov, *Ferrara Symp.* 1970, p. 501.

where Z represents the degree of inhibition, against the electron density at the N atom calculated by a method due to Hückel. In experiments on iron in N HCl containing 0·04 mol% of *para*-substituted compound, straight lines are obtained with (1) methyl, (2) ethyl, (3) propyl, (4) butyl and (5) amyl as the substituent group. Other researches by Grigoryev[1] refer to inhibition of the reaction of iron and zinc in HCl and H_2SO_4 by benzaldehyde derivatives containing various substituents, whose polar properties are defined by the Hammett constant. In the case of zinc the effect was found to increase with the nucleophile properties of the substituent group. Grigoryev has also studied the effect of such compounds on the cathodic production of hydrogen and its diffusion through thin iron membranes—a matter important in connection with hydrogen embrittlement (chapter XI). Here also the effect increases with the value of the Hammett constant. The potential of cathodically polarized metal in an acid solution containing an aniline derivative increases with the value of $|\sigma|$, the movement being in the same direction whether the sign of the Hammett constant is positive or negative. If the metal is iron or nickel, the potential moves in a different direction from that observed when the metal is copper or lead. In the case of nickel, the direction is altered if the acid contains As_2O_3, a substance which is known to promote the movement of hydrogen into the metal, probably by preventing it from being evolved as gas.

The electrochemical behaviour of iron in N H_2SO_4 in presence or absence of various ring-substituted benzoic acids has been studied by Japanese investigators working in the USA.[2] The logarithms of the corrosion rates plotted against the Hammett constant, σ, fall fairly well on a straight line. The inhibition provided by the ring-substituents fall in the order:

$$H < m\text{-OH} < m\text{-NH}_2 < p\text{-OH} < p\text{-NH}_2$$

The different effect produced by the same group in two different positions will be noticed; there are countless analogous examples in purely chemical reactions.

Although here the *meta*-compounds produce a different effect from the *para*-compounds, this is not always the case. A comparison of the inhibitive effects of three substituted cinnamates on the corrosion of cast iron by supply water carried out at Teddington[3] showed no significant difference; the three salts tested were sodium *m*-nitrocinnamate, sodium *p*-nitrocinnamate and sodium *o*-nitro*hydro*cinnamate.

Szklarska-Smialowska[4] has studied the effect of various substituents in thiophene on the inhibition efficiency; these included nucleophilic substituents like CH_2OH, $2CH_3$, $3CH_3$ and electrophilic substituents like $2Cl$, $2COOH$, $3Br$, $2COCH_3$, $2CHO$ and $2COOH$. The results (Fig. 7) show that $\log(i_0/i_R)$, where i_0 and i_R are respectively the corrosion rates obtained in

1.   V. P. Grigoryev and O. A. Osipov, *Ferrara Symp.* 1970, p. 473; V. P. Grigoryev and V. V. Ekilik; see *J. appl. Chem.* 1970, **20**, i, 23.
2.   A. Akiyama and K. Nobe, *J. electrochem. Soc.* 1970, **117**, 999.
3.   'Chemistry Research' 1951, Table 1, p. 8 (Dept. sci. ind. Res.).
4.   Z. Szklarska-Smialowska and M. Kaminski, *Tokyo Cong. Ext. Abs.* 1972, p. 217.

presence of thiophene itself and of its derivative containing the substituent group, is a rectilinear function of $|\sigma|$, and here both positive and negative values of σ move $\log (i_0/i_R)$ in the same direction. This movement in the same direction, irrespective of the sign of σ, was also found during a study[1] of the effect of thiophene derivatives upon the corrosion of steel in N H_2SO_4; thiophene itself is a poor inhibitor, the substitution of one H atom by any substituent, irrespective of its electrophilic or nucleophilic character, increases the inhibition efficiency by an amount proportional to the Hammett

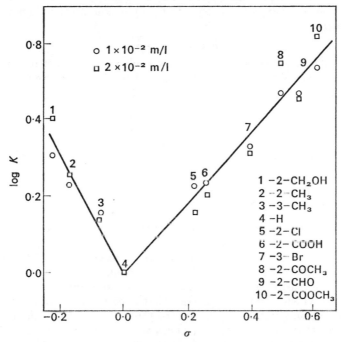

Fig. 7   Log K as a function of Hammett's constant (σ). K is i_0/i_R where i_0 is the corrosion rate in presence of thiophene (without substituent) and i_R is the rate in presence of a substituted thiophene (Z. Szklarska-Smialowska and M. Kaminski).

constant. It is believed that the inhibitor molecules lie flat on the metal surface, and that the adsorption results from interaction of the π-electrons of the thiophene rings with metal surface atoms. Nucleophilic substituents increase the density of π-electrons in the ring and accordingly improve protection. Electrophilic substituents decrease the electron density but nevertheless also improve protection—due, it is suggested to an increase of dipole moment.

Cases have arisen in chemistry where the Hammett equation has failed to explain the facts, and it has become evident that a substituent can do other things than confer or abstract an electric charge. It may cause a steric

1.   Z. Szklarska-Smialowska and M. Kaminski, *Corr. Sci.* 1973, **13**, 1, 557 (2 papers).

hindrance, preventing or slowing down a chemical change (because, for instance, the chains thrust out from the ring make it geometrically difficult for the molecules to get into the required position), or it may cause strain in the bonds joining the carbon atoms; where possible these bonds set themselves at the same angles as are present in diamond, the hardest and mechanically most stable form of carbon, but substituent chains may make such an angle geometrically impossible.

Taft[1] has considered cases where these effects are important and where electrical effects exert little influence. If electrical effects can be neglected, the equation to be expected is one closely analogous to the Hammett equation, namely

$$\log (k/k_0) = \delta E$$

where δ is a reaction constant depending solely on the nature of the reaction and independent of the substituent group, whilst E is a steric constant which is peculiar to a substituent group. Taft's determinations have been used in

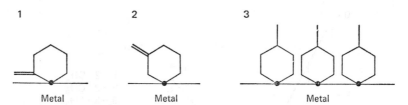

interpreting the experimental data of corrosion inhibition, but it is perhaps too early to assess the results.

There may be other geometrical effects of inserting substituent groups quite independent of charge transference. Indeed, explanations due to steric effects were proposed in early work, before the importance of electrical transfer became evident. It was suggested that a molecule so constituted as to push out a chain parallel to and close to the metallic surface might shield a greater area than one in which the chain was thrown out at an angle. Thus for equal masses of adsorbed matter, ① might be a more efficient than ②. However, it could be suggested that where the chains are thrown out at right angles to the surface, as in ③ a larger number of molecules per unit surface could be packed in close array; in that case, although more material would be demanded, more efficient inhibition might well be provided than in cases ① and ②. A Swedish suggestion,[2] based on rather similar lines, suggests that attachment of an inhibitor atom may fulfil a double function. The non-polar part of the molecules may provide a diffusion barrier preventing the inward movement of aggressive substances towards the metal and also preventing the dissolution of products formed from the metal. But there is another effect—not always sufficiently taken into account. If there is a strong bond between the polar end and the metal, the act of attachment releases a considerable amount of energy, and clearly the energy drop to be

1. R. W. Taft, *J. Amer. Chem. Soc.* 1953, **75**, 4534, 4538 (2 papers).
2. S. Friberg and H. Müller, *Ferrara Symp.* 1970, p. 93.

obtained when metallic ions escape during a corrosion process is made smaller by that amount; thus the escape becomes less probable and the corrosion rate is greatly reduced.

Synergistic Effects. It has been mentioned on pp. 94, 99 that two or more inhibitors acting together can produce a greater effect than would be expected from the effects observed when they act separately. It is not always possible to suggest an explanation; possibly extension of the theoretical reasoning based on the Hammett and other concepts may provide this in future years. Today it seems best merely to state some additional facts. In perchloric acid iron has been found by Murakawa[1] to be protected if certain amines are combined with Br^-. He believes that primary amines are adsorbed on the outside of a Br^- covering, whereas tertiary amines are adsorbed on the surface along with Br^- anions; these two modes of adsorption are called by him 'overlap adsorption' and 'co-adsorption'. Synergistic effects have been observed on low-carbon steel in $2N\ H_2SO_4$ for pyrrole combined with Cl^-.[2] In the case of zinc[3] in contact with H_2SO_3, hexamine behaves best as an inhibitor if combined with sodium oxalate, potassium ferricyanide or sodium pyrophosphate. A mixture of hexamine and sodium oxalate will give inhibition against $4N\ H_2SO_4$ or a mixture of H_2SO_4 and H_2SO_3. The advantage of combining an anodic and cathodic inhibitor is today generally realized, and has been emphasized in Polish studies on iron, nickel, zinc and lead in various acids.[4]

Synergistic inhibition is important in the oil industry. Horvath[5] reports that in presence of Cl^- or SH^- the inhibition by dicyclohexylamine is effective. tf Cl^- and SH^- should be absent, the addition of the amine actually increases Ihe current flowing within the passive range.

Changes of Properties remaining after Inhibition Treatment. The after-effects of the treatment of steel in a solution containing a sulphonate inhibitor have been studied by Borning and Martin,[6] who combined ellipsometry and potentiostatic methods to follow the development of oxide-films during polarization. They found that a pre-adsorbed film of dinonylnaphthalene sulphonate was rapidly desorbed under anodic polarization, but was stable if conditions were displaced by 0·1–0·2 V in the cathodic direction. The initial corrosion rate of bare steel was 4 mpy (mils per year), but a pre-adsorbed sulphonate film reduced this to 1 mpy.

Theory of Inhibition

General. It is not easy to present a unified picture of inhibition, since different inhibitors retard attack in different ways. The reader, however,

1. T. Murakawa, *Corrosion Engg.* 1968, **17**, No. 5.
2. R. M. Hudson and C. J. Warning, *Corr. Sci.* 1970, **10**, 121.
3. W. McLeod and A. R. Rogers, *Mat. Prot.* 1968, **8**, (4) 25.
4. B. Duš and Z. Szklarska-Smialowska, *Ferrara Symp.* 1970, p. 315.
5. J. Horvath, A. Rauscher, L. Hackl and F. Marta, *Ferrara Symp.* 1970, p. 851.
6. B. J. Borning and P. Martin, *Corrosion (Houston)* 1971, **27**, 315.

may care to consider a point of view which appears to explain the action of inhibitors commonly used today.

Organic Inhibitors. Several of the complicated substances used to suppress attack by acid almost certainly act by altering the electron density in the metal, and thus retarding either the cathodic or the anodic reaction; this explains the fact that substituent groups introduced into the molecule close to the site of attachment (generally a N or S atom) greatly influence the efficiency of inhibition, and that the effect varies both with the character and the position of the substituent. However, in many cases, a secondary product is produced on the metal surface—sometimes by reduction—and this product makes it possible to transfer the metal into liquid free from inhibitor and still obtain considerable protection; the attack is slower than that which would be suffered by metal which had never been in contact with the inhibitor, but faster than that which would occur if the liquid still contained inhibitor. In many cases, the identity of the substance left on the surface has been established.

The distribution of the blocking material is often far from uniform—at least in the case of azelates and benzoates, which inhibit in a relatively neutral range of pH. This is not surprising, since the energy needed for the escape of atoms (as cations) into the liquid must vary at different points; if, for instance, owing to internal tensile stresses, the mean distance between atoms is increased, or if, as at a micro-projection, an atom has fewer neighbours than it would possess on a truly smooth surface, or again if, as at a grain-boundary, structure is disorganized, escape will become specially easy, and a relatively thick block of obstructive material will be needed to make the probability of escape small. Where the situation is unfavourable to escape, no protective layer may be needed, so that the total amount of attached material—averaged over the whole surface—may be less than one monolayer. It was a tone time argued that, in cases where the amount of attached material present had been shown to be roughly a monolayer, this indicated an adsorbed film rather than the formation of a three-dimensional block; such an argument would seem to be unsound, and, in an increasing number of cases, the material producing the three-dimensional block has been identified with some familiar compound. Clearly the first stage in the process will be adsorption, but especially where retardation of attack continues even when the liquid ceases to contain inhibitor, the part played by three-dimensional matter is very important.

Inorganic Inhibitors. Many of the inhibitors used in near-neutral liquid are simple inorganic substances, which may act either at cathodic or anodic points. A magnesium salt solution into which iron is introduced can produce magnesium hydroxides by the cathodic reaction; since, however, anodic attack will not immediately be stopped, the Fe^{2+} formed may interact with the $Mg(OH)_2$ and a mixture containing both iron and magnesium is produced; this may be a bright green in colour whilst Fe^{2+}, Fe^{3+} and Mg^{2+} are all present, but later becomes brown (paler than ordinary rust) through further uptake of oxygen.[1] Zinc salts seem to behave similarly, producing a layer containing both zinc and iron. Calcium, if present as $Ca(HCO_3)_2$, will be thrown down as $CaCO_3$, and retard attack, especially if traces of organic matter are present.

The anodic inhibitors are often the sodium (or potassium) salts of weak acids, which provide OH^- by hydrolysis and this leads to an oxide-film formed by the anodic reaction, which is quickly slowed down; the fact that an oxide rather than a hydroxide is formed is easily explained by the mechanism suggested elsewhere (1968 book, p. 97). A little thought will show the advantage of using a salt of a weak acid to provide OH^-; it would be possible to obtain any desired pH value by adding a very small quantity of, say, sodium hydroxide or a much larger quantity of sodium carbonate. In the first case, the supply of OH^- may become locally exhausted before blockage at a susceptible point has been achieved; in the second case, a fresh supply will be produced by continuing hydrolysis. There is also the chance that the second anion present may play a part in the protection. Certainly a solution containing silicate behaves differently from one containing carbonate or phosphate, but the mechanism of inhibition by silicates (sometimes very efficient) calls for further investigation.

In the case of inhibition of attack on iron, produced by phosphates, electron diffraction studies[2] have shown the film to be essentially oxide when Na_3PO_4 has been used, but to contain particles of $FePO_4.H_2O$ after inhibition by Na_2HPO_4. The appearance of a ferric compound is instructive. The first anodic reaction on iron is $Fe - Fe^{2+} + 2e$, but if the Fe^{2+} remains in contact with the metal for a sufficiently long time, the reaction $Fe^{2+} = Fe^{3+} + e$ may follow; often Fe^{2+} will be removed by convection, and in some researches the electric current has corresponded exactly with that calculated for the production of Fe^{2+} alone. A high CD will favour the production of Fe^{3+}, both by increasing the chance that a given Fe^{2+} will be oxidized before it moves away from the metallic surface, and also because a high CD will raise the potential, by polarization, to a level where the possible value of Fe^{3+}/Fe^{2+} is higher than at low current densities. Thornhill[3], studying corrosion at a scratch-line on iron specimens covered with filter paper soaked in $0.1N$ $NaHCO_3$ solution, obtained evidence of the formation of Fe^{3+}; here the anodic area was small—which would increase the CD—and movement in the liquid was restricted by the filter paper.

Returning to the question of inhibition by phosphate solutions, it should be noted that the solubility of Fe_3O_4 is low; in the presence of an extremely small amount of Fe^{3+} the liquid will become supersaturated with Fe_3O_4 whilst it is still unsaturated with respect to $Fe(OH)_2$ or FeO. Thus when an iron specimen carrying an air-film is placed in Na_3PO_4 solution, a small amount of Fe^{3+} formation at discontinuities in the film will make the solution

1. U. R. Evans, *J. Soc. chem. Ind.* 1928, **47**, 55T.

2. J. E. O. Mayne and J. W. Menter, *J. Chem. Soc.* 1954, p. 103. Cf. radioactive tracer results by J. G. N. Thomas, *Brit. Corr. J.* 1970, **5**, 41.

3. R. S. Thornhill and U. R. Evans, *J. Chem. Soc.* 1938, p. 614; esp. Figs. 6, 7 and 8, p. 620.

supersaturated towards Fe_3O_4; since the air-formed film consists of cubic oxide, nucleation is provided, and Fe_3O_4 will be deposited, healing the gap (later it may be converted to γ-Fe_2O_3 by anodic removal of Fe, but this does not affect the principle involved). If, however, a solution of Na_2HPO_4 is used, the lower OH^- concentration may render the $FePO_4.H_2O$ the 'expected' product, in the sense that the solution becomes supersaturated with respect to it before it is saturated with Fe_3O_4. Here, however, no nucleus for the deposition of $FePO_4.H_2O$ is present and at many points Fe_3O_4 will still be deposited. If nucleation takes place, $FePO_4.H_2O$ crystals will appear—explaining why both compounds are found together in the film. It is evident that oxygen is necessary for complete suppression of attack, and that is found to be the case; however, in absence of oxygen, the cathodic reaction will be the production of hydrogen (probably in solution), not the reduction of oxygen, so that corrosion will be very slow.

Perhaps the most efficient of anodic inhibitors are the *chromates*, although they have two disadvantages; they are toxic, and there is the danger of intense localized corrosion if added to (say) a chloride solution in a quantity just insufficient to prevent attack altogether, so that the dangerous combination of large cathodic area and small anodic area arises.

The demonstration (this volume, p. 98) that the substance composing the protective film contains both Fe and Cr is consistent with the early view that the film results from the anodic reaction; anodic attack would normally produce Fe^{2+} ions, which in (say) a sulphate solution would pass into solution. However, Fe^{2+} is incompatible with CrO_4^{2-} and the deposition of a solid containing both Fe and Cr in contact with the metallic surface must be expected to start at once.

The presence of Cr in the film produced by chromate suggests an analogy with the film on stainless steel. Chromium is unstable in the divalent state, and thus a film containing trivalent chromium suffers reductive dissolution at a weak point less easily than a film of pure Fe_2O_3. Stainless steel withstands dilute H_2SO_4 containing oxygen, because oxygen will suffer reduction in preference to Cr^{3+}. However, if the oxygen concentration is kept very low, the film is destroyed and corrosion sets in (see this volume, p. 208). There is one difference between the two cases. On stainless steel, a crack (arising, perhaps, from internal stress) will generally heal itself, since the alloy contains chromium. In the case of iron or mild steel, it is necessary to keep chromate in the liquid if healing is to occur. Thus, although steel placed in a natural water dosed with chromate will remain unchanged for a long period, it will start to rust if removed to the same water free from chromate, since stress will sooner or later cause cracking. Presumably, however, alternate cracking and healing will gradually use up the internal stress; if so, it should be possible after a time to reduce the concentration of chromate without any rust appearing, although not to dispense with it entirely; Darrin's work suggests that such a reduction of concentration is indeed possible, but should be spread out over some months (1960 book p. 155).

The *nitrites* seem to have an advantage over chromates and benzoates in being more tolerant towards chloride in small amounts, but they are more dangerous if excess of chloride is present; sometimes they produce pitting and perforation, under conditions where chromates would give local but often self-stifling attack and benzoates would produce fairly general corrosion.[1] A nitrite solution is somewhat unstable, and performance is difficult to predict, especially if certain bacteria are present which can cause oxidation to nitrate. Remarkably little study seems to have been carried out on the mechanism of inhibition by nitrites, which, as inhibitors, are much more efficient than nitrates. Perhaps the most likely explanation is that the cathodic reduction of NO_2^- to NO occurs *more* easily than that of NO_3^- to NO_2; if so, the high current at the cathodic parts generated when first the metal comes in contact with a nitrite solution will require to be balanced by a high current density at anodic points, raising the potential into the passive range, so that attack soon ceases. Such an explanation is analogous to that advanced by the Pražák brothers for the passivation of iron by concentrated nitric acid (1960 book, p. 889). It may seem strange that NO_3^-, which contains more oxygen than NO_2^-, should be reduced less readily, but it has to be remembered that NO, the reduction product of NO_2^-, forms stable complexes with the cations of iron and some other metals, so that the drop of free energy accompanying the reduction of NO_2^- may be quite high. Such an explanation would appear consistent with the fact that some non-ferrous metals are not protected by nitrites. However, the question requires study—which may well lead to a different explanation.

Cohen[2] has detected ammonia in a by-product when a nitrite solution acts on iron. He considers that NO_2^- ions are adsorbed through the two O atoms, since the distance O—O in NO_2^- is not very different from that of Fe—Fe in metallic iron; OH^- is also adsorbed, and the removal of NH_3 leaves a monolayer of FeO:

Inhibitors which are not themselves oxidizing agents require the presence of oxygen. This has long been believed, but Cohen's delicate weight change measurements greatly clarify the situation. Iron placed in neutral phosphate solution free from oxygen steadily loses weight; when oxygen is present, there is loss of weight at the same rate for 30 hours, and then the loss (which has reached about $0·35$ mg/cm^2) suddenly ceases, the weight becoming constant. Evidently a protective film has been formed, and this is confirmed by the fact that the potential, which had previously been falling, suddenly rises from $-0·5$ V to $+0·1$ V; if oxygen is absent, it remains low at about $-0·55$ V.

1. A. D. Mercer, I. R. Jenkins and J. E. Rhoades-Brown, *Brit. Corr. J.* 1968, **3**, 136.
2. M. Cohen, *J. electrochem. Soc.* 1974, **121**, 191C.

Measuring Methods. The development of experimental methods for measuring accurately the effect of different inhibitors is obviously an important matter. In Polish work,[1] three methods (gravimetric, polarization resistance and Tafel-curve extrapolation) have been used, and excellent agreement obtained. This is satisfactory, but in other researches a lack of agreement between duplicate specimens tested by the same method has caused some concern. In cases where corrosion, if it occurs at all, starts locally, a scatter of results is to be expected, since the presence of an inhibitor may reduce the number of starting points on the specimen to a number so small that Bernoulli's principle (1960 book, p. 914) fails to operate; in such a situation, good agreement between duplicate experiments is not to be looked for, and anyone who claimed to have obtained it might render himself open to suspicion.

Nathan[2] makes reference to a series of experiments where at high concentrations or at very low concentrations, good reproducibility was obtained, in the sense that the scatter extended over a narrow range; at an intermediate range of concentration, the data were so widely scattered as to make meaningless the value of any average figure obtained for the concentration. An empirical relationship between the standard deviation σ, and the mean, ε, was found to be obeyed

$$\sigma = 1{\cdot}33\varepsilon(1 - \varepsilon)$$

and it will readily be seen that if either ε or $(1 - \varepsilon)$ is small compared to unity, their product is likely to be much smaller than if both are reasonably large. This explains the apparently better reproducibility when the inhibition is either very good or very bad. But before applying these arguments to a practical problem, the reader may do well to ask himself whether a small absolute value of σ is what he is seeking, or whether the value of the standard deviation expressed as a fraction of the mean value, σ/ε, is a rational measure of reproducibility for his purpose.

Impedance measurements have been developed in France in the study of inhibitors. Epelboin[3] and his colleagues have reached the conclusion that the corrosion rate is closely related to the transfer resistance (the limit of the Faradaic impedance at infinite frequency). They have tested the question by studying iron in $0{\cdot}5M$ H_2SO_4 to which has been added various small amounts of propargylic alcohol. Impedance methods show better correlation with values obtained by direct measurement of weight loss than do methods based on polarization resistance or double-layer capacity.

Further References

Conferences. The Report (published 1971) containing the papers presented in the Ferrara Symposium of 1970, should be studied with care. Several of the papers have been quoted elsewhere in this chapter, but refer-

1. B. Duš and Z. Szklarska-Smialowska, *Ferrara Symp.* p. 315; esp. Table 2, p. 323.
2. C. C. Nathan, *Ferrara Symp.* 1970, p. 829; esp. p. 835.
3. I. Epelboin, M. Keddah and H. Takenouti, *J. appl. Electrochem.* 1972, **2**, 71.

ence must here be made to a classification of inhibitive processes contributed by H. Fischer. The Tokyo Congress of 1972 also furnished valuable contributions to the subject; apart from papers quoted in this chapter, Okamoto's plenary lecture on the passive film on stainless steel has bearing on the subject of inhibition. A plenary lecture by J. E. O. Mayne, delivered at Ferrara to the Fourth European Symposium on Corrosion Inhibitors in Sept. 1975, should receive study when available in print.

Papers. 'Untersuchungen des Mechanismus der Schutzwirkung'.[1]
'Inhibition of Acid Corrosion of Iron by Nitrites'.[2]
'An electrochemical–statistical study of mild steel corrosion inhibition in oxygen-containing environments'.[3]
'Contribution to filming amine theory. An interpretation of experimental results'.[4]
'Activation of passivated steel in sodium chromate containing sodium chloride'.[5]
'Corrosion behaviour of steel in solutions containing both inhibitive and aggressive ions'.[6]
'The Hammett Equation'.[7]
'Sodium diethyl dithiocarbonate as a corrosion inhibitor for brass'.[8]
'Corrosion Inhibitors for Zinc'.[9]
'Inhibitive effect of an onium compound on the dissolution of steel in hydrochloric acid'.[10]
'Action of benzimidazoles on the corrosion of 63/37 brass in sodium hydroxide solution'.[11]
'Inhibition of the corrosion of iron by benzoate and acetate ions'.[12]
'Significance of the redox potential in the inhibition of the corrosion of iron by non-oxidizing inhibitors in the pH range 5–13'.[13]
'Acetylenic corrosion inhibitors'.[14]

1. I. L. Rosenfeld, E. K. Osche and W. G. Doroschenko, *Ferrara Symp.* 1970, p. 65.
2. V. Carassiti, F. Zucchi and G. Trabanelli, *Ferrara Symp.* 1970, p. 65.
3. S. A. Legault, S. Mori and H. P. Leckie, *Corrosion (Houston)* 1971, **27**, 418.
4. R. H. Hausler, L. A. Goeller, R. P. Zimmerman and R. H. Rosenwald, *Corrosion (Houston)* 1972, **28**, 7.
5. K. S. Rajagopalan and K. Venu, *Corrosion (Houston)* 1971, **27**, 506.
6. V. K. Gouda and S. M. Sayed, *Corr. Sci.* 1973, **13**, 841.
7. C. D. Johnson (Book), Cambridge Univ. Press, 1973.
8. A. S. Shah and M. Trivedi, *Werkstoffe u. Korr.* 1974, **25**, 521.
9. M. N. Desai, S. S. Rana and M. H. Gandhi, *Anticorrosion* Dec. 1973.
10. B. Sanyal and K. Srivastava, *Brit. Corr. J.* 1974, **9**, 103.
11. N. K. Patel, S. C. Makwana and M. M. Patel, *Corr. Sci.* 1974, **14**, 9.
12. J. E. O. Mayne and C. L. Page, *Brit. Corr. J.* 1974, **9**, 223.
13. J. E. O. Mayne and S. Turgoose, *Brit. Corr. J.* 1975, **10**, 44.
14. R. J. Tadeschi, *Corrosion (Houston)* 1975, **31**, 130.

CHAPTER VI

Bimetallic Contacts and Crevice Corrosion

Preamble

Picture presented in the 1960 and 1968 volumes. Rapid electrochemical corrosion can take place where two dissimilar metals are in contact. As explained in chapter IV, the situation is most serious when the area of the noble metal is greater than that of the less noble metal; if the total current flowing is determined by the oxygen reaching a large cathode, the current will then be strong, and the attack, concentrated on the small anode, may be very intense.

It is clearly important to decide the relative polarity of the various metals. The idea at one time established itself that the table of normal electrode potentials could provide information as to which of two metals would behave as anode and which as cathode. Unfortunately, however, this table is of little value for practical conditions. The table provides the potentials of oxide-free metal surfaces placed in a solution containing normal activity of the ions of the same metal. Such a situation practically never occurs in an industrial or engineering situation. Some metals, like aluminium, normally carry a highly protective oxide-film; the liquids which concern the industrialist or engineer do not usually contain appreciable amounts of ions of metals used for containing vessels; certainly these ions, if present, will not reach normal concentration. Attempts have been made to provide an empirical table, showing the potentials of the metal in contact with sea-water, but even this has limited value. Such numbers do not represent reversible values, and the EMF set up by two metals immersed in sea-water will not be equal to the difference between the two numbers as recorded in the table; it will generally depend on the supply of oxygen to the cathodic metal. Particularly unfortunate is the idea which once prevailed, that a combination was safe if the difference between the potentials, as indicated on the table favoured, was less than 0·25 V. Such an assumption has no logical basis.

As a result of this situation, a table was published by HM Stationery Office in 1956, providing purely empirical information (1960 book, p. 197). The table showed whether a given combination, when used in practice, had been found to produce (1) practically no attack, (2) slight attack, (3) serious attack, or (4) disastrous attack. Obviously such a statement represents an oversimplification of the position; a given combination can behave very differently under different service conditions; a number of notes were therefore included, designed to warn the reader about possible exceptions to the rules suggested. The information provided was greatly welcomed—as was shown by the rapid selling of the document; although the number of copies printed in 1956 was large, a second edition was called for in 1958, and a third edition in 1963.

Another situation where intense attack may take place exists at a crevice. This represents a place where the supply of oxygen is scanty, and where the replenishment of inhibitor (if present in the body of the liquid) may be insufficient to prevent attack. The situation is made worse by the fact that the attack, if it occurs, may not be noticed until it is in an advanced state; in addition, the combination of small anode and large cathode can cause the attack to be very severe. In contrast, corrosion at spots on the external part of a metallic surface will soon be noticed and the corrosion, which will usually spread out over the surface, will not become intense. If the situation is one where a sparingly soluble corrosion product is precipitated within the crevices, this may act as a wedge levering the sides apart and possibly causing the inception of a crack; on the other hand, if the material is sufficiently strong to resist cracking, the solid matter may block the path of the current and slow down the attack.

Other factors can be responsible for special corrosion in a crevice; under atmospheric conditions crannies may remain remain wet after rain, long after the external face has dried up.

Serious attack may also occur at joints between metallic members. At one time riveting was largely used for joining and clearly presented possibilities for serious localized attack. Even if the material of the rivet was identical with that of the plate, the deformation involved in beating it out would produce a highly stressed situation which, in many materials, would be favourable to anodic attack. In cases where liquid had access to the space between the rivet stem and the plate, crevice corrosion might arise. The most serious cases of rivet corrosion were reported where the metal used for the rivet was different from that of the plate. If it was cathodic, the attack set up on the plate around the edges of the rivet would generally not be very serious; where it was anodic, the dangerous combination of small anode and large cathode would arise, with the usual disastrous results.

In the case of soldered joints, the same principle holds good; solders which are anodic in relation to the material to be joined should be avoided. In the case of welded joints, there will often be less chemical difference between the weld metal and the plate, but the structure produced locally after heating to the melting-point followed by rapid cooling may be different from that of the main part of the metal. What may be more serious is the fact that the part of the material composing the plate close to the weld will itself have been strongly heated and rapidly cooled. The structure left behind will vary with the distance from the weld, but there is a risk, with some materials, of producing a line of material anodic to that of the unchanged part of the plate. Apart from that risk, the welding may leave behind serious internal stresses. Sometimes it is advisable to anneal after welding, but that may not deal with all sources of trouble.

Apart from the facts mentioned, corrosion due to substances in the fluxes used both in soldering and welding was at one time frequently reported. This source of trouble soon became appreciated, and attention has been given to the development of safe fluxes.

Developments during the period 1968–1975 and Plan of the new Chapter. Surprisingly little fresh research has been published during the period under review. It would be pleasant to draw the conclusion that difficulties have been overcome, but unfortunately that is not the true explanation. However, some useful papers have appeared. As regards crevice corrosion, several authorities have pointed out that other factors besides the limited supply of oxygen and inhibitor should be taken into account. On the the subject of welding, work has been published on precautions needed in connection with steel structures, and special attention has been given to the case of stainless steel, which is extremely liable to suffer corrosion at welds, unless precautions are taken; the so-called 'knife-edge attack' continues to be the subject of investigation. As regards materials used for soldering or brazing, attention has been given to the dezincification of brazed joints.

The arrangement of the chapter is similar to that adopted in the 1968 book, sections being devoted to galvanic corrosion, crevice corrosion and corrosion at joints.

Galvanic (or Bimetallic) Corrosion

General Situation. The problem of providing information about bimetallic corrosion was discussed in 1971 by D. N. L.[1] He pointed out that the Stationery Office Report, which had long provided guidance on the subject,was out of print, and that a Committee set up to provide an up-to-date document found difficulty in producing a generally acceptable statement; the first draft was considered too long by some of those to whom it was sent for an opinion, whilst a later version was considered too brief. It is understood that there were other points of contention. Earlier disagreements have been resolved, and discussion of the draft is continuing; it is hoped that the document,[2] which is badly needed, will soon be published.

Meanwhile, considerable interest in the matter is being shown in other parts of the world. A book by Rosenfeld,[3] besides reproducing an American table due to LaQue and the British (Stationery Office) table, provides an account of Russian work and views. There is, for instance, an account of work on aluminium–magnesium alloys in contact with other metals, with synthetic sea-water flowing over the junction at 10 m/s at 35°C; considerable corrosion was produced where the contact material was a low-alloy steel, bronze or copper; but stainless steel was especially dangerous, causing a loss of thickness of 1·5 mm/yr when the area ratio was 1 : 1. In practice, the steel hull of a ship will function as an anode if it is in contact with stainless steel or a copper alloy. When the area ratio is 1 : 1, the loss of thickness may reach 0·63 mm/yr. Stainless steel can intensify corrosion of aluminium, mild steel and copper alloys, but often itself suffers corrosion when in contact with copper

1. D. N. L., *Brit. Corr. J.* 1971, **6**, 1.
2. 'Corrosion at bimetallic contacts and its alleviation' with introduction by L. Kenworthy, Chairman of the Committee responsible for the production (British Standards Institution), in preparation.
3. I. L. Rosenfeld, 'Corrosion and Coating of Metals'. An English translation of pp. 150–79 has been prepared by the Building Research Station: Library Communication No. **1567** (1970).

alloys, titanium or hastelloy. Small areas of brass or other copper-based alloys in contact with a large area of stainless steel can suffer serious corrosion; but welded joints of an alloy containing 25% chromium, 20% nickel, with the rest iron, has caused no serious corrosion to plates of low-alloy steel after several years in slowly moving sea-water. In all these cases the rate of movement is important.

An important paper by Kaesche[1] on the pitting of aluminium and intergranular corrosion of its alloys also contains valuable information about bimetallic effects. His curves show that in moderately stirred 0·1M NaCl solution, attack upon aluminium is increased by contact with platinum more than with iron, but that copper accelerates corrosion even more; evidently copper provides a surface on which the cathodic reduction of oxygen proceeds easily. If there is no contact with a relatively noble metal, the rate of attack is very slight, and becomes negligible if the solution is pre-saturated with oxide—as shown by Lorking and Mayne.[2] On filmed aluminium the very low electronic conductivity of the film prevents it from acting as cathode.

The information provided by Kaeshe emphasizes the point that the position of a contact metal in the potential series is no guide to the danger which such contact introduces. The idea that the difference between the potentials of two metals in contact provides some index of the perils of the combination still lingers, although examples which disprove it were provided early in the century;[3] for instance, the corrosion of steel partially immersed in 0·1M NaCl is increased by contact with copper much more than with nickel, but contact with lead, which stands between copper and nickel in the series, causes less acceleration than either.

In an important paper Mansfeld[4] has examined the effect of the anode/cathode area ratio in causing corrosion. The so-called catchment principle represents only one of three possible cases. It is valid when the cathodic process on both metals is controlled by diffusion; in that case, intensity of attack is proportional to the ratio of the cathodic to the anodic areas, as shown in 1924 by Whitman and Russell.[5] But if the anode becomes polarized, the galvanic current may not represent the dissolution rate of the anode.

In another paper Mansfeld[6] recommends a combination of galvanic current records with weight loss measurements as the most promising method for examining galvanic corrosion; the current provides information about the variation of corrosion rate with time, whilst the weight loss represents a measure of average corrosion rate over a long period. The paper in question refers to work on aluminium alloys coupled to molybdenum stainless steel carrying a coating, but the recommendation probably has more general significance.

1. H. Kaesche, *Williamsburg Conf.* 1971, p. 516, esp. Fig. 2.
2. K. F. Lorking and J. E. O. Mayne, *Brit. Corr. J.* 1966, **1**, 181.
3. U. R. Evans, *J. Soc. chem. Ind.* 1928, **47**, 73T.
4. F. Mansfeld, *Corrosion (Houston)* 1971, **27**, 436.
5. W. G. Whitman and R. P. Russell, *Ind. Engg. Chem.* 1924, **16**, 276.
6. F. Mansfeld, *J. electrochem. Soc.* 1973, **120**, 231C.

E

An important paper by Waber[1] on the analysis of size effects should receive attention; he makes acknowledgement to an early paper of the subject by Agar and Hoar,[2] which still deserves study today. Unless the principles of dimensional analysis are taken into account, the use of 'models' in experimental work (the word 'model' seems to make a special appeal today, although the concept is not new) can lead to wrong conclusions; this may account for the mistrust felt by some engineers—not without justification—in the conclusions based on laboratory experiments. Attention may be directed to Waber's 'Fig. 25' which illustrates three cases where the corrosion distribution is controlled by (1) anode width, (2) liquid depth and (3) cathode width, respectively.

The Zinc–Iron Couple. It has long been known that the sign of the zinc–iron couple reverses above a certain temperature. This is a serious matter in connection with galvanized water pipes, where in cold water a zinc coat will generally provide protection to iron even at areas where the iron base is exposed; in very hot water such protection fails, and it is found that now zinc is cathodic towards iron. It was suggested by Gilbert[3] that the film formed on zinc at low temperatures was $Zn(OH)_2$ or perhaps a basic salt, whereas at high temperatures it would become ZnO, which is a reasonably good electronic conductor, and might therefore act as a cathode towards exposed iron.

The Teddington investigators,[4] however, consider that Gilbert's explanation is an over-simplification, since reversals occur at room temperature in presence of HCO_3^-. The fact that in many waters a basic carbonate is formed, as shown in Swiss work, also suggests that some modification of Gilberts' theory may be needed. The behaviour of zinc-coated steel has been studied in fresh waters containing HCO_3^- and Cl^-. The relative concentration of these ions will decide whether or not the corrosion product is the basic carbonate, $Zn_5(OH)_6(CO_3)_2$, which possesses a scaly character, as established by Feitknecht;[5] if formed as a continuous film in physical contact with the zinc surface, such a substance would be expected to provide protection. The composition-range of the water within which this compound is deposited on the zinc surface has been studied experimentally by Grauer and his colleagues.[6] The range is more extensive than would be expected on the assumption of thermodynamical reversibility (in absence of nuclei, the particular substance which is the least soluble and therefore the 'stable phase' in the thermodynamical sense, is often *not* the substance which is, in fact, deposited; some other compound, really metastable, appears). In cases where there is a complete coat of zinc covering the steel, the appearance of the protective, scaly substance, under conditions where it would not be the stable phase, is, of course, something to be welcomed. But where there are discontinuities in the covering, the unexpected avoidance of zinc corrosion may be an unfavourable

1. J. T. Waber, *Williamsburg Conf.* 1971, p. 221; esp. Fig. 25 on p. 233.
2. J. N. Agar and T. P. Hoar, *Disc. Faraday Soc.* 1947, **1**, 158.
3. P. T. Gilbert, *J. electrochem. Soc.* 1952, **99**, 16.
4. G. Butler, P. E. Francis and A. S. McKie, *Corr. Sci.* 1969, **9**, 715.
5. W. Feitknecht and H. R. Oswald, *Helv. chim. Acta* 1966, **49**, 334.
6. R. Grauer, U. Gut and K. Blaser, *Corr. Sci.* 1970, **10**, 489.

occurrence. It is known that where the steel basis of galvanized material is exposed, as at a cut edge, rusting often prevented by cathodic protection; this demands anodic attack on the zinc. If the water in contact with the metal is too hot, the cathodic protection may fail, and rusting sets in. Grauer believes that the main cause of the failure is the deposition of scaly material upon just those points of the zinc which would otherwise represent the sites of anodic attack; thus the steel receives no cathodic protection. This is, however, only true of high temperatures, since the crystal form of the basic carbonate depends greatly on the conditions. Earlier work with Feitknecht[1] had shown that the deposit formed in cold water contains a defect structure, shown up by X-rays; this faulty deposit does not prevent attack on the zinc and consequently there is cathodic protection of the steel.

The crystalline form depends not only on temperature, but also on the pH value of the solution. When precipitation is produced by a CO_3^{2-}/HCO_3^- buffer, thin rectangular platelets are formed; when the precipitant is Na_2SO_3, the platelets are small and irregular; at still higher pH values, the structure is fibrous, and there is departure from the stoichiometric composition. The fibrous material is unlikely to provide a continuous protective film over the zinc, and will therefore not prevent cathodic protection to the iron. Such observations seem to go far to explain why the polarity of the zinc–iron couple depends on temperature and on the composition of the water; but it cannot be stated that there is universal agreement on the matter.

The Iron–Copper Couple.   Bimetallic corrosion involving copper can be important in hot-water systems. As pointed out elsewhere (1960 book, p. 205), the combination of copper tubes and galvanized iron tanks or cisterns can cause trouble, without electrical contact, if the water is one capable of dissolving small traces of copper and redepositing it in metallic form on the zinc or iron. Different points of view have been conveniently brought together at a symposium,[2] organized by the British Joint Corrosion Group in 1970. The deposit of metallic copper may provoke anodic attack on the steel, as a result of the bimetallic cell iron–copper. Perhaps of greater importance is the fact that copper will catalyse the Schikorr reaction, converting dissolved ferrous hydroxide to magnetite and hydrogen (1960 book, p. 439); the magnetite may sometimes block the pump used in a domestic central-heating system, whilst the hydrogen collects in the radiator and needs to be blown off at intervals to prevent gas-locking. If a small amount of benzotriazole is added to the water, it will lock up the copper as a complex, thus diminishing redeposition, and also preventing catalytic enhancement of the Schikorr reaction; it will probably also act as an inhibitor towards the ordinary attack upon the iron.

Probably the increased reports of trouble in recent years is connected with the use of thin material. Instead of the old-fashioned cast iron radiators and relatively thick galvanized water tanks, we depend today largely on welded sheet.

1.   R. Grauer and W. Feitknecht, *Corr. Sci.* 1967, **7**, 629; esp. p. 632.
2.   *Brit. Corr. J.* 1971, **6**, 23–48, esp. J. B. Cotton and W. R. Jacob, p. 42.

Blanchard[1] describes laboratory experiments designed to reproduce practical experience of galvanized tubes in hot-water systems with special reference to the effect of copper. His results confirm existing beliefs, but they may not apply to all types of water. At ordinary temperatures galvanized iron remained unchanged, apart from a little 'white rust'; at elevated temperatures the familiar type of rust appeared; pustules of rust were formed at 60°C, whilst at 80°C protection practically ceased to exist. It is stated that the temperature should not exceed 60°C if virulent attack is to be avoided. The inclusion of a single tube of copper in a series of tubes constituting a circulating system caused pustules of rust under conditions where in the absence of copper there was little or no corrosion. Although a copper tube produces that effect, the presence of copper up to 1% in the galvanized coat can be tolerated.

It has generally been supposed that the copper is deposited in metallic form on the surface and serves to promote the cathodic reaction. The view has been expressed, however, that it enters the layer of zinc oxide, conferring on it electronic conductivity; in that case the oxide-covered zinc may act as a cathode towards steel, much in the same way as steel covered with mill-scale, which is mainly magnetite, can act as a cathode towards bare steel (at high temperatures, zinc oxide can become a fairly good electronic conductor even in the absence of copper). Such a possibility has been discussed by Pfeil,[2] who quotes earlier work by Kaesche. Pfeil provides an interesting diagram showing the fluctuation of current passing between zinc and iron after the addition of $CuSO_4$ to the water; the unsteadiness continues for 20 days, with the current passing sometimes in one direction and sometimes in another. It should be noted, however, that in this experiment N/100 $NaHCO_3$ was used as the 'water'.

The Aluminium–Iron Couple.   Considerable interest is taken in the aluminium–iron couple. This combination has caused serious damage to a new bridge near the coast in South Africa; the case is described by Copenhagen.[3] Aluminium alloy balustrades were used on a concrete bridge and came into contact with steel reinforcements at their entry points. The cell set up produced a foam-like corrosion product, and in some cases cracks due to pressure developed. All the balustrades had to be replaced before the formal opening of the bridge. The phenomenon was reproduced in the laboratory.

It has been found that small amounts of tin can enhance the current generated by the couple aluminium–iron. A study of the situation by Pryor[4] shows that the increase is only observed if tin is present in the oxide phase and this will only be the case when it was originally present in solid solution in the original metallic aluminium. The presence of minor constituents influences the chance of finding tin in solid solution. Bi, Zr, Mg and Ag, which expand the Al lattice, stabilize the AlSn solid solution; these elements

1.   F. Blanchard, *Corrosion (Paris)* 1971, **19**, (1) 29.
2.   E. Pfeil, *Deutsch. Forschungsgesellschaft für Blechverarbeitung* 1969, **20**, Nr. 2, p. 23.
3.   W. J. Copenhagen and J. A. Costello, *Mat. Prot. Perf.* 1970, **9**, (9) 31.
4.   D. S. Keir, M. J. Pryor and P. R. Sperry, *J. electrochem. Soc.* 1969, **116**, 319.

maintain or even increase the danger of a high galvanic current flowing between Al and steel. In contrast, Si, Zn, Cu and Mn contract the lattice and decrease the danger of a galvanic current. When Si and Fe are both present in aluminium (each is a common impurity), the Si-content exerts a greater influence on the current than the Fe-content.

Precautions to be taken at contacts between light alloys and steels on ships are reviewed in a recent report.[1]

An interesting case of bimetallic corrosion is described by Lee Craig.[2] An aluminium gas-pipe, placed in a steel culvert, developed leaks, which were found to be due to the fact that the contractor had used steel bolts to keep the pipe centred; the matter was rectified, and no trouble has occurred over a period of 16 years.

The Cadmium–Iron Couple. The galvanic relationship between cadmium and iron has some scientific interest. Judged on the basis of the normal electrode potentials, cadmium should be more noble than iron and should act as a cathode. In fact, it often acts as an anode. This is because the polarization of the system

$$Fe \leftrightarrows Fe^{2+} + 2e$$

is much greater than that of the corresponding system

$$Cd \leftrightarrows Cd^{2+} + 2e$$

which can behave as a nearly reversible electrode. Iron is a 'transition element', and polarization on such metals is high. If cadmium is immersed in a salt solution containing dissolved oxygen, the mixed potential set up when anodic attack on the metal is just balanced by cathodic reduction of oxygen will be more negative than the mixed potential of iron immersed in the same liquid; when the two are joined by a wire, a current will flow with the cadmium acting as anode.

The matter has been investigated by Yahalom.[3] He states that cadmium commonly acts as an anode towards steel, and that a cadmium coat will generally protect steel at a gap, as observed by Hoar in early work.[4] In some cases reversal occurs. In Yahalom's experiments there was no reversal in pure NaCl or in pure NaNO$_3$, but it was noticed in NaHCO$_3$ and even in a mixture of NaHCO$_3$ and NaNO$_3$. The reversal may be due to the building up of anodic insulating on the cadmium, and can be related to the buffering action of NaHCO$_3$. It occurs more quickly in concentrated NaHCO$_3$ solution than in dilute solution.

Crevice Corrosion

General. The usual explanation offered for intense corrosion observed in a crevice is that at such a place there is little replenishment of oxygen and inhibitor. This is probably sound reasoning, but another feature of a crevice,

1. 'Recommended Practice for the Protection and Painting of Ships' (British Ship Res. Assoc.), 1973; esp. pp. 175–9.
2. H. Lee Craig, *Williamsburg Conf.* 1971, p. 600; esp. p. 636.
3. L. Zanker and J. Yaholom, *Corr. Sci.* 1969, **9**, 157.
4. T. P. Hoar, *J. electrodep. Soc.* 1938, **14**, 33; esp. Fig. 4(a), p. 39.

the small volume of liquid contained in it, may influence the result. At any anodic point some acidity is likely to be produced owing to the hydrolysis of the salt formed by the anodic reaction. In general, however, this will become dispersed in a large quantity of liquid, and the drop of pH may not be great. In a crevice the amount of liquid available for dispersion is small, and the change in pH may be much greater; the acidity will stimulate attack. Vermilyea and Tedmon[1] have put forward a theory for the concentration and potential variation in a crevice; predictions based on it are in reasonable accord with experimental results obtained on a model.

A special type of crevice attack deserves attention. This is the meniscus formed at the water-line on a partly immersed vertical surface, In a simple NaCl solution, the meniscus zone is cathodic and escapes attack, because here there is ample oxygen replenishment, giving alkali, which is an inhibitor. If, however, the solution already contains an inhibitor (e.g. if it is a $NaCl–Na_2CO_3$ mixture), the lower part of the surface is well protected, and the meniscus crevice, which is unfavourably placed for the renewal of inhibitor, becomes anodic. Since the anodic area is small, perforation of a thin sheet specimen may occur in a time which produces no serious loss of thickness when a plain NaCl solution is used. The matter was investigated by Peers[2] in 1953, and is recalled here for the sake of completeness.

An interesting case has been recorded by Schikorr,[3] in which both crevice corrosion and a bimetallic effect are involved. He found that when a bare iron wire and a galvanized iron wire are twisted together, the corrosion is less than when two bare iron wires are twisted together, The zinc is more corroded than the iron, which receives cathodic protection. Two galvanized wires twisted together corrode still less.

Crevice corrosion set up at points of contact with a glass support have been studied by Butler.[4] This is a phenomenon commonly met with in experimental work, and may pose problems for the designer. It was found that the occurrence was very erratic; this is not surprising, since corrosion will only be set up if a sensitive point happens to be included in the small area rendered inaccessible to oxygen or inhibitor by the geometry. The matter received some study in early work by Mears.[5]

A study of corrosion at crevices in iron due to McCafferty[6] deserves study. It adopts as starting point the important work of Rosenfeld and Marshakov[7] which, by a regrettable oversight, received no notice in the 1968 volume of this book. Rosenfeld (whose results are fully confirmed by McCafferty) had shown that crevice corrosion usually proceeds in two stages; first the metal within the crevice suffers anodic attack as a result of the restricted oxygen supply, the external (aerated) surface forming a large cathode; later, cor-

1.   D. A. Vermilyea and C. S. Tedmon, *J. electrochem. Soc.* 1970, **117**, 437.
2.   A. M. Peers and U. R. Evans, *J. chem. Soc.* 1953, p. 1093.
3.   G. Schikorr, *Metall* 1967, **21**, 804.
4.   G. Butler and J. G. Beynon, *Corr. Sci.* 1967, **7**, 385; esp. Fig. 5, p. 392.
5.   R. B. Mears and U. R. Evans, *Trans. Faraday Soc.* 1934, **30**, 417.
6.   E. McCafferty, *J. electrochem. Soc.* 1974, **121**, 1007; also NRL Report **7781** (1974) Naval Research Laboratory, Washington.
7.   I. L. Rosenfeld and I. K. Marshakov, *Corrosion (Houston)* 1964, **20**, 115t.

rosion products accumulate within the crevice, and give rise to acidity. McCafferty has dealt with two cases. First, he has studied isolated crevices without an external (aerated) surface; here he found that the anodic corrosion rate was determined by the limiting current for oxygen reduction, and was less than that measured on an open sample; the limiting cathodic current for oxygen reduction (and hence the corrosion rate) was independent of pH and Cl^- concentration. Then he studied crevices connected electrically to an external (aerated) area of iron (or platinum). Here the internal current increased with Cl^- concentration, and with the area of the cathode (the aerated surface), but was independent of crevice dimensions. His experiments were particularly concerned with solutions containing both CrO_4^{2-} and Cl^-; at constant Cl^- the corrosion current decreased with increasing CrO_4^{2-} concentration; if the CrO_4^{2-} was insufficient for protection, the corrosion current increased with increasing Cl^- concentration in the early stages, but later became independent of it. Attack within the crevice was not uniform, being most severe near the crevice mouth.

The paper by Rosenfeld and Marshakov, mentioned above, deals with two other types of localized attack. The attack set up at contact with a dielectric in an acid solution has long been the subject of discussion. It cannot be attributed to differential aeration, since it can be produced in absence of oxygen. Rosenfeld relates it to the fact that, at places where acid will not easily be replenished, the pH will rise; now, within a certain range of pH, corrosion rate *increases* as pH rises—a fact brought out in earlier writing by Rosenfeld and also by Larkin and Iofa;[1] it is probably connected with the increased adsorption of OH^- (see this book p. 153). The rise of corrosion rate with pH value has been confirmed by other investigators, and the intense localized attack at crevices formed by contact with dielectrics thus receives an explanation.

As regards the water-line attack in a solution containing Cl^- and a quantity of inhibitor insufficient to prevent corrosion completely, Rosenfeld rightly attributes this to the failure of the inhibitor ions to be replenished at the crevice constituted by the meniscus. This is, in fact, the conclusion reached by the study of water-line attack carried out in 1953 by Peers and the author;[2] their paper states that 'The results support Schikorr's suggestion that water-line attack is due to slow replenishment of inhibitor at the meniscus. It is thus a form of crevice corrosion—analogous to that produced at contact with a glass rod.' Rosenfeld, in his paper of 1964 gives a reference to Peers' work, but seems to suggest that a different interpretation of water-line attack was favoured at Cambridge. Actually there is little difference of opinion on this matter.

One of the dangers of crevice corrosion is that it may not become visible to the eye until it has reached an advanced stage. Thus an electrochemical detection method—capable of giving a warning signal on a remote instrument —is welcome. The method chosen may depend on a change in the polarization resistance, dE/di, or simply on the corrosion potential. The subject is dis-

1. L. A. Larkin and Z. A. Iofa, *J. Physicheskoi-Kimii* 1960, **34**, 1470.
2. A. M. Peers and U. R. Evans, *J. Chem. Soc.* 1953, p. 1093.

cussed by Jones and Greene,[1] who recommend a combination of methods in seeking evidence of the presence or absence of localized corrosion.

A cell for testing crevice corrosion has been described by Atterly.[2] A small electrode representing the anode is placed in a crevice, and outside the crevice is arranged a large cathode; the current measured represents the corrosion rate. It is stated that correlation with field experiments in real tanks has been established. It was found that the current (and therefore the corrosion rate) is sometimes increased by the addition of an inhibitor to the water, which prevents the establishment of local cells due to minor anodic points on the mainly cathodic area. It is concluded that 'some common inhibitors may accelerate the pitting process and thus be dangerous'.

Crevice corrosion is particularly likely to occur on materials which normally develop protective films except at points where oxygen cannot be replenished. These materials will be discussed in turn.

Stainless Steel. Crevice corrosion is perhaps the most serious danger attending the use of stainless steel. Wilde[3] considers it more serious than

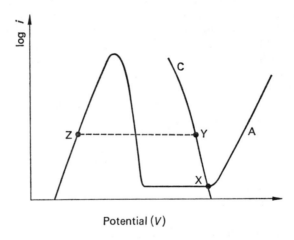

Fig. 8   Crevice corrosion on stainless steel. On the freely exposed part of the surface the situation is represented by point X, the intersection of the anodic and cathodic curves (A and C). If a crevice exists, an iR drop is needed, represented by YZ; although the large cathodic area outside the crevice remains in the passive region (at Y), the crevice is in the active region (at Z) (T. P. Hoar).

pitting. The electrochemical mechanism involved has been discussed by Hoar.[4] The anodic (A) and cathodic (C) polarization curves are shown in Fig. 8. On a freely exposed surface the intersection of the two curves is placed at

1.   D. A. Jones and N. D. Greene, *Corrosion (Houston)* 1969, **25**, 367.
2.   P. Atterly, *Ferrara Symp.* 1970, p. 421.
3.   B. E. Wilde, *Williamsburg Conf.* 1971, p. 342; esp. p. 350.
4.   T. P. Hoar, Athens internat. Cong. on Marine Corrosion and Fouling, 1968.

X, situated in the passive section, and there is little or no attack. If there is a crevice inaccessible to oxygen, the resistance (R) of the path between the anode in the crevice and face outside will demand a potential drop iR, where i is the current flowing. Thus, although the cathodic potential on the face may remain in the passive region, the anodic potential in the crevice may now be at Z, in the active region (YZ represents the iR drop). Corrosion therefore takes place. Hoar points out that on titanium the cathodic reaction is not the production of oxygen but the liberation of hydrogen, and this can take place within the crevice; thus no ohmic drop is needed, and the potential is situated within the passive range; consequently titanium and its alloys are relatively immune from crevice corrosion in most situations.

Lennox and Peterson[1] report that cathodic protection 'can be effective in not only preventing crevice corrosion of stainless steel but also stopping it once it has started'. This statement is based on laboratory experiments with NaCl solution and depends on observation of potential and pH changes. If under typical marine conditions, crevice corrosion can be arrested by zinc protectors, the method has obvious attractions.

The importance of geometrical factors in the crevice corrosion of stainless steel is discussed in an Italian paper.[2] The effect of oxygen and oxygen carriers, such as iron salts, at the cathodic areas outside the crevice can greatly influence results. Russian work[3] has indicated that the pH value within a crevice can be as low as 1·5, even when the liquid outside is NaCl solution at pH 4. These low figures at the anodic area within the crevice may be compared with those obtained in stress corrosion cracks (this volume, p. 327) and in pits (this volume, p. 72). It was found that the presence of molybdenum in stainless steel increases the resistance to crevice corrosion.

The design of stainless steel screw joints has been discussed by Bauer,[4] with special reference to the danger which can arise from the introduction of spring components.

The occurence of pitting and crevice corrosion on stainless steel exposed to sea-water has been discussed by Defranoux.[5] He proposes an electrochemical screening test for predicting the influence of alloying constituents and heat-treatment, including the effects of welding. Work by Peterson[6] shows that flowing sea-water causes much more severe attack than motionless sea-water, evidently owing to the better renewal of oxygen at the cathodic surface outside the cavity in the first case. Here also a low pH value (below 2·0) was found at the anodic surface, but this, it is thought, cannot arise from the hydrolysis of a ferrous salt; no Fe^{3+} was found, and the acidity is attributed to the hydrolysis of chromic salts. It is stated that cathodic protection, using

1. T. J. Lennox and M. H. Peterson, *Williamsburg Conf.* 1971, p. 173.
2. G. Bombara, S. Sinigaglia and G. Taccani, *Electrochimica Metallorum* 1968, **3**, (1) 81.
3. 'Corrosion and Protection of Structural Alloys': English translation by A. D. Mercer 1968, (National Lending Library, Boston Spa), Part I, p. 36.
4. C. O. Bauer, *Werkstoffe u. Korr.* 1972, **21**, 463.
5. J. Defranoux, *Corrosion (Paris)* 1970, **18**, 359.
6. T. J. Lennox, R. E. Groover and M. H. Peterson, *Mat. Prot.* 1968, **8**, (5) 41; 1970, **9**, (1) 23.

zinc or steel sacrificial anodes, is effective. Vreeland and Bedford[1] show that different types of stainless steel respond in different degrees to cathodic protection.

Aluminium. Certain aluminium alloys are susceptible to corrosion in crevices, which sometimes takes the form of exfoliation;[2] the exposed aluminium corrodes along laminar planes resulting in loss of metal in flakes or chunks. Exfoliation is often difficult to detect, as it proceeds below the surface or under paint coats. The danger can be avoided by filling crevices with suitable sealants; zinc chromate paints provide only temporary protection, since the chromate is in due course leached out.

Titanium. The extensive adoption of titanium in chemical plants has been largely due to its resistance to corrosion, and it is disappointing to find that (despite what has been stated above) crevice corrosion can occur in certain situations, notably in presence of chloride. The matter has been discussed by Griess,[3] who attributes it to differential aeration—an explanation favoured also by Cotton.[4] Crevice corrosion has been observed on heat-exchangers handling hot glycerol containing NaCl. An alloy containing 0·15% Pd was found to be resistant up to 140°C. The phenomenon has been reproduced in laboratory experiments which confirm the good effect of palladium. Japanese work[5] has shown that the pH value in a crevice on titanium can be as low as 0·7.

Crevice corrosion on titanium and its alloys has received authoritative discussion in papers presented at Williamsburg.[6]

Corrosion at Welds and Brazed Joints

General. Wranglén[7] has contributed some advice about the choice of weld type. 'From the corrosion point of view, butt welds are preferred to fillet welds . . . Intermittent welds are unsuitable both from a mechanical and a corrosion point of view . . . Welds should always be even and well shaped, without pockets and crevices. The welds should always be cleaned so that slag and spatter are removed.'

Ferritic Stainless Steel. Booker[8] has studied the corrosion of brazed stainless steel joints in domestic tap-water. Joints, produced with the standard quaternary Ag–Cu–Zn–Cd alloy, sometimes fail after a relatively short time—apparently by the preferential dissolution of a Cu–Zn–Fe phase which is present along the interface. In laboratory experiments it was found easy

1.   D. C. Vreeland and G. T. Bedford, *Mat. Prot.* 1970, **9**, (8) 31.
2.   *Metal. Progress* 1969, Sept. p. 72.
3.   J. C. Griess, *Corrosion (Houston)* 1968, **24**, 96.
4.   J. B. Cotton, *Brit. Corr. J.* 1972, **7**, 59.
5.   K. Shiobara and S. Morioka, *Tokyo Cong. Ext. Abs.* 1971, p. 133.
6.   *Williamsburg Conf.* 1971. Papers by A. Cerquetti, F. Mazza and M. Vigano (p. 661), L. W. Gleekman (p. 669), J. B. Cotton (p. 676).
7   G. Wranglén, 'Introduction to Corrosion and Protection of Metals' (Institut för Metallskydd), 1972, p. 150.
8.   R. A. Jarman, J. W. Myles and C. J. L. Booker, *Brit. Corr. J.* 1973, **8**, 33.

to separate the braze fillers from the stainless steel after 30 to 60 days. It is believed that the water gains access to the interface by dissolution of one of the phases making up the lamellar eutectic structure of the braze filler, rather than by crevice corrosion. Further investigation is promised.

Austenitic Stainless Steel. Knife-edge attack along the junction between weld metal and the base metal is still a problem, despite the experience of thirty years and the introduction of material with a much lower carbon content than was available some years ago. The causes are discussed by Čihal.[1] The matter is complicated, but one of the main causes of knife-edge corrosion of stabilized stainless steels has been shown to be the production of dendritic titanium carbides at grain-boundaries on superheating; these are eaten away by nitric acid.

German work[2] has indicated that the addition of 0·15% nitrogen to a 17/13/4·5 Cr/Ni/Mo steel can delay the precipitation of intermetallic phases and chromium carbides, so that plate up to 30 mm thickness can be welded without precipitation or cracking; the weld metal has the same ductility and resistance to corrosion as the base material. The steel is stated to be resistant to pitting, crevice corrosion and stress corrosion cracking, and there have been successful applications in the textile industry, shipbuilding, chemical industry and in plants for disposal of radioactive waste.

Herbsleb[3] states that the practice of grinding weld seams on Cr–Ni steel increases susceptibility to stress corrosion cracking, but shows that this danger can be met by a final pickling operation; he makes the further suggestion that possibly sand-blasting, which would produce compressive stress, might be effective.

Some disastrous cases of weld trouble in stainless steel have been described by Class.[4] He states that certain materials which behave admirably in the non-welded condition break down catastrophically in welding. He attributes the failure to avoid trouble largely to lack of cooperation between the various 'disciplines' which should be able to help one another. His paper describes many examples from practice, and discusses the alloys of aluminium, copper and nickel, as well as stainless steel.

A paper by Bianchi and his colleagues[5] on the pitting of austenitic stainless steels contains some important practical advice about the pickling of welded parts. They point out that a stainless steel heated at about 300°C develops a pale yellow covering which is 'hardly preceptible' to the eye, but which renders the material highly susceptible to pitting in the presence of Cl^-. Thus the pickling operations usually applied to remove oxide from definitely coloured regions must be extended to areas carrying this almost invisible film.

1. V. Čihal and J. Ježek, *Brit. Corr. J.* 1972, **7**, 76. V. Čihal, *Williamsburg Conf.* 1971, p. 502.
2. A. Bäumel, E. M. Horn and G. Siebers, *Werkstoffe u. Korr.* 1972, **23**, 923.
3. G. Herbsleb, *Werkstoffe u. Korr.* 1973, **24**, 867.
4. I. Class, *Werkstoffe u. Korr.* 1970, **21**, 559.
5. G. Bianchi, A. Cerquetti, F. Mazza and S. Torchio, *Williamsburg Conf.* 1971, p. 399; esp. p. 408.

A potentiostatic study of corrosion at welds in austenitic stainless steel has been carried out at the Welding Institute, Abington.[1] Trouble is most likely to occur under marginally or moderately oxidizing conditions. The composition and microstructure of the weld metal determine behaviour, and the welding process, *as such*, is found to have no significant effect. The carbon and molybdenum contents of the alloy influence results, whereas Cr, Ni, Nb and Ti have no significant effect on corrosion behaviour. Carbon above 0·08% can produce susceptibility to grain-boundary attack, particularly in multi-pass welds. Molybdenum, on the other hand, can prevent intergranular cracking, probably by retarding the precipitation of Cr-rich carbides, especially below about 750°C; it also reduces the peak CD at the active loop, which is beneficial. Mo does, however, lower the breakdown potential, and induces susceptibility to attack at the fusion boundary over a wide range of potential, including conditions where the rest of the weld would be expected to be passive; this occurs in moderately oxidizing environments. Preferential attack on ferrite is important, but occurs only under marginally oxidizing conditions, where it seems that the only way to avoid attack is the use of ferrite-free weld metal.

A research on the effect of weld microstructures on hydrogen induced cracking in transformable steels deserves mention.[2] Hydrogen embrittlement manifests itself in the form of cracking in the heat-affected zone and also in the weld metal; it depends, amongst other things, on the acicular microstructure formed during the decomposition of austenite. In the heat-affected zone twinned martensite is the most embrittled microstructure. The embrittlement is not simply the result of great hardness, but is related to twinning; indeed in a low-carbon steel, drastic embrittlement has been met with without significant increase in hardness. Hydrogen embrittlement in steel does not result in cracking of a specific morphology, but exploits the crack nucleation mechanism which is easiest in a given microstructure.

Finnish work[3] has shown that corrosion can occur near the heat-affected zone of welded mild steel, anodically polarized in artificial sea-water under potentiostatically controlled conditions. The corrosion rate of that zone is found to increase with the Mn-content of the steel. These results have importance for ships operating in icy waters, where the ice may remove the protective paint.

Brazed Joints.   Dezincification at brazed joints has been discussed by Campbell.[4] In the days when brazing materials containing zinc were in common use, severe dezincification was often met with at joints, even in waters where duplex ($\alpha\beta$) brass fittings gave no trouble; the trouble was, as usual, due to the combination of a large copper cathodic area and a small anodic area. It is now considered advisable to avoid brazing materials containing zinc; alloys containing copper, silver and phosphorus are available.

1.   T. G. Gooch, J. Honeycombe and P. Walker, *Brit. Corr. J.* 1971, **6**, 148.
2.   T. Boniszewski and F. Watkinson, *Met. Mat.* 1973, **7**, 90, 145 (2 papers).
3.   A. Saarinen and K. Onnela, *Corr. Sci.* 1970, **10**, 809.
4.   H. S. Campbell, *Water Treatment and Examination* 1971, **20**, 11; esp. p. 20.

Further References

Books. A book on welding for engineers should receive attention. It contains, for instance, information about modern brazing alloys.[1]

Papers. 'Application of electro-chemical hysteresis methods in assessing the corrosion characteristics of alloys' (with a section on 'evaluation of the influence of crevices').[2]

'Corrosion at welds in chemical apparatus of stainless steel.[3]

'The electrochemical behaviour of weld seams'.[4]

'Prevention of crevice corrosion by polar inhibitors'.[5]

'Resistance of welds to stress corrosion cracking and corrosion fatigue'.[6]

'Stress corrosion cracking of welds'.[7]

'Behaviour of Couples of Aluminium and Plastics reinforced with Carbon Fibre in Aqueous Salt Solutions'.[8]

'Galvanic corrosion of bare and coated Al alloys coupled to stainless steel 304 or Ti–6 Al–4'.[9]

'Galvanic Corrosion of Mild Steel when coupled to other materials'.[10]

'Korrosionsprognose für den Bereich der Schweisshant'.[11]

'Corrosion Testing of Weldments'.[12]

'Effects of Ni and Mo on Galvanic Currents between austenitic stainless steels'.[13]

'Inexpensive zero-resistance ammeter for galvanic studies'.[14]

'Polarity Reversal in the Zinc–Mild Steel Couple'.[15]

'Use of microelectrodes in the study of stress corrosion in aged Al–7·2 wt % Mg and Al–4·4 wt % Cu Alloys'.[16]

'Bimetallic corrosion effects on mild steel in natural environments'.[17]

'Localized Corrosion in Mild Steel'.[18]

'Crevice Corrosion of Metals and Alloys'.[19]

'Pitting and Crevice Corrosion of carbon steels in inhibited acid solutions'.[20]

1. H. Udin, E. R. Funk and J. Wulff, 'Welding for Engineers' (Wiley), 1954, pp. 306–9.
2. E. D. Verink, *Tokyo Cong. Ext. Abs.* 1971, p. 437.
3. J. Bolan, *Werkstoffe u. Korr.* 1968, **19**, 35.
4. H. Lajain, *Werkstoffe u. Korr.* 1972, **23**, 537.
5. T. K. Grover and B. Sanyal, *Labdev. J. Sci. Tech.* 1970, 8A, 99.
6. H. Gräfen, *Werkstoffe u. Korr.* 1972, **23**, 527.
7. T. G. Gooch and D. Willingham, *Met. Mat.* 1969, **3**, 176.
8. A. R. G. Brown and D. E. Coomber, *Brit. Corr. J.* 1972, **7**, 232.
9. F. Mansfeld and E. P. Parry, *Corr. Sci.* 1973, **13**, 605.
10. K. E. Johnson and J. S. Abbott, British Steel Corporation Report CEL/CH/23/73.
11. A. Bäumel, *Werkstoffe u. Korr.* 1972, **23**, 546.
12. M. Henthorne, *Corrosion (Houston)* 1974, **30**, 39.
13. J. J. Eckenrod and H. E. Trout, *Corrosion (Houston)* 1974, **30**, 24.
14. J. Wolstenholme, *Brit. Corr. J.* 1974, **9**, 116.
15. J. A. von Fraunhofer and A. T. Lubinski, *Corr. Sci.* 1974, **14**, 225.
16. P. Doig and J. W. Edington, *Brit. Corr. J.* 1974, **9**, 88.
17. K. E. Johnson and J. S. Abbott, *Brit. Corr. J.* 1974, **9**, 171.
18. G. R. Wallwork and B. Harris, *Williamsburg Conf.* 1971, p. 292.
19. I. L. Rosenfeld, *Williamsburg Conf.* 1971, p. 373.
20. G. Davolio and E. Soragni, *Williamsburg Conf.* 1971, p. 270.

CHAPTER VII

Anodic Corrosion and Passivation

Preamble

Picture presented in the 1960 and 1968 volumes. In the type of electrochemical corrosion discussed in chapter IV, the current flowing, and consequently the corrosion rate, could not exceed a certain value determined by the rate of renewal of oxygen at the cathode; thus, under any given conditions, there would be a limit to the rate of destruction of metal. If, however, current is provided from an external source, no limit would at first sight seem to be imposed. Fortunately, however, this idea of very high corrosion rates due to an external current is unduly pessimistic. If the current is high, the anodic potential will rise into a range where other reactions besides corrosion become possible; in most cases, the metal will cover itself with an oxide-film, becoming passive; at still higher potentials, the current may be expended largely in the production of oxygen gas. Evidently the subject of passivity produced by anodic action is an important one, and much experimental work has been performed; sometimes disagreement has arisen regarding the interpretation. Some authorities have attributed passivity to a monolayer of oxygen or possibly oxide; others think that, at least in some places, a considerable thickness of three-dimensional oxide is necessary to stop attack. The difference of opinion is largely due to the use of the word 'passivity' in different senses. There is little doubt that when the potential, raised by the current, enters the passive range, there may be only a monolayer or less of oxygen on the surface. To obtain a lasting change in properties (and this is what Faraday and Schönbein had in mind when they first used the word 'passivity'), a greater thickness is needed.

Much of the early experimental work on the subject—although accurately carried out—was difficult of interpretation, since the circuit used was neither exactly galvanostatic nor exactly potentiostatic. The improvement of instrumentation has enabled the modern worker to obtain results which are easier to reproduce and to interpret.

Anodic treatment can be used in two ways in the struggle against corrosion. Metals like aluminium, which form an oxide that is a poor electronic conductor, can be covered with a protective film by a process known as 'anodization'; the thickness of film is roughly proportional to the EMF employed, and the film remains after the anodic treatment has ceased; the metal is now more resistant to corrosion than if it only carried the ordinary thin air-formed film. In contrast, metals like iron, which form oxides with relatively high electronic conductivity, will not develop coats of any considerable thickness; after the oxide-film has appeared, the current, at sufficiently high potentials, will be devoted to the evolution of oxygen gas. However, in such cases, corrosion can often be averted by holding the potential at a value in the passive range. This process is known as 'anodic protection', and in suitable cases has been applied with success. The method can, however, have its dangers, since if the potential is not raised sufficiently, the anodic current flowing will accelerate corrosion instead of stopping it. In general, this type of protection can only be depended on whilst the current is flowing; the passivity may remain for a time after the current has ceased, but is likely to break down sooner or later.

Anodic corrosion—as opposed to anodic protection—can sometimes be usefully employed. Anodic polishing is used to produce an extremely bright surface without mechanical abrasion; if anodic treatment is carried out in a suitably chosen 'polishing bath', appreciable current will continue to flow, when the potential is held in the passive range. Even in an ordinary bath, a small amount of current continues to flow when the potential is in that range, and this is accompanied by a certain very slow attack on the metal. In a polishing bath, the current flowing may be much higher; the attack on the metal will be more rapid, but will be concentrated preferentially on the micro-prominences, so that the surface becomes smooth on an atomic scale. The process is extensively used today. Another practical application of anodic corrosion is the process known as electrocutting.

Developments during the period 1968–1975 and Plan of the new Chapter. One of the features of the period under review has been the development of instruments which allow curves relating current and potential to be obtained under conditions which are strictly potentiostatic or strictly galvanostatic. The classical work of W. J. Müller suffered in value to some extent owing to the fact that conditions were not strictly galvanostatic, although closer to a galvanostatic than to a potentiostatic situation. Perhaps for that reason his work, which provided a wealth of information, has today been largely forgotten; a summary of part of it is reproduced as an appendix to the present chapter, since it is still useful in solving the problems of today. A more recent development has been the use of ellipsometric methods simultaneously with electrical measurements during a polarization research; this has led to an improved understanding of the situation.

The new chapter opens with a general discussion of potentiostatic and galvanostatic procedures, after which attention is given to the behaviour of individual metals. Here again, for convenience, a start is made with iron, followed by iron–chromium alloys, including stainless steel; other metals are then considered. A discussion of the mechanism of anodic behaviour follows, introducing Müller's work on salt passivity. The industrial applications of the principles, including anodic protection, electrocutting, electromachining and electropolishing, receive brief attention. The chapter ends with a discussion of the breakdown of the passive film, including the influence of stress and of chloride additions.

Experimental Methods

Potentiostatic Procedure. Much of the information published during the period 1968–1975 has been expressed in iV curves, generally obtained by a *potentiostatic* procedure. The type of curve obtained is shown schematically in

Fig. 9. The potential may be fixed at a given value, and the current read after a steady state has been reached; then the potential is raised by a certain amount, and the current again read after the same time interval. It is generally agreed that the potential rise should be the same at each step, and that the period over which potential is held constant should be the same on each occasion. The appropriate choice will vary with the metal and solution

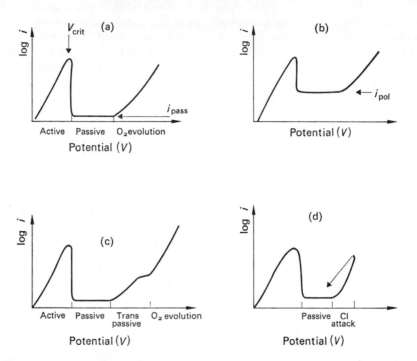

Fig. 9 Relation between anodic current and potential (schematic). (A) The current (i) increases with rising V until the critical volatage for passivation (V_{crit}) is reached, when it suddenly falls to a small value (i_{pass}); only when the voltage is sufficiently high for O_2 evolution does the current start to increase again. (B) In a 'polishing solution', the current flowing in the film-forming region (i_{pol}) is much larger. (C) Where the metal (such as Cr) forms a soluble higher oxide, a 'transpassive region' may occur over which considerable current will flow even though the potential is too low for oxygen evolution. (D) If Cl⁻ is present, the current may start to rise at a potential too low for oxygen evolution; on reducing the potential, a hysteresis effect is observed.

under study. If the time during which the potential is held constant is too short, the curve obtained will not represent the steady state; if the time interval between steps is too long, the composition of the solution will be changing (this is, of course, particularly true of the active and oxygen evaluation ranges, but there is some alteration of composition with time even when the metal is passive).

In the simplest case (Fig. 9(A)) the current (and corrosion rate) will increase with potential, until the critical potential for passivation (V_{crit}) is reached, when it falls abruptly and remains at a very low value (i_{pass}) over the whole passive range; in order to express the situation graphically, it is convenient to plot log i (not i) against V. In the passive range, the small current passing is partly devoted to very slow corrosion and partly to film formation—as pointed out by Vetter,[1] who has examined the situation both before and after the steady state has been established. When the potential has been raised sufficiently, oxygen is evolved, accounting for most of the current passing, although generally some very slight corrosion continues.

In certain solutions a relatively thick film is formed, and the current flowing in the film forming range of potential is much larger (i_{pol}); as suggested in Fig. 9(B). These are the solutions used in electropolishing—a subject to be considered later.

On chromium and some alloys containing it, there is a rise of current at a potential too low for oxygen evolution; this is connected with the formation of soluble compounds of hexavalent chromium, and the potential range involved is called the *transpassive range* (Fig. 9(C)). Vetter finds that the oxidation state of the layer formed in that range increases as the potential rises; he regards the layer as an oxide of variable composition.

If the liquid contains Cl⁻, the curve obtained (Fig. 9(D)) on iron or iron alloy generally rises sharply at a *breakdown potential* situated at a level lower than that at which oxygen would be evolved; the current passing is devoted to corrosion. On reducing the potential, there is considerable hysteresis; the current (and corrosion rate) only becomes small at a *protection potential* which is definitely lower than the breakdown potential. The significance of the protection potential has been discussed by Verink and Pourbaix.[2]

Instead of increasing the potential in steps, it is possible to arrange a slow continuous increase; here again, it must not be too fast (or the curve obtained will not represent a steady state); but it must not be too slow (or the solution may change in composition). This procedure is known as the *potentiokinetic* method. The manner in which iV curves alter with the stepping rate has been demonstrated in a paper from Rotherham.[3] This method has been criticized by Herbsleb[4] who writes: 'In several practical examples, the only reliable measuring method . . . is a series of potentiostatic point-by-point measurements at stationary potentials, combined, if necessary, with a gravimetric determination of the corrosion rate. Non-stationary, so-called potentiokinetic, measurements can cause considerable errors . . . .'

A bibliography of potentiostat design has been provided by Gabe;[5] West[6] has provided data regarding the prices of different models, which range from £50 for a 'home-made' type to £1350 for one of the more elaborate commercial

1. K. J. Vetter, *Electrochim. Acta* 1971, **16**, 1932; K. J. Vetter and F. Gom, *Werkstoffe u. Korr.* 1970, **21**, 703; W. J. Plieth and K. J. Vetter, *Ber. Bunsen Ges.* 1969, **73**, 1077.
2. E. D. Verink and M. Pourbaix, *Cebelcor Rapp. tech.* **191** (1971).
3. D. Shaw, E. Fletcher and J. S. Wilde, *Brit. Corr. J.* 1961, **4**, 249.
4. G. Herbsleb and W. Schwenk, *Corr. Sci.* 1969, **9**, 615.
5. D. R. Gabe, *Brit. Corr. J.* 1972, **7**, 236.
6. J. M. West, *Brit. Corr. J.* 1970, **5**, 65.

designs. A discussion by France[1] of the application and limitation of controlled potential in corrosion testing deserves study.

Galvanostatic Procedure. A cheaper and simpler method of obtaining iV curves depends on the use of a *galvanostatic* procedure; the current is set at a given value and the corresponding potential recorded; then the current is raised to a new level, the potential again taken, and so on. The simplest galvanostatic circuit consists of a very high EMF applied through a very high resistance; if the EMF is large compared to the potential variation within the cell, the current flowing should remain constant, but this requirement may not be easy to meet. In obtaining curves, galvanostatic methods are suitable for studying the active range, but they provide no information about the passive range where the current passing is extremely small; for the study of passivation the potentiostatic or potentiokinetic procedures are better.

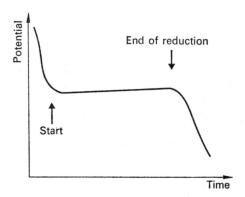

Fig. 10   Use of cathodic reduction under galvanostatic conditions to estimate the thickness of a film. The time during which the potential remains at the level corresponding to reduction, multiplied by the current flowing, gives the coulombs consumed, and from that the film thickness can be calculated by the application of Faraday's law.

A galvanostatic circuit is useful for measuring the thickness of the film after it has been formed by anodic action. It is destroyed by cathodic reduction; the duration of the period at which the potential remains at the level corresponding to the reduction process (Fig. 10) reveals the number of millicoulombs required for reduction—whence the film thickness is obtained by Faraday's law.

Much of the early work on anodic passivity was carried out by methods which were closer to galvanostatic than potentiostatic procedures but did not ensure strict constancy of current. Fig. 11 shows W. J. Müller's results for iron, nickel and chromium. In every case a current was applied sufficient to cause passivity, so that the potential rose from the initial value (shown by the left-hand curve) to the final (passive) value (shown by the right-hand curve); the change took place slowly if the current was weak, and quickly if it was strong; the times required for the change are shown by the figures placed on

1.   W. D. France, *Cebelcor Rapp. tech.* **178** (1970).

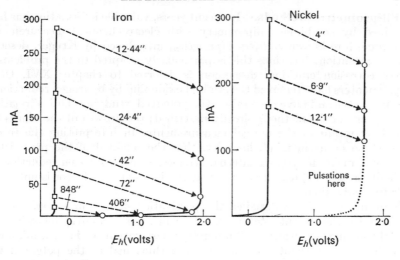

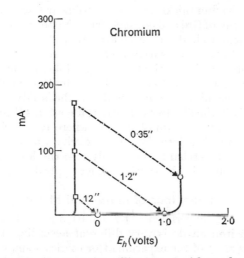

Fig. 11 Curves connecting current (in milliamperes) with anode potential (hydrogen scale) for Fe, Ni and Cr in N H_2SO_4. Active values are shown by squares and passive values by circles; transition to passive conditions is shown as broken lines, and the times needed for the transition are shown in seconds by the numbers placed on these lines (W. J. Müller).

the broken lines connecting the pair of points representing the situation in the active and the passive state respectively. Had the situation been truly galvanostatic, these broken lines would have been horizontal; in fact they slope down to the right, showing that a drop of current was accompanying the rise in potential.[1] The general situation is not, however, affected.

1. W. J. Müller, *Monatshefte für Chem.* 1927, **48**, 293. The early work is summarized and discussed by U. R. Evans, *Chem. Ind.* 1927, p. 1220.

Ellipsometric Methods. In recent years, valuable information has been obtained by combining ellipsometry with electrochemical measurements. The procedure known as *tribo-ellipsometry*, introduced by Kruger, deserves special attention; but since this is particularly adapted to the problems of stress corrosion cracking, discussion is deferred to chapter XVI. Other experimenters have removed the original oxide-film by hydrogen reduction in the apparatus and have then raised the potential, studying the $i–V$ relationship by one of the methods already described; the growth of the film is followed by means of the optical measurements. In interpreting the results it must be borne in mind, however, that the removal of the pre-existing film by abrasion may leave internal stresses, whilst hydrogen reduction (or cathodic treatment) may leave a hydrogen charge; either factor is unfavourable to passivation.

Sato and Kudo[1] have studied the passivation of iron in a borate buffer solution by the potentiostatic procedure mentioned above, following the film thickening by means of ellipsometry; passivation set in at a minimum film thickness of about 9 Å and the film thickened as the potential was increased, reaching 50 Å at the potential where oxygen evolution set in; over a certain range, the film thickness was a rectilinear function of potential. The ellipsometric studies of Ord and de Smet,[2] showing the formation of amorphous and crystalline material, also deserve attention.

Bockris[3] has studied passivation and grain growth on iron by means of transient ellipsometry; his method makes possible measurements 0·01 s after the start of the experiment. Smith and Mansfeld[4] have described an ellipsometric–elecetrochemical cell designed for the study of film growth on titanium, initially in the film free condition; it contains three cavities, and the specimen starts in the uppermost cavity, where it is etched in an H_2SO_4–HF–HNO_3 mixture, then into the middle cavity where electrolyte is forced over it to remove the etchant, and finally into the lower cavity where it is subjected to ellipsometric observation under applied potential.

Behaviour of individual Metals

Iron. Important experiments by Gibbs and Cohen[5] show that the behaviour of an iron anode is very different according as it is completely oxide-free at the start of the experiment or carries a small amount of oxide. Iron of high purity was subjected to anodic action in 0·15N Na_2SO_4 with the pH raised to 8·4 by the addition of NaOH. In certain experiments (curve C, Fig. 12) the specimen received preliminary cathodic treatment in a borate buffer (pH 8·4) at 10 $\mu A/cm^2$, with argon passing over the solution; this was thought to remove all oxide. In other experiments, after the cathodic reduction the specimen was exposed to air for 15 minutes before the solution

1.   N. Sato and K. Kudo, *Electrochim. Acta* 1971, **16**, 447.
2.   J. L. Ord and D. J. de Smet, *J. electrochem. Soc.* 1971, **118**, 206.
3.   J. O'M. Bockris, M. A. Genshaw and V. Brusic, *Symposia Faraday Soc.* 1970, **4**, 177.
4.   T. Smith and F. Mansfeld, *J. electrochem. Soc.* 1972, **119**, 663.
5.   D. B. Gibbs and M. Cohen, *J. electrochem. Soc.* 1972, **119**, 416.

was introduced into the cell. These specimens, which carried oxide-films, gave an entirely different type of curve (curve A), the current passing within the passive range being now very small. Even a small amount of oxide on the specimen was sufficient to produce curves of this general character. The authors comment that 'the passive region at high potentials, extensively studied by Freiman and Kolotyrkin, is seen to be entirely absent for electrodes which have been completely cathodically reduced before oxidation. This passive region evidently exists only if an air-formed film is initially present on the surface. . . . Even a very thin initial film . . . is able to give rise to such a region of passivity.'

The comment might be added that the trace of air-formed oxide needed to give the type of curve obtained by the Russian authors (and indeed by most

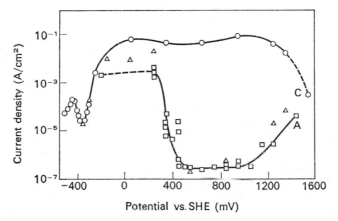

Fig. 12   Anodic polarization curves. A specimen carrying an oxide-film produced by previous air exposure gives a different curve (A) from a specimen freed from oxide by cathodic treatment (C) (D. B. Gibbs and M. Cohen).

other experimenters) probably serves to provide nuclei; if so, the curves published by these various experimenters should not be regarded as devoid of significance.

There has been considerable difference of opinion on the question of whether the film present on passivated iron consists of one layer or two. In 1962, work in Cohen's laboratory,[1] based on cathodic reduction and chemical analysis, led to the conclusion that two layers were present—a Fe_2O_3 film (with some defect structure) on the outer surface and a Fe_3O_4 layer next to the metal. This conclusion was reached when it was found that the cathodic reduction did not proceed at a single potential; the first part of the reduction proceeded at a relatively high potential and was taken to represent the reduction of the Fe_2O_3 layer to give Fe^{2+}, which passed into solution; subsequently there was reduction at a lower potential, which was taken to signify the reduction at low current efficiency of the Fe_3O_4 layer to give metal. More

1.   M. Nagayama and M. Cohen, *J. electrochem. Soc.* 1962, **109**, 781.

recent work by Sato and his colleagues[1] suggests a different interpretation. Ellipsometric measurements made during galvanostatic reduction in borate buffer solution show that the reduction of the whole film occurs in two stages. During the first stage, solid state reduction of Fe_2O_3 to a lower oxide occurs, and then in the second stage this lower oxide is reduced to metallic iron. The conclusion drawn by Sato from his observations is that the passive film consists of a single layer of Fe_2O_3 rather than a double film with two layers (Fe_2O_3 and Fe_3O_4). For films previously formed by one hour's anodic oxidation, the average composition was found to be Fe_2O_3, 1·28 FeOOH, which could also be written Fe_2O_3, 0·39 H_2O. The current efficiency of the cathodic reduction used in measuring the thickness of the film was only 80% in the first stage, whereas Cohen's results had seemed to give 100%. The loss of thickness per millicoulomb becomes much smaller during the second stage.

It would seem that these conclusions are correct for the films studied in Sato's laboratory. It is probable that films formed under some conditions really do contain two layers. Ord,[2] for instance, has presented optical evidence in favour of a two layer situation. A layer of Fe_3O_4 can be grown on the electrode whilst it is still in the active state, but the electrode becomes passive when an incomplete layer of Fe_2O_3 is formed at the interface between the Fe_3O_4 and the liquid. Further oxidation in the passive condition causes simultaneous growth of both layers; the inner one grows faster than the outer one.

Possibly both two layer and one layer views are correct in different circumstances. Asakura and Nobe,[3] studying an iron anode in N KOH, suggest that below $-1·1$ V (standard calomel electrode), there is logarithmic growth of a single oxide; above $-1·0$ V a duplex structure seems to be present.

The passivation of iron in Na_2CO_3 has been studied at Teddington.[4] There is some difference of opinion[5] as to the identification of the substances formed. The anodic behaviour of iron in a phosphate solution has been studied at Ferrara.[6] The film produced by anodic passivation in hot $Ca(NO_3)_2$ was found by work at Cambridge[7] to be Fe_3O_4; it underlies the original air-formed film, which is Fe_2O_3.

An Egyptian research[8] on the behaviour of steel in stagnant salt solutions, based on the tracing of anodic polarization curves under galvanostatic conditions, has led to the conclusion that 'the steady state potential of steel, its corrosion rate, as well as the CD needed for passivation, vary with the nature and concentration of the aggressive anions in stagnant salt solutions, in the range of concentration 10^{-6} to 10^{-2}M. The order of increasing aggressiveness is $NO_3^- < Cl^- < SO_4^{2-} < S^{2-}$. Additions of inhibitive anions ... to distilled water, at concentrations below the critical [value] for inhibition,

1. N. Sato, K. Kudo and T. Noda, *Corr. Sci.* 1970, **10**, 785.
2. F. C. Ho and J. L. Ord, *J. electrochem. Soc.* 1972, **119**, 139.
3. S. Asakura and K. Nobe, *J. electrochem. Soc.* 1971, **118**, 536.
4. J. G. N. Thomas, T. J. Burse and R. Walker, *Brit. Corr. J.* 1970, **5**, 87.
5. J. E. O. Mayne, *Brit. Corr. J.* 1970, **5**, 189.
6. F. Zucchi and G. Trabanelli, *Corr. Sci.* 1971, **11**, 141.
7. T. P. Hoar, M. Talerman and P. M. A. Sherwood, *Nature* 1972, **240**, 116.
8. V. K. Gouda and S. M. Sayed, *Brit. Corr. J.* 1973, **8**, 71.

cause a reduction in the corrosion rate of steel in the following order: benzoate < chromate < nitrate.'

There is little doubt that the composition of the anodic film depends on the potential at which it is being formed. This is shown by work in Bockris's laboratory[1] based on coulometric and ellipsometric methods. The evidence from coulometry is that at sufficiently positive potentials, the film is Fe_2O_3; in the lower range, it is likely to be $Fe(OH)_2$. The ellipsometric data suggest that in the region between 300 and 400 mV, the thickness remains roughly constant; this may be regarded as the region in which the film is being oxidized to a higher state. Near the end of the paper it is stated that 'earlier views in which the material before the break is [regarded as] adsorbed O are inconsistent with the present ellipsometric data, which indicates a phase oxide. If blocking of kink sites or adsorbed O^{2-} were the mechanism, one would not need a monolayer for passivation.'

An important research from Cohen's laboratory[2] showed that it was possible to deposit films on a platinum anode during electrolysis of an $FeSO_4$ solution. The films are mainly γ–FeOOH, but contain small amounts of Fe^{2+}, SO_4^{2-} and water; the amount of SO_4^{2-} decreases with increase of pH. The implications of these observations for the anodic passivation of iron are important. If anodic corrosion at a potential situated in the active zone has taken place before the potential is raised into the passive range, much $FeSO_4$ (or other ferrous salt) will be present in the electrolyte when passivation starts, and the passive film is likely to contain γ–FeOOH as well as Fe_2O_3. Thus anodic behaviour depends on the procedure. A slow or stepwise potential rise, allowing the production of much Fe^{2+} in the liquid, will cause a modification of the composition of the film and probably of its thickness. The incorporation of foreign anions will affect the stability of the film finally formed.

Further information about possible composition changes in the anodic film is likely to be provided by work now being conducted by Hickling,[3] who writes:

'Enormous ingenuity has been expended in seeking to explain the Flade potential of iron and the discrepancy between its value and the redox potential of iron oxides as calculated from free energy data. Such speculation has assumed that iron oxides will undergo cathodic reduction at the calculated potentials, and that a different value for the Flade potential of activation of passive iron is to be attributed to special characteristics of the passive film. Work now in progress on the electrolytic reduction of iron oxides in aqueous sulphuric acid solutions using the oxides mixed with graphite powder in the form of paste electrodes suggests that the basic assumption is incorrect and that both Fe_2O_3 and Fe_3O_4 undergo cathodic reactions at a high positive potential in the region of the Flade potential. This has led me to investigate the behaviour of a solid magnetite electrode in such solutions from the same

1. J. O'M. Bockris, M. A. Genshaw and V. Brusic, *Symposia Faraday Soc.* 1970, **4**, 177; esp. pp. 188 and 191.
2. J. L. Leibenguth and M. Cohen, *J. electrochem. Soc.* 1972, **119**, 987.
3. A. Hickling, *Electrochim. Acta* 1973, **18**, 635.

point of view, and the results are very simple and remarkable; it is almost incredible that they have not been previously reported. An electrode made by fusing ordinary Fe_3O_4 powder gives a well-defined static potential in N H_2SO_4 of 0·58 ± 0·01 V, and it behaves nearly reversibly on electrolysis at low current densities. It will sustain current densities of 10 $\mu A/cm^2$, both anodic and cathodic, with only slight displacement of potential. Thus cathodic reduction over a period of one hour occurred at 0·57 V, while the Flade potential of iron in the same solution was directly measured as 0·56 V. Furthermore variation of acid concentration displaced the magnetite potential in the same direction and by roughly the same amount as the Flade potential. It seems impossible to evade the conclusion that the two potentials are the same. . . .

'The activation of passivated iron has generally been interpreted as a reductive dissolution of a protective oxide film in which electrons set free by the ionisation of exposed metal provide the necessary cathodic current. The present results demonstrate directly that if the final stage involves the cathodic reduction of Fe_3O_4 this will occur at a potential which is the same as the Flade value. Thus there is no particular problem of the Flade potential. The real problem is why Fe_3O_4 undergoes cathodic reduction at a potential some 0·7 V more positive than the thermodynamic Fe_3O_4/FeO value and what is the nature of the reaction occurring. This problem is now being investigated. Preliminary results suggest that there is a simple initial step in the cathodic reduction of both Fe_2O_3 and Fe_3O_4 which effects a change in the relative population of Fe^{3+} and Fe^{2+} ions in an inverse spinel oxide lattice.'

Iron–Chromium Alloys, including Stainless Steel. A series of papers by Frankenthal[1] has largely cleared up our understanding of the passivity of iron–chromium alloys. Electrochemical studies, accompanied by microscopic observations, on an alloy containing 24% Cr subjected to anodic polarization, show that the film varies according to the potential applied. There are two types of film. The primary film is stable only within a few millivolts of the primary activation potential; its formation or destruction is a reversible process. If, after the primary film has been formed, the potential is lowered into the active range, the film disappears. The thickness of the primary film at the primary activation is less than the equivalent of one atom to each surface metal atom. The secondary film, which is slowly formed at more positive potentials, grows to a greater thickness than 10 Å; with increasing potential and sufficient time it becomes very resistant to reduction.

These two facts, (1) that the formation of the thin primary film is reversible, being destroyed when the potential is no longer held in the passive range, and (2) that the secondary thicker film can be brought down into the active range without immediate destruction, agree well with certain views expressed several years ago.[2] It was then suggested that, whereas a monolayer of adsorbed oxygen may put a stop to corrosion, it requires something considerably thicker (at least at the sensitive points) if the properties of the material are to

1.   R. P. Frankenthal, *J. electrochem. Soc.* 1967, **114**, 542; 1969, **116**, 580.
2.   U. R. Evans, *London Cong.* 1961, p. 3; esp. p. 8.

remain unchanged after the passivation treatment has been discontinued; in other words, for passivity in the Faraday–Schönbein sense of the word, a monolayer does not suffice.

The stability of the secondary film is responsible for a curious result observed if the thickness is estimated by measuring the number of milli-coulombs needed for its reduction. Because the film ultimately becomes in-capable of being reduced, there is an illusory reduction in the apparent thick-ness (as measured) after a long time of passivation; the true film thickness of the secondary film is greater than the apparent thickness as suggested by the coulombmetric method. It is believed that the secondary film, which is many monolayers thick, must be an oxide, because a structure consisting of separate layers of anions and cations, each carrying a charge density of several millicoulombs/cm^2, would be unstable; if such a structure were momentarily produced, interdiffusion of ions to form an oxide would quickly occur. In contrast, the thin primary film is best regarded as an adsorption product.

The growth of the passive film produced on stainless steel by anodic treat-ment has been studied by Bulman and Tseung.[1] Coulombmetric studies show that the thickness increases with the logarithm of the time. Thus the experi-mental results of Berwick (this volume, p. 208) are confirmed, but the inter-pretation of the logarithmic relationship suggested in Berwick's paper is not accepted. Indeed most of the explanations put forward for the logarithmic law, including that of Cabrera and Mott (see 1960 book, p. 825), are regarded with disfavour; Tseung considers that the best explanation of logarithmic growth on stainless steel is the place-exchange mechanism of Sato and Cohen.[2] A second research showed that the thickness obtained was in the range 10 to 50 Å; the material used initially carried an air-formed film 19 to 25 Å thick; this was removed by cathodic treatment, and then on anodic treatment a new film was produced. The optical measurements are considered to point to a two layer film similar to that formed on iron.

Uhlig[3] has found that the amount of oxygen needed to raise to the passive level the potential of Fe–Cr alloys of various Cr-contents between zero and 16·7% is independent of the chromium content, and is sufficient to represent a monolayer if a roughness factor of 8 is assumed. It is concluded that thick oxide films are not needed for primary passivity. It is found that in 3% sulphuric acid the critical CD needed for passivity falls off gradually as the Cr-content rises; it is stated that "passivity is established in a more stable form when the Cr-content reaches 12% or more'; that is the content found in stainless cutlery steels.

Frankenthal[4] emphasizes the fact that even a small fraction of a monolayer provides a high degree of protection. He considers that the transfer of ions from the metal surface into the film occurs at different rates for different types

1.   G. M. Bulman and A. C. C. Tseung, *Corr. Sci.* 1972, **12**, 415; 1973, **13**, 531 (2 papers).
2.   N. Sato and M. Cohen, *J. electrochem. Soc.* 1964, **111**, 512.
3.   H. H. Uhlig and G. E. Woodside, *J. phys. Chem.* 1953, **57**, 280.
4.   R. P. Frankenthal, *Electrochim. Acta* 1971, **16**, 1845.

of site, such as kinks, ledges and terraces; this is because an atom at each type of site has a different number of nearest neighbours, and therefore has a different activation free energy. Another important factor is that there are different energy changes connected with adsorption on the different sites. However, Genshaw and Sirohi[1] express disagreement with opinions which they attribute to Frankenthal. They argue that if passivation occurred by the blockage of kink sites, it should occur at about 0·01% coverage, since all the kink sites would then be blocked. In fact, they find passivation to set in at about 30% coverage. It should be noted that they were working with un-alloyed chromium, whereas Frankenthal was particularly interested in iron–chromium alloys. Like several other investigators, they find that the film becomes thicker as the anodic potential is raised. They consider the refractive index to support the idea that the film is Cr_2O_3. Their observations are consistent with the idea that oxide formation and metal dissolution occur simultaneously.

The dissolution kinetics of stainless steel in dilute sulphuric acid under anodic conditions is considerably influenced by cathodic pre-treatment. Wilde[2] finds that the magnitude of the effect is different for ferritic and austen-itic steels, owing to greater diffusivity of hydrogen in a body-centred cubic lattice. Pre-treatment in the (higher) trans-passive region of potential also affects behaviour.[3]

The anodic behaviour of the stainless steels depends considerably on composition. A 17/7 Cr/Ni precipitation hardening steel tested by Bruno[4] using potentiodynamic [potentiokinetic] scanning in acid media shows two peaks (Fig. 13), provided that the time interval between the primary cathodic activation process and the application of the scan is less than 6 minutes. If the interval is longer, one of them is found to disappear. The two peaks corres-pond to the active–passive transition relationship of the two phases—namely austenite and δ-ferrite.

Engells[5] has studied the effect of introducing nickel into Fe–Cr alloys. He finds that nickel reduces the current needed to produce passivity; but in the passive range the presence of nickel seems to have little effect on the curves. A dark film is observed on the surface of actively corroding specimens near the potential corresponding to the highest CD; this dissolves in the acid when the specimen becomes passive; if the CD in the passive state is not very high, the surface shows a metallic lustre. Although unalloyed iron remains passive at all potentials up to that at which oxygen is evolved, alloys containing Cr dissolve in the trans-passive region; nickel passes into solution as Ni^{2+}, iron as Fe^{3+} and chromium as Cr^{6+}.

The results of important Japanese investigaitons into the structure, com-position and thickness of the film produced on 18/8 Cr/Ni stainless steel by

1.   M. A. Genshaw and R. S. Sirohi, *J. electrochem. Soc.* 1971, **118**, 1558.
2.   C. D. Kim and B. E. Wilde, *Corr. Sci.* 1970, **10**, 735.
3.   Y. Hisamatsu and T. Yoshii, *J. Japan Inst. Mat.* 1970, **34**, 1207; 1971, **35**, 151.
4.   R. Bruno and F. Calderone, *Brit. Corr. J.* 1972, **7**, 174.
5.   K. Osozawa and H. J. Engell, *Corr. Sci.* 1966, **6**, 389.

anodic action in H_2SO_4 have been brought together and interpreted by Okamoto;[1] the methods used have included infra-red spectroscopy, radio-active tracer technique, ellipsometry and electron spectroscopy. An important finding has been that the character of the film is different according as the potential used has been below or above 0·4 V (sat. calomel electrode); in the lower range the film is relatively rich in Cr and contains much combined water (the content of which is measured by the use of tritiated water as tracer). At 0·4 V there is a remarkable increase in the Cr^- and Ni-contents (relative to iron) in experiments where the passivation had lasted 60 minutes; longer

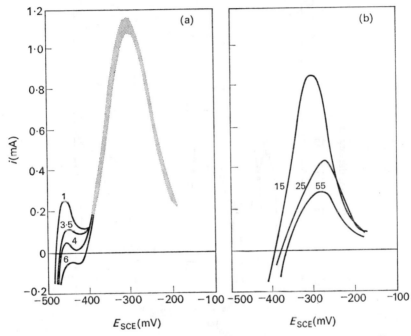

Fig. 13   Anodic polarization curves. Influence of free corrosion time (shown in minutes by numbers on curves, upon the anodic potentiodynamic curves of stainless steel (pickled surface) in 0.5M H_2SO_4 (R. Bruno and F. Calderone).

times decrease the Cr/Fe ratio. Films formed in a potential region above 0·4 V contain all three elements and the content of bound water is lower.

Perhaps surprisingly, the iV curves present no break at 0·4 V, as shown in Fig. 14, where the curve for stainless steel is compared with those for Fe, Cr and Ni. It will be noticed that the small current passing in the passive range is very low for Cr, being only about 1/100 that passing for Fe; but whereas for iron the current remains at a reasonably low level until the potential of O_2 evolution is reached, the current for Cr rises sharply at a much lower potential, doubtless owing to passage into the liquid as a soluble hexavalent compound (generally written CrO_3); the curve for Ni also rises at the same potential, and

1.   G. Okamoto, *Tokyo Cong.* 1972, Plenary Lecture.

in this (the trans-passive range) both elements pass freely into solution. Thus the film is mainly chromium oxide below 0·4 V, contains Cr, Ni and Fe up to about 0·95 V and is largely iron oxide above that level.

The effect of alloying additions in the anodic dissolution of stainless steel in media containing Cl⁻ has been studied by Tomashov.[1] The liquids used contained H_2SO_4 and HCl in different proportions. Small specimens corroded uniformly; large specimens became pitted, and the CD increased with time instead

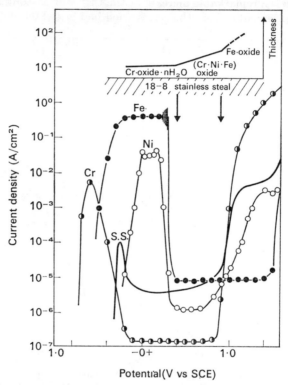

Fig. 14 Comparison of polarization curves for Fe, Cr, Ni and 18/8 stainless steel. Schematic sketch at the top suggests the composition and thickness of the passive film on stainless steel (G. Okamoto).

of falling. It was found that at any given potential the CD was decreased by the presence of V, Si, Mo and, especially, Re.

The polarization behaviour of Fe, Ni, Cr and seven different Fe–Ni–Cr alloys in concentrated NaOH at temperatures between 25°C and the boiling-point has been studied by Staehle.[2] The Fe–Ni–Cr alloys, and also iron and nickel, show an active–passive transition correlating well with the predictions

1. N. D. Tomashov, O. N. Markova and G. P. Charnova, 'Corrosion and Protection of Structural Alloys', translated by A. D. Mercer; 1968, p. 1.
2. A. K. Agrawal, K. G. Sheth, K. Poteet and R. W. Staehle, *J. electrochem. Soc.* 1972, **119**, 1637.

of the Pourbaix diagram. Cr exhibits a trans-passive behaviour, with an inflexion corresponding to the CrO_3^-/CrO_3^{2-} equilibrium.

An ellipsometric study of the passivation of Fe–Ni, Fe–Cr and Fe–Ni–Cr alloys has been carried out by Staehle.[1] The thickness of the film produced was found to lie between 11 Å and 55 Å, increasing with the potential used and also with the pH value. The character of the film was studied by cathodic reduction, and it was found that the chronopotentiogram obtained from films produced at pH 4 differed from those obtained from films produced at pH 6, 8·4, 10 and 12; the former showed two distinct arrests, whereas the films produced at the higher pH values showed no distinct arrests (except in the case of an alloy containing 10% Ni passivated at pH 10). The polarization curves of certain Cr–steels display a second anodic maximum after sensitization under appropriate conditions; this has been attributed to attack along the boundaries between α and γ phases. A possible explanation is based on impoverization in Cr; microprobe analysis and model experiments provide supporting evidence.

Iron–Silicon Alloys. The passivity of the Iron–silicon alloys has been studied by Omurtag and Doruk.[2] When the silicon content is below 14%, behaviour is controlled by a passive film consisting of iron oxide; at 14% silicon and above, the iV curve shows a change of shape, suggesting the growth of an SiO_2 film having a low conductivity. The corrosion rate in 35% H_2SO_4 falls off sharply at about 14% silicon.

The behaviour of the alloys under anodic conditions is of some interest. So long as the film is iron oxide, which has good electronic conductivity, anodic polarization produces evolution of oxygen. SiO_2, being a poor electronic conductor, will cause a thickening of the film rather than oxygen evolution.

Another study of the anodic behaviour of Fe–Si alloys in H_2SO_4 has shown[3] that the current which passes in the passive state at first rises with the silicon content, reaching a maximum at 6% Si; 12% Si decreases considerably the current passing in the passive state. Pure iron develops a black colour due to Fe_3O_4 and later a brown film of Fe_2O_3; alloys with 3% and 6% Si behave similarly. The alloy with 9% Si develops a milky white film. No visible surface discoloration is observed on alloys with 12% and 15% Si. The authors consider that the visible deposits are 'mechanical films and not those responsible for true passive behaviour'.

Cadmium. There has been some disagreement regarding the formation of films on cadmium in concentrated alkali; some authorities have favoured a solid state reaction leading to an oxide-film, whilst others believe that there has been dissolution as Cd^{2+} followed by precipitation of $Cd(OH)_2$. Results reported by Okinaka[4] suggest that in the active range the anodic formation of

1. K. N. Goswani and R. W. Staehle, *Tokyo Cong. Ext. Abs.* 1972, p. 99.
2. Y. Omurtag and M. Doruk, *Corr. Sci.* 1971, **10**, 225.
3. W. B. Crow, J. R. Myers and J. V. Jeffreys, *Corrosion (Houston)* 1972, **28**, 77.
4. Y. Okinaka, *J. electrochem. Soc.* 1970, **117**, 289.

$Cd(OH)_2$ takes place entirely through dissolution followed by precipitation. But the passivity observed at higher potentials cannot be attributed to complete coverage with $Cd(OH)_2$. The microscope shows clearly that coverage is incomplete, and the results point to the formation of a very thin continuous film of CdO which is only very slowly dissolved by alkali.

Aluminium. The porous films formed anodically on Al have been studied by Dekker.[1] The aluminium specimens were first covered with a porous layer by anodizing in oxalic acid and then further anodized in boric acid. The transport number for the positive ions during the first anodizing process was 0·33, but during the treatment in boric acid it was 0·24. The average pore diameter cell diameter and thickness of the barrier layer were found to vary linearly with the anodizing voltage; the thickness of the cell walls was somewhat thinner than the barrier layer.

The morphology of films obtained in borax, H_2SO_4 and CrO_3 baths was studied by Neufeld,[2] using scanning and conventional electron microscopy. There was good agreement between the results obtained after different film-stripping techniques had been used, and also between different methods of examination. The pore structure was found to change with time; the general trend was for a fine structure, with many holes, to give place to a coarser structure with fewer but larger pores. Another research based on radio-tracer methods[3] showed that, as a result of dissolution, the pores become wider near the pore mouth, where the products can diffuse away more easily.

Japanese work[4] has shown that in weak acids anodic treatment of aluminium generally causes pitting, but in stronger acids (sulphuric, oxalic, formic and tartaric) porous and uniform films are formed under appropriate conditions, although at low temperatures and low concentrations, pitting can still occur.

The mechanical properties of anodic films have been studied by Leach[5] with special reference to the effect of the movements of anions and cations. The relative mobility of an ionic species may depend on the total ionic flux. During the formation of the film, the current is carried by both anions and cations. At low rates of oxidation, most of the current is carried by the O^{2-} ions; when the field is increased, there is an increase in the flux of Al^{3+}, and this is associated with a decrease in the compressive stress produced in the oxide. Leach and Neufeld have confirmed an observation originally made by Bradhurst that if aluminium is subjected to a tensile stress, application of a suitable anodic current is accompanied by a lengthening of the wire. This was attributed to enhanced plasticity of the oxide-film in presence of the anodic current. The results were correlated with a model due to Nabarro and Herring, according to which grain-boundaries under tension have a higher concentration

1.   A. Dekker and A. Middelhoek, *J. electrochem. Soc.* 1970, **117**, 440.
2.   P. Neufeld and H. O. Ali, *Trans. Inst. Met. Finishing* 1970, **48**, 175.
3.   J. W. Diggle, T. C. Downie and C. W. Goulding, *J. electrochem. Soc.* 1969, **116**, 1347.
4.   Y. Fukuda and T. Fukushima, *Tokyo Cong. Ext. Abs.* 1972, p. 67.
5.   J. S. L. Leach and P. Neufeld, *Corr. Sci.* 1969, **9**, 225.

of vacancies than grain interiors, whilst under compression the reverse is true. The presence of a concentration gradient causes diffusion, and this is greatly enhanced by an electric field.

The anodizing of aluminium under galvanostatic conditions has been studied by Yahalom and Hoar[1] in H_2SO_4, Na_2SO_4 and ammonium borate. They found that when the field is sufficiently high, films formed in H_2SO_4 grow similarly to those in neutral solution. The breakdown voltage, and the corresponding film thickness, depend on the nature of the anion and not on the acidity; they are almost independent of CD, both in acid and neutral solutions. The observations are explained in terms of the introduction of anions into the oxide-film, and its ultimate disruption by mechanical force when the thickness reaches a value determined by the type and amount of anion introduced. The defect structure of thin films, which is different from that observed in thick films, has been studied by Pryor.[2]

The effect of anions on the anodic production of films on aluminium has been studied by Ammar.[3] The procedure was to anodize aluminium in 3% ammonium borate until the potential reached 100 V (normal hydrogen electrode) and then to add a solid sodium salt in an amount calculated to produce the concentration which it was desired to study. When the salt added was sodium carbonate, phosphate, succinate or oxalate, the potential showed a small decrease lasting less than 2 minutes, and then started to rise again. The addition of sulphate, sulphite or thiosulphate caused a fall to a constant value.

The film-growth on aluminium immersed in high purity water has been compared with that of a very dilute salt solution by Goddard.[4] The addition of 10 ppm SO_4^{2-} renders the film-growth slower than that in water. This is attributed to adsorption of SO_4^{2-} on the selected sites where reaction would otherwise occur.

The morphology of aluminium films obtained in sulphate solution has been studied by Wood and O'Sullivan.[5] Electron microscopy was carried out on films produced at various concentrations, pH values and temperatures. At low pH values, local film dissolution competes with film forming reactions to produce typical porous films, but as the pH is raised towards neutrality, pitting by metal dissolution becomes progressively an important competitive process, probably starting at the base of the pores in the film. In porous films it is found that the pore diameter, cell diameter and barrier layer thickness all vary with the formation voltage. The initiation and development of pores are probably due to field-assisted dissolution of oxide, with a possible minor contribution from Joule heating. At pH 6·7 the film is smooth and replicates the metal surface, but films are rougher at lower pH.

The mechanism of nucleation and electrochemical transport processes in

1.   J. Yahalom and T. P. Hoar, *Electrochim. Acta* 1970, **15**, 877.
2.   M. J. Pryor, *Oxidation Met.* 1971, **3**, 523.
3.   E. A. Ammar, *Corr. Prev. Control* 1972, **19** (2) 8.
4.   H. P. Goddard and E. G. Torrible, *Corr. Sci.* 1970, **10**, 135.
5.   G. C. Wood and J. P. O'Sullivan, *Electrochim. Acta* 1970, **15**, 1865. See also J. W. Diggle, T. C. Downie and C. W. Goulding, *Electrochim. Acta* 1970, **15**, 1079.

oxide formation during the anodic oxidation of aluminium has been studied by Czokan.[1]

A useful review of anodizing processes for aluminium, including baths for hard anodizing, and baths containing malonic and maleic acid, has been supplied by Tajima.[2]

Anodic treatment and sealing of aluminium alloys are important in connection with the external trim of motor cars.[3] Bright anodizing of aluminium alloys has been discussed by Jakobsen and Hulgren.[4]

An account of exposure tests lasting 10 years on anodized aluminium specimens on a railway triangle in East London, where the industrial atmosphere was very severe, has been provided by Liddiard and his colleagues.[5] They state that all the anodized specimens developed signs of pitting and lost reflectivity. The under-sides of the specimens (exposed slanting on a frame at 25° facing south) were attacked more than the upper sides; this was attributed to condensation, not to an accumulation of corrosion products or to deposits; periodical washing with distilled water did not prevent pitting. It was found that during an initial period (short in the case of unprotected aluminium) there was no corrosion; then attack set in. The rate often diminished with time unless stress was present or exfoliation occurred. Anodizing lengthened the initial period during which there was no attack; but when once the film had broken down, attack at the site of the breakdown proceeded in the same way as on unanodized material. Although all the specimens suffered pitting, resistance was found to improve with increase of film-thickness and with greater purity of the aluminium.

Titanium. The anodic passivation of titanium is interesting, since passivity can be reached in presence of a high Cl^- concentration. Levy and Sklover[6] have studied the anodic behaviour of titanium and its alloys in HCl. Increase of HCl concentration is found to increase the critical CD needed for passivity; it also raises the small dissolution current which still flows in the passive range. Increase of temperature increases the critical CD needed for passivity and the current flowing in the passive range. Commercially pure titanium requires a smaller CD for passivity than the alloys studied.

Many of the papers published on titanium refer to the practical use of titanium anodes in industry—for instance, in the manufacture of hypochlorites by the electrolysis of a chloride solution. Generally a small amount of a noble metal is employed, either as an alloying constituent, or on the surface, so as to keep the potential in the passive range. Platinized titanium anodes

1. P. Czokan, *Trans. Inst. Met. Finishing* 1973, **51**, 6; see also comments by R. W. Thomas, p. 10.
2. S. Tajima, Section of book on 'Advances in Corrosion Science and Technology', by M. G. Fontana and R. W. Staehle (Plenum Press, New York), 1970, Vol. I, pp. 229–362.
3. H. M. Bigford and R. W. Thomas, *Instn mech. Engrs Proc.* 1968, **182**, Part 3J, pp. 41, 47, 48.
4. O. Jakobsen and E. Hulgren, *Trans. Inst. Met. Finishing* 1968, **46**, 218.
5. E. A. G. Liddiard, G. Sanderson and J. E. Penn, *Trans. Inst. Met. Finishing* 1971, **49**, 200.
6. M. Levy and G. N. Sklover, *J. electrochem. Soc.* 1969, **116**, 323.

have been used in preparing a sterilizing solution by the electrolysis of sea-water. Sometimes, if the temperature is lower than 5°C, there is a loss of platinum. A research by Marshall and Millington[1] suggests that this is due to the high anode potential; at any given CD a low temperature will increase polarization and consequently produce a high potential; it is only indirectly responsible for the loss. At 25°C the platinum loss increases rapidly with the anode potential, but careful control of the anode potential should keep it small. The results of this research may be important for other industrial processes in which platinized titanium is used. In some cases, the presence of traces of platinum in a product can be objectionable, making it unstable; thus, quite apart from the money cost of the wasted platinum, traces in the product should be avoided.

Tomashov[2] has compared the dissolution of different titanium alloys in the passive state. Nb, Al and Mo increase the corrosion; Cr and Mn decrease it; Sn has no effect on alloys with the α-structure. In a titanium alloy containing 15% Mo, the dissolution rate in H_2SO_4 increases on passing from the pure β-structure (obtained on hardening) to the $(\beta + \omega)$- and $(\beta + \alpha)$-structures obtained on annealing after hardening. In the active state, Ti–Cr alloys corrode much more slowly in the one-phase state obtained after hardening than in the two-phase state obtained on subsequent annealing. Additions of Pd favour passivity.

Chromium. A study of the anodic behaviour of Cr in N H_2SO_4 has been carried out by Heumann.[3] Anodes of large grained Cr showed, after the completion of passivation, two further reactions at 0·2 and 0·7 V respectively (normal hydrogen electrode). The first of these was ascribed to the oxidation of the hydrogen present, whilst the second represented the oxidation of Cr^{3+} ions in the electrolyte.

The passivation of Cr has also been studied by Genshaw and Sirohi,[4] using an ellipsometric method.

Molybdenum. The anodic dissolution of Mo has been studied by Johnson[5] in solutions of KCl acidified with HCl. The apparent valency of the Mo passing into solution is approximately 6, but becomes lower as the solutions become increasingly acid. The molybdenum surface becomes covered with an oxide of a composition approximately Mo_2O_5. The anodic dissolution is controlled by the rate of oxidation of this oxide to MoO_3. The resistance to corrosion is ascribed to the 'Mo_2O_5 film', and not to an insoluble chloride, as has sometimes been suggested.

Tungsten. Johnson[6] has also studied the behaviour of tungsten. When placed in a sulphate solution, a tungsten specimen rapidly forms a film of

1. C. Marshall and J. P. Millington, *J. appl. Chem.* 1969, **19**, 298.
2. N. D. Tomashov, G. P. Chernova and G. A. Aynyan, *Tokyo Cong. Ext. Abs.* 1972, p. 107.
3. Th. Heumann and V. Sorajic, *Cambridge Conf.* 1970, paper 32.
4. M. A. Genshaw and R. S. Sirohi, *J. electrochem. Soc.* 1971, **118**, 1558.
5. J. W. Johnson, M. S. Lee and W. J. James, *Corrosion (Houston)* 1970, **26**, 507.
6. J. W. Johnson and O. L. Wu, *J. electrochem. Soc.* 1971, **118**, 1909.

F

composition roughly W_2O_5. Spontaneous corrosion then soon ceases (except at high pH values, where W_2O_5 is soluble). Upon anodic polarization, the appreciable conductivity of W_2O_5 allows further electrochemical oxidation to WO_3; hydration converts WO_3 to H_2WO_4, which has a very low conductivity, and is only slightly soluble in acids, but appreciably soluble in bases.

The colour changes observed on anodic polarization of tungsten in an acid solution follow the sequence

$$W \rightarrow \underset{\text{brown}}{WO_2} \rightarrow \underset{\text{blue}}{W_2O_5} \rightarrow \underset{\text{blue}}{W_4O_{11}} \rightarrow \underset{\text{green}}{W_5O_{14}} \rightarrow \underset{\text{yellow}}{WO_3}$$

Technetium. The anodic behaviour of technetium has been studied by Cartledge.[1] At overvoltages sufficiently high to prevent hydrogen evolution, the formation of surface oxides occurs at a rate of about 10^{-6} amps/cm^2, until a potential level is reached at which the formation of soluble pertechnic acid becomes possible.

Mechanism of Anodic Behaviour

Anodic Corrosion of Iron. The equation generally provided for the dissolution of an iron anode,

$$Fe = Fe^{2+} + 2e$$

merely represents the final change; it is clear that the reaction proceeds in steps, since the equation does not explain the fact that the pH value and the presence of anions affect the rate. In 1957, Bonhoeffer and Heusler[2] found that the dissolution rate was slower in acid than in relatively alkaline solution; Heusler[3] put forward a catalytic mechanism to explain this rather surprising state of affairs. Bockris and Kelly[4] suggested an alternative (non-catalytic) mechanism, and more recently two other mechanisms have been advanced.[5,6] It is likely that different mechanisms operate under different conditions. Useful surveys of the subject[7, 8, 9] are available, having special reference to acid chloride or acid sulphate solutions. The four proposed mechanisms are:

(1) *Heusler Mechanism*

$$Fe + H_2O \leftrightharpoons Fe.H_2O_{ads}$$
$$Fe.H_2O_{ads} \leftrightharpoons FeOH^-_{ads} + H^+$$
$$FeOH^-_{ads} \leftrightharpoons FeOH_{ads} + e$$

1. G. H. Cartledge, *J. electrochem. Soc.* 1971, **118**, 1752.
2. K. F. Bonhoeffer and K. E. Heusler, *Z. Elektrochem.* 1957, **61**, 122.
3. K. E. Heusler, *Z. Elektrochem.* 1958, **62**, 582.
4. J. O'M. Bockris, D. Drazic and A. R. Despic, *Electrochim. Acta* 1961, **4**, 325; 1962, **7**, 293. E. J. Kelly, *J. electrochem. Soc.* 1965, **112**, 124.
5. W. J. Lorenz, Y. Yamaoka and H. Fischer, *Ber. Bunsen Ges.* 1963, **67**, 932. F. Hilbert, Y. Miyoshi, G. Eichkorn and W. J. Lorenz, *J. electrochem. Soc.* 1971, **118**, 1919, 1927 (2 papers).
6. E. McCafferty and N. Hackerman, *J. electrochem. Soc.* 1972, **119**, 999.
7. S. Barnartt, *J. electrochem. Soc.* 1972, **119**, 812.
8. R. J. Chin and K. Nobe, *J. electrochem. Soc.* 1972, **119**, 1457.
9. L. Felloni, *Corr. Sci.* 1968, **8**, 133.

$$\text{FeOH}_{ads} + \text{Fe} \leftrightharpoons \text{Fe(FeOH)}_{ads} \text{ [catalyst]}$$
$$\text{Fe(FeOH)}_{ads} + \text{OH}^- \longrightarrow \text{FeOH}^+ + \text{FeOH}_{ads} + 2e \text{ [rate-determining]}$$
$$\text{FeOH}^+ + \text{H}^+ \leftrightharpoons \text{Fe}^{2+} + \text{H}_2\text{O}$$

(2) *Bockris-Kelly Mechanism*

$$\text{Fe} + \text{H}_2\text{O} \leftrightharpoons \text{Fe.H}_2\text{O}_{ads}$$
$$\text{Fe.H}_2\text{O}_{ads} \leftrightharpoons \text{FeOH}^-{}_{ads} + \text{H}^+$$
$$\text{FeOH}^-{}_{ads} \leftrightharpoons \text{FeOH}_{ads} + e$$
$$\text{FeOH}_{ads} \longrightarrow \text{FeOH}^+ + e \text{ [rate-determining]}$$
$$\text{FeOH}^+ + \text{H}^+ \leftrightharpoons \text{Fe}^{2+} + \text{H}_2\text{O}$$

(3) *Lorenz–Yamaoka–Fischer Mechanism*

$$\text{Fe} + \text{H}_2\text{O} \leftrightharpoons \text{Fe.H}_2\text{O}_{ads}$$
$$\text{Fe.H}_2\text{O}_{ads} + \text{X}^- \leftrightharpoons \text{FeX}^-{}_{ads} + \text{H}_2\text{O}$$
$$\text{FeX}^-{}_{ads} + \text{OH}^- \longrightarrow \text{FeOH}^+ + \text{X}^- + 2e \text{ [rate-determining]}$$
$$\text{FeOH}^+ + \text{H}^+ \leftrightharpoons \text{Fe}^{2+} + \text{H}_2\text{O}$$

(4) *McCafferty–Hackerman Mechanism* (proposed only for 6N chloride solution in the range of pH where H^+ promotes the reaction)

$$\text{Fe} + \text{H}_2\text{O} \leftrightharpoons \text{Fe.H}_2\text{O}_{ads}$$
$$\text{Fe.H}_2\text{O}_{ads} + \text{X}^- \leftrightharpoons \text{FeX}^-{}_{ads} + \text{H}_2\text{O}$$
$$\text{FeX}^-{}_{ads} + \text{H}^+ \leftrightharpoons \text{FeX}^-\text{H}^+ \text{ [complex]}$$
$$\text{FeX}^-\text{H}^+ + \text{H}^+ \longrightarrow \text{FeX}^+ + 2\text{H}^+ + 2e \text{ [rate-determining]}$$
$$\text{FeX}^+ \leftrightharpoons \text{Fe}^{2+} + \text{X}^-$$

McCafferty and Hackerman, studying iron in 6N chloride solution, found the behaviour to be different in three distinct ranges of pH; in these three ranges the 'reaction orders', indicating the influence of H^+, are -1, 0 and $+2$ respectively. This means that in the first range (where the H^+ activity is lowest), the corrosion current representing the reaction rate *decreases* as H^+ increases; then in the second range, it becomes *independent* of H^+ activity; finally, in the third range, the current *increases* as H^+ increases, but at a doubled rate. They explain their results by assuming that mechanism (3) operates in the first range, whilst their own mechanism (4) operates in the third range. However, at lower Cl^- concentrations the situation is different, doubtless because there is less tendency to form complex anions.

Several investigators have endeavoured to decide between the rival mechanisms by measuring the Tafel slope, which will be different in different cases; but the measurements have shown discrepancies. Barnartt thinks that these may have been due to impurities in the electrolyte or the metal. He himself has studied the matter using iron samples of different purities, and finds that the steady state Tafel slope drifts with time, except when zone-refined iron is used. He thinks that the Tafel slope is not a good criterion for the mechanism.

Classical work on Salt Passivity. From 1927 onwards, W. J. Müller studied anodic passivation under stagnant conditions, mostly in sulphate solution. He showed that the first change is generally the production of a layer of crystalline hydrated sulphate. This produced what he called *Bedeckungspassivität* ('covering passivity'). The formation of the salt crystals did not immediately put a stop to corrosion, but it diminished the area of uncovered metal, so that the same current now represented an increased current density, and this raised the potential into the passive range where the production of an oxide-film became possible. Under conditions where the anode is a rotating disc, or where the solution is stirred, or where the anode presents a vertical surface so that convection currents are set up, it is unlikely that a salt layer can be produced, except on lead where the sulphate is sparingly soluble; Müller used a horizontal surface carefully protected against disturbance. Modern workers have favoured conditions of rapid movement in the liquid, and Müller's work has not been followed up; under conditions of movement a much higher CD must be applied if passivity is to be produced, and probably the oxide-film will then be formed directly. Lately, however, papers have appeared describing the formation of a salt film on a metallic surface. In some cases, the authors appear to be unaware of Müller's work; this is true even of one paper from a German speaking country, and it is fairly certain that few corrosion students in English speaking countries are acquainted with Müller's researches. Yet to describe the new papers on salt passivity—as it is now called—and leave the reader unaware of Müller's extensive researches, would surely be wrong.

The same problem occurred in 1937. At that time, Müller had published several papers in the *Sitzungsberichte Akademie Wien*, a journal which rarely reached libraries in English speaking countries; the papers contained much elementary but tedious mathematics—not always free from errors; for various reasons the work remained unknown to most students of Corrosion. In hopes of bringing to notice these valuable researches, an effort was made about 1937 to prepare a brief and accurate statement of Müller's essential argument.* However the outbreak of war two years later frustrated this attempt to make Müller's work generally known. The statement, therefore, is reprinted almost unchanged as an Appendix to the present chapter. The reader must be lenient in his judgement of what was written over 35 years ago, but he will probably agree that Müller's researches, which placed the mechanism of salt passivity on a quantitative basis—without any instrumentation more advanced than a polarizing microscope—represented a remarkable achievement.

Recent work on Salt passivity. Italian work[1] on the behaviour of nickel in concentrated H_2SO_4 has revealed the presence of a salt film; X-ray

---

* At that time I was greatly indebted to Prof. H. A. Miley, who was a mathematician before he was a metallurgist, for scrutinizing this and indeed all the mathematical sections of my 1937 book. It is believed that certain errors noticed in the early Austrian papers invalidate only subsidiary arguments, and that the treatment in the statement as printed in the 1937 book contains no inaccuracies— U. R. E.

1.   G. Gilli, P. A. Borea, F. Zucchi and G. Trabanelli, *Corr. Sci.* 1969, **9**, 673.

study suggested that it is β–$NiSO_4.6H_2O$. A Swiss study[1] of lead in $M\ H_2SO_4$ shows that the passivity is entirely due to a film of $PbSO_4$, whilst that of Cd in $0.1M\ NaHCO_3$ is connected with a $CdCO_3$ layer. A German investigation[2] of the kinetics of passivation of zinc in concentrated $ZnSO_4$ solution has led to the following conclusion: 'Even in the case in which from a supersaturated solution a salt film is growing on the metal anode, there are enough pores to let the metal ions pass directly into the solution. But before these pores are shut by deposited salt, a direct reaction occurs between the metal and the anion, in order to form a solid salt layer at very low current density. By marking Zn with ^{65}Zn and $SO_4{}^{2-}$ with ^{35}S, we were able to show in a saturated $ZnSO_4$ solution that after the electrode was protected with hydrated $ZnSO_4$ deposited from the solution, initially about 70% of the zinc was dissolved and the layer deposited from the solution was growing. But with reduced current intensity no ^{65}Zn is transported into the solution and a direct transformation of the zinc into non-aqueous sulphate occurs. A very thin passivating water-free layer of $ZnSO_4$ with a specific resistance of about 10^{11} ohm-cm is formed.'

Industrial Application of Anodic Treatment

Anodic Protection. Ross[3] has discussed the relative merits of cathodic and anodic protection. He points out that, in some cases, especially in acid liquids, the efficiency of cathodic protection is low, because the current is mostly used in electrolysing the 'environment'. If the material is one which can build up a stable film, anodic protection may be preferable—especially under strongly acidic and oxidizing conditions. The problems of 'current attenuation' are less serious with anodic than with cathodic protection, since the passive film has considerable resistance; when the nearer points on the surface have become coated, the current is diverted to the remote region; as a result, the 'throwing power' is greater for anodic film deposition than for cathodic plating.

Certain requirements must be met, however, if anodic protection is to be effective. The electrical installation must be capable of providing a large current when the occasion demands it. When the process is started, a high CD is needed to produce a passive film; when once this has been formed, a low CD will suffice to maintain it; but, if the film should be broken by mechanical action or other cause, it is imperative that a high current should be available for its repair; otherwise there may be localized attack.

The theory and practice of anodic protection have been discussed by Walker and Ward.[4] They point out that although first demonstrated by Edeleanu in the UK in 1954, it has attained limited application in that country, whereas in the USA and USSR it is more widely used. A list of industrial applications is provided and it is noted that the CD needed to produce

1.   R. Grauer and E. Wiedner, *Corr. Sci.* 1971, **11**, 943.
2.   K. Schwabe and U. Ebersbach, *Amsterdam Cong.* 1969, p. 709.
3.   T. K. Ross, *Chem. Engr* 1971, No. **247**, p. 95.
4.   R. Walker and A. Ward, *Met. Rev.* 1969, **137**, 143.

passivity often exceeds by 3 orders of magnitude that needed to maintain it.

An account of the anodic protection of carbon steel in the digestors used for alkaline sulphide pulping has been provided;[1] also experience of the protection of titanium in the rayon industry.[2]

An acknowledged authority has written that 'it has taken 8 years to persuade engineers that anodic protection is safe'. Perhaps some readers will feel that the inventors have done well to accomplish the persuasion so swiftly; many innovations have remained neglected over much longer periods; moreover, considering the catastrophic effects that a failure might produce, the caution shown by the engineer is not altogether to his discredit.

Electrocutting and electromachining. Whilst in anodic protection, the metal is kept in the passive range, it is necessary in cutting or machining to produce anodic attack at high CD over a limited area without the metal becoming passive. The choice of anion is important, and has been discussed by J. P. Hoare,[3] who finds that NaCl gives wild cutting, $NaNO_3$ slow cutting, $NaClO_3$ excellent cutting, whilst $Na_2Cr_2O_7$ allows no cutting at all, since it produces a passivating film. Cutting appears to take place in the transpassive range, and this is conveniently obtained in a $NaClO_3$ solution.

Electropolishing. An ellipsometric study of copper undergoing electropolishing has been carried out at Philadelphia.[4] Polishing was carried out in 65% H_3PO_4, the copper anode being in a vertical position. The results confirm the observation of Hoar and Farthing (1960 book, p. 233) that a compact solid film is needed for polishing, and that a random removal of atoms leads to a surface which is smooth on an atomic scale (a preferential removal would leave an 'etched' surface). Actually it seems that two films are present during the process, an inner film at least 40 Å thick consisting of copper phosphate (possibly mixed with Cu_2O), and outside that a layer of viscous liquid consisting of highly concentrated electrolyte; this is about 2000 to 3500 Å thick. The outer layer can be removed without contact with air if the electrolyte is replaced by glycerine, which has a surface tension lower than that of the electrolyte; the ellipsometric measurements could then be carried out without introduction of an error. In contrast with a previous belief that the film thickness increases with potential, the Philadelphia investigators find that between 0·84 and 1·30 V the thickness and properties of the film remain constant. It is considered that the anodic current creates vacant cation sites at the outer film–liquid surface, and these vacancies migrate under the field to the metal surface, where they accept dissolving metal ions.

The relationship between anodic passivity and brightening has been

1. W. P. Banks, M. Hutchison and R. M. Hurd, *Cebelcor Rapp. tech.* **194** (1971).

2. L. S. Evans, P. C. S. Hayfield and M. C. Morris, *Amsterdam Cong.* 1969, p. 625.

3. J. P. Hoare, *J. electrochem. Soc.* 1970, **117**, 142; *Nature* 1968, **219**, 1034.

4. M. Novak, A. K. N. Reddy and H. Wrablowa, *J. electrochem. Soc.* 1970, **117**, 733.

clearly brought out by Hoar,[1] whose ideas are set out in Fig. 6 (p. 70). The curve for brightening is similar to the curve for passivation, but the current flowing over the horizontal part is very much larger. This is in fact essential. Hoar writes: 'It is evident that the film necessary to give anodic brightening must have a relatively high cation conductivity and must also allow easy passage of cations across the metal–film and film–solution interfaces— because dissolution has to occur from the metal through the film to the solution at a high rate and at quite low anode potentials. The film cannot, in fact, be simple oxide, because oxide-films have low ion conductivity at ambient temperatures and lead to passivity or, at high potentials only, to film-growth. Many recent papers on anodic brightening refer misleadingly to the "oxide" film. . . .

'Few attempts have been made to examine the composition of anodic brightening films. However, Williams and Barrett[2] found PO_4^{3-} in the residual films on copper anodically polished in aqueous O-phosphoric acid. . . . Sulphur and phosphorus have been found in anodic oxide-films formed on zirconium. . . . We think it reasonable to suggest that the thin, compact solid films responsible for anodic brightening are in general not simple oxides, but are oxides contaminated with significant amounts of the anion from the solution.

'Support for this hypothesis comes from consideration of the differences of composition of solutions giving anodic passivation and those giving anodic brightening. Generally, passivation occurs in the more dilute, brightening in the more concentrated, solutions of any particular acid or salt. Thus nickel passivates in dilute sulphuric acid, but brightens in 10M acid. . . . Brightening conditions are produced, it appears, by a sufficiently high anion/water ratio— which strongly suggests that the compact solid film present during brightening must have a sufficiently high anion/O^{2-} ratio.'

Perhaps a few remarks may be added. Simple geometrical considerations suggest that if a rough specimen is exposed to anodic attack in a liquid which produces no film, the prominences will be preferentially removed unless they consist of material more 'noble' than the body of the specimen. In material of uniform composition, the difficulty in obtaining a smooth surface is not the removal of prominences, but the avoidance of the development of pits or crystal ledges. For, when once a given atom has been removed, the probablity of the removal of each of its former neighbours becomes greater than that of a typical atom elsewhere, since each of these neighbours is in contact with a smaller number of other atoms than would be the case for a typical atom. Thus there is a danger of starting a micro-pit, bounded by crystal faces; as anodic attack continues, the acid formed by hydrolysis will help local attack, and the surface, instead of becoming smooth, becomes increasingly irregular. If, however, the electrolyte is one which covers the whole surface with a film through which ions (or cationic vacancies) can move under an electric field, the situation is different. As before, prominences will be most readily removed,

1. T. P. Hoar, D. C. Mears and G. P. Rothwell, *Corr. Sci.* 1965, **5**, 279, esp. Fig. 1, p. 281.
2. E. C. Williams and M. A. Barrett, *J. electrochem. Soc.* 1956, **103**, 363.

and if somewhere there happens to exist a very small volume less noble than the rest (perhaps a region where through internal stress the interatomic distance is unnaturally great, so that atoms require less energy for removal), this volume will be preferentially removed. But when once its removal has been completed, there is nothing to cause further preferential digging down into metal at that point; any acid is produced on the external surface of the film. On the contrary, it is likely that the irregularity produced by this preferential attack will be rectified, since the film will have become locally thicker at the point where attack had been occurring and the movement of vacancies through the film will be slower than elsewhere. It is believed that if the film substance has a suitably high ionic resistance, the effect of the greater local thickness will more than overcome the opposing effect of the loosening of the structure and that attack at the point in question will become slower than on the main part of the surface. The result should be to keep the surface smooth.

Breakdown of the Passive Film

Effect of Stress. An attractive suggestion regarding the mechanism of breakdown is made by Sato.[1] He points out that the high electric field imposed on the film under the conditions prevailing will produce electrostriction and consequently compressive stresses, which could cause film breakdown. There is a critical thickness above which deformation or breakdown becomes possible from the standpoint of energy. This is itself influenced by the concentration of Cl^- and other anions.

Effect of Chloride. It was pointed out in chapter IV that the addition of small amounts of Cl^- to a specimen held in the passive range of potential can cause breakdown of the protective film at certain places, and lead to pitting. It was pointed out in chapter V that normally reliable inhibitors fail to prevent corrosion in presence of chlorides. The mechanism of the breakdown has been deferred to the present chapter since many of the researches on the subject have depended on anodic polarization. It should be noticed that such methods have been questioned by some acknowledged authorities, who point out that the pitting and inhibitor failures met with in service occur in situations where no anodic current is applied from an external source. But, however much sympathy is felt for that criticism, it must in fairness be acknowledged that researches based on potentiostatic procedure, in which the specimen is held at a potential where considerable current will flow if and when the passive film breaks down, have greatly increased our understanding of the subject, and deserve careful study.

Film breakdown on Iron and Mild Steel. A study at Nottingham by Leach and his colleagues[2] on the breakdown of passive films by Cl^- ions on

1. N. Sato (with K. Kudo and T. Noda), *Electrochim. Acta* **16**, 447, 1683, 1909 (3 papers).
2. V. Ashworth, P. J. Boden, J. S. L. Leach and A. Y. Nehru, *Corr. Sci.* 1970, **10**, 481.

mild steel placed in moving liquid under potentiostatic conditions disclosed three stages. During the first stage, there was an increasing number of small pits of constant size. During the second (short) period, the pits increased in size. During the third period, there was breakdown with intense local attack, leading to a green product, and with hexagonal cavities below it. The current passing, which had been small during the first period, rose rapidly during the second, but the rise became slow again during the third.

Work has been carried out at Washington by Kruger and his colleagues,[1] using ellipsometric observation under potentiostatic control. The breakdown of the film was indicated by a sudden rise in current at time t_1. The main optical change observed arrives later at time t_2 (but another research carried out with a different wavelength suggests that possibly some change may occur between t_1 and t_2). The behaviour depends both on the potential maintained, and the composition of the solution. On pure iron in sodium tetraborate (pH 8·4), film-growth at 0·24 V (relative to the normal hydrogen electrode) was the same whether Cl^- was present or not. At higher potentials, the film grew thicker and faster if Cl^- was present, and severe pitting then occurred. In some experiments, passive films were grown in this potential region in the absence of Cl^-, and then sodium chloride was added. After a time depending on the potential and chlorine concentration, breakdown of the film occurred, signalled by a sudden increase in the anodic current density, followed by a rapid growth of porous non-protective oxide. The breakdown of passivity was attributed to (1) the adsorption of Cl^- at the oxide–solution interface followed by penetration, (2) an anodic oxidation of Fe^{2+} to give a porous plug of γ-FeOOH, (3) rapid dissolution of iron through the conducting paths thus formed, resulting in localized pitting, (4) precipitation of γ-FeOOH at the oxide–solution interface.

Shepherd and Schuldiner[2] have studied the effect of adding very small quantities of Cl^- to an NaOH solution made up with extremely pure, chloride-free water. When the Cl^- added was very low (not more than 0·01 $\mu g/cm^3$) there was no visual corrosion or change on the iron surface, and no iron could be detected in the solution. When the NaCl added was sufficient to give 2 $\mu g/cm^3$ or more, the pattern of the iV curves changed from that of Fig. 15(A) to that of Fig. 15(B). The maximum of the curve at K was always found at about −0·2 V (NHE). At 2000 $\mu g/cm^3$, the CD at K was 100 times the value obtained when Cl^- was almost absent. Along the limb EF the CD involved in O_2 production vastly exceeded that connected with other reactions, and this part of the curve was not affected by the amount of Cl^- present. At 2000 $\mu g/cm^3$ the iron became pitted after two cycles, with a considerable quantity of dark brown corrosion product. The current i_K flowing at K was considered to be a good measure of the extent of corrosion, and it was found that

$$i_K = 1·2 \times 10^{-7} C_{Cl}^{0·4}$$

1. J. Kruger and J. R. Ambrose, *Amsterdam Cong.* 1969, p. 698. H. T. Yolken, J. Kruger and J. P. Calvert, *Corr. Sci.* 1968, **8**, 103.
2. C. M. Shepherd and S. Schuldiner, *J. electrochem. Soc.* 1972, **119**, 572.

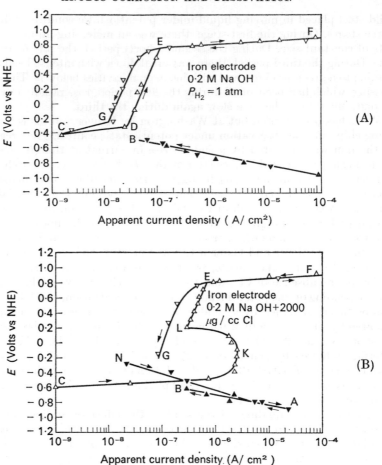

Fig. 15   (A) Polarization curves for iron in 0·2M NaOH at 25°C. Anodic current during increasing potential, △, and during decreasing potential ▽. Cathodic current during increasing potential ▲ and during decreasing potential ▼. (B) Effect of addition of 2000 g/cm³ of Cl– to the 0·2M NaOH (C. M. Shepherd and S. Schuldiner).

Breakdown on Stainless Steel. Hoar and Jacob[1] subjected 18/8 Cr/ Ni stainless steel to anodic treatment in either 0·01M H_2SO_4 (pH about 2) or in phosphate buffer (pH 7) under conditions of slow flow, and at a potential within the passive range; when the anodic CD had become constant, the liquid was replaced by a similar liquid containing a small amount of Cl^- or Br^-, with the flow conditions remaining the same. At first the CD remained unaltered, but after a certain induction period of length τ the current abruptly rose. The value of τ depended on the Cl^- (or Br^-) concentration, the potential and the temperature; it could vary between a few seconds

1.   T. P. Hoar and W. R. Jacob, *Nature* 1967, **216**, 1299.

and several hours. A modest rise in temperature from 25° to 40°C lowered the value of τ by over 2 orders of magnitude, suggesting a high activation energy, which calculation shows to be 60 kcal/mol. It was found that log τ, plotted against log C_{Cl} where C_{Cl} is the Cl⁻ concentration, produces a straight line; if $1/\tau$ is an appropriate measure of the breakdown process (whatever that is), the rate of breakdown is proportionate to $(C_{Cl})^n$, where n lies between 2·5 and 4·5. It is concluded that the process depends on the joint adsorption on the oxide-film of 3 or 4 Cl⁻ (or Br⁻) ions, producing an adsorption complex of high energy, which will separate and pass into the liquid far more readily than an ordinary Fe^{2+} or Fe_{aq}^{2+} ion—at least at certain susceptible points; at such points, the passage into solution of complexes will leave the film thinner, and if the potential remains constant, the field will increase, and the movement of ions outwards at the point in question becomes more rapid; thus the velocity of the local breakdown, once started, increases explosively. Hoar and Jacob describe the phenomenon as 'catalytic'; some authorities object to the use of that term, but the argument itself does not seem to have been seriously challenged.

The pitting of alloys containing 5·6% and 11·6% chromium has been studied by Stolica.[1] He found that curves produced under galvanostatic conditions showed an induction period depending on the Cl⁻ concentration, and a reaction period characterized by oscillations; a minimum concentration of Cl⁻ ions was necessary for attack; he observed the pits to be of different forms, and (in contrast with some other observers) states that they are hemispherical in the early stages, becoming roughened and irregular later.

Breakdown on Nickel. A research on nickel in the polycrystalline and monocrystalline state with a potential held in the passive range has been carried out by Tokudo and Ives.[2] In sulphuric acid containing Cl⁻, pits started at scratches and grain boundaries. Susceptibility to pitting on polycrystalline nickel, mechanically polished {111} faces, and electropolished {111}, {100} and {110} faces, was found to fall in the same order as their reactivity under conditions of active dissolution. It was concluded that on nickel the sites for pitting are identical with the sites for active dissolution. On the {110} face of a single crystal, mechanically polished, which had received a scratch, made with a blade, the pit started at one end of the scratch and developed to an almost hexagonal shape.

Breakdown on Aluminium. The manner in which Cl⁻ promotes pitting has been studied in detail by Pryor at New Haven.[3] He believes that it is due to a defect structure set up where Cl⁻ ions exchange with O^{2+} ions on the Al_2O_3 lattice, electrical neutrality being maintained by the passage of the appriate number of Al^{3+} ions from the oxide into solution. This leads to the production of vacancies at cation sites, and thus facilitates the passage of Al^{3+} ions outwards through the film. Although in the absence of oxygen rapid pitting only occurs above a certain critical potential, he found it in presence

1. N. Stolica, *Corr. Sci.* 1969, **9**, 205, 455 (2 papers).
2. T. Tokuda and M. B. Ives, *Corr. Sci.* 1971, **11**, 297.
3. M. J. Pryor, *Williamsburg Conf.* 1971.

of oxygen at a much lower range of potential. In fact, he considers that under service conditions all pitting of aluminium and its alloys is produced in a region well below the so-called critical potential.

Metzger[1] has shown that even in the absence of Cl^- breakdown events do take place on an aluminium surface; he attributes the stability of aluminium, not to immunity from film-breakdown but to the efficiency of repair; the function of Cl^- is to inhibit repair, not to initiate breakdown. Aluminium, previously electropolished in perchloric acid, was subjected to anodic attack in H_2SO_4, to which was added Cl^- at 0·01M concentration; this did not interfere with the growth of the film, which continued for $2\frac{1}{2}$ hours before stable pits were formed; in the earlier part of the treatment, many small, shallow pits were produced but soon repaired themselves.

The importance of the electric field strength in pitting is brought out by Videm.[2] Many previous researches had been carried out under conditions allowing simultaneous film-growth and breakdown. In Videm's experiments a pre-formed film of known thickness and structure was used, permitting calculation of field strength. It was found that Cl^- had no effect on the leakage current in the period preceding breakdown. Just before the sharp rise of current which was to indicate local breakdown, pulses were observed in the current passing; these became frequent after breakdown. Videm believes that the first step is the absorption of Cl^-, not uniformly, but distributed in a manner determined by the field strength, and therefore greatest at just those places where the barrier film is thinnest. The chloride absorbed into the film converts the part near the surface into an ionic conductor and thus increases the strength of the field existing in the oxide underneath. This in turn increases the rate of Cl^- absorption, so the process proceeds in a manner recalling a chain reaction.

Comparison between different metals. An interpretation of the influence of Cl^- ions on anodic behaviour has been put forward by Vijh,[3] using solid state concepts. He finds that Cl^- is deleterious to passivation on Al, Cr and Fe–Cr alloys, but that it does not affect passivation on Mo, W or Zr. He explains the difference by the high metal–metal bond energies of the three last-named metals, and also the high lattice energies of their oxides, which makes it more difficult for metal ions it leave the lattice of the corresponding metal oxide.

Concepts of Breakdown Potential, Pitting Potential and Protection Potential. It has been mentioned that some investigators consider that pitting will only set in if the potential is raised above a certain value at which the passive film breaks down, but that both Pryor and Wood state that in practice it is often obtained below the value in question. Japanese work[4] shows that certain iron–chromium alloys, after anodic treatment in H_2SO_4 in the (high) trans-passive range of potential, will develop pits in H_2SO_4 con-

1. J. Zahavi and M. Metzger, *Tokyo Cong. Ext. Abs.* 1972, p. 115.
2. K. Videm, *Tokyo Cong. Ext. Abs.* 1972, p. 121.
3. A. K. Vijh, *Corr. Sci.* 1971, **11**, 161.
4. Y. Misamatsu and K. Ichikawa, *Tokyo Cong. Ext. Abs.* 1972, p. 127.

taining NaCl when the potential is reduced to the upper limit of the active range. Hoar (this volume, p. 160) considers that the concept of a 'breakdown potential' may have to be modified, perhaps by inclusion of a time factor—a point of view with which Pourbaix[1] concurs. Pourbaix distinguishes between the *Pitting Potential* and the *Protection Potential*. If pitting is to start, the potential must be raised above the Pitting value; but, when once started, a pit will not be stopped from growing by merely reducing the potential just below that value. In order to stop the growth of a pit, the potential must be reduced further, namely into the range below the Protection Potential. In the potential range between these two values, pits which have already started will continue to grow, but new pits will not be started. The idea of a Protection Potential is particularly helpful in discussing practical problems. Wilde,[2] who has emphasized the utility of the concept, draws attention to the fact that the Protection Potential is not a unique property of the material, but will vary with prior pit propagation, which in turn affects the amount of acid produced by the hydrolysis of the metallic chloride in the pit; Pourbaix has expressed agreement.

Varying opinions about the significance of the 'pitting potentials' were expressed at the Williamsburg conference.[3] Some authorities were critical, but most seemed to agree that, if the concept is used at all. two levels must be recognized, namely the value above which pitting will start and that below which it will cease.

Rajagopalan[4] has studied the breakdown of passivity in a solution containing an inhibitor and also Cl^- ions; it occurs when the specimen is raised anodically above the breakdown potential, which depends on the nature and concentration of both ions; for instance, in phosphate–chloride solutions it becomes more negative with increase of Cl^- and more positive with increase of PO_4^{3-}. Time–potential curves produced under galvanostatic conditions fall into three categories in which:

(1) the potential rises and then falls permanently to a negative value; this effect is obtained when $NaOH$ or Na_2CrO_4 (in addition to NaCl) is present; one or more spots remain active after the potential has fallen to the negative value.

(2) potential oscillation occurs; this effect is obtained with Na_2HPO_4 or $Na_2B_4O_7$; at each fall, the pits previously formed by breakdown heal up, but fresh pits are formed when the potential rises again.

(3) the potential rises and remains steady at a positive value; this occurs

1. M. Pourbaix, *Corrosion (Houston)* 1969, **25**, 267; 1970, **26**, 431; *Cebelcor Rapp. tech.* **179** (1970); **191, 199** (1971); **203** (1972); confirmed and extended by E. D. Verink, T. S. Lee and A. L. Cusumano, *Corrosion (Houston)* 1972, **28**, 148.
2. B. E. Wilde, *Corrosion (Houston)* 1972, **28**, 283. B. E. Wilde and E. Williams, *J. electrochem. Soc.* 1969, **116**, 1539; 1970, **117**, 775; 1971, **118**, 1057; *Electrochim. Acta* 1971, **16**, 1971.
3. *Williamsburg Conf.* 1971; see esp. R. P. Frankenthal and M. B. Ives, p. 311; Z. Szklarska-Smialowska, p. 312; B. E. Wilde, p. 342; G. Bianchi, A. Cerquetti, F. Mazza and S. Torchio, p. 399.
4. K. S. Rajagopalan, K. Venu and G. Venkatachari, *Ferrara Symp.* 1970. p. 711.

with NaNO$_2$ or sodium benzoate; the film thickens, but does not break down
—a circumstance attributed to the 'mutual protective effect, whereby the
spot suffering attack protects the adjacent area' (1960 book, p. 114).

Further References

Cotton[1] provides important information about anodic protection in industry, which should be studied in his paper. One example may be quoted. The
final stage in the manufacture of concentrated sulphuric acid (cooling from
115°C) has traditionally been carried out in cast iron vessels. A joint British–
Canadian research showed that for air-cooling stainless steel and for water-
cooling a nickel-based alloy can be safely used if anodic protection is provided; this leads to considerable reduction in cost and space, and provides
a purer acid with an iron-content below 5 ppm. Since 1970, thirty commercial plants based on this procedure have come into operation in different parts
of the world.

Papers. 'Passivation of chromium containing Iron and Nickel-base
alloys in Aqueous Solutions at 289°C.'[2]

'Investigation of the second anodic current maximum on the polarization
curves of commercial stainless steels in sulphuric acid'.[3]

'Stability of the Anodic Film on Aluminium in relation to the nature of
some anions'.[4]

'Effect of chloride on growth of an anodic film'.[5]

'Some questions concerning the anodic potentiostatic phase dissolution of
heterogeneous alloys.'[6]

'Outdoor corrosion performance of anodized and electrolytically coloured
aluminium for architectural use'.[7]

'Investigation of electrolytic colouring of porous anode films on aluminium,
using electron microscope'.[8]

'Accelerated corrosion test to determine the continuity of thin oxide films
on aluminum'.[9]

See also the important paper by Revie, Boker and Bockris, quoted on p. 15,
ref. 4.

1. J. B. Cotton, *Brit. Corr. J.* 1975, **10**, 66.
2. D. A. Vermilyea and M. E. Indig, *J. electrochem. Soc.* 1972, **119**, 39.
3. L. Felloni, S. S. Traverso, G. L. Zucchini and G. P. Cammarota, *Corr. Sci.* 1973, **13**, 773.
4. I. A. Ammar, S. Darwish and E. A. Ammar, *Werkstoffe u. Korr.* 1973, **24**, 200.
5. J. Zahavi and M. Metzger, *J. electrochem. Soc.* 1973, **121**, 268.
6. A. Piotrowski and A. Slimak, *Werkstoffe u. Korr.* 1974, **25**, 262.
7. J. Patrie, *Trans. Inst. Met. Finishing* 1975, **53**, 28.
8. A. J. Doughty, G. E. Thompson, J. A. Richardson and G. C. Wood, *Trans. Inst. Met. Finishing* 1975, **53**, 33.
9. N. J. McDevitt and W. L. Baun, *J. electrochem. Soc.* 1975, **122**, 523.

APPENDIX TO CHAPTER VII

W. J. Müller's Theory of Anodic Passivity

Case of Vertical Electrodes. Very interesting are the relations prevailing where a layer of insoluble matter can be produced over the anode and passivity can set in. The current passing provides no guide to the corrosion rate, since most of it will now be utilized in some other reaction (e.g. production of oxygen). It is naturally of great importance to know the conditions under which the passive state may be expected to appear. In examining this question Shutt and Walton[1] used a vertical gold electrode under conditions of violent stirring, and found a limiting current density, ω_0, below which passivity was *never* produced, however long the experiment was continued. Above the limiting current density, numerous experiments with potassium chloride and hydrochloric acid led to a simple relation between the current density ω, and the time needed for passivation, t_p.

$$t_p(\omega - \omega_0) = Q$$

where Q was a constant, which in the stronger solutions was proportional to the chlorine ion concentration. The limiting current density, ω_0, was found to be proportional to the chlorine ion concentration over a wide range of conditions. For sulphate solutions, Shutt and Walton obtained the same linear relation between ω and $1/t_p$, but both ω_0 and Q were much smaller; here the value of Q represented approximately the quantity of electricity required for the formation of a unimolecular layer of auric oxide (Au_2O_3).

In normal hydrochloric acid, the straight-line graph connecting ω and $1/t_p$ obtained with vigorous stirring coincided at the higher current densities with that obtained with relatively little movement in the liquid; at low current densities, the two curves diverged, the more stagnant condition greatly reducing the time of passivation, t_p. The addition of anions like $(SO_4)''$, $(NO_3)'$ or $(HPO_4)''$ had surprisingly little effect on the time of passivation in chloride solutions, whilst an increase of $(OH)'$ concentration did not seriously affect the limiting current density in N potassium chloride, until a pH value of 10·8 was reached, after which point any further increase in alkalinity produced an enormous reduction in the value of the limiting current density.

Shutt and Walton explain their results by assuming that adsorption of chlorine ions must take place on the gold surface before gold can pass into the liquid. The velocity of adsorption being proportional to the chlorine ion concentration in the liquid, it follows that the limiting rate at which gold can be dissolved (i.e. the limiting current density) will be proportional to the chlorine ion concentration. If this current density is exceeded, the adsorption process of chlorine ions cannot keep pace with the corrosion, and hydroxyl ions begin to play a part in the anodic reaction, leading to a film of some

1. W. J. Shutt and A. Walton, *Trans. Faraday Soc.* 1932, **28**, 740; 1934, **30**, 914; 1935, **31**, 636. See also W. J. Müller and E. Löw, *Trans. Faraday Soc.* 1935, **31**, 1291.

oxide (or hydroxide) on the surface and consequent passivity. The pH value (10·8 in N potassium chloride), beyond which t_p drops suddenly, represents the condition where the direct adsorption of OH′ ions can compete with that of Cl′ ions, thus greatly facilitating passivity.

Armstrong and Butler[1] have obtained the same linear relation between ω and $1/t_p$ for a gold anode in unstirred chloride solutions, but explain it somewhat differently.

Case of Horizontal Electrode.   During the period whilst an anode is becoming passive, the value of the current diminishes steadily with the time. W. J. Müller and his colleagues[2] have studied this decline, using a horizontal anode protected against accidental stirring. Here, the liquid just above the anode surface will soon become saturated with the salt constituting the anodic product; supposing, now, that this salt commences to crystallize out at certain points, and extends outwards over the surface *as a layer of uniform thickness*, y, the relation between i and t can be calculated:

If the current were *solely carried by anions*, then Faraday's Law would predict the deposition of $K_s i$ dt of the salt on the anode surface in time dt, where K_s is the electrochemical equivalent of the *salt* (i.e. its equivalent weight divided by Faraday's number). Actually, the fact that the current is partly carried by *cations* reduces the amount deposited to (say) $XK_s i$ dt where $X < 1$. If we assume that the movements of ions across the metal–liquid interface is the same as across any other plane perpendicular to the current-path in the liquid, then

$$X = \frac{v}{u+v}$$

where u and v are the mobilities of the metallic ion and the anion. The element of surface covered up will be

$$\mathrm{d}F = XK_s i\, \mathrm{d}t/sy \tag{1}$$

where s is the density of the salt formed. If the total area of the anode is F_0, and the part covered up is F, the resistance of the pores threading the layer is

$$R_1 = \frac{y}{\kappa(F_0 - F)}$$

where κ is the specific conductivity of the saturated liquid in the pores. If e is the external EMF applied to the cell, and ε_C and ε_A the local potentials at cathode and anode, Ohm's Law states that

$$e + \varepsilon_C - \mathrm{e}_A = i(R_0 + R_1)$$

where R_0 is the resistance of the free liquid. Writing E for $e + \varepsilon_C - \varepsilon_A$, we have

$$E = i\!\left(R_0 + \frac{y}{\kappa(F_0 - F)}\right)$$

1. G. Armstrong and J. A. V. Butler, *Trans. Faraday Soc.* 1934, **30**, 1173.
2. W. J. Müller, *Z. Elektrochem.* 1934, **40**, 110, 570, 578; 1935, **41**, 83, 641; 1936, **42**, 166, 366; 1937, **43**, 407, 561.

whence
$$-\mathrm{d}F = \frac{y}{\kappa R_0} \cdot \frac{E/R_0}{(E/R_0 - i)^2} \, \mathrm{d}i \tag{2}$$

combining with (1) we get

$$\frac{K_s X \kappa R_0^2}{E s y^2} \mathrm{d}t = - \frac{1}{i(E/R_0 - i)^2} \mathrm{d}i$$

Integration gives

$$t = C + \frac{sy^2}{\kappa K_s X R_0}\left[-\frac{1}{E/R_0 - i} + \frac{1}{E/R_0} \ln \frac{E/R_0 - i}{i} \right] \tag{3}$$

At $t = 0$, $F = 0$, i has the value i_0 such that

$$i_0 = \frac{E}{R_0 + y/\kappa F_0}$$

Hence
$$C = \frac{sy^2}{\kappa K_s X R_0}\left[\frac{R_0(R_0 + y/\kappa F_0)\kappa F_0}{yE} - \frac{1}{E/R_0} \ln \frac{y}{R_0 \kappa F_0} \right]$$

By adopting rough numerical values for the y and κ, Müller shows that under the conditions of the experiments, $y/\kappa F_0$ must be small compared to R_0, and that the second term within the square bracket must be small compared to the first; hence to a first approximation

$$C = \frac{sy R_0 F_0}{K_s X E}$$

Consequently, altering the base of logarithms, equation (3) becomes

$$t = C - A\left(\frac{1}{i_0 - i} - \frac{2 \cdot 3}{i_0} \log \frac{i_0 - i}{i} \right) \tag{4}$$

where
$$C = \frac{ys F_0}{K_s X i_0} \quad \text{and} \quad A = \frac{sy^2}{\kappa K_s X R_0}$$

This relation between the current i and time t has been tested experimentally by Müller and his colleagues in a number of cases, for instance with iron, copper, nickel, zinc, lead and cadmium anodes in sulphuric acid solution. It is found that, during the early part of the passivation process, reasonably constant values of A are obtained, which indicates that the layer of solid sulphate preserves a fairly uniform thickness. After a certain point (at which practically the whole surface has been covered up with the film and only minute pores remain) the value of A commences to change, indicating that the film is commencing to thicken. In other words, the *sideways extension is over*, and *growth at right-angles to the surface is setting in*. If during this second stage, the pore area is assumed to be constant whilst the thickness y increases with the time, Müller argues that

$$\mathrm{d}y = \frac{K_s X}{s F_0} i \, \mathrm{d}t \tag{5}$$

Now it follows from Ohm's Law that

$$E = iR_0 + \frac{iy}{\kappa F_0}$$

which gives on differentiation

$$dy = -\kappa EF_0 \frac{di}{i^2}$$

Combination with (5) leads to

$$dt = -\frac{\kappa EF_0^2 s}{K_s X} \cdot \frac{1}{i^3}\, di \tag{6}$$

Integration between the times t_1 and t_2 with corresponding current values i_1 and i_2 gives

$$t_2 - t_1 = B\left(\frac{1}{i_2^2} - \frac{1}{i_1^2}\right) \tag{7}$$

where

$$B = \frac{\kappa EF_0^2 s}{2K_s X}$$

This relation also has been tested for several cases (e.g. copper in saturated copper sulphate, lead in sulphuric acid, iron in normal sodium sulphate), and it is found that, plotting $1/i^2$ against the time, remarkably straight curves are obtained during the second stage of the passivation process.

The two growth laws are well illustrated by the curves of Müller and Machu for the falling off of current upon a lead anode in sulphuric acid (of 'accumulator concentration'), reproduced in Fig. 11. The curve marked i indicates

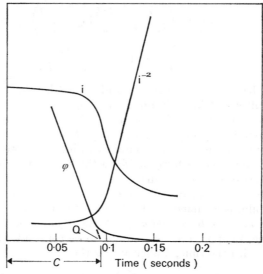

Fig. 16   The falling off of current with time at a lead anode in dilute sulphuric acid (W. J. Müller and W. Machu).

the current at different times; it will be noticed that the period of rapid drop lasts only a fraction of a second and naturally a special oscillograph was needed to obtain a trace of this kind; the short time of passivation is due to the low solubility of lead sulphate; similar curves for an iron anode extend over many minutes. In the curve marked ϕ, the function

$$\phi = \frac{1}{i_0 - i} - \frac{2\cdot 3}{i_0} \log \frac{i_0 - i}{i}$$

is plotted against time, and the straightness of this curve is a proof of the validity of equation (4), representing the law of sideways extension. Almost as soon as this law fails (shown by the departure from straightness of the curve connecting ϕ and t), the depth-law begins to be obeyed, as shown by the straightness of the curve connecting i^{-2} with time. The point, Q, where the ϕ-curve cuts the horizontal axis clearly gives the value of C, a magnitude which forms a convenient measure of the time of passivation.

Müller's work proves, therefore, that passivation usually consists of two stages:

(1) extension of the layer *sideways*, the thickness being nearly constant,
(2) growth of the layer *in depth*, the area represented by pores remaining nearly constant.

Nature of the Obstructive Layers.   From the values of C and A

$$C = \frac{ysF_0}{K_s X i_0} \quad \text{and} \quad A = \frac{sy^2}{\kappa K_s X R_0}$$

it is possible to eliminate y, and obtain an expression for the specific conductivity of the liquid threading the pores

$$\kappa = \frac{C^2}{A}\left(\frac{i_0}{F_0}\right)^2 \cdot \frac{K_s X}{sR_0} \tag{8}$$

Müller found that, for an iron anode, the number obtained experimentally (468×10^{-4} ohm^{-1}) agreed well with the conductivity of a saturated solution of iron sulphate heptahydrate (470×10^{-4} ohm^{-1}). This suggests that the solid layer which causes the current to die away is none other than the familiar 'green vitriol'—a conclusion supported by the polarizing microscope.

Since passivity depends on the separation of a sparingly soluble body on the anode, it is natural that passivity will set in most quickly when the solubility of the anodic product is low. As already stated, the time needed for a lead anode to become passive in sulphuric acid is much smaller than that needed for an iron anode. Evidently quantitative measurements of the time of passivation would possess interest, but as the passivity layer accumulates gradually, some definition of the term must be given. Shutt and Walton, in the work on gold already quoted, adopted the time needed to reach a potential at which chlorine gas was freely evolved. Müller adopts a different criterion. A study of his data on the connection between i and t indicates that the diminution of the current becomes sudden at a time equal to the numerical value of the constant C. This constant therefore provides a suitable measurement of the time of passivation, t_p. It depends, naturally, on the initial value of the current density, and Müller has found that, for many systems, straight curves are obtained when log i_0 is plotted against t_p. Clearly t_p will be reduced if the greater part of the surface of the metal is already covered up with a film, when the anodic treatment starts, since this will increase the true current density on the parts left uncovered. Müller showed that under certain conditions, the time of passivation of an iron anode in normal sodium sulphate was about 0·01 second; this short time was due to the presence of the invisible air-formed oxide film, for if, before the experiment, the iron was freed from this film by a special treatment (anodic polarization followed by activation through touching with a zinc wire) the time of passivation was multiplied by about 20 000. Measurements of times of passivation thus become important as a test for the presence of an invisible film, and in some cases may provide an idea of the area of the metal still exposed at pores in that film. Müller[1] considers that a passive behaviour must be expected when the pores represent less than 0·01% of the total area, and an active behaviour when they represent more than 1%.

Apparent Potential Changes on Passivation and Activation. The passage between activity and passivity is also conveniently followed by measuring the potential-drop at the electrode surface by means of a tubulus held in close contact with that surface, and leading to a calomel or hydrogen electrode. For iron in the active state, the potential is strongly negative towards hydrogen; during the process of 'passivation' the potential rises, and when the discharge of oxygen begins it is strongly positive. During the first part of the passivation process, whilst the surface is being covered up with salt, the apparent rise of potential is simply that needed to force the current through the resistance of the pores threading the layer of salt. Müller's experimental verification of equation (4) (which assumes that the genuine anodic potential, ε_A, is constant throughout the passivation period) affords proof that (in many cases, at least) the metal, under its blanket of salt, continues to preserve the potential proper to its environment. The apparent rise in the potential is mainly due to the current flow, and would largely disappear when the current was turned off. When, owing to the great diminution of the pores, the true current density has reached a certain value (probably 40–100 amps/cm²), other changes become possible, such as the formation of ions of higher valency, or of an oxide-film, or of free oxygen. These changes call for a *genuine* rise in the potential, and that rise will remain, for a time at least, when the current is turned off. With an iron anode in acid solution, the return of the potential to the 'active' value occurs very soon after the turning off of the current, and the potential-tumble usually corresponds with the restoration of active properties—i.e. the power to dissolve freely at low current densities instead of evolving oxygen. The addition of alkaline substances or oxidizing agents defer or even prevent the potential-tumble to the active values, and likewise the return of active properties.[2]

1. W. J. Müller, *Z. Elektrochem.* 1934, **40**, 119; *Korr. Met.* 1932, **8**, 253; 1935, **11**, 31.
2. See also early work by C. Fredenhagen, *Z. Phys. Chem.* 1903, **43**, 1; E. P. Schoch and C. P. Randolph, *J. Phys. Chem.* 1910, **14**, 719; G. R. White, *J. Phys. Chem.* 1911, **15**, 768.

Buried and Immersed Metal-work

Preamble

Picture presented in the 1960 and 1968 volumes. The types of corrosion discussed in chapter IV refer to complete or partial immersion in water or salt solution. Special factors are introduced when metal is buried in the soil; one of the most important is the action of bacteria, which allows rapid corrosion to take place in the absence of oxygen. Under immersed conditions, oxygen is needed for rapid attack by most near-neutral liquids, although rapid attack by acids, and sometimes by alkalis, can occur in absence of oxygen, hydrogen gas being evolved in some cases; this type of attack is considered in chapter IX.

Chapter VIII deals with the problems of soils, with particular reference to 'cathodic protection', which often provides a solution of the corrosion problem for the pipes or metal-work included in the protective scheme, but may sometimes cause enhanced attack on other buried systems which have not been included. The chapter also deals with corrosion under marine conditions. Here complications may be caused by the attachment of living organisms—animal or vegetable. Consequently the painting system used generally has to consist of an inner coat intended to prevent corrosion, and an outer coat intended to deal with 'fouling'. Some anti-corrosive paints tend to be softened or loosened by alkali, and if a cathodic protection system is in use, such paints must be avoided.

Special factors are introduced in the corrosion of steel reinforcements in concrete. Under favourable circumstances the alkalinity due to lime liberated during the setting of concrete prevents attack; but if the concrete is too thin or too porous, the lime may be converted to calcium carbonate, and attack on the reinforcement may then become possible. This also applies to cases where a layer of concrete has been laid over a flat steel surface; if the concrete is too thin or too porous, serious trouble may occur. If chloride is present in the concrete, the dangers are enhanced. Chloride may be added intentionally to accelerate setting, or to prevent freezing; it may also reach the steel accidentally under marine conditions. The dangers are further enhanced if the steel is under tensile stress, as in a pre-stressed system.

Developments during the period 1968–1975 and Plan of the new Chapter. During the period under consideration there has been much useful information published regarding all the problems discussed in this chapter. The importance of biological corrosion has become generally appreciated, and the amount of research devoted to it has been considerable. Marine problems have also received much attention, particularly the advantages of applying cathodic protection; this practice, however, has made necessary the choice of a paint capable of withstanding the action of cathodic-ally produced alkali. Finally the behaviour of steel in concrete, with or without chloride, has rightly received much experimental study, since the effect of breakdown of steel reinforcement could have disastrous results.

The chapter opens with a study of biological corrosion in soils and above ground. Protective methods for buried metal-work independent of cathodic protection are then discussed, after which cathodic protection methods receive attention. A section devoted to corrosion under marine conditions follows, with the application of cathodic protection methods (marine paints, however, are discussed in chapter XIV). Various problems of water treatment, including those arising at desalination plants, then receive attention. The remainder of the chapter is devoted to the corrosion of steel in concrete —the general situation being first discussed, followed by a consideration of protective measures and the influence of stress.

Biological Corrosion

Soils. A useful survey has been provided by Iverson.[1] He states that in 1954 biological corrosion, largely directed against buried pipes, was costing the USA 500 to 2000 million dollars a year. In one group of American oil wells, sulphate reducing bacteria were responsible for more than 77% of the corrosion. In Great Britain it has been estimated that at least half of the failures of buried metal are due to microbiological causes. Soil that has been removed for the burial of pipes and replaced is more likely to cause trouble than the original soil, for various reasons. One is that the decomposition of organic matter by aerobic organisms produces the nutrients necessary for the multiplication of anaerobic organisms. These aerobic organisms can themselves produce corrosion at a steel surface accessible to oxygen. In one peaty soil in England it was found that the H_2S produced by anaerobic sulphate reducing bacteria at a low level was diffusing upwards and was being oxidized to H_2SO_4; this produced severe corrosion to bitumen covered gas-pipes and water-mains. In addition, iron bacteria, which promote oxidation of Fe^{2+} to Fe^{3+}, can cause accumulations of rust. However, the main cause of the trouble is that soils containing sulphates and sulphate reducing bacteria may electrochemically convert iron to FeS and H_2, especially at pH 7—a level favourable for the growth of *desulphovibrio*. If the FeS is deposited on the iron, so as to form a protective layer, the rate of attack may decrease; but such protection is only temporary, and after the film becomes detached, the rate increases again, probably owing to an Fe/FeS couple. Differential aeration is probably an important factor in microbiological corrosion, since many organisms tend to adhere to surfaces, and maintain very low oxygen concentration on the small parts covered. It is suggested that the original theory of von Wolzogen Kuhr (removal of polarizing H from the surface) can account for only part of the high corrosion rate met with in soil. There is rather less evidence that the removal of cathodic hydrogen is connected with the reduction of SO_4^{2-}. One highly probable mechanism which would account for the high corrosion rate ob-

1. W. P. Iverson, *Gas J.* 1968, **334**, 363.

served in the field may be the depolarizing action of iron sulphide when in contact with the iron.

A useful literature survey has been provided by Costello.[1] He discusses in turn:

(1) *thiobacilli*, which oxidize sulphur or sulphite to H_2SO_4; these organisms can survive in 10% to 12% H_2SO_4.

(2) *ferrobacilli*, which can oxidize Fe^{2+} to Fe^{3+}; the latter in turn oxidize sulphur compounds (e.g. pyrites, if present) to H_2SO_4. These iron bacteria can produce growths on the insides of water-pipes, and corrosion is set up by differential aeration.

(3) *sulphate reducing bacteria*, which are discussed at some length. The Pourbaix diagram is applied to show that, in presence of sulphur, FeS can be produced at potentials more negative than oxide; in the absence of oxygen and presence of sulphur, the EMF of the cell Fe/H can be high.

The mechanism of sulphate reduction has received authoritative discussion from Booth and his colleagues.[2] Studies of cathodic polarization of steel suggest that, besides the utilization of hydrogen by the bacterial hydrogenase system, depolarization by solid FeS at the cathode makes a contribution. They provide general confirmation of the von Wolzogen Kuhr mechanism, and state that an additional negative potential (0·1 V) is needed for cathodic protection in the presence of bacteria. They have studied also the action of certain strains of sulphate reducing bacteria on aluminium, where experiments carried out in presence of the bacteria produced up to 100 times the corrosion rate measured on the sterile controls.

The mechanism of the corrosion set up by FeS is being studied at Manchester.[3] Chemically prepared FeS is used, and it is found that 88 mg of FeS causes the corrosion of about 10 mg of iron. Continued high corrosion rates are observed if (and only if) the FeS is regularly replenished, and the corrosion rates are proportional to the quantity of FeS added. There are two possible ways in which FeS could stimulate attack. Current may flow between anodic points, where iron passes into solution as Fe^{2+}, and cathodic points, where hydrogen is produced and is absorbed into the lattice of the FeS, or removed by the bacteria—if present. Alternatively, the FeS particles may provide the cathodic areas, the hydrogen being formed upon them. It is not easy to decide between the two possibilities. Later experiments[4] have confirmed the belief that ferrous sulphide produced by the bacteria is the major corrosive agent, and that the concentration of soluble iron is the determining factor; a high corrosion rate is not maintained if the Fe^{2+} concentration is diminished. Apparently the importance of sulphate reducing bacteria is connected with their ability to produce FeS.

1. J. A. Costello, *Internat. biochem. Bull.* 1969, **5**, (3) 101.
2. G. H. Booth (with L. Elford, D. S. Wakerley and A. K. Tiller), *Brit. Corr. J.* 1968, **3**, 242; *Corr. Sci.* 1968, **8**, 549, 583.
3. R. A. King and D. S. Wakerley, *Brit. Corr. J.* 1973, **8**, 41.
4. R. A. King, J. D. A. Miller and D. S. Wakerley, *Brit. Corr. J.* 1973, **8**, 89.

The relative susceptibility of steel and cast iron to microbiological corrosion has been investigated.[1] Where the iron content of the environment is very low, the cast iron is found to be more susceptible to corrosion than any of the steels studied; but there is little difference between the corrosion rate of cast iron and steel specimens suspended in an iron-rich medium.

von Wolzogen Kuhr recognized that the oxidation of cathodic hydrogen by nitrate reducing bacteria might lead to corrosion similar to that caused by sulphate reducers. Cases of this have now been reported,[2] with a strain of an organism which possesses a hydrogenase system and is able to use nitrate as an acceptor under anaerobic conditions. Mild steel specimens suspended in liquid containing KNO_3, small amounts of $FeSO_4$ and sodium thioglycollate, as well as nutrient broth, corroded, when the organism (*E. coli*) was present, 6 times as fast as specimens exposed under sterile conditions. The findings support the original theory of von Wolzogen Kuhr; but it is felt that, in view of the diverse nature of the nitrate reducing bacteria and their frequency in soils, the corrosion caused by such organisms should be studied in greater detail.

Other bacteria can serve to produce nitrates. Two species exist, one of which oxidizes ammonia to nitrite, whilst another converts nitrite to nitrate. These reactions may be important in connection with problems of pollution control.[3]

It has been mentioned that corrosion in undisturbed soils is far less serious than in back-filled excavations. It was noticed in 1962 that sulphate reducing bacteria appeared to be absent in undisturbed soils into which pilings had been driven (1968 book, p. 123). A more recent American observation[4] confirms that in undisturbed soils corrosion is less than below a back-fill, and this is attributed to lack of oxygen replenishment. A French paper[5] describes a method of comparing the efficiency of various inhibitors against bacterial corrosion; the method is based on an estimation of the amount of H_2S produced by the reduction of sulphate.

Formulae for the strength remaining in corroded pipes have been provided by Marvin.[6] If this strength can be determined from geometry and material specification, better criteria as to the need for the replacement of pipes will be available; comparison has been made between the strength predicted from the formulae and that obtained experimentally by rupture tests. Some variation of the research formulae is suggested.

A summary of the extensive work carried out in the USA on soil corrosion has been provided from an Indian source.[7] Indian experience of cathodic protection is also summarized. One petroleum pipe-line 730 miles long crosses 70 rivers. The resistance of the soils traversed varies from 1000 to 1 000 000

1. D. D. Mara and D. J. A. Williams, *Brit. Corr. J.* 1972, **7**, 139.
2. D. D. Mara and D. J. A. Williams, *Chem. Ind.* 1971, p. 566.
3. G. Knowles, *Chem. Ind.* 1970, p. 697.
4. J. R. Rossum, *J. Amer. Waterworks Assoc.* 1969, **61**, 305.
5. E. Lagarde and A. Malderez, *Corrosion (Paris)* 1967, **15**, 276.
6. C. W. Marvin, *Corr. Abs. (Houston)* 1972, **11**, 67.
7. S. K. Gupta, B. Sanyal and J. N. Nanda, *Labdev. J. Sci. Tech. (Kanpur)* 1968, **6A**, 1, 51 (2 papers).

ohm-cm. Sacrificial anodes were used only in soils with a resistance less than 3000 ohm-cm.

The behaviour of welded titanium tubes in soils has been described by Romanoff.[1] The tubes, including the heat-affected zone and the weld, were found to remain unaffected by corrosion in all the soils included in the test. There was no pitting, and no apparent attack on the metal. The welding was carried out automatically by machines, which provided a structurally sound weld.

Protection of Buried Metals

Protective Methods independent of Cathodic Protection. Coal-tar enamels are largely used for the protection of underground pipe-lines. It would be satisfactory to find a coating which would prevent corrosion even when no cathodic protection was applied. Unfortunately this is not easy. The requirements have been enumerated by Kemp.[2] He regards impermeability to water, high electrical resistance and good adhesion as important, as well as resistance to bacteria, which exist in the pipe-line ditches, and can destroy some organic coating materials; in general, coal-tar enamel is found to be resistant to them. The coating must also resist attack by alkali which may be formed if cathodic protection is applied, and also loosening by cathodically produced hydrogen which may be formed at a holiday.

A specification was published in 1967[3] covering coal-tar coatings for hot application to iron and steel. Four types were recognized:

(1) *Unmodified coal-tar* applied by hot dipping and generally without a primer, so as to give a thin coating (0·1 mm maximum).

(2) *Filled unmodified coal-tar*, for application where no extremes of temperature are likely to occur. This is considered specially suitable for flood-coating materials which have been previously primed.

(3) *Modified coal-tar* for purposes similar to type (1); modification improves the physical properties.

(4) *Filled modified coal-tar*, for a wide range of application on previously primed materials.

The primers may be based on coal-tar or on a material such as chlorinated rubber. The filled coatings should be subjected to impact tests and also peeling tests.

This information is useful, but there is very little mention of corrosion in the specification. Organizations representing corrosion specialists do not appear to have collaborated in framing the recommendations. The result has been unfortunate. Some engineers have tended to assume that the application of any of these four types justifies the claim that protection has been provided which fulfils the accepted specification; in the absence of any clear statement as to when the more elaborate methods or materials should be used, there has been a tendency to assume that the simplest and cheapest

1. B. T. Sanderson and M. Romanoff, *Mat. Prot.* 1969, **8**, (4) 29.
2. W. E. Kemp, *Mat. Prot.* 1970, **9**, (6) 14.
3. *Brit. Standard Spec.* **4164** (1967).

one will suffice. This assumption is probably not, in general, justified. Some years after the appearance of the specification, a pile of tar-coated pipes was noticed by the author on the margin of a road, awaiting burial; already they were rusting along scratch-lines left by rough handling received in transport. Presumably, in such a case, if the place of burial chanced to lie in the anodic zone of a stray current system, corrosion set up at these gaps in the black coating would be more severe than if no coating had been applied at all. It is, of course, not certain whether in this particular case the coating used had indeed conformed to the specification mentioned; but it would seem that either the specification was not being used, or that the specified coating, when used, did not resist abrasion and produced a situation favourable to intensified attack.

A protective method used by the Welsh Gas Board[1] for a super-grid has been described. The pipe sections are welded, and the exposed metal surface is coated with a bituminous primer and then wrapped with polythene tape, allowing a 50% overlap. This is stated to provide a corrosion free joint of high dielectric strength.

Cathodic Protection. The choice of cathodic protection systems has been discussed by Palmer and Chapman.[2] They point out that magnesium alloy anodes provide only 50% current efficiency, whereas some zinc alloys provide 95%; they consider these alloys to be a major advance, and hold that the low voltage is an advantage, being calculated to avoid serious damage to paint. (It should be noticed, however, that the EMF provided by zinc alloys would be too low for some purposes.) They provide a table of comparison of the performance of magnesium, zinc alloys and aluminium alloys.

The relative advantages of different systems are also discussed by Berkeley.[3] He is dealing mainly with the problems of an effluent treatment plant, but much of his information is of general interest. He considers that magnesium with 50% efficiency is useful when poorly coated surfaces are present, that aluminium alloys with about 40% efficiency are useful in flowing hot water, whilst zinc with 95% efficiency is useful in cases where the resistance does not exceed 500 ohm-cm.

Berkeley also discusses impressed current methods, and compares the claims of different anode materials. He states that impregnated graphite is in general use, that iron containing 14% silicon should not be used in presence of Cl⁻ (otherwise its utility is fairly general), but that iron containing 3% molybdenum as well as 14% silicon can be used even when Cl⁻ is present, that platinized titanium, although expensive, is much used, and that a Pb–Sb–Ag alloy is effective in sea-water. He points out that cathodic protection is not a cure for all evils, nor is it always the cheapest method of solving a problem; very often, however, it is an attractive investment when used in conjunction with protection by coatings. It is probably unwise to consider cathodic protection of any structure against a water having a re-

1. *Corr. Prev. Control* 1967, **14**, (11) 15.
2. C. A. S. Palmer and D. A. Chapman, *2nd internat. Corr. Conf.* 1969.
3. K. G. C. Berkeley, *Chem. Ind.* 1971, p. 287.

sistance higher than 15 000 ohm-cm. In the so-called 'zero hardness waters' cathodic protection will generally be unsuitable; apart from their low conductivity, these waters do not build up a retentive film containing Ca and Mg compounds on the surface; such a film is often the real basis of the success provided by cathodic protection.

The system known as 'deep ground-beds' is discussed by Peabody.[1] The placing of ground-beds vertically below the surface to be protected, rather than on one side, has, he says, captured the imagination of many corrosion engineers; it reduces the problems due to stray current, and may also lessen expense connected with the purchase of rights-of-way. Deep ground-beds have been in use at least since 1941. Sometimes expendable materials are used for the anodes. One form of anode is an iron cylinder, but since iron is consumed at the rate of 20 lbs per ampere-year, a substantial quantity must be provided to give the bed a long operating life. The upper part of the cylinder used is coated with a protective layer; the lower part is largely bare, although a quarter or more of the circumference is covered with an armoured strip of first class material. The use of deep ground-beds for the cathodic protection of unwrapped oil pipe-lines has also been discussed by Hatley.[2] They may be buried at a depth exceeding 50 ft.

It is not unusual to prescribe a minimum current in a cathodic protection scheme. The strength depends upon the corrosion rate which would occur in the absence of protection; this can be expressed as a current, and, according to one view, the protective current should be 1·5, 2·0 or 3·0 times the corrosion current, according as the resistivity is below 5000 ohms, between 5000 and 10 000 ohms, or more than 10 000 ohms. However, Pourbaix[3] considers the control of potential more satisfactory than that of current. In the case of steel, this should be its 'immunity potential', which is about $-0·6$ V relative to the standard hydrogen electrode. He states that the electrode potential, rather than the CD, has now been generally accepted as the criterion of control.

If cathodic protection is to be controlled by potential regulation, a serious source of error must be avoided. The reference electrode will generally be placed on the ground above the pipe, and in general the measurement will not yield the potential difference between pipe and soil, since, if current is flowing, there will be an iR drop which vitiates the result. Various methods are available to eliminate (or at least reduce) this error, and reference should be made to a Swedish paper[4] in which they are compared. The conclusion reached is that 'the new earth megger proved to be very simple and reliable, and the method is well suited to iR-drop measurements on cathodically protected installations demanding a low total current'. Accuracy is less good when the current is high, and under such conditions the 'recorder method' is to be preferred.

The adsorption of hydrogen by cathodically protected steel pipes has been

1. A. W. Peabody, *Mat. Prot.* 1970, **9**, (5) 13.
2. H. M. Hatley, *Anticorrosion* 1968, **15** (6) 11.
3. M. Pourbaix, *Cebelcor Rapp. tech.* **205** (1972), p. 13.
4. B. Wallén and B. Linder, *Brit. Corr. J.* 1973, **8**, 7.

discussed by Hackerman.[1] For a given pipe-to-soil potential, the rate of hydrogen penetration is decreased by oxygen and by arsenic, but increased by the presence of sulphide.

American authorities[2] consider that stray currents from d.c., mainly electric railways, are still the major source of underground corrosion, but that developments in electronic devices may be expected to be helpful in solving the problems of control. The legal aspect of pipe-line corrosion, especially in regard to cathodic protection, is discussed by Hatley.[3]

A statement regarding the protection of Gas Council pipe-lines in the UK provides the following information.[4] 'Cathodic protection is always applied, as 100% protection can never be achieved or maintained by coatings alone. . . . Particular consideration is given . . . to points near or crossing electrified railway systems.'

McCaffrey[5] has discussed the effects of over-protection on pipe-line coatings; this may cause disbonding of coal-tar epoxy and coal-tar enamel coatings, but extruded polyethylene shows better resistance.

A somewhat disturbing report on the effects of cathodic protection comes from Denmark.[6] Certain district heating distribution system suffered from leakage due to corrosion. The installation of a cathode protection system was effective in reducing the frequency of leakage, but produced failures due to stress corrosion cracking, set up by the cathodically produced alkali which became concentrated by evaporation. It is stated that, unless it is found possible to control the potential so as to avoid the level where stress corrosion occurs, it may be necessary to dismantle the cathodic protection system.

The question as to whether a.c., when present, can increase corrosion has long been a subject for argument. An Australian view[7] is that, whilst the traditional belief that it causes additional corrosion equal to only 1% of that caused by d.c. may be realistic for steel, it is certainly untrue for aluminium, where the figure can be as high as 50%. Caution should, therefore, be observed in the use of aluminium pipes underground. Induced a.c. voltages may be dangerous. Tulloch[8] states that the effect of a.c. will vary with conditions. The damage may be 45% that of the equivalent d.c., but in other cases there may be no effect at all. Under stagnant conditions, where the rate of removal of the cation is slow, redeposition during the cathodic half-cycle may nearly compensate for corrosion during the anodic half-cycle. (But this happy state of affairs is not always achieved; some metals, notably aluminium, will not be redeposited.)

Pourbaix points out[9] that cathodic protection as used at paper works is

1. P. E. Hudson, E. S. Snavely, J. S. Payne, L. D. Fiel and N. Hackerman, *Corrosion (Houston)* 1968, **24**, 189.
2. L. H. Schwalm and J. G. Sandor, *Mat. Prot.* 1969, **8** (6) 39.
3. H. M. Hatley, *Anticorrosion* 1971, **18** (11) 6.
4. *Brit. Corr. J.* 1971, **6**, 6.
5. W. R. McCaffrey, *Mat. Prot. Perf.* 1973, **12** (2) 10.
6. B. Doulson, L. C. Henriksen and H. Arup, *Brit. Corr. J.* 1974, **9**, 91.
7. B. McCaffrey, *Australasian Corr. Engg* 1971, **15** (8) 9.
8. D. S. Tulloch, *Corr. Prev. Control* 1967, **14** (11) 13.
9. M. Pourbaix, *Cebelcor Rapp. tech.* **166** (1969).

really due to the alkali formed by the cathodic reaction, and not directly to the shift of potential. Hamner,[1] discussing the cathodic protection of a paper-making cylinder, recommends short cathodic pulses at relatively high CD (0·2 amp/cm^2). The double layer formed may, for a time, keep the surface potential low enough to avoid corrosion, even when the current is not flowing. One benefit of the system is the removal of fibres and filler residues from cylinder surfaces—probably as a result of the hydrogen generated. The quality of the paper is thus improved, and it has been possible to discard two of the 'doctor blades' (scrapers).

Corrosion under Marine Conditions

General. An authoritative survey of marine corrosion has been provided by Rogers.[2] An account of 16-year testing of marine materials exposed to sea-waters under tropical conditions has been provided by Southwell and Alexander.[3] The effect of water velocity has been studied by Danek.[4] He finds that alloys fall into three groups:

(1) Alloys with excellent resistance at all velocities. This group includes Ti alloys and a Ni–Cr–Mo alloy.

(2) Alloys which are resistant at high and moderate velocities, but which suffer localized corrosion at low velocities and in crevices. This includes most Ni alloys and stainless steels.

(3) Alloys resistant at low velocities, but which suffer corrosion-erosion at medium and high velocities. This includes copper-base alloys.

The effect of alloying elements on the corrosion of steel by flowing water has received study.[5] Fully annealed steels suffer more rapid corrosion in 3% NaCl than normalized steels. In the former, the elements Al, Si, P, Cr and Mn are beneficial, whilst C is detrimental. In the latter Al, Cr and Cu are beneficial.

A German paper[6] discusses the effect of bacteria in brackish water and sea-water; it only becomes important above 10°C. Bacteria other than sulphate reducers may be important; the fact that sulphate reducing bacteria have attracted so much attention is probably to be ascribed to the ease with which FeS is detected. It is stated that during the summer months bacterial action is inhibited by the fouling of the iron surface; in the autumn, when fouling organisms are killed by lower temperatures, bacterial cathodic action increases again.

An account[7] of the condition of the ships entrapped in the Great Bitter Lake near Suez in June 1967, as a result of political events, may arouse interest. By March 1970 one group of ships had attained the same corrosion

1. N. E. Hamner, *Mat. Prot.* 1969, **8** (2) 48.
2. R. H. Rogers, 'Marine Corrosion' (Newnes), 1968.
3. C. R. Southwell and A. L. Alexander, Report to Naval Research Laboratory, Washington, 1969.
4. G. F. Danek, *Corr. Abs. (Houston)* 1968, **7**, 116.
5. M. Kowaka and H. Nagano, *Corr. Abs. (Houston)* 1971, **10**, 301.
6. I. Ehlert and M. Pantke, *Werkstoffe u. Korr.* 1972, **23**, 196.
7. R. Juchniewicz and A. Banas, *Brit. Corr. J.* 1971, **6**, 234.

potential, independently of the paint used and the date of the last dry-docking; in this group corrosion at the water-line was more severe than that lower down. Other groups of ships had received some cathodic protection; these showed different potentials, and apparently the underwater corrosion had been lessened.

The result of 16-year tests on ferrous materials exposed to sea-water and fresh water under tropical conditions are described by Southwell and Alexander.[1] Their curves show how misleading can be attempts to predict behaviour from tests extending over relatively short periods, such as one year. The pitting of a steel containing 2% and 4% of nickel was less than that of mild steel after 1 year but much deeper after 8 to 16 years. Steel containing 3% or 5% Cr suffered lower losses than carbon steel at first, but after 4 years the situation was reversed. The same authors[2] describe tests on non-ferrous materials in the Panama Canal Zone lasting 16 years. They found that aluminium behaved well in marine environments, but suffered intense random pitting in a fresh-water lake. On Ni–Cu alloys the rate changes rapidly in the early stages, and short tests would certainly give inaccurate results.

A later report by Southwell[3] brings out the important point that the growth of organisms may, in some circumstances, prolong the corrosion life of steel structures in sea-water. When a clean steel surface is exposed to sea-water, a very high corrosion rate (up to 16 mpy*) is observed; this is due to corrosion of the oxygen absorption type. Then as a mat of fouling organisms is produced, the corrosion rate declines, falling after about a year to a steady value which may be 3·0 mpy in tropical sea-water or 2·5 mpy in temperature regions. This is due to the fact that the organisms exclude oxygen from the surface, but a slower type of anaerobic corrosion due to sulphate reducing bacteria continues. It is suggested that 'if these bacteria can be selectively controlled, whilst maintaining an adequate fouling cover, then very low corrosion rates for bare steel immersed in sea-water should be attainable'. There would seem to be great possibilities for greatly reducing wastage due to marine corrosion.

Influence of Minor Constituents of Sea-water. It was once common practice in laboratory testing to use N/10 NaCl as a substitute for sea-water. Such a practice can lead to misleading results. Under conditions of partial immersion, N/10 NaCl attacks steel more quickly than sea-water, and the difference increases as the period of immersion becomes more prolonged; in one set of tests (1960 book, p. 165), N/10 NaCl was found to corrode 1·9 times as quickly as sea-water in 2 days, but 3·3 times in 128 days; the Ca and Mg present in sea-water helps in building up a protective film. In contrast, under atmospheric conditions, the presence of sea-salt as

1. C. R. Southwell and A. L. Alexander (Naval Research Laboratory, Washington) (1969) Report, Part 9.
2. C. R. Southwell and A. L. Alexander, *3rd Internaval Corrosion Conference* 1969, p. 19A.
3. C. R. Southwell, J. D. Bultman and C. W. Hummer, *NRL Report* 7672 (Naval Research Laboratory, Washington), 1974.
* mpy stands for 'mils per year'; 1 mil = 10^{-3} inch.

dust on the surface may lead to more rapid corrosion than pure NaCl, because $MgCl_2$ present in the film of moisture on the surface may prevent evaporation; thus a surface carrying sea-salt may remain wet under conditions where one carrying pure NaCl would dry up.

Attempts have been made to provide formulae for 'synthetic sea-water' to be used in laboratory tests; this has sometimes been a NaCl solution with ions Mg^{2+}, Ca^{2+} and SO_4^{2-} added. Unfortunately there has generally been no attempt to introduce some of the other minor ingredients which can influence the corrosion rate; of these organic compounds derived from sea-weed would probably slow down corrosion, but certain compounds present in polluted estuaries, such as cysteine (1960 book, p. 479) would sometimes produce intense localized attack. Perhaps even more serious is a failure to consider the effect of the very small amount of iodine present in sea-water. Liss[1] has shown that this exists both as iodate and iodide; the ratio of IO_3^- to I^-, as measured, is about 20. Apparently there will be two redox systems O_2/OH^- and IO_3^-/I^-, available for the cathodic reactions in the corrosion of steel by natural (as opposed to synthetic) sea-water. It is generally assumed that the cathodic reaction, which often determines the corrosion rate, is the reduction of O_2 to OH^-; this reaction is known to proceed sluggishly. The reduction of IO_3^- to I^- would probably proceed much more readily, provided that IO_3^- does not become exhausted. The matter has not been studied experimentally, but it is possible that measurements of corrosion by NaCl solution with and without IO_3^- might provide valuable information.

Cathodic Protection under Marine Conditions. Special problems arise when cathodic protection is applied to ships, mainly because the cathodically formed alkali can either soften or loosen the paint. Paints free from saponifiable substances are therefore to be preferred. A coal-tar epoxy paint has been approved for ships in the British Navy fitted with impressed current systems.[2] It is stated to have a better adhesion than a vinyl paint, and that a thickness of 10 mils is obtained with only three coats. Inert pigments such as red iron oxide are recommended. Alkali attack is stated to be the main cause of the breakdown of paint systems.[3] It is recommended that impressed current systems should have an automatic control; for sacrificial systems zinc is being extensively used, as being less likely to cause troubles due to 'over-protection'.[4]

Paints for use in connection with the cathodic protection of ships in the Canadian Navy are discussed by Anderton.[5] He states that replacement of a red lead vinyl paint by an Al-pigmented vinyl paint has resulted in an improvement of adhesion. Experience in the Australian Navy[6] showed that

1. P. S. Liss, J. R. Herring and E. D. Goldberg, *Nature Phys. Sci.* 1973, **242**, 108.
2. R. Holland, *Brit. Corr. J.* 1969, **4**, 113; see also comment by F. R. Boynton, same page.
3. A. F. Routley, *Paint Tech.* 1967, **31** (4) 28.
4. *Corr. Prev. and Control*, 1969, **16** (12) 12.
5. W. A. Anderton, *3rd Internaval Conf.* 1969, p. 7.1.
6. P. J. Knuckey, *3rd Internaval Conf.* 1969, p. 21.1.

a zinc alloy with 0·3 Al and 0·05 Cd has been extremely satisfactory as sacrificial anode. An impressed current system provides virtually complete absence of rust and of pitting on the underwater hull area. Dome shaped anodes of lead containing 1% Ag and 6% Sb have proved reliable, but are heavy. Platinized anodes have been tried, but the platinum coating showed poor adhesion after $5\frac{1}{2}$ years.

Dutch tests[1] on paints used in the zone above the water-line show that specimens coated with paints which have no saponifiable constituents were in good condition after one year cathodic polarization. Where the paint was saponifiable, there was complete removal over a large area above the water-line. When the lower coat was unsaponifiable and the second saponifiable, the coating above the water-line remained in good condition, but there was now some damage on the submerged portion.

For impressed current systems, platinized platinum anodes, despite the expense, are now largely used. Cerny[2] states that these are stable in river-water containing Cl^- over a wider range of potential than in sea-water or pure salt solution. He attributes this to inhibition by other constituents present, possibly sulphates. However, the effect depends on the composition of the water, the working conditions and the construction of the anode. In Cerny's research 100% of the specimens were unaffected when Cl/SO_4 was 5/20, 88·9% when it was 5/10, but only 50% when it was 5/3·5. Tests carried out at Brixham, Devonshire,[3] show that electro-deposition was the only practical method of application capable of controlling thickness for a wide range of anode shapes; the coat is hard and not easily damaged, but can sometimes be detached by bending. It is held[4] to be bad practice to use platinized titanium anodes under stagnant conditions, since chlorine will be evolved and NaOCl formed, which can produce a bad effect on condenser tubes.

The use of aluminium alloy sacrificial anodes would have many attractions if the current efficiency could be made high, since the atomic weight is much lower than that of zinc, whilst each atom provides three electrons as opposed to two for zinc. Some of the alloys recommended contain mercury, which imposes problems due to toxicity. Pearson[5] describes tests with a sacrificial anode containing Al, Zn, Mg and In; prolonged tests on a raft in Poole Harbour are stated to have given satisfactory results, and the alloy is said to be 'currently in commercial use'.

Blakemore[6] reviews some recent developments in cathodic protection. One is the trailing anode used for the protection of ships. This generally takes the form of an aluminium ribbon coiled on a drum mounted at the stern of the vessel, and continually fed out as the immersed portion is consumed. When the ship is docked, the system is switched off and the ribbon wound

1.   J. H. de Vlieger, *Verfkroniek* 1971, **44**, 102.
2.   M. Cerny, *Werkstoffe u. Korr.* 1970, **21**, 610; also *Amsterdam Cong.* 1969, p. 721.
3.   D. M. Brasher, *Brit. Corr. J.* 1971, **6**, 197.
4.   W. E. Heaton, *Brit. Corr. J.* 1971, **6**, 193.
5.   A. W. Pearson, *Brit. Corr. J.* 1971, **6**, 55.
6.   J. S. Blakemore, *Australasian Corr. Engg* April 1972, p. 7.

in. He states that trailing anodes of platinum have also been used with success.

Hudson[1] provides interesting information about the cost of anodes for ships. On British ships zinc anodes are preferred for general purposes. Of 750 ships covered by the Chamber of Shipping's report of 1969, 558 were fitted with zinc anodes, 161 with aluminium anodes, and 31 with magnesium anodes. According to estimates given by an individual shipowner, the cost of supplying and fitting anodes to a 50 000 tonne ship would be £540 for aluminium, £700 for zinc and £810 for magnesium. The calculation assumes that aluminium or zinc anodes would need replacement after 4 years and magnesium anodes after only 2 years; in this respect, therefore, magnesium is at a disadvantage. He adds that the zinc anode systems for the Royal Canadian Navy are now designed to give an average life of 8 to 10 years.

Regarding impressed current systems, several types of anode can be used, including silicon iron, titanium or tantalum that has been electroplated with platinum, silicon iron (unplated) or lead–silver alloy. Silver–lead anodes appear to be the most economical, and it is hoped that, through improvement in the resins used for the insulating shields, modern designs should be capable of lasting a full 15 years before replacement.

Cathodic protection of stationary marine structures has been discussed by Fitzgerald.[2] Case histories include that of a dock in Chile, two wharfs on the New England coast, and highway bridges in the south-east part of the USA.

The problems of off-shore structures are becoming increasingly important. Suitable designs for sacrificial protection with zinc or aluminium, or with an impressed current system, are provided by Lehmann,[3] who also discusses the relative advantages of the various systems. Further information is provided by Hanson and Hurst,[4] who advise cathodic protection in the submerged zone, protective wrappings in the splash zone, and paint coatings in the atmospheric zone. In regard to the choice of materials, they state that magnesium anodes require frequent replacement; but, since the number needed is small, the loss due to storms is low. Zinc was at one time regarded as unreliable, but its problems were solved in the late fifties. Two aluminium alloys are mentioned, preference being given to the aluminium–mercury alloy, which provides a consistent efficiency of about 95%. (It might be added, however, that there is a widespread feeling against the use of mercury in the sea, as it tends to accumulate in fish.) The problems of the cathodic protection of mooring buoys and chains are discussed by Drisko,[5] with special reference to the Californian coast. In early systems it was noticed that some parts were not completely protected because of poor electrical continuity between certain links in the chains. Accordingly a new system with cast zinc anodes and a steel cable woven through the links has been

1. J. C. Hudson, 'Recommended Practice for the Protection and Painting of Ships' (British Ship Research Association), 1972, pp. 126, 128, 132.
2. J. H. Fitzgerald, *Mat. Prot. Perf.* 1972, **11** (5) 23.
3. J. A. Lehmann, *J. Metals* 1970, **22** (3) 56.
4. H. R. Hanson and D. C. Hurst, *J. Metals* 1970, **22** (4) 46.
5. R. W. Drisko, Note N-1045 (Naval Civil Engineering Lab., Port. Huememe, California).

installed. This has been found to give complete protection to buoys for at least five years.

Cathodic protection of the piles of marine ferro-concrete gantries has been found to require a CD of 0·3 to 0·5 mA/dm², according to the result of Russian tests.[1]

Automatic potential control rectifiers for cathodic protection in sea-water are described by Ferry.[2]

Problems of Water Treatment

Corrosion by softened water. The report of the Hoar Committee includes some comments on the effect of softening on the corrosive properties of supply water.[3] The use of new sources to meet increasing demand for water has led to problems. Lime-softening is cheap and effective for treating river-water in order to reduce hardness and remove organic pollution, but the almost complete removal of temporary hardness makes the treated water liable to corrode ordinary brass fittings. If the chloride content of the water is low, alternative but more expensive treatments are available. The additional cost of changing the water treatment would be much less than the replacement of brass fittings by some material unaffected by the water, and would be in the public interest. However, for water with a high chloride content, no practical alternative to lime-softening is at present available; expenditure on research into alternative methods of water treatment could be justified from a national standpoint.

Corrosion in desalination plants. The necessity of obtaining supplies of fresh water from the sea has involved corrosion problems in the desalination plants. An American report[4] states that dissolved oxygen has the greatest effect on the corrosion rate of a copper alloy in hot sea-water; velocity is not an important factor. Increased time of residence in a circulating system diminishes the corrosion rate when the oxygen concentration is 20 to 100 parts per billion, but increases the rate at a lower oxygen concentration. At the lowest oxygen concentration studied (below 5 parts per billion), the alloys tested suffered very little corrosion. Another American paper,[5] describing tests at San Diego, reaches the conclusion that chemical treatment of the water produced is the most satisfactory method of controlling corrosion under present circumstances, but that more research is needed.

Bom,[6] describing a desalination plant in Holland, states that about 90% of the material used is 90/10 Cu/Ni alloy, but there are small amounts of 70/30 Cu/Ni, stainless steel, nickel, Ni cast iron and Monel alloy.

1.   B. A. Zamanov and F. I. Samedova, *Corr. Control Abs.* 1970, No. 11, p. 28.
2.   R. Ferry, *Mat. Prot.* 1968, **7** (8) 26.
3.   'Report of the Committee on Corrosion and Protection'. Chairman: T. P. Hoar. (H.M. Stationery Office), 1971, p. 128.
4.   Metals for desalination plants: tests organized by the US Office of Saline Water. See F. Schreiber, O. Osborne and F. H. Coley, *Mat. Prot.* 1968, **7** (10) 20.
5.   E. I. Crossley and F. O. Waters, *J. Amer. Waterworks Assoc.* 1970, **62**, 188.
6.   P. R. Bom, *Brit. Corr. J.* 1970, **5**, 258.

The specification of an alloy for tubing in a desalination plant of the distillation type is an important matter, since it will greatly affect the capital cost. Italian tests[1] on tubes with sea-water at 50°C flowing through them at rates up to 4 m/s have yielded the result that aluminium brass displays better corrosion resistance than either 70/30 or 90/10 Cu/Ni alloy (containing iron) for rates up to 3 m/s; this is true whether the sea-water is 'clean' or contaminated with ammonia. In the presence of S^{2-} ions, the 70/30 alloy is superior to aluminium brass.

Corrosion of Steel in Concrete

General. The problems connected with steel reinforcements in concrete, and those which arise where a concrete layer has been used for protecting a steel surface, continue to arouse interest, and there is a welcome increase of scientific thinking. An encyclopaedia article,[2] after recalling the well known reason for using reinforced concrete—namely that concrete resists compression well whilst steel resists tension—brings out the additional point that steel is one of the few materials which expands with temperature in nearly the same degree as concrete. The assumption is often made that the steel cannot slip within the concrete, which shrinks during the setting process, and therefore grips the steel firmly, provided that the thickness of the concrete is sufficient; it should never be less than $\frac{1}{2}$ inch thick, nor less than the diameter of the steel bars. The best shape appears to be a cylinder with a rough surface. Reinforcing bars should, it is stated, be rusty and free from grease. An authoritative handbook[3] also recommends a film of rust, on the grounds that it causes an improvement of the bond, but the warning is given that when the reinforcing bars have become very badly rusted, the cross-sectional area may have been reduced so much that the bars are unsuited for use. It is pointed out that rusting is accompanied by an increase in volume; presumably if rust formation occurs after the concrete has come into service, a modicum of rusting will improve the bond, but a large amount may cause cracking or spalling of the concrete.

It should be pointed out that, whilst a uniform coating of rust may have practical advantages, local patches of rust might be dangerous through producing differential aeration cells with large cathodic and small anodic areas —always a dangerous situation. The danger would seem to be greatest when an inhibitor has been added to the concrete, since, although it will reach the unrusted part of the surface, it may fail to reach the steel below the rusty patches; it must be remembered that most inhibitors will interact with, and be destroyed by, ferrous salts, which will probably be present in the rust.

A sensible treatment of the subject comes from the Battelle Institute.[4] The increased use of NaCl or $CaCl_2$ for the removal of ice and snow from

1. I. Giuliani and G. Bombara, *Brit. Corr. J.* 1973, **8**, 20.
2. *Chambers' Encycl.* 1966, **3**, 837.
3. 'Concrete Construction Handbook', J. J. Waddell (McGraw-Hill), 1968.
4. A. B. Tripler and W. K. Boyd (Battelle Mem. Inst.), *Anticorrosion* 1969, **16** (4) 20.

roads, has caused corrosion of steel reinforcing rods in cases where good concreting practice has not been followed, or where traffic pounding has destroyed the concrete envelope. Differential aeration cells are a frequent cause of damage. A high salt content may increase the effect, not only by counteracting the passivity which would otherwise be set up by the alkali liberated during the setting of the cement, but also by lowering the solubility of oxygen. The Battelle authors consider that protective coatings are the best method of counteracting the trouble, and regard zinc, nickel, or asphalt–epoxy coatings as the most suitable, although further field tests would seem advisable. A report from Brussels[1] attaches importance to differential carbonation, but inequalities of the chloride content at different places may also set up corrosion cells. Chloride, which is often added intentionally to accelerate the acquisition of compressive stress, may become concentrated at certain points owing to unequal evaporation, porosity, or permeability. Differential humidity and differential aeration are regarded as possible causes of trouble.

The principles of fracture mechanics have been applied by Raharinaivo[2] to the problems of pre-stressed concrete.

A case is mentioned where pipes placed at a junction between concrete and glass wool (stated to have an acid reaction) suffered attack. The view is expressed that if the pipe had been buried wholly in cement, there would have been no corrosion, whilst if it had been wholly in the glass wool, there would have been less attack than occurred with burial at the junction. Other cases described are attributed to sulphate reducing bacteria. For instance, a pipe buried in sand in contact with clay on the upper surface had been attacked on the clay covered part; sulphate reducing bacteria and also sulphides were found. If the pipe had been surrounded by the same soil over the whole surface, there might have been no attack at all.

It is well known that the ratio Cl^-/OH^- is important in deciding whether passivity will break down. Experiments by Hausmann[3] suggest that added chloride does not cause corrosion if the ratio does not exceed 0·6. These experiments, however, appear to have been carried out with bare steel rods in an aqueous solution containing a gravel of limestone pieces $\frac{1}{4}$ to $\frac{3}{4}$ inch in size; the reader should consider whether this simulation is valid for the conditions in which he is interested.

An inspection of reinforcements in Indian structures, 20 to 40 years old, provides some interesting information.[4] Deterioration was least where the roof had a good slope, and was worst in servants' quarters, kitchens, latrines and bathrooms. Experiments on the creepage of solution into hollow spaces shows that the rate was 33 700 times as fast through a 1/4/8 mix as through 1/2/4; the diffusion of Cl^- and SO_4^{2-} is much faster in the former than in the latter mix.

1.   A. V. de Seabra and M do R. T. Cravo, *Cebelcor Rapp. tech.* **202** (1972).
2.   A. Raharinaivo, *Corrosion (Paris)* 1972, **20**, 276.
3.   D. A. Hausmann, *Mat. Prot.* 1967, **6** (11) 19.
4.   K. S. Rajagopalan, N. S. Rengaswamy and T. B. Balasubramanian, *J. sci. ind. Res.* 1969, **28**, 382.

Protective Measures. The addition of inhibitors to concrete has been proposed by many writers, but questioned, not without justification, by others. A discussion from a Hungarian source[1] states that nitrite cannot be recommended, since it becomes completely oxidized to nitrate in two to three years. An authoritative discussion of the subject has been provided by Treadaway and Russell.[2] They point out that concrete normally has a pH value of 11·5 to 12·5, and will keep the steel passive unless the cover is too thin (in which case CO_2 may destroy the alkali), or unless Cl^- can reach the steel. Some engineering designs require a thin cover, whilst others involve aerated concrete; in such cases, steel reinforcements must be protected either by coating with a bituminous composition (or with epoxy paint) or should receive hot-dip galvanizing. It is considered that in some cases the protection can be increased by incorporating corrosion inhibitors. Nitrites have been used in Russia, but produced bad pitting at certain Cl^-/NO_2^- ratios. The use of sodium benzoate has been patented, but this is stated to lower the compressive strength. In a new research a mixture of nitrite and benzoate has been tried. Nitrite is consumed during the passivation process, and becomes less effective with time; moreover the deep pitting reduces the cross section. Benzoate is an adsorption inhibitor, and its consumption rate is low. Both inhibitors reduce strength, and for that reason are best applied as a slurry to the steel before it is embedded in the concrete mass. The attack produced by Cl^- in presence of benzoate is stated to be more uniform than in nitrite, and the corrosion rate tends to become slower after some weeks.

Gouda[3] has studied the behaviour of steel both (1) in an alkaline solution, and (2) embedded in concrete. He finds that in aqueous alkali free from Cl^- the critical pH above which stable passivity is established is 11·5. That value rises when Cl^- is present in concentration C, according to the equation $pH = n \log C + K$, where K and n are constants (n is 0·83); this relation ceases to be valid at pH 13·5. The results obtained on steel embedded in concrete give results which are not always the same as those obtained in aqueous solution. Steel embedded in concrete made with Portland cement or a mixture of Portland and blast furnace cements becomes passive, provided that distilled water or supply water have been used; if sea-water has been used, the steel corrodes severely. Passivity in concrete is not impaired by sulphate up to 8%, but in experiments with a cement extract, the passivity failed at 0·2% Na_2SO_4. Additions of benzoate, chromate, nitrite, phosphate or stearate were found to be efficient in preventing the corrosion of steel reinforcements; the concentration necessary was found to be higher for mixtures containing blast furnace slag cement than for neat Portland cement. It is important to use the proper concentration, which must be determined under field conditions; if the addition is insufficient, there can be local intensification of the attack.

1.   I. Medgyesi, *Amsterdam Cong.* 1969, p. 591.
2.   K. W. J. Treadaway and A. D. Russell, *Highways and Public Water* 1968, **36**, 19, 40 (2 papers).
3.   V. K. Gouda (with W. Y. Halaka), *Brit. Corr. J.* 1970, **5**, 198, 204 (2 papers).

An Indian paper[1] describes a case where mild steel pipes carrying steam had suffered severe corrosion where they were in contact with a lagging material containing asbestos and magnesia. Tests suggested that protection could be obtained by means of barium potassium chromate.

Zinc coatings for Reinforcing Wires.   A digest made available by the British Building Research Station[2] may help the practical man. It is pointed out that so long as concrete remains alkaline the steel does not corrode (at least in absence of chloride); but that if CO_2 or SO_2 destroys the alkalinity, there is a risk; the corrosion may be of the differential salt concentration type. If chloride has to be added, it should be dissolved in water (rather than introduced as flake into the dry materials), so as to keep the distribution uniform. Zinc coatings have been used successfully to avoid corrosion; hot-dip galvanizing is generally the most economical form. If spraying is used, shot-blasting must precede it, whilst if zinc-rich paint is adopted, the surface should be relatively free from rust. Tests on bond strength seem to show that zinc-coated surfaces provide a better bond than smooth steel, and possibly slightly better than rusted steel; evidence about steel covered with mill-scale is contradictory. (It may be suggested that this is due to the fact that there are many kinds of mill-scale.) The action of alkali on zinc produces H_2, which may adversely affect the bond; but the risk is reduced by dipping the galvanized wire in a chromate bath.

The influence of zinc coatings on iron in contact with cement has been discussed by Kaesche.[3] He points out that zinc undoubtedly increases the evolution of hydrogen, but whether this sinks into the steel as H atoms, causing embrittlement, depends on circumstances. The presence of sulphides, which prevent the union of H atoms into harmless H_2 molecules, is detrimental. Zinc and galvanized iron are less sensitive to Cl^- than is bare iron. On zinc, Cl^- only becomes harmful between 0·1 and 1·0 mol/l, whereas on bare iron it is harmful between 0·001 and 0·01 mol/l.

An American paper reporting Norwegian work[4] confirms the effect of chromate in improving the bond strength on galvanized bars. In one concrete, which was low in chromate, the bond strength obtained with galvanized bars was greatly inferior to that obtained with ungalvanized steel; a chromate addition to the freshly made concrete, and especially a pretreatment of the galvanized bars in chromate solution, was found to eliminate the reduction in bond strength.

A research carried out by Everett and Treadaway[5] compares different methods of applying the zinc coat, including galvanizing, sherardizing and zinc-spray. They found that after three years there was no rust-staining, splitting or spalling, and no significant loss of zinc. Although after the first

1. S. K. Gupta, R. Kumar and B. Sanyal, *Labdev. J. Sci. Tech. (Kanpur)* 1969, 7A, 146.
2. *Building Research Station, Digest* No. **109** (1969).
3. H. Kaesche, *Werkstoffe u. Korr.* 1969, **20**, 119.
4. A. Hofsoy and I. Gukild, *J. Amer. Concrete Inst.* 1966, **66** (3) 174.
5. L. H. Everett and K. W. J. Treadaway, *Building Res. Station, Current Paper* 3/70 (1970).

year there were local dark patches, probably representing points of corrosion, where aggregate particles had been pressed into close contact with the zinc surface, such areas showed no significant loss of zinc even after two years. It is considered that the importance of hydrogen has been exaggerated, and its evolution is often suppressed by the chromate normally present in British concrete. With pre-stressed steel, the hydrogen might perhaps cause embrittlement, and this point still requires investigation. Some of the cases where zinc has been condemned by workers overseas, on the ground that it causes a loss of bond, refer to the use of a cement not commonly employed in the UK.

A method for the estimation of Cl^- in concrete, capable of being used by those who have had little or no chemical training, is greatly to be welcomed; it is stated that satisfactory results have been obtained by operatives on a construction site. The test—due to Smith and Banfield[1]—is designed for situations where ready-mixed oncrete receives additions of $CaCl_2$ to prevent freezing.

Effects of Stress. There has been much discussion regarding the corrosion of pre-stressed wires in concrete. The currents set up in such systems have been studied in Australian work,[2] and the distribution of anodic and cathodic areas appears to alter with time; the initial distribution seems to be determined by differential aeration, but the later one is connected with local leaching away of alkali. Another Australian paper, particularly concerned with concrete dams,[3] states that the wires may corrode in natural waters which are high in Cl^-, but that the concrete itself may be attacked if unwise cathodic protection is used with a view to protecting the wires. It is concluded that 'a proper study of the position seems opportune'. The behaviour of pre-stressing wires in concrete vessels is also the subject of a paper.[4]

An Austrian laboratory study[5] of stressed iron wire immersed in saturated $Ca(OH)_2$ deserves attention, but it should be remembered that behaviour in aqueous solution is different from that of metal buried in concrete. Potentiodynamic and potentiostatic methods have been used. In saturated $Ca(OH)_2$ the wire becomes passive, but the time needed for complete passivation increases with the stress applied; the critical CD for passivity also increases with the stress applied, owing to the increased concentration of defects. Additions of K_2CrO_4 bring about a reduction of the critical CD for passivity, and also of the current flowing in the passive state. $CaCl_2$ additions surprisingly reduce the current flowing in the active region. This is ascribed to the displacement by Cl^- ions of the adsorbed OH^-, which is a catalyst for the iron dissolution. (See this volume, p. 153.) Although in pure saturated

1.   A. V. Smith and P. R. Banfield, *Chem. Ind.* 1971, p. 1227.
2.   B. W. Cherry and N. L. Miller, *Australasian Corr. Engg* 1972, **16** (3) 15.
3.   *Brit. Corr. J.* 1971, **6**, 91.
4.   P. J. Hollingum, *Anticorrosion* 1972, **19** (4) 4.
5.   H. Grubitsch, H. Miklautz and F. Hilbert, *Werkstoffe u. Korr.* 1970, **12**, 485.

$Ca(OH)_2$ there is no stress corrosion cracking, such cracking can be produced when 2% $CaCl_2$ is present.

Authoritative information from the (British) Building Research Station[1] starts by saying that occasional failures of pre-stressed steels have been attributed by different authorities to

(1) contact with concrete of two different compositions,
(2) stress corrosion cracking due to nitrates in the soil,
(3) $CaCl_2$ in the concrete with cold-drawing on the site,
(4) H_2S produced by CO_2 acting on CaS in blast furnace slag,
(5) sulphides generally, causing H_2S and hence embrittlement.

There is an account of tests lasting 29 months at Beckton. After short exposure tests there was appreciable corrosion in concrete containing 2% $CaCl_2$ and still more with 5%; there was only superficial corrosion in absence of $CaCl_2$. Pitting started after 12 months on some of the specimens where $CaCl_2$ was present; this is regarded 'as a serious hazard', consequently the 'quality' of the concrete and grouting should be maintained and chloride contamination avoided. Carbonation of lime to $CaCO_3$ was complete within 4 months in concrete with a cement/aggregate ratio of 1 : 9, but had penetrated less deeply where the ratio was 1 : 4. Many steel specimens were virtually uncorroded when they were removed. When the wires were tested to fracture under tensile stress after their removal, the pits tended to be the origin of the fracture, but this was of a ductile character with visible necking. There was no evidence of crack formation, nor does it appear that Cl^- promotes stress corrosion cracking in high-tensile steel. There was little difference between concrete made with ordinary cement or blast furnace cement, but the degree of the corrosion depended on the quantity of chloride present.

Further References

Code of Practice.   'Cathodic Protection' (British Standards Institution). CP *1021* (1973).

Papers.   'Protective film formation on ferrous metals in semicontinuous cultures of nitrate reducing bacteria'.[2]

'Corrosion of mild steel by iron sulphides'.[3]

'Zinc ribbon anodes underground'.[4]

'Selection of anode systems for cathodic protection in natural waters'.[5]

'Cathodic protection of sub-sea pipelines'.[6]

'Effect of over-protection on pipe-line coatings'.[7]

1.   K. W. J. Treadaway, *Brit. Corr. J.* 1971, **6**, 66.
2.   S. A. Ashton, R. A. King and J. D. A. Miller, *Brit. Corr. J.* 1973, **8**, 132.
3.   R. A. King, J. D. A. Miller and J. S. Smith, *Brit. Corr. J.* 1973, **8**, 137.
4.   G. U. Kurr, *Mat. Prot. Perf.* 1973, **12** (2) 17.
5.   G. I. Russell and J. Banach, *Mat. Prot. Perf.* 1973, **12** (1) 18.
6.   W. B. Mackay and A. T. G. Edmonds, *Corr. Prev. and Control* 1973, **20** (3) 6.
7.   J. B. Lankes, *Mat. Prot. Perf.* 1973, **12** (3) 10.

'Research into the properties of materials for underwater structures in the North Sea'.[1]

'Bases experimentales de l'Étude électrochimique du comportement des Métaux en présence du béton'.[2]

'Pipeline voltage hazards due to proximity of overhead high voltages'.[3]

'A guide to plastic pipe materials'.[4]

'Placement of reference electrode and impressed current anode. Effect on cathodic protection of steel in a long cell.'[5]

'Scanning electron microscope study of the hot water corrosion of zinc'.[6]

'Tubes et Conduites'.[7]

1. K. J. Kievits and H. Slebos, *Werkstoffe u. Korr.* 1972, **23**, 1075.
2. P. Longuet, P. Peguin and A. Zelwer, *Corrosion (Paris)* 1972, **20**, 155.
3. E. A. Cherney and D. J. Carrigan, *Mat. Performance*, July 1974, **13**, 10.
4. *Corr. Prev. and Control* 1974, **21** (3) 7.
5. E. Schaschl and G. A. March, *Mat. Performance*, June 1974, **13**, 9.
6. H. Grubitsch and H. Harmer, *Werkstoffe u. Korr.* 1973, **24**, 1.
7. Eight papers with Introduction by B. le Boucher, *Corrosion (Paris)* 1973, **21**, 7–64.

CHAPTER IX

Acid Corrosion and Resistant Materials

Preamble

Picture presented in the 1960 and 1968 volumes. The types of corrosion discussed in chapter IV required the presence of oxygen; those discussed in chapter VII required an electric current from an external source. Cases are, however, known where corrosion can proceed without oxygen and without an applied current. The base metals, such as zinc and iron, will dissolve readily in an acid solution, with evolution of hydrogen; the more noble metals will be attacked readily by oxidizing acids such as nitric acid. In the latter case, the reaction is autocatalytic and, for that case, becomes much slower if the acid is in motion, because the substances forming the intermediate stages are removed; conversely, in the stagnant conditions prevailing at crevices, the reaction is particularly rapid.

The hydrogen evolution type of corrosion has some historical interest, because it is greatly influenced by the presence of certain impurities in the metal. It is well known that the reaction

$$2H^+ + 2e = H_2$$

proceeds much more readily on some surfaces than on others. On pure zinc, the hydrogen over-potential is high, and substances in solid solution may hardly affect it. If, however, a substance present in solid solution passes into the acid and is then reprecipitated as a second phase, this may provide a cathode on which hydrogen is readily evolved; the over-potential is then much lower and consequently the effective EMF driving the corrosion process is much higher. As a result, impure zinc containing lead and iron placed in acid evolves little hydrogen at first; but, when once these impurities have been reprecipitated as a dark sponge, the attack becomes much more rapid, hydrogen being evolved readily from the sponge. If the sponge is brushed away from part of the surface, the hydrogen evolution from the cleaned part becomes slow again. Although the sponge generally consists mainly of lead, it is probably the small amount of iron deposited on the extensive lead surface which really forms the cathode on which the hydrogen is evolved. Thus the corrosion rate of impure zinc in a non-oxidizing acid such as dilute H_2SO_4 is slow at first and then becomes fast. The action of nitric acid on copper also starts slowly and increases with time, but this has nothing to do with hydrogen over-potential; as already mentioned, it is connected with the building up of products needed for the autocatalytic cycle. On aluminium, the oxidizing properties of nitric acid serve to build up a protective oxide-film, which, like other sesquioxides, is only slowly dissolved by acid. In concentrated nitric acid, the attack is so slow that aluminium vessels can be used for storage and transport; the dilute acid permits more attack, but this is much slower than that produced on such noble metals as copper and silver.

Developments during the period 1968–1975 and Plan of the new Chapter. Only a small amount of work has been published during the period in question on the attack of acids on non-resistant materials. Considerable effort, however, has been devoted to the development of acid-resistant alloys, but much of the new information provided has arisen from a study of polarization curves, and has already been presented in chapter VII; it will not be repeated in the present chapter.

It has been thought well to re-name the chapter 'Acid Corrosion and Resistant Materials'. The chapter starts with a comparatively short section on non-resistant materials and their reaction with acid, and then deals with resistant materials, such as stainless steel. A perusal of recent papers on this subject has suggested that today many investigators are only slightly acquainted with the work of Pryor on reductive dissolution, and are apparently unaware of Berwick's application of the principle to explain the resistance of stainless steel towards acid. It has seemed worthwhile to provide a summary of both sets of research as an Appendix to this chapter.

Non-resistant Materials

Iron in Acid. The effect of temperature on the corrosion of Fe in HCl has been studied by Butler and Frost,[1] who obtained the corrosion rate by measuring the change in resistance of an iron wire corroding in acid; the Arrhenius diagram obtained on plotting the logarithm of that rate against the reciprocal of the temperature, consisted of two straight lines intersecting at about 40°C.

An Egyptian study[2] of the effect of acid concentration, carbon content and annealing temperature on the corrosion rate on steel in HCl will arouse interest, since the carbon content of the materials used extended from 0·14% to 1·13%; it thus embraced regions where ferrite and cementite are present as separate phases. A detailed study of the manner in which different phases present in steel behave during acid attack has long been awaited. Staehle and his colleagues,[3] using electron transmission microscopy, have now provided a wealth of information; their paper is illustrated by nearly a hundred electron micrograms. In different cases attack can be directed on the ferrite, the carbide or the interface between the two; again it can be general. The mode of dissolution of pearlite can be made to vary between different forms merely by change of the anion. The observed fact that the carbide can be dissolved at −1000 mV in HCl at pH 4 may arouse surprise, but it is in conformity with the appropriate Pourbaix diagram; there has been a tendency to assume that the attack on carbide is necessarily a slow change—even under conditions where it is thermodynamically possible; it now becomes evident that the reactivity of carbide is high compared to that of

1. E. B. Butler and J. J. Frost, *Corrosion (Houston)* 1971, **27**, 241.
2. A. A. Abdul Azim and S. H. Sanad, *Corr. Sci.* 1972, **12**, 313.
3. C. J. Cron, J. H. Payer and R. W. Staehle, *Corrosion (Houston)* 1971, **27**, 1.

carbon. The authors state that the attack on carbide is probably rationalized in terms of the reduction of carbide to iron and methane (or possibly methanol).

The effect of various elements on the corrosion of steel in $0 \cdot 1N$ H_2SO_4 is discussed by Cleary.[1] P and C are detrimental to corrosion resistance; Cu is mildly beneficial and Si slightly harmful; S, Cr and Ni are found to have no effect. The distribution of iron carbide exposed is important.

Further information is provided by Japanese work.[2] This refers to carbon steel in 1% H_2SO_4 and stainless steel in boiling 5% H_2SO_4. Although P and S decrease the resistance, the detrimental effect of P is suppressed by the presence of Cu and S; Pb, Sn, As and Sb improve corrosion resistance in presence of Cu and S.

Attack by organic acids on steel has been studied by Gupta.[3] The corrosion first declines with the number of carbon atoms and then increases again. Thus propionic acid produces intense attack at the water-line, butyric acid appreciably less; caproic, capric and lauric cause considerably less corrosion, but myristic and palmitic acids attack more quickly.

The attack upon Fe, Zn, Ni and Cu by non-aqueous monocarboxylic acids has been studied by Heitz.[4] If oxygen is absent, the reduction of the acid to the corresponding alcohol would be the 'expected reaction' in the sense that it would be accompanied by a greater drop of free energy than the evolution of hydrogen. However, as is so often the case, the result actually obtained is decided by kinetic and not by thermodynamical considerations; in Heitz's experiments, no alcohol or aldehyde could be detected by gas chromatography.

Calculation of the corrosion rate by extrapolation of the Tafel lines shows good agreement with direct analytical determinations, and points to an electrochemical mechanism. The overall reaction is controlled to the extent of two-thirds by the cathodic partial reaction, and only one-third by the anodic step. This result is analogous to results noted in reactions in several aqueous solutions. An interesting observation is that the ohmic drop between the statistically distributed anodic and cathodic reaction sites does not influence the reaction rate; that is true even of a solution of butyric acid containing lithium butyrate, which has a high resistivity, similar to that of distilled water. This rather surprising state of affairs is explained by the fact that the overall resistance is influenced by the reduction in size of the anodic and cathodic sites but also by the corresponding decrease in the distance between them.

Zinc in Acid. Krug and Borchers[5] have studied the local ring-pitting which occurs at the places where bubbles are produced when a non-oxidizing

1.   H. J. Cleary and N. D. Greene, *Corr. Sci.* 1967, **7**, 821.
2.   A. Takamura, K. Shimogori and K. Arakawa, *Amsterdam Cong.* 1969, p. 466.
3.   S. K. Gupta, *Labdev. J. Sci. Tech. (Kanpur)* 1968, **6A**, 113.
4.   E. Heitz, *Amsterdam Cong.* 1969, p. 533; *Werkstoffe u. Korr.* 1970, **21**, 360, 457 (with M. Hukovic and K. H. Maier).
5.   H. Krug and H. Borchers, *Werkstoffe u. Korr.* 1971, **22**, 309.

acid acts on zinc. It is regarded as a case of crevice corrosion, related to an impoverishment of H^+ ions in the constricted layer between the bubbles and the specimens. At first sight it might seem that this would reduce the local corrosion rate, but an analogous case, where reduction of H^+ increases corrosion rate, has been mentioned on p. 152.

The behaviour of zinc in H_2SO_4–H_2CrO_4 or H_3PO_4–H_2CrO_4 mixtures has received study.[1] Small additions of H_2CrO_4 to H_2SO_4 accelerate the attack; larger amounts give a brown, partly hydrolysed film containing trivalent chromium. If, however, H_2CrO_4 is added to H_3PO_4, the corrosion rate sinks below the value obtained with H_3PO_4 alone.

Although considerable knowledge exists regarding the way in which impurities in zinc affect attack by acids, usually by altering the hydrogen over-potential, less is known about their effect on attack by alkali. Lee[2] has measured the hydrogen over-potential in 9N KOH on zinc containing small amounts of impurities uniformly distributed. Except for Cd, the impurities studied (Ca, Fe, Hg and Mn) shift the cathodic polarization curve in such a way as would predict increased corrosion; direct corrosion measurements by weight loss or other method do not seem to have been carried out. The fact that Hg reduces over-potential may cause surprise; the situation may be not entirely simple. Powers[3] brings out the point that a coherent ZnO film on zinc can catalyse H_2 production in concentrated KOH.

Resistant Materials

Iron–chromium alloys. The nature of the passive film formed on Fe–Cr alloys has been studied by McBee and Kruger.[4] They used very thin foils produced by reducing the thickness of sheet material by means of a mixture of 70% perchloric acid with 5 times the quantity of glacial acetic acid; such foils could be studied by transmission electron diffraction; ellipsometric methods were also used. It was found that alloys containing less than 5% Cr, like Cr-free iron, carry a film composed of a well oriented spinel; alloys with 12% Cr carry a poorly oriented spinel; those with 19% Cr carry a mainly amorphous film, whilst the alloy with 24% Cr has a wholly amorphous film.

A question which has aroused much interest is whether the resistance of Fe–Cr alloys suddenly improves at a certain critical composition, or whether it improves gradually over a range. If corrosion resistance is due—as is believed in some quarters—to the filling of the D-band in the atoms, one would rather expect a sudden improvement at a certain composition. Kaesche[5] provides curves in which i_{crit} is plotted against Cr-content; these show a straight line, with no special features at 12% Cr. Likewise Fot and Heitz[6]

1. A. M. Shams-el-Din and M. G. A. Khedr, *Metalloberfläche* 1971, **25**, 200.
2. T. S. Lee, *J. electrochem. Soc.* 1973, **120**, 707.
3. R. W. Powers, *J. electrochem. Soc.* 1971, **118**, 685.
4. C. L. McBee and J. Kruger, *Electrochim. Acta* (in the Press).
5. H. Kaesche, 'die Korrosion der Metalle' (Springer), 1966, p. 195 (esp. Abb. 106).
6. E. Fot and E. Heitz, *Werkstoffe u. Korr.* 1967, **18**, 529.

state that no sudden change in corrosion resistance could be found at a critical Cr-content.

Work by Tomashov and others[1] has been directed to the effect of various constituents on the corrosion of stainless steel by acids. They find that Ni, Mo and (especially) Re greatly diminish the corrosion rate in H_2SO_4; the same elements reduce the current passing in the passive range when the stainless steel is made an anode. The function of Ni and Mo is to retard the anodic reaction directly. Re, although actually promoting anodic dissolution, promotes also the cathodic process, thus serving to bring the potential up into the passive range even when no anodic current is applied from an external source; its behaviour is thus analogous to that of Pt. It is found that Ag and Pd decrease the corrosion of certain stainless steels containing Fe–Cr–Ni, but Pd actually increases that of a steel containing Fe–Cr–Ni–Mo–Cu.

The effect of sulphur in stainless steel is discussed by Henthorne.[2] This may be present as an impurity (less than 0·03%) or added intentionally to improve machinability (about 0·3%). Sulphurized stainless steel which is low in manganese resists corrosion better than steels high in manganese, since manganese sulphide is less stable than sulphides rich in chromium.

The causes of resistance of stainless steel to acid are discussed in the Appendix to this chapter, where importance is attached to the fact that Cr is unstable in the divalent state, so that a film containing it is not destroyed by reductive dissolution.

An extensive study of the beneficial effect of passivating treatment on certain alloy steels has been carried out by Herzog.[3] Steels containing different amounts of Cr, Ni and Al have been studied both in the untreated and in the passivated condition. The passivating treatments used included 5 minutes' immersion in concentrated HNO_3; in some cases this was followed by treatment in boiling 3% CrO_3, or by cathodic treatment in boiling 3% CrO_3. Tests were carried out in marine, industrial and country atmospheres, and also under total or partial immersion in the laboratory. It was noted that, even in cases where preliminary passivating treatment reduced corrosion under conditions of total immersion, its effect was often lost under conditions of partial immersion, where water-line attack took place.

It is essentially the presence of chromium in the material which enables passivating treatment to bring advantages. Additions of Al or Ni were found to reduce corrosion under water-line conditions in 3% NaCl at 25°C, and resistance was further increased by a passivating treatment, but to no greater extent than was obtained on steel free from these elements. In contrast, when Cr was present in the alloy, the advantage of passivating treatment was much more pronounced.

1.   N. D. Tomashov and others. From Collection of Russian Papers No. 2, translated by A. D. Mercer and C. J. L. Booker (Nat. Lending Library, Boston, Spa), Vol. I (1967), pp. 1, 11, 121, 138, 146, 192, 380.
2.   M. Henthorne, *Corrosion* (*Houston*) 1970, **26**, 511.
3.   L. Backer and E. Herzog, *Corrosion* (*Paris*) 1967, **15**, 117.

In recent years, efforts have been made to reduce the carbon content of stainless steel, so as to extend serviceability, despite the increasing severity of conditions which are being imposed today. The manufacture of stainless steel with a very low carbon content has long aroused interest. A process based on the induction electric furnace has been described from an Italian source.[1]

In the case of castings, carbon tends to be taken up on the surface, so that the external layer comes to carry a much higher carbon content than the interior. Work in Fontana's laboratory[2] shows that this surface carbon uptake can be due to the use of moulds containing organic or carbonaceous material, and that it can be avoided by choosing a mould material free from carbon. The corrosion rate then declines; it has been shown that the corrosion rate of castings produced in green-sand or ceramic moulds may be only a quarter of those produced in a resin shell mould.

Electrochemical methods have been used by Jackson and van Rooyen[3] in evaluating the resistance of various stainless steels in presence of chloride. They depend on the determination of the critical CD and pitting potential, and also on the measurement of the Cl^- concentration above which the current flowing (at a certain fixed potential) suddenly increases, indicating a breakdown of passivity. The results are found to correlate well with the rust-staining observed in a marine atmosphere.

The pitting resistance of ferritic stainless steels in HCl is improved by the addition of Mo, especially in the case of materials produced *in vacuo*.[4] Material with 25/2 Cr/Mo is superior to 17/3. The addition of Mo also prevents crevice corrosion. Other information regarding the improvement of the resistance of iron with 28% Cr by the addition of Mo is provided in a paper from Pittsburg.[5]

It would be of great economic advantage to certain countries if, in stainless steel, a cheap austenite stabilizer could be used instead of the expensive nickel. Recent Egyptian work[6] suggests that considerable replacement of Ni by Mn is possible; as regards pitting resistance, an alloy containing 17·3% Cr, 5·3% Ni and 5·6% Mn is found to be comparable to one with 18% Cr and 9% Ni.

Nickel-based Alloys. A new series of alloys designed to resist hot H_2SO_4 of various concentrations is arousing interest.[7] Optimum resistance is found to be given by an alloy containing 9·25% to 9·75% Si, 2·5% to 3·0% Ti, 2% to 3% Cu and 2·75% to 3·25% Mo. The resistance depends on the structure. In the binary alloy with 9·2% Si, the maximum penetra-

1. V. Giacomazzi and A. Ricci, *Met. Ital.* 1967, **59**, 105.
2. W. A. Luce, M. G. Fontana and J. W. Cangi, *Corrosion (Houston)* 1972, **28**, 115.
3. R. P. Jackson and D. van Rooyen, *Corrosion (Houston)* 1971, **27**, 203.
4. A. P. Bond and E. A. Lizlovs, *Werkstoffe u. Korr.* 1970, **21**, 336.
5. C. R. Rarey and A. H. Aronson, *Corrosion (Houston)* 1972, **28**, 255.
6. A. M. Shams-el-Din, M. M. Badran and S. E. Khalil, *Werkstoffe u. Korr.* 1973, **24**, 290.
7. W. Barker, T. E. Evans and K. J. Williams, *Brit. Corr. J.* 1970, **5**, 76.

tion by boiling 26% H_2SO_4 occurs when the γ-phase content is just sufficient to provide continuous paths into the interior; the α–γ combination provides an efficient corrosion couple. However, in boiling 80% and 94% H_2SO_4 the α-phase is preferentially corroded. Another nickel alloy[1] containing 25% Cr, 10% Mo was originally developed for high temperature purposes such as flame-tubes of gas turbine engines. It is now being used in other situations, since it has been found to be resistant to acids. In 10% H_2SO_4 the thinning due to corrosion is only 0·6 mm per year, and in H_3PO_4 up to 80% it does not exceed 0·7 mm per year; however, 20% H_2SO_4 corrodes it 8 times as fast as the 10% acid. There is good resistance to pitting and crevice corrosion in oxidizing chloride solutions.

Titanium.   An authoritative paper on extended uses of titanium has been provided by Cotton.[2] He points out that neutral brines and most organic acids (except oxalic and formic) produce practically no corrosion on titanium, unless hydrogen is present, when, in some circumstances, a passive film may break down. Methanol is more reactive than ethanol. Complete dehydration may produce breakdown of passivity in circumstances where a small quantity of water prevents attack; breakdown may occur in completely dry chlorine or dry red fuming nitric acid.

Titanium anodes coated with Pt (or a Pt alloy containing 30% Ir) are now being used in cells producing sodium chlorate, and also in both the mercury and diaphragm types of cells producing chlorine. This departure has resulted in considerable economies over the traditional use of graphite anodes, since the power consumption is less, the cell maintenance easier, and the product purer.

As a cathode, titanium is now tending to replace 'oiled copper' or stainless steel for the production of the thin 'starter sheets' of copper needed in copper refineries. (On these starter sheets the bulk of the copper is cathodically deposited.) Titanium provides just enough adhesion for the deposition process, but allows the starter sheet to be stripped easily when the desired thickness has been attained.

The use of titanium in the oil and petrochemical industries is discussed by Kay.[3] One outstanding merit is resistance to chlorides, but titanium also resists sulphur compounds. Probably the largest application lies in organic syntheses involving chloride catalysts, where the Cl makes stainless steel unsuitable, whilst the high temperature and pressure rules out most non-metallic materials. Titanium is useful in the manufacture of acetaldehyde from ethylene using a copper chloride catalyst.

The behaviour of titanium in acidic, oxidizing, aqueous and alcoholic systems is discussed in a paper from Haifa.[4] Titanium is found to exhibit a wide passivity range in acid solutions containing Cl^- and NO_3^-. These media show an ability to form and repair protective films.

1.   A. C. Hart, *Brit. Corr. J.* 1972, **7**, 105.
2.   J. B. Cotton, *2nd internat. Titanium Conference, Boston* 1972.
3.   V. B. Kay, *het Ingenieursblad* 1970, **39**, 797 (in English).
4.   A. Alon, M. Frenkel and M. Schorr, *Amsterdam Cong.* 1969, p. 636.

Protection by Salt-like Oxides. An important discussion by Hoar[1] emphasizes the part played by 'salt-like oxides' in forming protective layers. He remarks that cations with high charge and small radius tend to form covalent bonding between metal and oxygen; the oxides are then sparingly soluble and also tend to form complex anions; Al, Sn, Mo and especially Si and B, provide examples. Many other cations have only a small tendency to covalent bonding with oxygen; examples are provided by Cu, Fe, Ni, Cd, Hg and Mg. The two groups sometimes combine to give salt-like oxides, and such alloys or phases display relatively high resistance to corrosion. Examples are provided by NiSn and $CuAl_2$; it is suggested that the protective films formed may be $NiSnO_2$ and $Cu(AlO_2)_2$ respectively. Hoar's views are summarized in Table III.

TABLE III

THE USE OF 'SALT-LIKE OXIDES' IN RESISTANT FILMS (T. P. Hoar)

Component with 'basic' oxide	Component with 'acidic' oxide	Alloy(s) more resistant than 'basic' component
Cu	Zn	Brasses
	Al	Aluminium bronzes
	Sn	Tin bronzes
	Si	Silicon bronzes
Fe	Cr	Chromium irons; stainless steels
	Sn	$FeSn_2$
	Si	Silicon irons
Ni	Cr	Nichrome, Nimonics
	Sn	NiSn
	Mo	Hastelloys etc.
Co	Sn	CoSn
	Cr	Vitallium
Ti	Mo	16% (wt) Mo alloy[2]
	Ta	5% (wt) Ta alloy[2]
Zr	Al	35–50% (wt) alloys[3]

Clearly if the combination of the basic and acidic oxides to form a salt is attended by a drop in free energy, the salt-like oxide will be a stable one, since this energy has to be supplied if the two metals are to pass into solution. Hoar states, 'Generally the solubilities and the dissolution rates of the oxides formed by cations that form strong covalent oxide links are smaller than those of oxides that do not. However, a strong degree of interaction between these "acidic oxides" and the more easily dissolved "basic oxides" may be expected to produce salt oxides wherein the "basic cations" are prevented from such easy dissolution. Thus binary alloys of the copper, iron, nickel group, with the aluminium, tin, molybdenum group, may be expected

1. T. P. Hoar, *J. electrochem. Soc.* 1970, **117**, 17C.
2. T. P. Hoar and D. C. Mears, *Proc. Roy. Soc.* (A) 1966, **294**, 486.
3. T. P. Hoar and O. Radovici, *Trans. Inst. Met. Finishing* 1964, **41**, 88.

to be more readily and more nearly completely passivatable than the pure metals, as is in fact found.' He gives several examples, including the inter-metallic compound NiSn, which can be formed, with almost exactly stoichio-metric composition, by electrodeposition. It exhibits passive behaviour over a wide pH range, and in most solutions becomes passive at a less positive potential than do either nickel or tin. There appear to be three different films lying over one another; these, he suggests, may be respectively $NiSnO_2$, $NiSnO_3$ and $NiSnO_4$; all three compounds would contain Ni and Sn in the same ratio as in the alloy. It is suggested that the bonding would be sub-stantially covalent as to the tin and electrovalent as to the nickel, and that there is stabilizing resonance with other possible structures.

Elsewhere, Hoar[1] discusses the conditions which must be fulfilled if a metal which is not 'noble' shall nevertheless resist corrosion in service. Most metals develop oxides or hydroxides which over a limited range of pH are 'insoluble'; for practical purposes this may be taken to mean that the solu-bility is less than $10^{-6}M$, but for many metals the range is a narrow one; for instance, $Zn(OH)_2$ dissolves as a cation if the pH is less than 8·5 and as an anion if it exceeds 10·7—so that the range is only 2·2. In contrast, Ta_2O_5 and Nb_5O_3 are 'insoluble' (in the sense adopted) over a whole range of pH from below −2 to above +16, so that protection may be expected under all conditions normally met with. An important case is that of titanium; TiO_2 is soluble as a cation when the pH is lower than 2·4 and as an anion when it exceeds 12·0, giving a range of resistant behaviour of 9·6—a satis-factory situation.

Hoar also provides a useful review of the behaviour of iron-based alloys containing Cr and Ni. He states that the so-called 'duplex' austenitic ferritic alloys can be made considerably stronger (mechanically) than the fully austenitic alloys and can in some cases be at least as corrosion resistant, especially against stress corrosion cracking and pitting in Cl^- environments. An alloy with 18·5% Cr, 4·7% Ni, 2·7% Mo, 1·5% Mn and 1·65% Si is at least 50% ferritic, with a 0·1% proof stress much higher than the classical 18/8 alloy and a greatly improved resistance to stress corrosion cracking. An alloy with 26% Cr, 5·25% Ni, 2·1% Mo, 3·4% Cu, 1% Mn and 0·6% Si has a much improved resistance to pitting in sea-water. Among alloys with 18% Cr, those containing only 2% Ni still retain moderately good corrosion resistance, whilst the nickel free 17% Cr alloy (used in automobile trim) and especially the 25% Cr alloy are resistant to many environments.

'Thus Ni in Fe–Cr–Ni steels is mainly a modifier of physical properties, which can be adjusted by heat and mechanical treatments, rather than a direct aid to corrosion resistance. However, very high Ni contents, 40% or so, confer near immunity to stress corrosion cracking in chloride environ-ments.'

1.   T. P. Hoar, *S.A. Engineer and Elect. Rev.* 1972, May, p. 31; June, p. 2.

Further References

An authoritative discussion of materials for chemical plants has been provided by Edeleanu[1] and deserves study. He states: 'No specialist can work in isolation. ... It is extremely simple to resolve a corrosion problem by specifying glass or some out-of-the-way unweldable alloy, but this is no help if it throws the problem from the corrosion specialist on to the engineer, who will then be faced with designing in glass, or the welding technologist, who will have to think of a clever way of welding something which is basically unweldable.'

A useful review of non-ferrous corrosion resistant metals and alloys for the chemical industry has been provided by Rama Char.[2] A paper on the 'control of HNO_3–HF stainless steel pickling baths with the aid of a permaplex membrane electrode and a lanthanum fluoride electrode'[3] will interest some readers. Barnartt[4] has discussed 'Tafel slopes for iron corrosion in acidic solutions'.

The beneficial effect of Mo in high-purity stainless steels containing 13% and 18% Cr (no Ni) has been studied by Lizlovs and Bond.[5] Potentiodynamic polarization curves in N HCl showed that 13% Cr steel without Mo only becomes passive when the applied anodic current density reaches the critical value of about 10^{-1} amps/cm^2, whereas if 3·5% Mo is present one-tenth of that value suffices. Immersion tests in 0·33M $FeCl_3$ acidified to give pH 1 produced severe general corrosion on 13% Cr containing no Mo or only 1% Mo. The other materials tested suffered pitting. Mo addition produced no improvement on 13% Cr steel, but on 18% Cr steel, a high Mo concentration did diminish the pitting-rate. Potentiostatic experiments in N HCl showed that 1% Mo could produce complete passivity in the 18% Cr alloy, but failed to do so if the Cr-content was only 13%.

The same authors[6] have studied the deterioration of 18/2 Cr/Mo ferritic stainless steel stabilized with titanium produced by heating at 475°C. The well-established embrittlement is accompanied by loss of corrosion resistance —notably in boiling formic acid—due to the precipitation of α' particles, both at grain-boundaries and in the matrix; these consist of a Fe–Cr phase with 60–80% Cr, and the depletion of Cr explains the change of behaviour.

1. C. Edeleanu, het Ingenieursblad 1970, **39**, 783.
2. T. L. Rama Char, Corr. Prev. and Control 1971, **19** (1) 8.
3. T. Eriksson and S. E. Lunner, J. Iron Steel Inst. 1973, **211**, 581.
4. S. Barnartt, Corrosion (Houston) 1971, **27**, 467.
5. E. A. Lizlovs and A. P. Bond, J. electrochem. Soc. 1975, **122**, 719.
6. E. A. Lizlovs and A. P. Bond, J. electrochem. Soc. 1975, **122**, 589.

APPENDIX TO CHAPTER IX

Early Work on Reductive Dissolution and its Application to the Behaviour of Stainless Steel

(mainly by M. J. Pryor and I. D. G. Berwick)

Direct Dissolution of Ferric Oxide. Pryor[1] prepared α-ferric oxide by precipitating a solution of ferric ammonium sulphate with aqueous ammonia, washing thoroughly with a large volume of boiling distilled water, drying and igniting in a platinum boat for 24 hours at different temperatures between 200° and 1000°C. He then studied direct dissolution in 0·1N HCl at 25°C. The dissolution is extremely slow and decreases with the temperature of ignition, as shown in Fig. 17. The dissolution rate increases with the acid

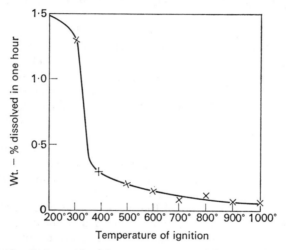

Fig. 17 Relation between ignition temperature of ferric oxide and percentage dissolution in 0·1M HCl at 25°C (M. J. Pryor and U. R. Evans).

concentration as shown in Fig. 18; it is quicker in HF and slower in H_2SO_4. Comparative experiments carried out in HCl under stagnant and agitated conditions indicated that movement of the liquid made only a small difference to the rate of dissolution; this indicates that the slow rate of dissolution cannot be explained by exhaustion of acid in the layer next to the oxide particles.

The dissolution rate was found to fall off with time. This suggests that there is some preferentially soluble material on the surface of the ferric oxide particles; when it has been removed, the dissolution rate declines. Pfeil (1960 book, p. 24) had found that material which is nominally Fe_2O_3 really contains more than the stoichiometric proportion of iron; evidently part of this excess iron must be in the ferrous condition, if electrical neutrality is to be

1. M. J. Pryor and U. R. Evans, *J. Chem. Soc.* 1949, p. 3330.

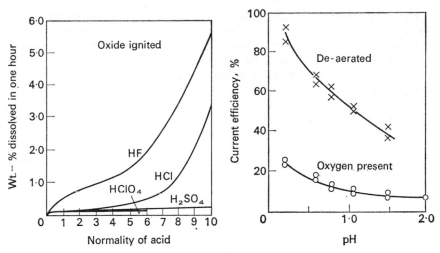

Fig. 18 Relation between normality of different acids and percentage dissolution of oxide ignited at 1000°C (one hour at 25°C) (M. J. Pryor and U. R. Evans).

maintained. Tests with di-2-pyridyl showed that ferrous iron was present in the acid extract. It has been shown that α-Fe_2O_3 is a semiconductor of the metal-excess type containing vacant sites in the oxygen lattice. For the maintenance of electrical neutrality, these oxygen defects must be associated with the Fe^{2+} ions in the lattice. It is probable that around defects the inter-atomic distances will be slightly different from those in a 'perfect' Fe_2O_3 lattice, making passage into the liquid easier. The relatively large amount of ferrous iron in the acid extract suggests strongly that dissolution takes place most readily at the lattice defects where the energy needed for dislodging ions will be smaller than elsewhere. This view is supported by the fact that oxide which has been pre-treated in acid and has become slowly dissolving can be restored to the relatively quickly dissolving condition by heating at 1000°C, which will allow lattice defects from the interior of the grains to move up to the surface and thus provide quickly dissolving material.

Reductive Dissolution of Ferric Oxide. If ferric oxide receives cathodic treatment in acid, we may reasonably expect to obtain ferrous ions and therefore rapidly dissolving conditions on the surface. This had previously been regarded as the reason why, despite the slow attack on powder which is nominally Fe_2O_3, films of Fe_2O_3 present on heat-tinted iron can be dissolved away in a few seconds. It has been shown that if the films are removed from the iron to a plastic basis, they can be exposed to acid for relatively long periods without dissolution. If, on the other hand, a heat-tinted iron specimen is dipped in the acid, the cell Fe | acid | Fe_2O_3 will be set up at every discontinuity in the film; the exposed metal will suffer anodic attack, and the oxide-film will be reduced to the ferrous condition, rapidly passing into solution as Fe^{2+} ions—thus explaining the rapid disappearance of the colours.

Pryor[1] demonstrated this mechanism by means of the cell shown in Fig. 19. The oxide used in these experiments was the same powder as had been used for the work on direct oxidation. The current efficiency was high at first if oxygen was excluded, falling off with time, but was low from the outset if the solution in the cell contained oxygen. That is evidently because, if oxygen is present, an alternative cathodic reaction is possible, namely the reduction of oxygen. It would thus appear probable that, if a relatively powerful oxidizing agent were present in sufficient amount, the whole of the current passing would be devoted to the reduction of that oxidizing agent, and the Fe_2O_3 would remain unreduced and would escape dissolution in the acid. It has long been known[2] that destruction of the tints on heat-tinted iron by 0·01M H_2SO_4 can be prevented by the presence of chromic acid.

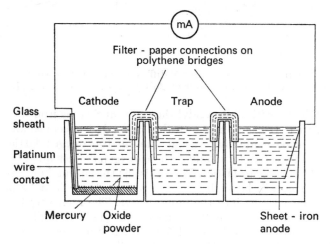

Fig. 19   Cell for studying the reductive dissolution of ferric oxide (M. J. Pryor and U. R. Evans).

Although on iron tinted to the first order red-brown colour the tints disappear within 2 s in acid of pH 1·0, they are unchanged after 24 hours at pH 4. The remarkable effect of pH on the time needed for autoreduction is shown in Fig. 20. The effect of pH is partly due to the fact that the equilibrium potential of the reductive dissolution reaction includes the term $(3RT/F \log_e [H^+])$, whereas the values for either of the two alternative reactions (the liberation of hydrogen and the reduction of oxygen) include the (smaller) term $(RT/F \log_e [H^+])$, so that the first reaction is hindered more by a rise in pH than the other two.

It is interesting to consider these facts on the basis of the Pourbaix diagram for iron, which shows a sloping line representing the equilibrium

$$2Fe^{2+} + 3H_2O \leftrightharpoons Fe_2O_3 + 6H^+ + 2e$$

1.   M. J. Pryor and U. R. Evans, *J. Chem. Soc.* 1950, pp. 1259, 1266, 1247 (3 papers).
2.   U. R. Evans, *J. Chem. Soc.* 1930, p. 481.

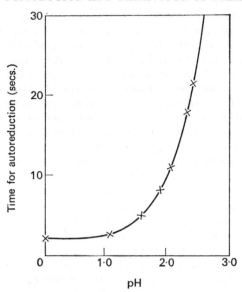

Fig. 20   Relation between pH value and time of auto-reduction of first order red-brown films (M. J. Pryor and U. R. Evans).

Below this line the reaction can proceed in the right-to-left direction, representing the destruction of the film by reductive dissolution and the loss of passivity. Above the line the reaction can proceed in the left-to-right direction; work in Cohen's laboratory (this book, p. 141) shows that such a reaction can play an important part in building up the film under conditions where an appreciable amount of Fe^{2+} has accumulated in the liquid (e.g. if a period of active dissolution has elapsed before conditions have become suitable for passivation).

The general conclusion from Pryor's work is that the destruction of ferric oxide films which occurs when heat-tinted iron is dipped into dilute acid is *not* due to *direct* dissolution as *ferric* ions, but to *reductive* dissolution as *ferrous* ions. The destruction occurs quickly and regularly when the films are thin, so that the colours change uniformly over the whole surface; probably electrons can pass through these thin films easily, permitting uniform cathodic action at all points of the outer surface. The thicker films are destroyed slowly and irregularly, probably because the reaction occurs preferentially around discontinuities.

Early work (published in 1930 and summarized on p. 222 of the 1960 book) had shown that iron (carrying either an invisible air-formed film or an early interference tint film produced by gentle heating) was rendered passive by anodic polarization in dilute H_2SO_4 at high CD, providing a potential capable of producing oxygen. When the current was stopped for a short time, and then re-started, the iron was found still to be passive and oxygen evolution was observed at once. If the cessation of current had been too long, the iron was found to have become active once more; when the current was re-started, it produced, not evolution of oxygen, but anodic corrosion of the metal; the loss of passivity was shown also by the electrical readings. In some experiments, during the cessation of current, an evolution of oxygen in bubbles was observed to continue; after a time bubbles ceased, and shortly afterwards the metal became active. This suggests that passivity is preserved as long as an excess of oxygen is present in the film; whilst that is the case, any anodic attack on the metal at a tiny gap in the film is balanced by the cathodic reduction of oxygen, and reductive dissolution of the film does not occur; only when the excess oxygen has been used up, does destruction of the film occur, leading to activity.

Recently Sato[1] has definitely demonstrated the excess of oxygen by an ellipsometric method; he considers that the passive film consists of an inner layer of anhydrous oxide (Fe_2O_3), which thickens with the potential, and an outer layer of hydrous oxide containing the excess oxygen.

It could be suggested that the iron in this outer layer may be in the hexavalent condition; the two layer picture here given by Sato is different from the two layer picture (Fe_3O_4 and Fe_2O_3 films) provided by Cohen—which Sato does not accept. On the other hand, Frankenthal[2] has expressed general agreement with Cohen's outlook. On nickel, Sato[3] believes that the film formed in the passive range is essentially NiO, but that in the oxygen evolution range a higher oxide, probably Ni_2O_3, is present; in that range, the film thickness increases as potential rises.

In 1952 Berwick[4] showed that if heat-tinted iron was charged with oxygen by anodic pre-treatment in acid for different periods before the current was turned off, the time which elapsed before passivity suddenly broke down (as shown by a sudden change in potential) increased with increase in the period of pre-treatment—up to a certain value, beyond which there was no further increase. The reproducibility between duplicate experiments was less than perfect, but the general trend of results confirmed the suggestions made above; so long as some of the excess oxygen survived, an alternative cathodic reaction was possible (the reduction of the excess oxygen), and consequently the reductive dissolution of the main Fe_2O_3 film did not take place. It is uncertain whether the 'oxygen charge' consisted of extra oxygen ions in the lattice, or was connected with a deficiency of iron below that suggested by the formula Fe_2O_3 (with Fe^{6+} present at some cation sites). The scatter of the results is attributed to periodic cracking of the film at randomly scattered points, which causes premature destruction by cathodic reduction.

Application to Stainless Steel. Although in the case of unalloyed iron a relatively powerful oxidizing agent (like chromic acid) is needed to provide an alternative reaction and prevent reductive dissolution over a long period, the state of affairs is different if the iron contains Cr as an alloying constituent, since in Cr the divalent state is unstable. Even if oxygen is present

1.   N. Sato, *Cambridge Conf.* 1970, paper 12.
2.   R. P. Frankenthal, *Cambridge Conf.* 1970, paper 3.
3.   K. Kudo and N. Sato, *Corr. Abs. (Houston)* 1972 (11) 88.
4.   U. R. Evans and I. D. G. Berwick, *J. Chem. Soc.* 1952, p. 3432.

at very low concentrations, it will be reduced at a potential where the reduction of Cr from the trivalent to the divalent state does not occur. This was shown by Berwick,[1] who saturated $0\cdot 1M$ H_2SO_4 with various gases and allowed it to act on 18/8 Cr/Ni stainless steel; the specimens had been degreased in xylene and anodically etched for 30 minutes in 5N H_2SO_4 at 15 mA/cm². The experiments were carried out for 72 hours, after which the specimens were reweighed and the solution was analysed for Cr, Ni and Fe. It was found that where the acid had been saturated with oxygen or air, the weight change was very small and the amount of metal found in the solution was also very small; sometimes it was undetectable. In the case of saturation

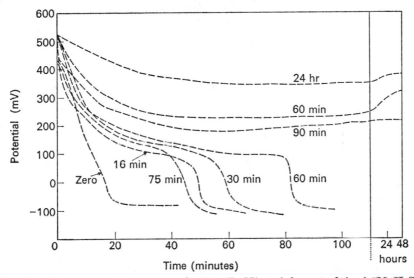

Fig. 21 Time–potential curves of 18/8 Cr/Ni stainless steel in $0\cdot 5M$ H_2SO_4 de-aerated with H_2 after previous exposure to oxygen-containing acid for various times (I. D. G. Berwick and U. R. Evans).

with N_2, H_2 or argon, the weight loss was high and the amount of all three metals found in the solution considerable. Almost all the iron was in the ferrous condition; very little Fe^{3+} was found; the Cr was found in the solution in the Cr^{3+} condition. The appearance of the specimens accorded with the gravimetric and analytical figures. All the test pieces immersed in acid containing no oxygen appeared etched and darkened; the specimens tested in acid treated with oxygen or air showed no sign of any corrosion.

Berwick also studied the movement of potential, which dropped suddenly when the material lost its passivity. If the stainless steel had received no special treatment, this drop occurred within about 10 minutes of immersion in de-aerated acid. If, on the other hand, it had been exposed to acid containing oxygen, the time needed for the drop could be prolonged, and if the preliminary treatment in the acid containing oxygen was sufficiently lengthy,

1. I. D. G. Berwick and U. R. Evans, *J. appl. Chem.* 1952, **2**, 576.

the drop seemed to be prevented altogether. The effect of pre-treatment is shown in Fig. 21.

Particularly interesting are Berwick's experiments using acid containing different amounts of oxygen. These are shown in Fig. 22. In this series the stainless steel was initially in the active condition, but quickly became passive if sufficient oxygen was present. If the amount of oxygen present was

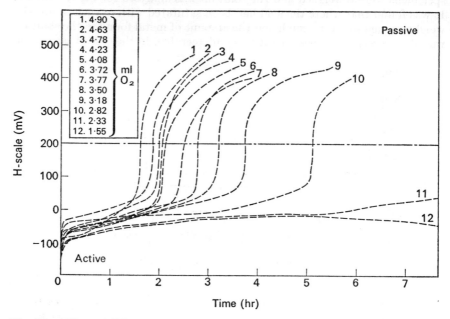

Fig. 22   Effect of different quantities of O_2 (in ml) on the potential of active 18/8 Cr/Ni stainless steel in stirred 0·5M H_2SO_4 (I. D. G. Berwick and U. R. Evans).

reduced, the time needed for the rise of potential into the passive region was lengthened, and with very small amounts, it did not occur at all during the experimental period. It was found that if $1/t$ (where t is the time needed for passivation) was plotted against V, the amount of oxygen introduced into the acid, straight lines were obtained, as shown in Fig. 23. Prolongation of these lines to cut the horizontal axis will clearly give a point at which $1/t$ is zero, so that t is infinity. In other words, it indicates the amount of oxygen needed to prevent corrosion indefinitely. A smaller amount is needed if the liquid is stirred than under stagnant conditions.

Berwick's paper provides an interpretation of the fact that a straight line is obtained on plotting $1/t$ against V, based on the assumption that the film is growing according to the direct logarithmic law; the interpretation is independent of the derivation of the logarithmic law—a matter about which opinions differ. On the other hand, had the film been thickening according to the parabolic law, a straight line would have been expected on plotting $t^{-1/2}$ against V.

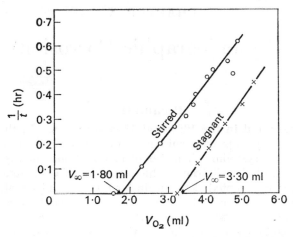

Fig. 23 Relation between oxygen addition and reciprocal of time of passivation (I. D. G. Berwick and U. R. Evans).

Final Conclusions. It would seem that this work provides a rational explanation of the reason why stainless steel resists dilute sulphuric acid under ordinary circumstances. The explanation generally given, namely that there is a protective film on the stainless steel, is clearly insufficient, since ordinary mild steel carries an invisible film. The difference between the behaviour of the films on the two materials is that the film on mild steel will suffer reductive dissolution even if oxygen is present, whereas that on stainless steel will escape reductive dissolution, if oxygen is present even at a low concentration. This, of course, explains why the passivity of stainless steel will generally break down at crevices or other places where there is no renewal of oxygen.

The papers of Pryor and Berwick contain an account of many series of experiments pointing in the direction indicated. Only a few of these series have been mentioned in this appendix, and even then drastically summarized. If the reader feels that the argument is sound, he would do well to study in the original journals, all the papers quoted.

CHAPTER X

Crystallographic Corrosion

Preamble

Picture presented in the 1960 and 1968 volumes. In the 1960 book it was felt necessary to provide guidance to readers who had little previous knowledge of the crystalline character of metals. Consequently an elementary explanation of the allotriomorphic structure of cast metals, with the crystal direction interrupted at the grain-boundaries, and sometimes at sub-grain-boundaries, was provided, along with a discussion of screw dislocations and edge dislocations; it was suggested that the reader already familiar with these things could do some skipping. In the 1968 book, this chapter was retained, being used for a discussion of stacking faults, dissociated dislocations, and the question as to whether the dislocations provide sites for pitting; the view was advanced that on copper pitting occurs mainly at dislocations, whereas on aluminium this is not always the case.

The preferential corrosion often met with along grain-boundaries was discussed, starting from the bubble raft pictures of Bragg and Nye, which show that, along the boundary separating two crystal grains of different orientation, there must be 'holes' of a size not present in the 'perfect' parts of each crystal. Such holes will clearly be too small to accommodate an atom of the main constituent, but if 'impurity' atoms are present, these will often tend to place themselves at the grain-boundaries, since such segregation can reduce the discontinuity, and thus—other things being equal—diminish the energy of the system. If the foreign atom is very small, it can enter the hole and largely fill it up; if the foreign atom is over-size, its presence near the hole is likely to reduce the unoccupied space, thus lowering the overall energy.

The presence of foreign atoms along a grain-boundary does not in itself amount to formation of a second phase. However, a second phase may form preferentially at a grain-boundary, where the size of a stable nucleus will probably be smaller than in the body of the grain, and where enrichment of solute may have occurred as a result of the grain-boundary segregation. Thus it is not surprising that materials which consist at high temperature of a single phase, but at room temperature of two phases, will, on slow cooling from a high temperature (or rapid cooling followed by reheating at an intermediate temperature) often deposit the second phase along the grain-boundaries rather than in the grain interiors. This precipitation will produce a changed composition in the part of the main phase which lies closest to the boundary. Thus austenitic stainless steel, after certain heat-treatments, deposits carbide containing chromium at the grain-boundaries, and leaves the main phase around the carbide particles denuded in chromium. The intergranular corrosion resulting is generally regarded as due to anodic attack on this zone, which contains too little chromium to become passive. It is possible that the carbide particles play a special role in the cathodic reaction, since, although the area which they present is small, they are likely to be more efficient cathodes than the oxide coated solid solution; moreover, they are situated very close to the regions upon which anodic attack is being directed. In other cases the second phase precipitated at the grain-boundaries may be anodic towards the rest of the material, so that its destruction follows. In other cases, again, the attack which appears to be directed upon the grain-boundaries when the situation is viewed under the optical microscope, is really due to attack on parts of the slip-planes close to the grain-boundaries where dislocations have piled up. Attack is sometimes directed on slip-planes crossing the grains, and then has a definitely transgranular character. It has been suggested, but not proved, that this is connected with a second phase deposited, in particles too small to be detected, at dislocations connected with the gliding which occurs on application of stress.

Developments during the period 1968–1975 and Plan of the new Chapter. Many of the papers published during the period have been concerned with the different oxidation rates or corrosion rates on different crystal faces, or have described studies of the crystal forms of pits. In the present volume these matters have been dealt with in the chapters devoted to oxidation or to the appropriate type of corrosion. A relatively small number of researches have dealt essentially with the crystallographic aspect, and these are collected in the present chapter. The main sections are headed 'behaviour of different crystal faces', 'crystallographic factors determining localized attack' and 'the effect of strain'.

Behaviour of different crystal faces

Iron. Interesting work from Iofa's labatory[1] concerns the effect of deformation on inhibition when different crystal faces are exposed. Although deformation was found to increase the corrosion rate of high purity iron by acid, and annealing at 750°C to decrease it, the population density of the dislocations increases with the deformation, so that inhibitors are better adsorbed. Annealing decreases the density of dislocations, and causes faces possessing smaller adsorption activity to be oriented at the surface. Earlier investigators had stated that the increase in dissolution rate on deformation only occurs if iron contains impurities, especially C, and had stated that iron containing only 0·001% C does not show an increase in the number of dislocations. However, Iofa's new experiments show that there is a significant increase in the population density of dislocations and in the associated adsorption in the case of iron with less than 0·001% C. This agrees with the work of Greene,[2] who found that wire made from zone-refined iron containing less than 0·001% carbon, dissolves in N H_2SO_4 at 30°C, when strained much faster than in the annealed state. The change appears to be due to an

1. Z. A. Iofa, V. V. Batrakov and Y. A. Nikiforova, *Corr. Sci.* 1968, **8**, 573; *Ferrara Symp.* 1970, p. 183.
2. N. D. Greene and G. Saltzman, *Corrosion (Houston)* 1964, **20**, 293t.

increase in the anodic reaction rate; there is only a minor effect on the cathodic evolution of hydrogen.

Nickel. Important work in Ives's laboratory[1] has compared the shape of pits obtained on different crystal faces on nickel. Nickel was passivated in N H_2SO_4 (or acidified $NiSO_4$) at various potentials, and NaCl was added until pitting started. The shape of the pits obtained was compared with that which would be predicted on various assumptions. If it is assumed that the atom which is preferentially removed is the one possessing fewest bonds (i.e. the one with fewest close neighbours), good agreement is obtained with observation in the case of {100} and {111} faces, but there is considerable discrepancy in the case of {110} faces. A new assumption, suggested by Kolotyrkin,[2] has been tested. He took into account aggressive anions (in this case Cl^-) and assumed that removal depends on the accessibility of aggressive ions through the spaces left by previous atom removal. This appears to explain satisfactorily the observed morphologies, and fits in with results of other investigators using body-centred cubic metals. The morphological changes vary with the reaction rate, which is itself dependent on the degree of passivation, the potential and acidity. The proposed mechanism provides a satisfactory explanation.

Research on nickel in the polycrystalline and monocrystalline states with a potential held within the passive range has been carried out by Tokuda and Ives.[3] In N H_2SO_4 containing Cl^-, the pits were observed to start at scratches and grain boundaries. Susceptibility to pitting on polycrystalline nickel, on mechanically polished {111} faces, and on electropolished {111}, {100} and {110} faces, was found to fall in the same order as their reactivity under conditions of active dissolution. It was concluded that on nickel the sites for pitting are identical with the sites for active dissolution.

Copper. Interesting work by Markovac and Petrocelli[4] on copper adds considerably to our knowledge. In earlier research, Jenkins and Bertocci[5] had found that, when oxygen was completely excluded, no difference could be detected between the equilibrium potentials of faces with different orientations. When, however, O_2 was allowed to leak into the system, dissolution of copper occurred, with potential differences between differently orientated faces, producing morphological changes on the various electrode surfaces. In the new work, oxygen was not completely excluded, and the morphological changes were studied. The conclusion reached was that 'the topographies for all specimens were consistent with the stepwise ledge model of metal crystal dissolution'.

Silver. Vetter[6] has measured the exchange CD on low index single crystals of silver, and finds that it varies with the crystal face. When the

1.   T. Tokuda and M. B. Ives, *J. electrochem. Soc.* 1971, **118**, 1405.
2.   Y. M. Kolotyrkin, *J. electrochem. Soc.* 1961, **108**, 209.
3.   T. Tokuda and M. B. Ives, *Corr. Sci.* 1971, **11**, 297.
4.   V. Markovac and J. V. Petrocelli, *J. electrochem. Soc.* 1970, **117**, 256C.
5.   L. H. Jenkins and U. Bertocci, *J. electrochem. Soc.* 1965, **112**, 517.
6.   K. J. Bachmann and K. J. Vetter, *Z. phys. Chem. (neue Folge)* 1966, **51**, 98.

concentration of Ag^+ is 0·1M, the exchange CD is 1·6 amps/cm^2 for a {111} face, but 2·0 amps/cm^2 for a {100} face. In the case of 0·5M solution the numbers are 4·2 and 5·4 amps/cm^2 respectively.

The character of pits produced on films of silver grown epitaxially on NaCl or mica has been studied by Jaeger.[1] The etchant used was one of a class often adopted in studying the distribution of dislocations in solid silver, and contained NH_3, H_2O_2, CrO_3 and HCl. It was found that the orientation of the pits, grooves and ridges did not depend on the proportion of the constituents, but the addition of Cl^- increased the slopes of the sides and made the features more conspicuous. Etching of the {111} or {100} films, produced pits whose sides consisted of close-packed steps. Shallow single pits with sides sloping at 9° to 15° to the surface were observed at isolated dislocations. Twins were found to dissolve faster than the matrix in the {111} films, but more slowly than the matrix in {100} films. Some of the dislocations did not generate pits; indeed, in the {111} films the mean concentration of pits (5×10^7 per cm^2) was sometimes more than one order of magnitude lower than the concentration of dislocations.

Aluminium.   Metzger[2] has provided a polarigram showing the character of corrosion on different faces in 16% HCl; this is 41% higher near {110} than {100} and 71% higher near {111}. Pitting is attributed to a weak heterogeneity connected with the segregation of iron or other impurity, and is not necessarily the result of thermodynamic stability of special crystal faces.

Crystallographic factors determining localized attack

Iron.   Work by Ives[3] has gone far to clear up certain matters which had previously been the subject of dispute. Etch pitting has long been used in endeavours to pinpoint the sites of dislocations, but endeavours to prove that, with some particular etchant, there is a one-to-one correspondence between dislocations and pits have been unconvincing. The new work shows that the removal of the air-formed oxide-film by annealing pure iron in H_2, followed by etching below a H_2 atmosphere, completely alters the etching behaviour. Instead of pits being formed at flaws in the film (this is usually determined by stresses in the film, but sometimes by grain-boundaries), their position is determined by singularities in the structure of the basic metal. The fact that, under normal etching conditions (i.e. when oxide is present), the pit density increases with the time of etching, rules out the possibility of a one-to-one correspondence. On a surface freed from oxide by reduction, two types of pitting appear:

(1) Deep pits with conical sides apparently formed at places where aged dislocations intersect the surface.

(2) Shallow, irregular pits, attributed to dislocation networks parallel to, and just below, the surface.

1.   H. Jaeger, *Acta Met.* 1971, **19**, 621, 637 (2 papers).
2.   O. P. Arora and M. Metzger, *Amsterdam Cong.* 1969, p. 435.
3.   G. M. Spink and M. B. Ives, *J. electrochem. Soc.* 1971, **118**, 903.

In these experiments pure iron was used with an etchant containing HCl + H_2O_2 in water. It should be noticed that since pure iron was used, the effect of inclusions such as sulphides would not be brought out.

Nickel. Work by Schatt and Worch[1] has shown that on nickel the number of pits is generally smaller than the number of dislocations cutting the surface. This discrepancy is understandable if we assume that the dislocations are decorated in different degrees with various atoms, of which only certain types, when present in sufficient number, can lead to localized attack. The object of the new work was to characterize the dislocation which leads to pitting in single crystals. Microanalysis showed that dislocations carrying precipitations of Ni_2S_3 provide the starting point of pits. It was suggested that 'clean' dislocations might generate pits if the electrolyte contained sulphide ions.

Russian work[2] on high purity nickel obtained by electro-beam melting shows that the polarization curves vary with the crystal face, and are also affected by deformation and annealing. Face {111} proved the most difficult to oxidize, as shown by the high value of the active solution current; {110} was the most easily oxidized with a considerably lower current value. It is thought that behaviour is determined by structure rather than by impurities.

Zinc. French studies[3] of corrosion figures produced on pure zinc by Finkeldey solution (CrO_3–Na_2SO_4) suggest that local corrosion is often set up at points where lead particles exist in the zinc. The number increases with the Pb-content, and in the case of Zn which has been very slowly cooled, the corrosion figures are replaced by less numerous globules of coalesced Pb.

Effect of Strain

The so-called 'Electrode Potential' of a strained Metal. Clarke[4] has performed a service by clearing up a matter on which there has been much loose thinking. When metal is strained within the elastic range, the additional strain energy persists only as long as the stress is applied; the process is mechanically reversible. If, however, the strain exceeds the elastic limit, the process becomes mechanically irreversible, and residual strain energy remains after the removal of the applied stress. Such strained metal immersed in a solution of its own salt will always take up a 'mixed potential'; corrosion is proceeding and the potential measured does not represent any reversible reaction. Attempts to calculate residual strain energy by measuring the potential of the strained electrode lack a sound thermodynamic foundation. When a clean electrode of copper, possessing residual strain energy, is immersed in $CuSO_4$ solution (free from O_2), the deformed parts may suffer anodic attack, whilst Cu is deposited on the cathodic areas. The electrode may be in 'mass equilibrium', in the sense that it is not changing in weight, but it is certainly not in thermodynamic equilibrium.

1. W. Schatt and H. Worch, *Corr. Sci.* 1971, **11**, 623.
2. L. N. Yagupolskaya and B. A. Movchan, *Amsterdam Cong.* 1969, p. 473.
3. N. and P. Dreulle, *Rev. Met., Mem. Sci.* 1969, **66**, 245.
4. M. Clarke, *Corr. Sci.* 1970, **10**, 671.

Effect of strain on potential during flow of current. Despite what has just been said, studies of the effect of strain on apparent potential, as measured by recognized techniques, are not without interest. For instance, different types of copper wire strained in a flowing electrolyte[1] produce potential–time curves of four different forms; all types of wire, however, give the same strain–time curves when tested with high speed photography. The photographic strain–time curve relates to the overall strain of the copper wire, whereas the potential results from the strain on the copper oxide film and its consequent rupture.

The apparent effect of micro-strain on the potential and anodic corrosion of copper has also received study.[2] The quasi-stationary potentials of cold-worked copper were found to show a rectilinear relationship to the strain energy. Discrepancies between the results of X-ray measurements and electrode potentials are ascribed to the fact that X-ray studies yield information about the volume or bulk strain, whereas electrode measurements yield information about the surface strain.

Czech work[3] has compared the potential of steel cleaned by blasting and with that of undeformed steel. The difference is found to be about 50–70 mV at a CD of 10 mA/cm^2, as related to the geometric surface; the potential on the undeformed steel is the more noble.

1. K. J. Roberts and L. W. Shemilt, *Trans. Inst. chem. Eng.* 1969, **47**, T204.
2. D. Lewis, D. O. Northwood and C. E. Pearce, *Corr. Sci.* 1969, **9**, 779.
3. M. Pražák and M. Havrda, *Amsterdam Cong.* 1969, p. 130.

H

CHAPTER XI

Hydrogen Cracking and Blistering

Preamble

Picture presented in the 1960 and 1968 volumes. It was pointed out in chapter IX that certain metals are attacked by acid in the absence of oxygen; in such cases, the cathodic reaction is often the evolution of hydrogen. Certain impurities in the acid interfere with the evolution of hydrogen—which may at first sight appear as something to be welcomed. That is not necessarily the case. If, for instance, H_2S present in the acid poisons the reaction which would lead to the formation of molecular hydrogen, it may leave atomic hydrogen in the metal, which will then diffuse inwards, producing mischief. The simplest explanation of the action of H_2S is that it becomes adsorbed, particularly at the points where hydrogen bubbles would normally be formed. The combination of two H atoms to form one H_2 molecule probably requires a pair of atoms adsorbed at contiguous sites. If most of the sites are occupied by attached sulphur, this will rarely occur, and a high concentration of atomic H will be built up, so that diffusion inwards is favoured. If now within the metal sub-microscopic cavities exist, the walls of which carry nothing likely to prevent the combination of pairs of H atoms, molecular hydrogen will be formed in those cavities. Calculation from the over-potential values produced by the H_2S indicate a concentration of atomic hydrogen so high that, if at a cavity it forms molecular hydrogen, the pressure produced when equilibrium between H_2 and 2H has been established, will be extremely high. Laboratory experiments have, indeed, confirmed pressures as high as 200 to 300 atmospheres. Perhaps the most convincing proof of these high pressures is the formation of blisters in cases where the internal evolution of hydrogen takes place at points a short distance below the surface. This may happen during the pickling of steel to remove mill-scale. If there are inclusions just below the surface, blisters arise at the sites of these inclusions. Quite likely, the inclusions, consisting of some brittle substance, have fractured when the steel has been rolled, so that true cavities are already present in which the molecular hydrogen can develop pressure; however, if the adhesional force between inclusion and metal is small, the evolution of hydrogen between inclusion and metal would require comparatively little nucleation energy, even in the absence of a pre-existing cavity. The forcing up of the surface layer of the steel into a blister can be likened to the blowing of a bubble, and obviously it takes considerable pressure to blow a bubble in solid steel. Probably the formation of blisters during the pickling of steel containing inclusions will occur most easily if the acid contains H_2S as an impurity, or if H_2S is formed by the action of (relatively pure) acid on sulphide particles at the surface. However, blistering would not be impossible in pure acid and with steel free from sulphide, since even then appreciable over-potential is required for hydrogen evolution; thus some diffusion into the steel is possible.

Clearly at internal cavities, or inclusions, situated far from the surface, the blowing of bubbles will be impossible, but the consequences of entrapped hydrogen at high pressure may still be serious. For instance, if the steel is subjected to stress otherwise insufficient to cause fracture, cracking may be started owing to the combination of the applied stress and the local stresses occasioned by the entrapped high pressure hydrogen.

Unfortunately, the situation is extremely complicated, because several factors potentially capable of causing or assisting fracture are present, and their effects are not additive. As already stated, H_2S at the surface, by preventing the harmless evolution of hydrogen as gas, may cause diffusion of atomic hydrogen into the interior; that can cause high stress at points where molecular hydrogen is evolved. Apart from that, however, the atomic hydrogen in the metallic phase may alter the mechanical properties in a manner which may be either favourable or unfavourable. (It will probably increase the strength, but reduce the ductility, thus favouring the inception of cracks.) In some materials, the sulphur may itself diffuse into the interior, perhaps producing a sulphide network, and thus reducing ductility. It may be difficult to distinguish between the hydrogen embrittlement which is the indirect effect of the H_2S from true sulphide cracking. Furthermore, if an external stress has been applied, it may be difficult to distinguish between these two types of cracking and the stress corrosion cracking discussed in chapter XVI.

This may seem to be a sufficiently complicated situation, but discussion is made still more difficult by suggestions of other effects for which there is no convincing evidence. For instance, some authorities have suggested that H_2S (or H_3As in cases where arsenical acid is used in pickling) has a positive catalytic effect in favouring the entry of hydrogen into the steel. It is difficult to see why hydrogen present as H_2S or H_3As should pass into the metal more readily than that already present as atomic hydrogen (*proton*). A more likely explanation of the effect of H_2S or H_3As is that, by adsorption on the surface, it renders the presence of two H atoms on contiguous sites less frequent, and thus prevents the evolution of gaseous hydrogen. It has also been suggested that the presence of a stress gradient in the metal increases the diffusivity of atomic hydrogen. This would not be impossible, but a more likely explanation of the fact that stress appears to favour hydrogen diffusion is that it opens internal cavities, thus favouring the conversion of 2H into H_2; this will lower the concentration of atomic hydrogen at points next to the cavity, so that a true concentration gradient is produced, thus explaining the more rapid diffusion. As the crack extends under the combined influence of externally applied stress and internally generated pressure, so the volume into which hydrogen can pass increases, and the concentration gradient is maintained.

Practical problems arising from hydrogen occur in many situations. They may arise in the pickling of steel to remove mill-scale, and interfere with the benefits otherwise to be expected from the so-called cathodic pickling. One of the troubles which arise from pickling by straight immersion in acid is that, besides removal of the scale, there will be attack on the metal. When the material undergoing pickling is made the cathode, this attack is lessened or avoided; but the material may be embrittled and may develop cracks, either in the pickling bath or after removal; thus the advantage of the diminished attack on the metal may be too dearly bought.

A situation where troubles connected with hydrogen cause anxiety arises in the so-called sour oil-fields—regions in which the water accompanying the oil contains H_2S. As a result, either hydrogen embrittlement, sulphide cracking or stress corrosion cracking may affect the metallic parts of the wells themselves, or the pipes conveying the oil from them.

Developments during the period 1968–1975 and Plan of the new Chapter. Considerable interest in hydrogen cracking has been shown during the period under review, but this mainly concerns situations where hydrogen is believed to contribute to the advance of a stress corrosion crack; many of those cases are dealt with in chapter XVI. Work on the effect of hydrogen under conditions where stress corrosion cracking is not involved is discussed in the present chapter. Headings include 'Entry of Hydrogen', 'Sites of Internal Precipitation of Hydrogen', 'Effect of Hydrogen on Ductility', 'Practically Important Cases of Hydrogen Embrittlement', 'Hydrogen induced Cracking at Welds', 'Avoidance of Hydrogen Embrittlement', 'Hydrogen Troubles due to Phosphating', 'Determination of Hydrogen in Steel'; there is a discussion of hydrogen cracking in iron–nickel alloys, maraging steels and stainless steels, also in non-ferrous materials.

Mechanism of Embrittlement in Steel

Entry of Hydrogen. An important research in Bockris's laboratory[1] on the cathodic production of hydrogen and its movement into the metal has shown that, below a certain critical over-potential, hydrogen permeation is a simple function of time; the same results can be obtained over and over again on same specimen; evidently the movement of hydrogen inwards is not causing damage to the metal. But above the critical value, the curve connecting permeation and time attains a new, characteristic shape, and the same results can no longer be obtained several times on the same specimen. It is believed that the critical over-potential is that at which micro-cracks commence to nucleate, spread and provide traps for molecular hydrogen. Calculated pressures are consistent with the assumption of hydrogen pressure developing in voids which would be expected to favour the spreading of cracks. The decay transients show two maxima; the first is believed to represent hydrogen in cavities and the second to be connected with hydrogen bonded to chemical impurities.

Against the view that hydrogen pressure in voids is important must be set the fact[2] that cracks grow when open to the atmosphere; in such a case

1. J. O'M. Bockris and P. K. Subramanyan, *J. electrochem. Soc.* 1971, **118**, 1114.
2. E. Strecker, D. A. Ryder and T. J. Davies, *J. Iron Steel Inst.* 1969, **207**, 1639.

hydrogen pressure could not be serving to reinforce the applied stress in producing an increased propagation rate of a crack.

Work on free-cutting steel[1] suggests that there is a correlation between micro-void volume and diffusivity. A balance, dependent on working process and temperature, is set up between the formation and elimination of voids; working introduces voids; high temperature annealing eliminates them. The view has been expressed that micro-voids serve to retard the diffusion rate of hydrogen, but do not retain the gas permanently.

It may be pertinent to recall the earlier work by Akimov.[2] He found that at very low CD the small amount of hydrogen evolved at a cathode adheres to the surface owing to adsorptive forces. After gradual accumulation, it may be slowly removed by diffusion into the metal, also into the solution, and thence into the gas phase. At higher CD, hydrogen may appear as bubbles which on reaching a certain size may overcome surface forces and detach themselves. The points on the cathode where H^+ is discharged do not coincide with those where hydrogen bubbles are formed.

Views expressed by Townsend[3] on the mechanism of hydrogen cracking produced by the presence of H_2S deserve attention. The susceptibility produced reaches a maximum near to ambient temperature and decreases at higher temperatures. Townsend's measurements show that the time to failure is shortened by cathodic and lengthened by anodic polarization. The results are held to support Petch's internal adsorption theory,[4] according to which H atoms diffusing through the lattice become adsorbed on the surfaces of embryonic cracks (i.e. dislocation pile ups); this reduces the energy required to create a new surface which would convert the pile-up into a crack. Townsend considers the internal pressure theory of Zappfe (1960 book, p. 419) to be useful in explaining the occurrence of blisters and laminations; but he thinks that it cannot be reconciled with the temperature dependence of susceptibility to cracking.

Sites of Internal Precipitation of Hydrogen. A research by Tetelman[5] suggests that the precipitation of hydrogen in molecular form on carbide particles or on inclusions is the cause of voids and micro-cracks, and helps the first two stages of brittle fracture, either by causing plastic deformation or by cleavage, according to the toughness of the steel in question and the shape of the nucleating particles. The three stages of brittle fracture are (1) crack nucleation, (2) slow crack growth and (3) rapid, unstable fracture.

Oriani[6] states that in cold-worked steel micro-crack surfaces are more important than dislocations for the trapping of hydrogen.

Internal pressure need not be always due to elementary hydrogen. Russian

1. G. M. Evans and E. C. Rollason, *J. Iron Steel Inst.* 1969, **207**, 1591.
2. G. V. Akimov, *Corrosion (Houston)* 1958, **14**, 463t; esp. pp. 466–7t.
3. H. E. Townsend, *Corrosion (Houston)* 1972, **28**, 39.
4. N. J. Petch, *Phil. Mag.* 1956, **1**, 331.
5. A. S. Tetelman, *Corr. Abs. (Houston)* 1968, p. 20.
6. R. A. Oriani, *J. electrochem. Soc.* 1969, **116**, 285C.

work[1] shows that pressures of CH_4 of the order of 10^6 atm. are thermodynamically possible in the pores of certain steels at high temperature and hydrogen pressures, and can lead to a breakdown of the steel. Formation of methane, an increase of volume and the cracking of the steel have been observed.

Interesting results[2] have been obtained on exposing steel carrying laminated oxide to a solution containing $FeCl_2$ and $NiCl_2$ at 316°C. It appears that blisters arise from the accumulation of hydrogen in spaces between an inner layer of fine magnetite and an outer layer of coarse magnetite; and this forces up the coarse magnetite which is mechanically weak. Blistering can only occur at places where hydrogen is formed more quickly than it can leave the oxide layers; the blisters are associated with local concentrations of Ni, which represent sites of low over-potential.

Effect of Hydrogen on Ductility. Various views have been advanced as to the way in which hydrogen causes deterioration of the properties of steel. A Japanese paper[3] attributes it to the formation of hydride-like compounds and to the weakening of the Fe–Fe bond; the trap theory is rejected. A Polish paper[4] suggests that a heat-treatment producing only slight locking of the dislocations allows the penetrating hydrogen to interact with them and change the structure; this leads to heavy embrittlement. In contrast, when a dislocation is locked by a saturated atmosphere of other interstitials, e.g. nitrogen, the hydrogen is only feebly attracted, and its effect on the ductility of the steel is less pronounced. In this way it is possible to explain the fact that 18% Cr ferritic steel which has been heated for an hour at 880°C *in vacuo*, quenched and cooled, and then aged for 4 to 5 weeks at room temperature, is seriously embrittled when hydrogen is introduced; if, instead of being aged, it is heated for 1·5 hours at 320°C, followed by slow cooling, the embrittlement by hydrogen is much less pronounced.

Another Japanese paper[5] discusses the diminished ductility of pipe-line steel after cathodic protection to prevent corrosion in a sour oil-field. In such situations high strength steels have been employed so as to transmit the maximum quantity of gas. Sometimes cathodic protection and coatings have failed to prevent stress corrosion cracking. New experiments show that with materials having strengths below 60 kg/mm^2 there is no serious problem; for stronger material the cathodic protection must be carefully controlled. The decrease in ductility caused by excessive CD is attributed to atomic hydrogen diffusing inwards and being precipitated as molecular hydrogen between the matrix and inclusions elongated during the rolling process.

If hydrogen cracking is connected with the production of high pressure hydrogen in voids, the factors leading to the presence of voids must be im-

1.   V. I. Alekseev, Y. I. Archakov, S. D. Bogolyubskii and L. A. Shvartsman, *Corr. Abs.* (*Houston*) 1971, p. 292.
2.   M. J. Longster and D. J. Arrowsmith, *Tokyo Cong. Ext. Abs.* 1972, p. 359.
3.   S. Yoshizawa and K. Yamakawa, *Tokyo Cong. Ext. Abs.* 1972, p. 175.
4.   E. Lunarska and M. Smialowski, *Tokyo Cong. Ext. Abs.* 1972, p. 177.
5.   E. Sunami, M. Tanimura and G. Tenmiyo, *Tokyo Cong. Ext. Abs.* 1972, p. 169.

portant. Baker and Charles[1] have studied voids connected with sulphide inclusions. They reach the conclusion that matrix voids are 'only found to be associated with the less deformable inclusions, that is single-phase sulphide rolled at 1200°C and the non-deformable duplex sulphides at the lower rolling temperatures. . . . The formation of voids appears to be due to the inability of the steel to flow round the non-deforming inclusion whilst simultaneously maintaining contact with it.'

There has been considerable disagreement as to whether elements like As, Se and Te act by preventing the harmless evolution of hydrogen as gas on the surface, or have some positive effect in promoting entry of hydrogen into the lattice. Shreir[2] has compared the action of six different catalytic poisons. He found that the permeation current decreases in the order As > Se > Te > S > Pb > Bi. The stability of hydrides (As > Se > Te) appears to be related to the decrease of poisoning action with increase of concentration.

Beck[3] quotes several cases where compressional stress has presented or slowed down hydrogen embrittlement. The compressional stresses introduced by shot-peening render high strength material resistant to the embrittlement which might otherwise be produced by electroplating. An interpretation is offered based on the energy of activation of hydrogen within the lattice and the building up of energy at dislocations.

The part played by molecular hydrogen is discussed by Bruch.[4] He considers that molecular hydrogen is formed in cavities and at lattice defects, and that the high pressure leads to flake formation at temperatures below 200°C. The results obtained by different methods of analysis have sometimes led to the assumption that the hydrogen occurs in two different types, and produces different metallurgical effects; Bruch feels that such conclusions should be judged with caution.

Practically important Cases of Hydrogen Embrittlement

Hydrogen uptake in manufacturing processes. Hydrogen embrittlement is generally attributed to hydrogen taken up in service. The effect of hydrogen present in the newly manufactured steel should not be neglected, but Scott and McCullagh[5] state that modern steels are generally low in hydrogen, and it is hydrogen pick-up from service environmental conditions which can be deleterious; they are writing of bearing material, but the statement is possibly true of steels generally.

The 'temper blueing' of low-alloy steel in steam[6] has given rise to embrittlement. This process is applied industrially to high tensile bolts, razor

1.   T. J. Baker and J. A. Charles, *J. Iron Steel Inst.* 1972, **210**, 680, esp. p. 689.
2.   T. P. Radhakrishnan and L. L. Shreir, *Electrochim. Acta* 1966, **11**, 1007; 1967, **12**, 889.
3.   W. Beck, P. K. Subramanyan and F. S. Williams, *Corrosion (Houston)* 1971, **27**, 115.
4.   J. Bruch, *J. Iron Steel Inst.* 1972, **210**, 153.
5.   D. Scott and P. J. McCullagh, *N.E.L. Report* **551** (1973) (East Kilbride, Glasgow).
6.   J. C. Wright and S. E. Webster, *J. Iron Steel Inst.* 1970, **208**, 680.

blades, pen nibs and certain parts of guns, typewriters, calculators, and cycle chains, partly as a matter of tradition and partly out of a desire to improve shelf life and provide some resistance against rusting. The blueing can be carried out in air, steam or CO_2 between 400° and 500°C. Delayed failures have been noted after tempering in steam at about 475°C—a condition which gives rise to an increased hydrogen content. Attempts to avoid this trouble by adoption of air-tempering instead of steam-tempering have produced unsightly discoloration. CO_2-tempering has proved more successful. Experiments have shown that the embrittlement produced by steam-tempering is largely due to the presence of some hydrogen in the steam, which prevents the escape of hydrogen introduced into the steel during an earlier austenizing process; if already the steel has an appreciable hydrogen content, the steam-tempering could increase this.

It has been suggested that, during cold-rolling, hydrogen can be taken up by the rolls. There is evidence that spalling is initiated by hydrogen injected into the roll as a result of a corrosion reaction with the soluble-oil coolant. Various coolants have been compared experimentally.[1]

Hydrogen induced Cracking at Welds. It has long been known that if steel contains hydrogen when it is welded, cracking can occur spontaneously after the austenite has changed to a hardened acicular microstructure. Research partly conducted at the (British) Welding Institute[2] provides reliable information as to the diverse behaviour of different materials.

The cracking occurs because hydrogen reduces the amount of plastic deformation which the metal is capable of withstanding under the influence of applied or internal stresses. When steel is cooled, internal stresses are produced by volume differences along the thermal gradients and by phase transformations. In large masses of steel containing hydrogen, such stresses lead to the well known defects known as hair-line cracking and flaking. In fusion welds, cracking of a similar nature can occur; there are here geometric stress raisers in addition to the stresses set up by transformations or thermal gradients; fit-up discontinuities in individual joints and other welding flaws cause stress concentration which can lead to hydrogen cracking where it would not otherwise occur.

Hydrogen cracking in steel welds will occur if there is insufficient attention to three factors, which require attention from three different classes of people:

(1) *Hydrogen content:* this is the responsibility of the manufacturer and the welding engineer, who must ensure the lowest possible moisture content of the welding materials.

(2) *Residual stress:* this is the responsibility of the designer and the welding engineer, who must avoid unnecessary stress concentration.

(3) *Susceptible microstructure:* this is the responsibility of the metallurgist,

1.   D. A. Melford, V. B. Nileshwar, R. E. Royce and M. E. Giles, *J. Iron Steel Inst.* 1972, **210**, 163.
2.   T. Boniszewski and F. Watkinson, *Met. Mat.* 1973, **7**, 90, 145 (2 papers).

who should ensure that, when a 'weldable' steel is being developed, the least susceptible microstructure in the heat-affected zone is aimed at. Some steels once in common use show extreme susceptibility to cracking in that zone, whilst others have little or no susceptibility.

The very susceptible steels include those with a medium carbon content (0·25–0·45% C) and low contents of alloying elements. The papers quoted provide particulars of the extent in which the various alloying elements produce types of structure likely to cause cracking.

Hydrogen troubles in the Oil Industry. Bates[1] has discussed the behaviour of steels of high yield strength in the sour oil-fields of West Texas. They were conducted on two large storage tanks containing oil with moderate and low H_2S contents respectively. The factors determining behaviour seem to be the yield strength, the metal, the welding and the H_2S content of the oil.

Avoidance of Hydrogen Embrittlement. Since electroplating is a serious source of hydrogen embrittlement, interest attaches to methods of obtaining a metallic coat without cathodic treatment. The process known as peen-plating is discussed by Everhat.[2] The parts to be plated are tumbled in a barrel containing a water slurry of fine particles of metal to be deposited along with glass spheres and a chemical promoter.

The hydrogen introduced by cathodic treatment is often removed by baking at comparatively high temperatures but this method has its disadvantages. It has been shown, however, that anodic pickling in HCl can avoid the trouble. Experiments by Lui and Rogers[3] on pin specimens immersed in $N/10$ HCl without application of current caused 60% of the pins to break. If cathodic treatment was applied 80% to 100% of them broke. When the pins were given anodic treatment at $-0·5$ to $+0·4$ V (saturated calomel electrode), none broke. Any non-metallic material left on the surface after the pickling could be removed by ultrasonic vibration in water.

The embrittlement caused by cadmium plating can be reduced if, instead of a cyanide bath, a bath containing di amino-n-butyric acid is used. Its superiority is attributed to lower adsorption. The subject is discussed by Beck.[4]

Italian work[5] compares the effect of various inhibitors on hydrogen embrittlement. It is found that various polyamines avoid the embrittlement often associated with phenyl-thiourea. Indian investigators[6] recommend the use of 2-mercapto-benzothiazole where mild steel has to come into contact with H_2SO_4 (as in pickling). Experiments at 40°C with and without the

1. J. F. Bates, *Mat. Prot.* 1969 (1) 33.
2. J. L. Everhat, *Corr. Abs.* (*Houston*) 1969, p. 235.
3. A. W. Lui and R. R. Rogers, *J. electrochem. Soc.* 1969, **116**, 130.
4. W. Beck, A. L. Glass and E. Taylor, *Plating*, 1968 July, p. 232.
5. G. Trabanelli, F. Zucchi and G. Gilli, *Ann. Univ. Ferrara*, Sect. 5, 1960–8, **2**, 93; see *J. appl. Chem.* 1969, **19**, ii, 354.
6. I. Singh and T. Banerjee, *Corr. Sci.* 1972, **12**, 503.

inhibitor show that even at concentrations as low as 0·08 g/l, 98% inhibition is obtained. The cathodic reaction is the one mainly retarded, and it would appear that the discharge of H^+ ions is suppressed, so that hydrogen absorption and permeability are suppressed also.

Hydrogen Troubles due to Phosphating. The phosphating process often used before painting may introduce hydrogen into high strength steels. Shreir[1] finds that the permeation is related to pH value, and that oxidizing agents added to the phosphating bath decrease the permeation; nitrites produce the greatest decrease. Hydrogen embrittlement due to phosphate treatment is discussed in a paper contributed by Andrew and Donovan;[2] only in the case of the strongest steels is trouble caused. The paper discusses specifications for baking treatment and describes studies of the hydrogen content by vacuum extraction.

Determination of Hydrogen in Steel. A report by Beck[3] contains a discussion of methods for the determination of hydrogen along with tests for hydrogen embrittlement and a discussion of the correlation between hydrogen content and hydrogen embrittlement. A paper by Shreir[4] distinguishes between hydrogen existing in the atomic, ionic and molecular states, and describes methods of determining these separately, as well as methods of studying internal stresses due to the presence of hydrogen. Swinburn[5] describes two new types of apparatus for the hot vacuum extraction and determination of hydrogen in steel. One is a very simple apparatus, capable of being operated by a relatively unskilled person, and producing duplicate results in 12 minutes. The other is semi-automatic and requires a minimum of attention. It is stated that these two types of apparatus have been used for six and two years respectively, and have cut down the time spent on hydrogen determination by 90%—apart from savings on the maintenance of the complicated and fragile apparatus used previously.

Hydrogen Cracking in Highly Alloyed Steels

Iron–Nickel Alloys. Uhlig[6] finds that alloys containing 10% to 19% of nickel with a low carbon content become much more resistant to hydrogen cracking after severe cold-rolling; when the carbon content is increased, the life of both cold-rolled and unrolled specimens is shortened, but the difference between the two is diminished. Studies of other additions show that Co, which produces a single-phase structure, is not damaging, but that Ti, Mo and Al, which give precipitates of intermetallic compounds, reduce the life. Hydrogen cracking is considered to be related to micro-voids at the surface of the precipitated particles. Cold-rolling presumably aligns such

1.   J. Sherlock and L. L. Shreir, *Corr. Sci.* 1970, **10**, 561.
2.   J. F. Andrew and P. D. Donovan, *Trans. Inst. Met. Finishing* 1970, **48**, 152.
3.   W. Beck, E. J. Jankowsky and P. Fischer, Naval Air Development Center, Warminster, Pa., Report No. NADC-MA-7140.
4.   L. L. Shreir, *Werkstoffe u. Korr.* 1972, **21**, 613.
5.   D. G. Swinburn, *J. Iron Steel Inst.* 1971, **209**, 620.
6.   J. A. Marquez, I. Matsushima and H. H. Uhlig, *Corrosion (Houston)* 1970, **26**, 215; esp. Fig. 3, p. 217.

nuclei, which accounts for the greater resistance of specimens stressed in a direction parallel to the rolling direction. The explanation is described as speculative; further experimentation is needed.

Maraging Steels. The embrittlement of Cr–Mo steels is discussed by Snape.[1] He uses a pre-cracked cantilever specimen in evaluating susceptibility to a liquid containing 3·5% NaCl and 0·5% acetic acid, saturated with H_2S. The results suggest that maraging steel with 12% Ni, 5% Cr and 3% Mo is more resistant to stress induced cracking than steels with no nickel, 1% Cr and 2% Mo, or 2% Ni, 1% Cr and 3% Mo. For these low-alloy steels without applied potential, the failure seems to be caused mainly by hydrogen embrittlement. Impressed anodic and cathodic potential have no effect on the time to failure on low-alloy steels, but for the maraging steel there is an immunity zone between −0·6 and −1·0 V.

The avoidance of hydrogen cracking in martensitic steel with high elastic limit can be brought about by depositing and diffusing suitable elements so as to produce on the surface an austenitic layer only slightly permeable to hydrogen; this can be Ni in the case of pure 18/8 stainless steel, or Cr in the case of maraging steel. The possibilities have been studied in Chaudron's laboratory.[2]

The hydrogen induced cracking of high strength steels has been attributed to trapping, but one writer[3] considers that trapped hydrogen is innocuous in maraging steels.

Stainless Steels. Holzworth[4] finds that hydrogen causes a loss of ductility in 304L stainless steel (18/10 Cr/Ni) to an extent which depends on the amount of martensitic phase present. Only a thin surface layer is damaged, because the hydrogen diffuses very slowly into the metal. The loss of ductility is related to the width and depth of surface cracks; narrow cracks are the most dangerous.

A study of 304 stainless steel[5] shows that ductility is lost by intergranular cavitation at voids on carbide particles. This favours bubble formation on the carbides, thus reducing the amount of grain-boundary sliding needed for the formation of voids.

Another study of austenitic stainless steel[6] shows that thin oxide-films consisting of $\alpha\text{-}Fe_2O_3$ provide excellent barriers against the infusion of hydrogen, which would otherwise take place during cathodic treatment in 0·1M NaOH under conditions of stirring. The effectiveness of the oxide-film as a barrier disappears when the thickness exceeds a certain value—probably owing to cracking of the film itself. Tensile stress in the substrate increases the diffusion rate; compressional stress has no effect.

1. E. Snape, *Brit. Corr. J.* 1969, **4**, 253.
2. M. C. Belo, G. P. Legry, J. Montuelle and G. Chaudron, *Amsterdam Cong.* 1969, p. 111.
3. H. R. Gray, *Corr. Abs. (Houston)* 1969, p. 83.
4. M. L. Holzworth, *Corrosion (Houston)* 1969, **25**, 107.
5. D. Kramer, H. R. Brager, C. G. Rhodes and A. G. Pard, *J. nuclear Mat.* 1968, **25**, 121.
6. M. R. Piggott and A. C. Siarkowski, *J. Iron Steel Inst.* 1972, **210**, 901.

Embrittlement in Non-Ferrous Materials

Titanium Alloys.   The embrittlement of Ti alloys has been discussed by Scully.[1] Evidence that the trouble is due to hydrogen is provided by the following facts:

(1) Pre-exposed unstressed specimens which exhibit cleavage to considerable depths when subsequently broken in air do not show this effect when there is an intermediate baking.

(2) Fractographic measurements on specimens anodically polarized in $CH_3OH-HCl$ show that the cleavage marking normally obtained on fracture in air is not shown if the dissolving front progresses at a rate comparable to H diffusion.

(3) Fractographic comparisons of broken hydride specimens provide supporting evidence.

It is true that cracking has been produced in CCl_4, which strictly speaking should contain no hydrogen; this, however, can be attributed to the traces of water which in practice are generally present.

The results tend to support the idea that, at slow strain rates, hydrogen embrittlement is operative. The important step is the nucleation of a hydride on an operative slip-plane, restricting the ductility of the grain and causing cleavage when the process is repeated. It is, however, agreed that 'much more work needs to be done'.

The effect of the initial hydrogen content present in a titanium alloy containing 8% Al, 1% Mo and 1% V has been studied by Gray.[2] It is found that reduction of the H-content in the alloy from 70 to 9 ppm does not influence resistance to the stress corrosion set up by heating when covered by a salt layer. Green[3] finds that hydrogen rather than Cl^- is the essential factor in causing stress corrosion cracking in Ti alloys in aqueous solution.

Zirconium Alloys.   The embrittlement of zirconium alloys has been discussed by Fontana[4] with special reference to the life of the cladding of nuclear fuel elements. He states that the cladding becomes unsuitable when the hydrogen content reaches about 400 ppm. The cracking of the oxide present on Zr alloys has been attributed to a second phase, believed to be hydride, at the metal–oxide interface.[5] The occurrence of hydride particles in zirconium alloys has been studied by Ells[6] and by Ambler.[7] It appears to be established that the degree of embrittlement greatly depends both on the morphology of the hydride and on its orientation relative to the stress system. In Zircaloy 2, grain-boundaries which make angles of 35–40° and 60–75° to the basal plane of the adjacent grain are preferred sites for the

1.   J. C. Scully and D. T. Powell, *Corr. Sci.* 1970, **10**, 719.
2.   H. R. Gray, *Corrosion (Houston)* 1972, **28**, 186.
3.   J. A. S. Green and A. J. Sedriks, *Corrosion (Houston)* 1972, **28**, 226.
4.   M. G. Fontana, *Corrosion (Houston)* 1971, **27**, 132.
5.   J. R. Moon and D. G. Lees, *Corr. Sci.* 1970, **10**, 85.
6.   C. E. Ells, *J. nuclear Mat.* 1968, **28**, 129.
7.   J. F. R. Ambler, *J. nuclear Mat.* 1968, **28**, 237.

hydride particles. The situation is somewhat complicated and the original papers should be consulted.

Marino[1] states that the α-phase in Zircaloy can become supersaturated with hydrogen, since the precipitation of hydride is inhibited to some extent by local stress resulting from the 10% to 20% volume increase which would occur if the hydride was formed.

Further References

Review. An important volume[2] devoted to the 'Cracking of High Strength Steels', produced by W. Beck and two colleagues, deserves close study. Individual chapters discuss (1) mechanical (dynamic and static) tests for embrittlement and cracking, (2) determination of hydrogen, (3) propagation of cracks due to hydrogen, (4) effect on fatigue strength, (5) effect on micro-surfaces, (6) the problem of distinguishing stress corrosion cracking from hydrogen cracking, (7) embrittlement in liquid metals, organic compounds and aqueous environments, (8) delayed cracking in weld metals, (9) sulphide cracking, (10) damage due to high pressure hydrogen, (11) embrittlement due to moisture during heat-treatment, (12) caustic cracking, (13) hydrogen embrittlement of cathodically protected steel, (14) chemical processing, (15) embrittlement due to plating, (16) cadmium plating formulations designed to reduce embrittlement, (17) methods of minimizing hydrogen embrittlement (removal of surface hydrogen, pre-straining, coatings, ageing, baking, introduction of compressive stresses, and use of inhibitors during pickling), (18) suggested mechanisms, (19) interpretation of certain factors and (20) role of CN^- and other anionic groups.

Papers· 'Autoradiographic analysis of microsegregation of hydrogen in metals'.[3]

'Lattice dilatation and hydrogen embrittlement cracking'.[4]

1. G. P. Marino, *Corr. Abs. (Houston)* 1969, p. 12.
2. 'Hydrogen Stress Cracking of High Strength Steels', W. Beck, E. J. Jankowsky and P. Fischer (Naval Air Development Center, Warminster, Pa., USA), 1971.
3. M. R. Louthan, D. E. Rawl and R. T. Huntoon, *Corrosion (Houston)* 1972, **28**, 172.
4. B. C. Syrett, *Corrosion (Houston)* 1973, **29**, 23.

CHAPTER XII

Boilers and Condensers

Preamble

Picture presented in the 1960 and 1968 volumes. Most of the previous chapters have been concerned mainly with corrosion at ambient or slightly raised temperature. Where, as in boiler systems, distinctly elevated temperatures are encountered, new factors demand consideration. Reactions take place more quickly at high temperature, and such effects as polarization, which may restrain corrosion at ordinary temperatures, cease to have that welcome effect in hot situations; against that can be set the fact that oxygen solubility becomes lower at high temperatures. Thus the problems of boilers require special treatment. Condenser tubes, which have also caused very serious difficulties in the past, are conveniently treated in the same chapter; here the introduction of new alloys has helped to solve the problems.

The situation has altered with time, largely owing to changes in methods of power production. Atomic energy has introduced new materials, like zirconium, with corrosion behaviours very different from that of the steel and copper alloys used at conventional power stations. But even in the latter, changes of design have reduced certain difficulties whilst introducing new ones. Thus less is heard of the caustic cracking at joints involving rivets; now that welding has replaced riveting, the danger has diminished. The introduction of oil-firing has brought a new menace connected with the presence of low-melting vanadium compounds in ash.

Early experience with boilers working at pressures lower than those customary today, showed that it was possible to reduce corrosion troubles by partial removal of oxygen. It was perhaps natural to believe that, if only oxygen could be eliminated, corrosion would cease. Accordingly, efforts were directed to the removal of oxygen, and it became possible to keep the oxygen concentration below 0·1 ppm. Whilst this was a remarkable achievement on the part of those who designed the water treatment processes, the final result was disappointing; boilers continued to corrode. It is easy to be wise after the event, but the situation might well have been foreseen. The reaction of iron with (oxygen-free) water to produce hydrogen and an iron oxide is a change which is thermodynamically possible; the only reason why in most boilers such a reaction did not lead to trouble is that the oxide-scale normally served to isolate the steel from the water. In a boiler operating under conditions leading to cracking of the scale (such as sudden temperature fluctuations), trouble must be expected. The experiments of Holmes and Mann at Leatherhead have shown that sudden reduction of the temperature from 300° to 150°C every 14 minutes (followed by a fresh rise to 300°C) produced corrosion 50 times as fast as that observed in experiments at constant temperature; these laboratory conditions are more severe than those present in service boilers, but the experiments serve to show what will happen if the scale fails to protect. However, the same experiments indicate that other factors are probably operative in boiler troubles. Pits in boilers are usually found to contain chloride. Apparently the accumulation of chlorides in a pit is governed by the principles discussed in chapter IV, but the Leatherhead experiments suggest that pitting only occurs when the steel is made locally anodic by some special factor—such as local overheating, which may be produced either by steam-blanketing or by deposits capable of restraining heat escape. If the hot spot is anodic towards the rest (and a rise of temperature does not invariably shift potential in the anodic direction), the combination of small anode and large cathode is likely to produce a dangerous situation.

Developments during the period 1968–1975 and Plan of the new Chapter. During the period under review a start has been made on the production of Pourbaix diagrams representing equilibria at high temperatures; these are likely to be extremely useful for the understanding of reactions proceeding in boilers and heat-exchangers; some references are given at the opening of the chapter, but clearly these must be consulted in the original papers. Work has continued on the reaction of iron with high temperature water, and particularly on the morphology of the double magnetite layer formed in high pressure boilers after long periods. The results of these researches are summarized in the present chapter. Other sections deal with 'the control and cleaning of boilers', 'materials for very high temperatures' and 'the effect of copper compounds on boiler corrosion'. An important matter which has recently attracted attention has been the effect of the direction of heat flux on the corrosion rate; this is briefly reviewed. Attention is called to possibilities of using lithium hydroxide in the place of sodium hydroxide for maintaining an alkaline reaction in boiler waters. Condenser problems then receive study, including the effect of iron in cooling water and in condenser alloys and new work on the dezincification of brass; the cleaning of tubes is also discussed. Finally a short section makes reference to experimental work published on the problems of nuclear power plant, including studies of the influence of radiation on corrosion rates; a reader should consult the original papers before basing decision on the necessarily summarized information provided in this book; also, if he does not himself possess specialized knowledge on the subjects involved, he would do well to consult someone with expert qualifications.

Boilers

Pourbaix Diagrams at High Temperatures. The regions of a Pourbaix diagram indicating conditions under which immunity, corrosion or passivation may be expected will vary with the temperature. Although diagrams for 25°C are available for practically every metal, corresponding information was not, until fairly recently, obtainable for higher temperatures. Clearly it is important for the understanding of events in boilers that diagrams should be constructed showing the state of affairs at higher temperatures, and a start has been made to providing what is needed. For instance, work at Nottingham[1] has given us Pourbaix diagrams for iron corresponding to 30°, 50°, 75° and 90°C, whilst Townsend[2] has carried the information up to 200°C. A comparison of the diagram obtained at a high temperature with that at 25°C shows that the area corresponding to the formation of ferroates in presence of alkali is much larger at high temperatures than at low ones; that is important in connection with caustic cracking.

Staehle[3] has provided diagrams for nickel at 100°, 200° and 300°C. Butler[4] has published the corrosion potentials of iron and other metals at 20°C and 150°C in solutions containing oxygen. It is pointed out that the potentials of different metals differ much less at 150°C than at 20°C. The comforting conclusion is drawn that at high temperatures bimetallic corrosion will be less important. Undoubtedly the EMF's operating will be lower, but, since reactions take place more quickly at high temperatures, it must not be assumed that in all circumstances the currents flowing will be less; however, the fact that the oxygen solubility is lower at high temperatures suggests that, where the rate of oxygen supply controls the current, bimetallic corrosion may indeed become less menacing as the temperature rises.

For the understanding of processes occurring at elevated temperatures and pressures, special measuring apparatus is demanded. A. H. Taylor and Cocks[5] describe two systems for the measurement of potentials; one incorporates an external reference electrode designed for use when it is desired to express the measurements on the hydrogen scale, and the other has an arbitrary internal reference electrode and is designed specially for the measurement of corrosion rates. Jones and Masterson[6] describe apparatus for the measurement of electrode processes above 100°C. There has been discussion of a reference electrode suitable for corrosion studies up to 289°C.[7]

Thermogalvanic Effect. The current flowing between two specimens of the same metal held at different temperatures is clearly a phenomenon important in connection with boiler corrosion. The most important case will arise where the hotter surface is the anode, since then at each 'hot spot', we shall obtain a small anode surrounded by a large cathode. There has been much contradictory evidence as to whether, in general, the hot region will be anodic or cathodic to the cold region. Russian work[8] appears to clear up the

1.   V. Ashworth and P. J. Boden, *Corr. Sci.* 1970, **10**, 709.

2.   H. E. Townsend, *Corr. Sci.* 1970, **10**, 343; also *Amsterdam Cong.* 1969, p. 477.

3.   R. L. Cowan and R. W. Staehle, *J. electrochem. Soc.* 1971, **118**, 557, esp. Figs. 3, 4 and 5, p. 559.

4.   G. Butler, P. E. Francis and A. S. McKie, *Corr. Sci.* 1969, **9**, 715.

5.   A. H. Taylor and F. H. Cocks, *Brit. Corr. J.* 1969, **4**, 287.

6.   D. de G. Jones and H. G. Masterson; see 'Advances in Corrosion Science and Technology' (editors M. G. Fontana and R. W. Staehle), Vol. I, pp. 1–49.

7.   M. E. Indig and D. A. Vermilyea, *Corrosion (Houston)* 1971, **27**, 312.   Cf. O. L. Biggs, 1972, **28**, 63.

8.   S. A. Kaluzhina, A. Ja. Shatalov and T. A. Kravchenko, *Amsterdam Cong. Ext. Abs.* 1969, p. 228. This paper did not appear in the final volume containing the papers delivered to the Congress.

matter to some extent. It is stated that the normal thermogalvanic effect (hot region anodic, cold region cathodic) is generally met with on active metals in an aggressive liquid which does not cause passivation. The reverse effect (hot region cathodic) appears on metals in a passivating medium.

A thermodynamic discussion of a thermal corrosion cell is provided in an Italian paper.[1]

Reaction of Iron in Water at High Temperatures. Berge[2] has performed a service by showing that the ideas of Bloom are reconcilable with those of Potter (1968 book, pp. 173–6). His views are based on the concept that the Schikorr reaction is reversible.

$$3Fe(OH)_2 \leftrightharpoons Fe_3O_4 + 2H_2O + H_2$$

He believes that the direct action of water in the pores of the scale on the steel produces Fe_3O_4 and H_2. When the hydrogen pressure increases, this may reduce the magnetite by the reverse Schikorr reaction, producing Fe^{2+} ions which diffuse outwards ($Fe(OH)_2$ is appreciably soluble). At places favourable to the direct Schikorr reaction these may now deposit magnetite. The magnetite is often deposited at the hot parts of the external surface, especially at places where nuclei of cubic oxide exist. Adherent deposits of magnetite have been produced experimentally by placing a steel plate in contact with magnetite powder in water and heating in an autoclave at 300°C for one week. Berge discusses the effect of pH, water movement and O_2 concentration, and it is stated that oxygen can inhibit corrosion under favourable conditions to some extent, but that under stagnant conditions oxygen is dangerous and likely to produce pitting. The argument explains why magnetite can be either protective or dangerous.

Berge[3] has also studied the formation of the two layer scale which Potter and Mann observed on steel exposed to alkaline solutions under pressure at 300°C, showing that similar double layer scale can be formed on carbon steel and stainless steel exposed to less aggressive solutions and at temperatures as low as 200°C. It is found that hydrogen, produced either by the action of the water on the steel or formed from $Fe(OH)_2$ by the Schikorr reaction, greatly facilitates the formation of the outer layer, by reducing the sparingly soluble Fe_3O_4 to yield a ferrous compound, so that the iron passes into solution, probably as $[Fe(OH)_3]^-$, the anion of sodium ferroate $Na[Fe(OH)_3]$. If the hydrogen is prevented from escaping, the reductive dissolution of the magnetite is sufficient to cause the scale to be non-protective. This is shown by experiments with capsules on the lines developed by Bloom (1968 book, p. 176); if the capsule containing 13% NaOH is heated at 290°C under conditions where the hydrogen can pass outwards through the steel walls and escape, the scale formed is thin (1 to 2 μm) and the inner layer practically invisible. If, however, the outer surfaces of the capsule have been copper-plated before the experiment, so that outward passage of the hydrogen becomes impossible,

1. R. Bruno and R. Cigna, *Ann. chimica* 1969, **59**, 739.
2. P. Berge, Lecture, 'Électricité de France', March 1, 1972.
3. P. Berge and P. Saint-Paul, *Comptes rend.* (Série C) 1973, **276**, 1747.

the scale formed is much thicker; the inner layer alone is 30 μm thick, and numerous pustules are formed upon it.

Morphology of the double Magnetite Layer formed in High Pressure Boilers after long periods. German work[1] confirms the view that the inner layers growing into the steel are compact, but that the state of the outer ones is influenced by geometry, temperature and flow velocity. In areas covered by water they are non-uniform and composed of an agglomeration of particles; in the steam areas they consist of coalesced crystals forming a compact layer with almost even thickness. In superheated steam there is believed to be solid state diffusion of O^{2-} anions and iron cations by way of vacancies and interstices. In the water zone, there is outward diffusion of iron cations, with inward penetration of water through pores.

Kirsch[2] has also obtained two strata of oxide, having predominatingly the spinel structure. The first is formed below the original surface of the steel, whilst the second is formed outside it, and is more porous and less protective than the first.

Laminated oxide on iron containing chromium. Work has been carried out at Leatherhead under Mann[3] on corrosion in high temperature $FeCl_2$ solution; the first results were somewhat erratic, but by using $NiCl_2$ (which deposits metallic Ni and produces $FeCl_2$), more consistent behaviour was obtained; probably the metallic Ni provides a reproducible surface for the production of hydrogen by the cathodic reaction. Iron containing 1% Cr, exposed to 0·1 mol/l $NiCl_2$ at 300°C for 66 h, develops an oxide-coat with a banded structure; the bands are composed of two types of oxide formed alternately, namely light bands consisting of fine-grained oxide rich in Cr, and dark bands consisting of coarser grains in which Cr is almost absent. The oxide morphology is strongly influenced by the Cr concentration, whilst other elements affect the corrosion rate; thus a steel with 1% Cr was found to become oxidized more slowly than pure iron containing 1% Cr, probably owing to the presence of Si.

Another paper by Mann[4] develops the mathematical basis of the experimental results. A parabolic rate law has been observed, and receives an interpretation based on a stoichiometry gradient in the oxide.

Control and Cleaning of Boilers. Warworth[5] discusses means of controlling blow-down by means of the conductivity of the water. Excessive blow-down should be avoided, since it means extravagant losses in heat and chemical treatment. A conductivity controller will signal when blow-down has become necessary in order to avoid foaming and carry-over. The controlling apparatus contains two electrodes. One of them is surrounded by a liquid of stable pH, which is separated by glass (of the character commonly

1. P. H. Effertz, H. Meisel and H. Christian, *Tokyo Cong. Ext. Abs.* 1972 p. 451.
2. P. Kirsch, *Werkstoffe u. Korr.* 1971, **22**, 527.
3. G. M. W. Mann and P. W. Teare, *Corr. Sci.* 1972, **12**, 361.
4. G. J. Bignold, R. Garnsey and G. M. W. Mann, *Corr. Sci.* 1972, **12**, 325.
5. P. Warworth, *Process Engineering*, Oct. 1969, p. 108; esp. Fig. 9, p. 113.

used in glass electrodes) from the liquid under test; the latter has access to the second electrode; as the composition of the two liquids becomes different, the EMF provided increases.

The cleaning of boilers is discussed by Corker.[1] He points out that particles of iron oxide from steam-pipes and superheaters can damage the throttle valve seats of turbines and blading. The use of HCl containing an inhibitor has become usual for cleaning, but is considered to cause stress corrosion cracking of austenitic steel due to residual chlorides. Citric acid is safer but more expensive. The cleaning of a boiler before it is put into service may be carried out by flushing, degreasing (with $NaOH + Na_3PO_4$), followed by an acid wash with inhibited HCl, and then with a weak acid wash to complex any $FeCl_2$ remaining on the surface. The superheater and the reheater is now brought into the circuit being washed, and the main treatment with citric acid takes place. The cleaning of boilers which are already in service can be carried out by the proprietary Citrosolv process, starting with 3% citric acid at pH 3·5 to 4 at 95°C, then raising the pH to 10 by means of ammonia, the temperature now being 60°C. Corker considers that a chelating agent like EDTA is excellent for removing iron and copper products; it should be injected at pH 9·2.

Another account of the cleaning of power station plant is provided by Daves and Murgrave.[2] They say that if the component parts can be cleaned at the works and thereafter kept free from corrosion, the pre-commissioning cleaning can be simplified or even omitted.

The removal of magnetite from power station boiler tubes which have been in service is more difficult than the cleaning of new tubes. Studies of the procedure favoured by the (British) Central Electricity Generating Board[3] suggest that HCl provides the best all-round performance, and should be chosen when preliminary evaluation is impossible. The risk of stress corrosion cracking of austenitic steels can be largely eliminated by 'good housekeeping' (e.g. water-wedging of superheaters). In most respects H_2SO_4 is as good as HCl, but there are problems regarding inhibition over long periods. Sulphamic acid is better than ammoniated citric acid, and merits further investigation where a weak acid must be used. It is stated that EDTA is of no practical use in acid water.

The cleaning of once-through boilers with chelating agents has been discussed by D. J. Turner.[4] In the old drum boilers, a separation of phases (water and steam) occur, and the steam, after being superheated, passed to the turbine. In the new, once-through, type of boiler there is no separation of phases, and any non-volatile matter in the boiler water can be deposited in the superheater or the turbine. EDTA, which had been used in low pressure drum boilers, cannot here be adopted, since it decomposes at operating temperatures above 300°C; moreover, the complexes formed by it are not

1.   J. M. Corker, *Chem. Ind.* 1969, p. 1329.
2.   I. Daves and N. S. Murgrave, *Chem. Ind.* 1969, p. 1335.
3.   J. Brown, D. G. Kingerley, V. Ashworth and W. J. Willett, *Chem. Ind.* 1969, p. 1369.
4.   D. J. Turner, *J. appl. Chem. Biochem.* 1972, **22**, 983.

sufficiently volatile. Potentially suitable agents are being considered for stability and corrosion risk, as well as for the stability and volatility of the complexes formed. Oxine (8-hydroxy-quinoline) would seem to be satisfactory, and is worthy of detailed examination. Its complexes are just sufficiently volatile to avoid deposition problems in a turbine, and probably just sufficiently weakly bound at low temperatures to permit of decomposition by means of ion-exchange. As regards corrosion, the morphology of the protective magnetite film changes in presence of strong chelating agents, but oxine, being relatively weak, is unlikely to increase corrosion. Other bidentate chelating agents with only one ionizable proton are considered. It is thought that oxine should also be able to prevent the formation of metal–oxide debris in high pressure boilers of the drum type.

Materials for very High Temperatures. Holmes[1] has considered the requirements which will have to be met if it is decided to increase the steam temperature at power stations from 560° to 660°C. It seems doubtful whether any existing materials really provide what is needed; mechanical properties of some corrosion resisting materials at high temperatures are inadequate, so that changes of design or the use of composite materials may have to be sought. Alternatively the corrosion resistance of existing high strength austenitic materials should be improved by minor element changes. The effect of nickel is not clear and further research is suggested; also the apparently excellent corrosion resistance of ferritic Fe–Cr alloys containing Al should be studied in tests lasting for a longer time.

A paper from a manufacturer source[2] discusses the present position and future trends. It is stated that for industrial boilers working below 1000 lbs/in², heat fluxes are likely to increase, and it will be necessary to reduce water hardness to below 1 ppm. Conductivity alarms will have to be installed. Central station and industrial boilers working above 1000 lbs/in² are today mostly performing well, but there are some occurrences not easy to explain; of two apparently identical boilers, one may give trouble and the other none. The writer expresses anxiety about high temperature cleaning in HCl and the danger of stress corrosion cracking in austenitic steels.

Effect of Copper Compounds on Boiler Corrosion. There has long been disagreement as to whether certain kinds of boiler trouble can be explained by the presence of copper in the water; cases are known where boilers have worked properly with waters containing copper, but this is not conclusive. Dutch writers[3] have reached the conclusion that 'copper and copper oxides must be avoided in boiler water'. On the other hand Indian authorities[4] state that $CuSO_4$, used to control algae growth in spray-ponds at power stations, does not increase corrosion in most cases. It is stated that 4 ppm $CuSO_4$ actually reduces the corrosion of mild steel, copper, brass, zinc

1.   D. R. Holmes, *Corr. Sci.* 1968, **8,** 603.
2.   B. R. Scriven, *Chem. Ind.* 1969, p. 1292.
3.   W. M. M. Huijbregts, G. A. A. van Osch and A. Snel, *Cebelcor Rapp. tech.* **195** (1971).
4.   V. K. Gouda and S. M. Sayed, *Corr. Sci.* 1973, **13,** 647, 653 (2 papers).

and cast iron, but increases that of galvanized iron. It is possible that the explanation of these contradictory statements can be found in the physical condition of metallic copper developed by interaction between the water and the iron; this may be influenced by minor constituents of the water.

Heat Flux and Corrosion. Japanese work[1] in an apparatus with twin-cells shows that, in the temperature range between 20 and 80°C, the corrosion rate of iron and steel in an acidified solution of sulphate, chloride or nitrate (0·5N) depends not only on the surface temperature of the specimen but on the magnitude of the heat flux and also its *direction*. Corrosion is stimulated when heat is flowing in the same direction as the transfer of metallic ions, but tends to be suppressed when the flow is in the contrary direction. In the temperature range of reaction control, heat flux may affect the number of active sites, whilst in the range of diffusion control a temperature gradient in the diffusion layer may affect the thermal diffusion rate of the products. Although the temperatures studied in this research do not represent boiler conditions, the principles emerging may be useful in interpreting observations in working boilers. Another Japanese research,[2] dealing with the effect of heat flux on corrosion by flowing HCl, would seem to have more direct application to certain service problems—namely those arising during the cleaning of boilers; here again corrosion rate is found to be increased or decreased according to the direction of the heat flux, which may influence the transfer rate of Fe^{2+}, H^+, H_2 gas and also dissolved O_2—as well as the rate of the electrochemical reactions.

Superheater Problems. German work[3] has shown that vigorous cold-work treatment, such as surface grinding, can raise the corrosion resistance of austenitic stainless steel to that of nickel-base alloys.

Treatment of Boiler Water

Use of Additives. The inhibitive power of hydrazine and sodium sulphite have been investigated by Gouda,[4] whose work, although not carried out under boiler conditions, deserves study. The good effect of hydrazine had usually been attributed to the removal of oxygen, or to the provision of an anodic reaction alternative to $Fe \rightarrow Fe^{2+} + 2e$. The new research suggests that hydrazine inhibits by adsorption on sites where otherwise anodic attack would start. If only some of these sites become covered with N_2H_4 molecules (as happens if the hydrazine concentration is low), general attack will be replaced by pitting. Higher concentrations allow coverage of all sites, and corrosion is practically prevented. The good effect of Na_2SO_3 had also been attributed to oxygen removal, but it is now found that, under conditions where deaerated water still allows slow corrosion, water containing Na_2SO_3 prevents corrosion entirely. Here again we seem to be dealing with

1. T. Ishikawa and R. Midorikawa, *Tokyo Cong. Ext. Abs.* 1972, p. 395.
2. T. Shinohara and Y. Suezawa, *Tokyo Cong. Ext. Abs.* 1972, p. 397.
3. S. Leistikow, *Tokyo Cong. Ext. Abs.* 1972, p. 355.
4. V. K. Gouda and S. M. Sayed, *Corr. Sci.* 1973, **13**, 653 (2 papers).

genuine inhibitive action, and the results suggest that it is not prevented by the presence of Cl^- or SO_4^{2-}.

Corrosion in steam condensate systems can be a financial burden in industry. Experiments at Trevose, Pa., USA[1] suggest that inhibitor formulations based on octadecylamine are the most effective of commercially available film-forming systems.

Possibilities of Lithium Hydroxide. A paper from Leatherhead[2] states that the attack on the internal surfaces of mild steel boiler tubes is often rapid and locally severe. Detection is difficult whilst the plant is on load, and the tube may fail, bringing the boiler unit and its associated plant to a standstill without warning. The cost of a single failure can be extremely serious. A correlation has been found between boiler failures and the level of the O_2 concentration, but this is probably related to an association found recently between the O_2 level and the concentration of metallic impurities. There is a connection between reaction rate and the solubility of iron in NaOH, which itself increases with the alkali concentration. If so, the use of LiOH might have possibilities; the excellent results obtained by Bloom may be due to the limited solubility of LiOH, but this might itself lead to an increased 'hide-out'. Investigations of possible adverse effects of LiOH are therefore being carried out.

Japanese work[3] confirms the view that LiOH is safer than NaOH. In LiOH, the attack is governed by a parabolic law, the constant being independent of concentration; in NaOH, a cubic law is obeyed up to 0·5N but a linear law above that concentration.

One fear expressed regarding the use of LiOH has been that it may cause stress corrosion cracking. New experiments reported by Jones and Humphries[4] suggest that at all potentials studied the susceptibility to cracking is lower in LiOH than in NaOH of equivalent concentration. The low susceptibility in LiOH is probably due to the pore-blocking action of $LiFeO_2$.

The films formed at 300°C in presence of LiOH have received study from Moore and Jones.[5] The oxide first formed is oriented with a morphology dependent on the grain orientation of the substrate, but later this breaks down and gives place to a fine-grained randomly oriented film at the base covered with a layer of Fe_3O_4 crystals grown from the solution. The breakdown probably arises from stress in the epitaxial oxide as the film thickens. The base film attains a limiting thickness when the rate of its formation at the metal–oxide interface no longer exceeds the dissolution rate at the solution–oxide interface.

1. J. J. Schuck, C. C. Nathan and J. R. Metcalf, *Mat. Prot. Perf.* 1973, **12** (10) 42.
2. H. G. Masterson, J. E. Castle and G. M. W. Mann, *Chem. Ind.* 1969, p. 1261.
3. O. Asai and N. Kawashima, *Amsterdam Cong.* 1969, p. 492.
4. D. de G. Jones and M. J. Humphries, *Tokyo Cong. Ext. Abs.* 1972, p. 171. Cf. R. L. Jones and E. W. Steinkeller, *Corrosion (Houston)* 1971, **27**, 353.
5. J. B. Moore and R. L. Jones, *J. electrochem. Soc.* 1968, **115**, 576.

Condensers

Localized Corrosion at hot spots. Boden[1] has carried out an important study of the corrosion of Cu and Cu-base alloys under conditions of boiling heat transfer. Thermogalvanic cells are set up, with the hot spots constituting small anodes surrounded by the less hot surrounding region as large cathodes; this naturally leads to intense localized attack. An unexpected feature is the fact that, although under boiling conditions dissolved oxygen would presumably be expelled into the steam bubbles, the cathodic reduction of oxygen seems to proceed readily at the boiling-point; in experiments on Cu in aerated 3% NaCl the reduction was equivalent to 70 $\mu A/cm^2$ at a temperature of 30°C; it becomes 1200 $\mu A/cm^2$ when the solution is boiling at the hot spot with a metal surface temperature about 102°C.

Choice of Materials. The increasing pollution of waters available for cooling purposes has raised some serious problems. Japanese experience[2] has favoured the use of alloys containing tin and aluminium. Trials in a model condenser show that tubes made from a copper alloy containing 8% Sn, 1% Al and 0·1% Si resist polluted water better than aluminium brass tubes, and are likely to prove economical in the long run (at Japanese prices prevailing in 1969). They develop a uniform white layer of SnO_2 which is stable even in the presence of H_2S, whereas aluminium brass develops a porous black scale of Cu_2S which is brittle and may lead to pitting at breaks, since Cu_2S is an efficient cathode.

Information[3] presented at the Lausanne Power Conference held in 1947 suggests that aluminium brass, cupro-nickel and stainless steel were then in common use. For aluminium brass tubes it was considered important to maintain correct storage conditions. Copper–nickel alloys were being used where the water was polluted or where erosion was likely to occur. The use of stainless steel was extensive in 1967, but precautions against pitting were necessary. The desirability of using the same material for tubes and tube-plates was emphasized; in some cases, it may be possible to clad them with the same material (e.g. a Cu–Ni alloy)—thus avoiding bimetallic corrosion.

Indian studies,[4] undertaken as the result of frequent tube failures at thermal power stations, emphasize the potential dangers of using certain 'neutralizing substances', such as cyclohexylamine, morpholine and ammonia, in condensers fitted with aluminium brass tubes. If such substances are present at high concentrations, the corrosion rate may be increased 5 times; polarization experiments suggest that they depolarize the anodic reaction.

Practical rules for the choice of condenser materials are provided by Eichhorn.[5] For fresh water containing less than 0·1% dissolved matter,

1.   P. J. Boden, *Corr. Sci.* 1971, **11**, 353, 363 (2 papers).
2.   S. Sato, *Amsterdam Cong.* 1969, p. 795.
3.   R. Gasparini and P. Sturla, *Corr. Abs. (Houston)* 1969, p. 129.
4.   K. Balakrishnan, P. Annamalai, B. Sathianandham, N. Subramanyan and
S. Sampath, *Corrosion (Houston)* 1969, **25**, 92.
5.   K. Eichhorn, *Werkstoffe u. Korr.* 1970, **21**, 553.

70/30 Cu/Zn brass (or 70/28 containing tin) can be used. For sea-water and brackish water, with flow-rates exceeding 1·4 m/s, aluminium brass or a copper–nickel alloy is needed. If the sea-water is seriously polluted, it is advisable to use an alloy containing iron; 1–2% iron in a 90/10 Cu/Ni alloy, or 0·4 to 1·0% Fe in a 70/30 alloy, will suffice.

The corrosion of valves in flowing warm sea-water has been studied by Mattsson.[1] Ordinary free-turning brass suffers dezincification and is unsuitable, unless it is in contact with iron which provides cathodic protection and greatly reduces the attack. For more general situations a special brass with 1·7% Al, 0·05% As, 65% Cu, 2·5% Pb, 0·8% Sn, with the balance Zn, resists dezincification. Gun-metal and phosphor-bronze suffer erosion corrosion, but not to such an extent as to disturb the functioning of the valves.

Use of iron in preventing the corrosion of brass. The addition of ferrous sulphate to cooling water for the production of a protective film on brass tubes has given satisfaction in several cases. Studies of the effect on a model condenser at a Japanese power station[2] have shown that aluminium brass tubes became covered with a smooth brown film when 0·01 ppm of Fe^{2+} was present in sea-water, and this gave complete protection even at 3·7 m/s flow-rate. When no Fe^{2+} was added, there was erosion at the inlet end, with a maximum corrosion rate of 0·6 mm/year.

The mechanism of the process has been studied in Italy.[3] Negatively charged particles of lepidocrocite are attracted by the electric field associated with the Cu_2O film (which has a positive zeta-potential). Theory predicts the possibility of reducing the amount of $FeSO_4$ added to the water without impairing protection.

North and Pryor[4] have studied the formation of a film on a Cu–Fe alloy. The alloy containing particles of α-Fe in a Cu matrix develops protective films in chloride solution after an initial period of relatively rapid attack; the protective film is a mixture of Cu_2O and lepidocrocite, arising from the oxidation (and subsequent hydrolysis) of Fe^{2+} formed by anodic attack on the iron particles, followed by electrophoretic deposition of the positive ferric hydroxide colloidal particles on the cathodic copper matrix. The lepidocrocite covering spreads out as patches which finally meet one another.

There is evidence that the spreading obeys the laws of expanding circles, as developed in an early paper.[5] Once formed over the entire surface, the lepidocrocite film inhibits further corrosion, probably by preventing the cathodic reduction of dissolved oxygen.

Pearson[6] provides a review of the role of iron in preventing the corrosion of Cu–Zn alloys in marine heat exchangers. The iron may come from a sacrificial anode, or it may be added to the water as a salt, or again it may come from iron present as an alloying constituent. It is suggested that iron

1. E. Mattsson and L. Svensson, *Brit. Corr. J.* 1972, **7**, 200.
2. S. Sato and T. Nosetani, *Corr. Abs. (Houston)* 1971, p. 271.
3. R. Gasparini, C. della Rocca and E. Ioannilli, *Corr. Sci.* 1970, **10**, 157.
4. R. F. North and M. J. Pryor, *Corr. Sci.* 1969, **9**, 509.
5. U. R. Evans, *Trans. Faraday Soc.* 1945, **41**, 365.
6. C. Pearson, *Brit. Corr. J.* 1972, **7**, 61.

becomes incorporated in the surface film, which ceases to be cathodic, so that local corrosion cells are not set up. Apparently iron does not improve the corrosion resistance of aluminium-bronze in sea-water, where the resistance is attributable to an Al_2O_3 film.

North and Pryor[1] have provided an interesting discussion of the fact that the oxide-film formed on 90/10 and 70/30 Cu/Ni alloys is more protective towards boiling NaCl solution than that formed on unalloyed copper. Essentially the oxide formed is Cu_2O in all cases, but whereas that formed on unalloyed copper can contain no other metal, the film formed on the alloys can contain nickel, and this may add to its resistance. It is well known that the oxide generally written Cu_2O contains, in fact, less copper than that formula would suggest; there are vacant cation sites. The increased resistance of the film produced on the alloys can be attributed either to the filling up of the cation vacancies with Ni^{2+} ions, or to the replacement of Cu^+ by Ni^{2+}. The first change destroys two positive holes per exchange and the second destroys only one. (A 'positive hole' is equivalent to the transformation of Cu^+ to Cu^{2+}.) In either case the electronic resistance is increased, as well as the ionic resistance, when cation vacancies disappear; both changes can contribute to an improvement in the corrosion behaviour.

Alternative Mechanisms of Dezincification. In the years when the dezincification of brass condenser tubes was a serious problem, two alternative explanations were favoured by different authorities. According to one, the anodic reaction consisted of the preferential extraction of zinc, leaving residual copper; according to the other, both zinc and copper were brought into solution by the anodic reaction, and then copper was redeposited as part of the cathodic reaction. Both proposed mechanisms led to the same final results, and under reversible conditions the driving EMF would be the same in both cases. However, under service conditions the current would be different in the two cases, and the corrosion rate would be different also; the first mechanism, which involves zinc finding its way from the interior of the alloy into the solution, would involve a large resistance term, whilst the second mechanism, involving copper dissolution at one point and redeposition at another, might well involve concentration polarization.

The dezincification of both α- and β-brasses has been studied by Heidersbach and Verink,[2] who express their results in terms of Pourbaix diagrams. On both materials they found that either of these two alternative mechanisms could take place. Each can occur over the appropriate range of potential and pH, but sometimes both are observed on the same specimen.

Modifications have been put forward for each of these mechanisms. Wagner and Pickering[3] have suggested that the preferential dissolution of zinc is caused by the diffusion of double vacancies into the alloy, producing pores which can fill themselves with electrolyte; this proposal overcomes some of the objections previously raised against the theory of preferential

1. R. F. North and M. J. Pryor, *Corr. Sci.* 1970, **10**, 297.
2. R. H. Heidersbach and E. D. Verink, *Corrosion (Houston)* 1972, **28**, 397.
3. H. W. Pickering and C. Wagner, *J. electrochem. Soc.* 1967, **114**, 698.

dissolution of zinc. The theory based on the simultaneous dissolution of both zinc and copper has been modified by Lucey;[1] he considers that the copper passes into solution at anodic points as Cu^+ ions, and that there is then disproportionation of CuCl, giving $CuCl_2$ and metallic copper.

A research which seems to reconcile these two modified views has been described by Rothenbacher.[2] He has studied the behaviour of 70/30 brass in 0·1M NaCl containing 0·01N H_2SO_4; in case it should be objected that sea-water is alkaline rather than acid, it could be replied that when once anodic attack has been set up, the pH value at the anodic points is likely to be brought down into the acid range. Rothenbacher finds that two layers of copper are produced. The outer one bears no relation to the structure of the brass, and is coarsely crystalline in character; this is probably formed by dis-proportionation of CuCl. Below it is found a fine-pored copper layer almost free from zinc, probably arising from the in-diffusion of double vacancies. The finely pored copper skeleton retains the shape of the original grains and also the twin boundaries; it is about 10^{-6} cm thick. Dezincification is favoured by plastic deformation; where dislocations cut the surface, the work of sepa-ration of the metal atoms from the lattice is diminished. These also act as sinks for vacancies, thus diminishing the length of the diffusion path and favouring the development of porosity.

Evidence in favour of the preferential extraction of zinc is provided by Pickering's X-ray study[3] of partly dissolved γ- and ε-brass. He finds that new phases appear, more rich in copper. N H_2SO_4 and various buffer solutions were used, the CD being controlled at 2–5 mA/cm².

Another research designed to distinguish between the various mech-anisms has been provided by Revie and Uhlig.[4] They studied the creep of α-brass surrounded by a deaerated acetate buffer solution. After an hour of creep in the absence of current, an anodic current of 0·9 mA/cm² was applied, and the extension rate was found to increase 8 times; the rate declined again when the current was discontinued. This result is regarded as supporting the Pickering–Wagner mechanism. The authors write: 'The divacancies diffuse into the alloy, enabling the sessile dislocations to climb, thereby causing additional plastic flow and an increased creep rate.'

Cleaning of Condenser Tubes.   In early days, when the flow-rate of water running through condenser tubes was much lower than that common today, one of the most disastrous types of corrosion was 'deposit-attack'; the deposition of debris at points on the tube surface set up the usual dangerous combination of small anodes and large cathodes; intense local corrosion was the result. Very great importance was, therefore, attached to the proper cleaning of the tubes. The adoption of higher flow-rates introduced new forms of trouble, such as impingement attack, but possibly rendered the old types less serious. Nevertheless, cleaning remains a most important matter

1.   V. F. Lucey, *Brit. Corr. J.* 1965, **1**, 9, 53 (2 papers).
2.   P. Rothenbacher, *Corr. Sci.* 1970, **10**, 391.
3.   H. W. Pickering, *Trans. electrochem. Soc.* 1970, **117**, 8.
4.   R. W. Revie and H. H. Uhlig, *Corr. Sci.* 1972, **12**, 669.

even today, and Indian research[1] on the removal of hard scale from condenser tubes, using HCl containing a commercial inhibitor, deserves attention. It is found that there is no appreciable attack on the metal, and the scale can be removed in 15 minutes by the 9% acid, and in 5 minutes by 25% acid.

Nuclear Power Plant

Influence of Radiation on Corrosion Rates. Several of the papers presented at the Amsterdam Congress of 1969 refer to the influence of radiation produced at nuclear reactors on corrosion and related changes. For instance, a research from Harwell[2] describes the effect of irradiation by fission fragments upon the oxidation of a ferritic steel containing 15% Cr, 4% Al and 0·86% Y in oxygen at 650° and 800°C. This material has been shown to have good resistance to oxidation and superior adhesion at temperatures around 1000°C, and compares favourably with other alloys; consequently it appears to have potential utility for components of gas-cooled nuclear reactors. Any effect of irradiation upon its oxidation behaviour is clearly important. Fission fragments have been shown experimentally to be a convenient radiation species for such studies, provided that the oxide-film thickness is smaller than the range of the fragments (about 10 μm). It must, however, be recognized that the associated chemical contamination of the growing oxide-film with the wide range of fission product elements and uranium oxides could possibly modify oxidation behaviour. It was found that irradiation with fission fragments had no influence on the oxygen uptake at 650°C, whilst at 800°C it actually reduced the attack produced in 48 hours. The amount of oxidation at 800°C increased with the time of exposure up to 1000 hours; thereafter there was no further significant weight gain up to 5889 hours. It is possible that the irradiation facilitates nucleation and growth of a more protective oxide-film, e.g. by enhancing the diffusion of beneficial alloying elements in the steel to the steel–oxide interface; another possible explanation, suggested by Leach, is that the effects arise from a change in the thickness of the protective layer.

Engell[3] has found that fast neutron irradiation increases the stress corrosion susceptibility of 19/10 Cr/Ni steel.

Hoyer[4] discusses radiolytic corrosion in cooling-water circuits. Here the concentration of copper derived from the ducts is much greater under irradiation than in blank experiments. Hydrazine additions provide some benefit, but the protective effect seemed to persist for only about one hour. Benzotriazole is the most efficient inhibitor. The question has been raised as to whether benzotriazole affects the exchange capacity of ion-exchange resins; it seems that experience extending over two years at one synchrocyclotron and one year at another indicates that there is no adverse effect.

1. V. K. Nigan, T. K. Grover and B. Sanyal, *Labdev. J. Sci. Tech. (Kanpur)* 1968, **6A**, 220.
2. M. J. Bennett, G. H. Chaffey and J. E. Antill, *Amsterdam Cong.* 1969, p. 155: cf. J. S. L. Leach, p. 162.
3. F. Schreiber and H. J. Engell, *Werkstoffe u. Korr.* 1972, **23**, 255.
4. F. E. Hoyer, *Amsterdam Cong.* 1969, p. 164.

Coriou[1] has studied inhibition of the corrosion of steel in nuclear plant circuits. He states that $NaNO_2$ is decomposed by γ-radiation, whereas Na_3PO_4 is stable. However, the latter does not prevent corrosion of steel under conditions of irradiation, although it will do so if sufficient NaOH is added to raise the pH value to 12. Borax is found to protect against irradiation at a concentration of 40 g/l, which gives a pH value of 9·5.

Zirconium Alloys. The oxidation of zirconium alloys has received study from Cox.[2] At the outset growth may follow a parabolic or even a cubic law, but at a certain thickness, which depends on the temperature, there is a sudden increase in corrosion rate, which now approximately follows a rectilinear law. The primary cause of the transition is the generation of a network of small pores (10–50 Å in radius) which penetrate to the oxide–metal interface. Large cracks due to temperature cycling and stresses due to grain growth are comparatively unimportant factors. The size distribution and character of the gaps has been studied under the electron microscope and also by means of mercury entry under pressure.

The uptake of hydrogen by zircaloy in concentrated LiOH has been studied by Kass.[3] It is usually rectilinear with time, and is greatest when oxygen is excluded. Kass provides photo-micrographs showing the distribution of the hydride.

A paper from Harwell[4] shows that radiation in a reactor may enhance the corrosion rate of zircaloy 2 and also that of zirconium containing 2·5% Nb by a factor of 10 or more. On the other hand, experience at Culcheth[5] seems to suggest that the alloy with 2·5% Nb is oxidized more slowly in the pile than outside. Further information regarding the effect of alloying constituents is provided by Lacombe.[6] His results show (1) an opening stage with a complex growth law, (2) parabolic growth, (3) a stage with cracking and peeling. Some metallic additions only enter the oxide phase in small amounts (or as inclusions) and do not greatly affect kinetics; these include Cu, Cr, Mo and Fe. Others (Al, Nb and Sn) accelerate oxidation, entering the oxide in amounts similar to those in which they are present in the alloy; they produce cracking, with the formation of cubic ZrO_2. The Wagner–Hauffe principle appears in this case to be unhelpful in explaining the effect of additions.

Johnson[7] has studied the effect of radiation induced corrosion on the introduction of hydrogen into zirconium alloys. This is a matter of some importance, since the formation of hydrides can lead to the embrittlement of such alloys. The conclusion reached is that hydrogen absorption can be tolerated in zirconium alloy reactor components under most reactor conditions; however, special configurations such as thick oxides or the hydrogen

1. H. Coriou, L. Grall and H. Willermoz, *Ferrara Symp.* 1970, p. 627.
2. B. Cox, *J. nuclear Mat.* 1969, **29**, 50.
3. S. Kass, *Corrosion (Houston)* 1969, **25**, 30.
4. R. C. Asher, D. Davies, T. B. A. Kirstein, P. A. J. McCullen and J. F. White, *Corr. Sci.* 1970, **10**, 695.
5. C. S. Campbell and C. Tyzack, *Brit. Corr. J.* 1970, **5**, 172.
6. B. de Geles, G. Beranger and P. Lacombe, *J. nuclear Mat.* 1969, **29**, 1.
7. A. B. Johnson, *Amsterdam Cong.* 1969, p. 168.

'windows' produced where there is bonding to Inconel can contribute to accelerated hydriding.

Problems in different Types of Plant. Various papers have been published dealing with problems arising in different designs of reactors. Brush and Pearl[1] deal with corrosion by neutral feed-water in boiling water reactors, and state that copper-based alloys can be used only if no oxygen is present, whilst carbon steel can be used under certain conditions of oxygen concentration; it is stated that stainless steel can be used regardless of oxygen content.

Work has been conducted in Sweden[2] on the oxidation of stainless steel and nickel-based alloys used for cladding fuel elements in nuclear reactors depending on high temperature steam for heat transfer. Oxidation at various temperatures (500° to 800°C) in steam with different oxygen contents has been studied on numerous alloys, but there is no simple relation between behaviour and alloy composition. At 300°C in water, materials with ground surfaces suffered oxidation more quickly than pickled specimens, but at 650°C in steam the reverse was true.

Catchpole[3] deals with the problems of 'Magnox stations', where the fuel is natural uranium clad in magnesium alloy. He states that in a CO_2 atmosphere at 350° to 450°C mild steel is oxidized to Fe_3O_4, which gradually thickens and breaks away. Preliminary investigations suggested that the effect was likely to be small and should cause no appreciable weakening due to metal loss. However, after a broken bolt had been found at a power station, a more searching investigation was carried out. This confirmed the original expectations as regards ordinary free surfaces, but at trapped surfaces between nuts and bolts, the increase of volume due to oxidation was found to be capable of leading to failure. By limiting the temperature to 360°C, with only occasional operation at 380°C, the situation was brought under control, but this seriously reduced the power output of the group of stations concerned.

Corrosion experience with Dragon heat-exchangers has been discussed by Gray and Salter.[4] Where the magnetite film formed by action of water is intact, corrosion falls off according to the parabolic law. This has been verified during studies carried out during the starting-up period of a new series of exchangers. Earlier series had suffered failure due to localized attack, but apparently in the latest pattern this has been eliminated by attention to design aimed at achieving flow stability. Corrosion experiments point to the danger of using volatile inhibitors such as morpholine sulphate, which can generate local acidity.

A study of the oxidation of rimming steels in CO_2 at 350–550°C and pressures up to 600 psig by Goodison and Harris[5] is of value in connection with

1. E. G. Brush and W. L. Pearl, *Corrosion (Houston)* 1972, **28**, 129.
2. S. Jansson, M. Hübner, G. Östberg and M. Pourbaix, *Brit. Corr. J.* 1969, **4**, 21.
3. S. Catchpole, *Chem. Ind.* 1972, p. 592; esp. p. 594.
4. P. S. Gray and W. B. Salter, *J. British Nuclear Energy Soc.* July 1971, p. 207.
5. D. Goodison and R. J. Harris, *Brit. Corr. J.* 1969, **4**, 146.

the type of plant where CO_2 is used for heat transfer. At 1 atm pressure a protective film is formed, but at high pressures, especially in presence of moisture and CO, this may break down locally, giving rise to oxide excrescences; whilst these excrescences are being formed, the growth law changes gradually from parabolic or cubic to rectilinear.

Problems of Fusion Reactors. Although several years may pass before controlled thermonuclear fusion is a practical possibility, authorities are becoming increasingly confident that the necessary conditions (plasma temperature, containment time and density) can be achieved. A new international research programme is to be started. It is believed that by scaling in size the plasma confinement system, conditions for fusion can be obtained which will ensure a net power gain. The energy crisis today facing the world makes it highly important that no time should be lost through delays arising from the deterioration of materials subjected to the action of neutrons or gamma-rays. It is satisfactory to note that such problems are already receiving attention; a paper by Holmes-Siedle[1] should be studied. This is particularly concerned with materials suitable for plant of the British TOKAMAK design, but there is some allusion to the alternative forms of plant being developed in other countries.

In the case envisaged, the energy gain arises from the reaction between deuterium and tritium to form ^4He; the tritium is to be bred within the reactor from neutrons acting on lithium. Materials likely to be damaged include insulators as well as metallic conductors, and many of the changes discussed should not, perhaps, be described as 'corrosion'; they include the production of gas bubbles within the material, which may alter mechanical properties, and void formation, which can produce catastrophic loss of strength. Problems calling for intensive research include 'the development of structural metals with minimum change in ductility, creep, stress-rupture, fatigue and swelling under high neutron fluences at high temperatures, and with high resistance to corrosion by the breeding material'.

Wright[2] quotes a 'provisional date' for the completion of a commercial reaction producing electrical energy by fusion as the 'mid-nineties', but states that two laboratories are to be built shortly in the USA 'for testing new materials essential to the building of fusion reactors'. He regards the decision to cut down British research on the fusion method as an error.

Further References

Papers. 'Corrosion in high-temperature water'.[3]
'Corrosion of Condenser Tubes due to carbonaceous film on the surface'.[4]
'Corrosion inhibitors for steam condensate systems'.[5]

1.　A. Holmes-Siedle, *Nature* 1974, **251**, 191.
2.　P. Wright, *The Times*, Aug. 3, 1975.
3.　G. Butler, *Anticorrosion* 1973, **20** (3) 6.
4.　R. Retief, *Brit. Corr. J.* 1973, **8**, 264.
5.　J. J. Schuck, C. C. Nathan and J. R. Metcalf, *Mat. Prot. Perf.* 1973, **12** (10) 42.

'Water-side Corrosion of Diesel engines'.[1]

'Growth of tubercles on iron-base alloys in elevated temperature water'.[2]

'The thermogalvanic corrosion of mild steel in alkaline solution'.[3]

'Central heating Corrosion—causes into prevention'.[4]

'Prevention of Condenser Inlet Tube Erosion-corrosion'.[5]

1. T. K. Ross and A. F. Aspin, *Corr. Sci.* 1973, **13**, 53.
2. R. W. Staehle and K. N. Shetty, *Williamsburg Conf.* 1971, p. 305.
3. V. Ashworth and P. J. Boden, *Corr. Sci.* 1974, **14**, 183, 199, 209 (3 papers).
4. P. S. Muetzel, *Corr. Prev. and Control* 1974, **21** (2) 11.
5. J. Lichtenstein, *Mat. Performance*, March 1974, **13**, 17.

CHAPTER XIII

Atmospheric Corrosion

Preamble

Picture presented in the 1960 and 1968 volumes. The types of corrosion hitherto discussed have concerned metals entirely unwetted (chapters II and III) or immersed, at least partially, in a liquid (chapters IV to XII). The so-called 'atmospheric corrosion' embraces a number of intermediate conditions. Outdoor exposure generally involves a metal surface being flooded by water during rainy periods, but drying up when the rain ceases. Indoor corrosion usually involves no visible wetting, but if dust settles on the surface, conditions comparable to those produced by a visible drop of liquid can be set up. The classical work of Vernon made it clear that various types of dust behave differently, but that when certain salts are present in or on the dust particles, spots of rust are produced; this will only happen if the relative humidity exceeds a critical value, which Bukowiecki has shown to be close to the vapour pressure of a saturated solution of the salt present. It would appear that in these cases the salt will take up moisture until a liquid phase is produced over a microscopic area. In such a case, conditions present analogies with those met with in the corrosion of a partly immersed specimen.

Although outdoor corrosion can be produced in pure, country air (especially in the case of steel), attack becomes far more serious if salt is present, or if the atmosphere contains sulphur dioxide—as is usually the case in industrial and urban districts where fuel containing sulphur is being burned. Indoor corrosion is also possible if the relative humidity is sufficiently high, and if either sulphur dioxide contamination or particles of sea-salt are present.

It has been explained that dry corrosion (direct oxidation) is electrochemical in the sense that ions and electrons are passing through the oxide-film in a direction roughly at right angles to the metallic surface. However, the use of the term 'electrochemical corrosion' is generally restricted to cases of wet corrosion where electrons will be passing through the metal between anodic and cathodic areas and in a direction roughly parallel to the metallic surface; this happens in most cases of corrosion by natural waters, salt solutions or acids. Much discussion has taken place as to whether atmospheric corrosion belongs to the first or second type. It could be argued that the corrosion set up by salt particles would have the same mechanism as that set up by drops of salt solution, where without doubt currents are passing between a cathodic zone at the periphery and an anodic region in the interior. But such arguments have not been universally accepted. Certainly in the case of corrosion set up by small amounts of sulphur dioxide, many authorities have, at least until recently, considered that there is little or no electrochemical attack—using the word to mean a flow of current between well separated anodic and cathodic points. Indeed, in the case of iron, the mechanism most generally favoured has been the so-called 'acid regeneration cycle'. According to this idea, sulphur dioxide and moisture are adsorbed, and the former is oxidized by atmospheric oxygen to give sulphuric acid, which attacks the iron, giving ferrous sulphate; this is now oxidized to produce ferric sulphate; the latter hydrolyses to produce ferric hydroxide (rust), liberating sulphuric acid, which now acts on further iron. Such a cycle would explain the fact, emphasized by Schikorr, that one molecule of sulphur dioxide can produce many molecules of rust. However, when once rust and ferrous sulphate are present on the surface, conditions became favourable to a mechanism of another character, involving anodic and cathodic reactions occurring at different points. At the cathodic points ferric rust ($FeOOH$) can be reduced to magnetic (Fe_3O_4), whilst at the anodic points iron passes into the liquid as Fe^{2+} ions. The magnetite formed cathodically can quickly be oxidized by air to give fresh ferric rust, in an amount greater than had been present before. The reactions can be written

$$Fe = Fe^{2+} + 2 \qquad \text{(anodic reaction)}$$
$$8FeOOH + Fe^{2+} + 2e = 3Fe_3O_4 + 4H_2O \qquad \text{(cathodic reaction)}$$
$$3Fe_3O_4 + 0{\cdot}75O_2 + 4{\cdot}5H_2O = 9FeOOH \qquad \text{(chemical re-oxidation)}$$

It will be observed that such a cycle increases the amount of rust ($8FeOOH$ becomes $9FeOOH$), with one atom of iron contributed by the metal.

This mechanism, commonly known as the 'electrochemical cycle', would also account for the fact that one molecule of sulphur dioxide can produce many molecules of rust. The current produced by the cell

$$Fe \mid FeSO_4 \text{ solution} \mid FeOOH \text{ (supported on Cu)}$$

has been demonstrated; the orange-brown $FeOOH$ changes to black Fe_3O_4, and when there is no $FeOOH$ left the current is found to die away. The rapid reconversion of freshly produced magnetite to the ferric state by the action of air has also been demonstrated.

There is little doubt that both the 'acid regeneration cycle' and the 'electrochemical cycle' can really operate. The question which has remained doubtful until recently is which of the two cycles proceeds most rapidly and can therefore account for the greater part of the rust formation. It may be mentioned, in passing, that, although the so-called electrochemical cycle involves separate anodic and cathodic areas, the movement of electrons will be taking place in a direction roughly at right angles to the metallic surface— so that even this mechanism has something in common with the dry oxidation produced when iron is heated in air.

It should be noticed that the electrochemical mechanism just suggested is possible on iron because the intermediate oxide, magnetite, is an electronic conductor, so that the passage of electrons is possible between the point where anodic dissolution is occurring and the cathodic points where rust is being reduced. With most non-ferrous metals, no oxide possessing good electronic conductivity exists, and the mechanism is likely to be different. This feature of the situation explains why zinc, which is attacked more quickly than iron in a salt solution, resists far better than iron under atmospheric conditions; indeed a coating of zinc is used to protect iron from atmospheric corrosion. It also explains why aluminium, which forms an oxide possessing insulating properties, suffers only very slow atmospheric corrosion, although the drop in free energy connected with its conversion to oxide or hydroxide is far greater than in the case of iron.

Developments during the period 1968–1975 and Plan of the new Chapter. During the period under review much experimental work has been devoted to the study of rust, particularly the composition and crystal structure of the various compounds formed under different conditions. Possibly of greater interest is the mechanism of atmospheric rusting, and considerable work has been carried out which appears to show that, as in the case of immersed corrosion, the mechanism is really electrochemical. The reactions assumed to be important in the 'acid regeneration theory' can, indeed, take place, but they only possess positive importance in the early stages; the electrochemical reactions can occur very much faster and are mainly responsible for the continuation of rusting.

Great interest continues to be shown in low-alloy steels, which, in certain parts of the world, can be used successfully in the unpainted condition. It would seem that the action of the alloying constituents can also be explained reasonably if the electrochemical mechanism is accepted.

The new chapter opens with a section on the composition of rust. Then the mechanism of rust formation is discussed—firstly the rusting set up by salt particles in dust, and then that set up by the presence of sulphur dioxide traces in moist air above a certain critical humidity; the evidence for the electrochemical mechanism is presented. It is believed that Taylor's recent experimental study of the electrochemical mechanism of atmospheric corrosion possesses an importance comparable to that of Hoar's work on the electrochemical mechanism of immersed corrosion published in 1932.

The distribution of pollution in the atmosphere and efforts now being made to reduce it—a matter which affects the health of Men as well as that of Metals—is then discussed at some length. The application of Taylor's work to the good behaviour of low-alloy steels receives brief treatment; this is somewhat speculative, and it cannot be claimed that, at the moment, there is general agreement; it is hoped, however, that accord will be reached in due course—doubtless as the result of further experimentation. Testing procedures are then reviewed and the chapter ends with a section on the atmospheric behaviour of non-ferrous metals.

Composition of Rust

General. Much interest has long been shown in detecting the various compounds present in rust`and distinguishing between the various forms possessing the same composition but different crystal structures. Without underestimating the value of such investigations, it should be pointed out that the position of the formation of a compound may be of greater importance than its composition or structure; a compound formed in physical continuity with the metal phase is most likely to be protective. Moreover, the protective character of a compound is likely to be affected more by the presence or absence of defects (such vacant sites or dislocations) than by the structure which would exist if defects were absent.

Main Constituents. A thorough study of rust produced in different types of atmosphere has been carried out by Keller.[1] In his earlier work he discovered that rust formed at Stuttgart contained α-FeOOH (goethite), γ-FeOOH (lepidocrocite) and Fe_3O_4 (magnetite); anhydrous Fe_2O_3 (haematite) was not found. In later work he noticed an amorphous phase, whilst rust formed in marine situations was generally found to contain β-FeOOH (akaganemet). This compound appears to be formed only if Cl^- or F^- is present, and in the first case a certain amount of OH^- is replaced by Cl^-; perhaps it should be described as a basic salt rather than a hydroxide. It is generally regarded as a dangerous constituent of rust, probably because Cl^- will often be liberated during its reactions.

Importance of Magnetite in Rust. Schwarz,[2] also working at Stuttgart, has continued his fruitful investigations. His latest paper records a study of three series of steel plates exposed (1) for one year in the Ruhr, (2) for one year near Stuttgart and (3) for 10 weeks near Stuttgart. After these exposures, half the specimens in each series were exposed to N_2 (free from O_2) at 97% RH. Such treatment produced black spots on the Ruhr plates, and also smaller spots on the one-year Stuttgart plates; the latter also were found to carry green crystals when examined under the microscope. The 10-weeks plates became uniformly black-brown. This treatment in damp N_2 free from oxygen was found to convert the FeOOH into magnetite. The effect of the conversion of FeOOH into magnetite was found to make the rate of rusting produced on subsequent exposure to damp O_2 (97% RH) *faster* than would otherwise be the case. The fact that the presence of magnetite enhances liability to rusting is to be expected from Taylor's results (this volume, p. 254); but the confirmation in the Stuttgart work is particularly welcome, since some previous workers have looked upon the formation of magnetite as something favourable to corrosion resistance. Since the volume occupied by magnetite is likely to be much less than the FeOOH from which it is formed, it is unlikely on any mechanism that magnetite would be specially protective. It is quite easy to abolish the colour of rust simply by exposure to wet conditions in absence of oxygen. Taylor found that rusted specimens placed in water or salt solution ($M/10$ $FeSO_4$, $M/10$ Na_2SO_4, $M/10$ $NaCl$ or $M/10$ NH_4Cl in tubes completely filled with the various liquids, without air-space) soon lost their rusty appearance and became black. Similar changes have been made the basis of commercial treatments which, it has been claimed, render a rusty surface suitable for painting. It is not suggested that these commercial processes are in all cases without value (without knowledge of the composition of the liquid used for the treatment, no judgement on that

1. P. Keller, *Werkstoffe u. Korr.* 1967, **18**, 865; 1969, **20**, 102; 1971, **22**, 32.
2. H. Schwarz, *Werkstoffe u. Korr.* 1972, **23**, 648.

point is possible); but the change of appearance (rust colour to black), although highly impressive, is not in itself any indication of an improvement of condition. Indeed, as Schwarz's work shows, the formation of magnetite renders the metal *more* susceptible, and *not* less susceptible, to rusting. It should be borne in mind that the objection to painting over rust is due to the presence of ferrous salts in the rust, and these may remain after the FeOOH has been converted to Fe_3O_4.

Atmospheric Corrosion of Iron and Steel

Effect of Salt Particles. The simple case of corrosion by a drop of NaCl solution placed on a horizontal iron surface has been studied in some detail (1960 book, p. 118), but it is by no means a perfect model for what is happening on a much smaller scale when a particle of sea-salt settles on a clean iron surface. In the latter case, a water phase is only possible at times when the humidity exceeds a certain value, so that in practice rusting may not be continuous. Still more important is the fact that the particle of sea-salt will contain Mg^{2+} as well as Na^+, and this will have two effects. The presence of the ions of $MgCl_2$, a hygroscopic body, will enable corrosion to take place at a much lower RH value than would be present if only pure sodium chloride was present. The vapour tension of saturated NaCl is given in the literature[1] as 75·1%, whereas that of $MgCl_2.6H_2O$ is only 32·7%. In fact laboratory studies have shown that in presence of particles of pure sodium chloride corrosion does become slow below about 75%, whereas in presence of sea-salt it is quite rapid round about 50%. However, the presence of magnesium can produce another effect, because the cathodic product will now be magnesium hydroxide, which on the arrival of Fe^{2+} ions from the anodic points, can be converted *in situ* into a form of rust which has some protective value because it is in physical contact with the metal. Thus we have two factors acting in opposite directions.

Informative studies of the rusting set up by NaCl crystals adhering to an iron plate have been provided by Norwegian investigators. Henriksen[2] dipped carefully prepared iron plates into NaCl solution containing acetone, dried them *in vacuo* and then exposed them to air at 92% RH for different periods between 1 and 64 hours; the NaCl used contained a suitable proportion of the isotopes ^{24}Na and ^{36}Cl, which show a considerable difference in half-life; this makes it possible to obtain the distribution of Na^+ and Cl^- on the same specimen. The intensity of the radiation given out by ^{24}Na is sufficient to produce blackening of a photographic film in 1 to 4 hours; in that time ^{36}Cl gives no perceptible blackening. After he had obtained the distribution of Na^+, Henriksen kept his specimen plates in dry air for 9 days to allow the ^{24}Na to decay. They were then again placed between photographic films to record the ^{36}Cl.

After exposure to humid air, the iron specimens were seen to be peppered with rust spots. The photographic film after recording of the Cl and Na distribution showed darkening in spots corresponding in position to those

1.   J. F. Young, *J. appl. Chem.* 1967, **17**, 241.
2.   J. F. Henriksen, *Corr. Sci.* 1969, **9**, 573.

rust spots. When the exposure to humid air had been only 1 hour, the Na$^+$ and Cl$^-$ were both fairly equally distributed over the area of a spot; but when the exposure had been longer, the Na$^+$ was found to be concentrated at the periphery of each rust spot and the Cl$^-$ at the centre (Fig. 24). This is what would be expected from the studies of corrosion by relatively large drops of

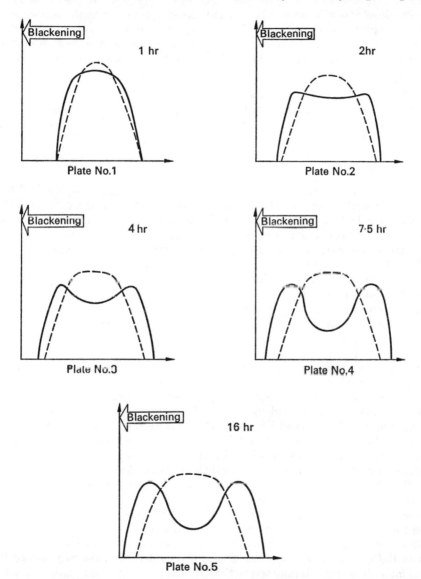

Fig. 24   Distribution of Na and Cl on corroded spots obtained by a radioactive tracer method, Curves ——— show blackening due to ^{24}Na, and curves - - - - show blackening due to ^{36}Cl (J. F. Henriksen).

NaCl solution, which develop a cathodic ring along the periphery, with an anodic area in the centre (1960 book, p. 118). The results certainly support an electrochemical mechanism.

Effect of Sulphur Dioxide. Schikorr[1] provides a useful summary of the known facts of rusting by moist air containing SO_2. He states that one molecule of SO_2 will convert 15 to 20 atoms of iron to rust in the winter, and 30 to 40 atoms in the summer; but since in winter 4 times as much SO_2 is present in the air as in summer, the rusting is about twice as fast. He interprets the facts by assuming the acid regeneration cycle, proceeding in three steps:

(1) adsorption of SO_2 on rust
(2) formation of $FeSO_4$ from SO_2, Fe and O_2;
(3) oxidation of $FeSO_4$ to rust and free H_2SO_4, which then attacks more iron, giving fresh $FeSO_4$, hence more H_2SO_4 and so on. The cycle can continue many times until the SO_4^{2-} is removed as sparingly basic ferric sulphate.

A laboratory research carried out at Cambridge by C. A. J. Taylor[2] has sought to decide between the claims of the acid regeneration cycle and the electrochemical cycle. He finds definitely that the reactions postulated in each of the two cycles can really take place, but that, when once rust and $FeSO_4$ are present, the electrochemical cycle proceeds very much faster than the acid regeneration cycle, and the latter contributes only a negligible positive effect to atmospheric corrosion. Shallow layers of ferrous sulphate solution were exposed to air in petri dishes at 25°C for various periods and then filtered; a weighed specimen of iron was placed in a glass tube filled with the filtrate. When the liquid used was a solution which had been exposed to air for 8 days, there was marked attack on the iron, producing a visible evolution of hydrogen gas and a detectable loss of weight. In contrast, freshly prepared ferrous sulphate solution produced practically no visible change on iron, and actually a slight *gain* in weight. If the air exposure had lasted only one day, the loss of weight was much smaller than if it had lasted eight days. It is evident that *oxidative hydrolysis*, liberating acid and rendering the liquid corrosive—as postulated by the supporters of the acid regeneration cycle—*does in fact take place*, but *very slowly*. In fact, the acid regeneration cycle requires days for its completion, whereas the electrochemical cycle can be repeated several times within a single hour.

But the acid regeneration cycle, although its *positive* effect is negligible, has an important *negative* effect. In some of Taylor's experiments, the rusty layer left sticking to the glass surface of the petri dishes after the air exposure on the ferrous sulphate solution was washed until the wash-water ceased to show the presence of Fe^{2+} or Fe^{3+}; it was then dissolved in HCl and tested with $BaCl_2$. Much SO_4^{2-} was found, and it is evident that the rust formed by oxidative hydrolysis is not merely FeOOH but also contains a basic sulphate.

1. G. Schikorr, *Werkstoffe u. Korr.* 1963, **14**, 69; 1964, **15**, 457.
2. U. R. Evans and C. A. J. Taylor, *Corr. Sci.* 1972, **12**, 227.

Thus there is removal of Fe^{2+} and SO_4^{2-} from solution, which will reduce the conductivity of the liquid joining anodic and cathodic points, and slow down the electrochemical reaction. This explains why a finite quantity of ferrous sulphate does not produce an infinite amount of rust—as might perhaps be expected if the equations suggested above were taken quite literally. It appears that the oxidative hydrolysis—once regarded as the most important factor in *producing* corrosion—is really important in *limiting* the amount of corrosion which the electrochemical cycle, acting by itself, would be expected to produce.

The claims of the acid regeneration and electrochemical mechanisms have been discussed by Duncan,[1] who states that 'no conflict exists between the recent results and acceptance of the electrochemical model as the main corrosion path'.

As a result of Taylor's work, the situation of iron developing rust in moist air is suggested by Fig. 25. Anodic attack on the iron at XX′ is balanced by

Fig. 25 Mechanism of atmospheric rust formation (schematic). The anodic attack on iron occurs at level of XX′ and the cathodic reduction of FeOOH to Fe_3O_4 (later reconverted to FeOOH, in larger amount) occurs at YY′.

cathodic reduction of the rust at YY′ to magnetite, which is quickly re-oxidized by air to the ferric state, producing a quantity of fresh rust additional to that previously present. Like all corrosion cells, this one has two limbs, the electrolytic limb being provided by the $FeSO_4$ solution in the channels threading the magnetite, and the electronic limb being provided by the magnetite, which allows the electrons liberated by the anodic reaction at XX′ to reach the level YY′ where they are needed for the cathodic reaction. If the humidity sinks too low, the $FeSO_4$ solution will dry up, and the electrolytic limb will fail; thus corrosion will cease. If magnetite were to be replaced by some other substance which is a poor electronic conductor, the electronic limb would fail and the corrosion would be slow.

The mechanism just suggested has been approved by de Miranda,[2] whose important review of the situation and experiments on the effects of alloying elements, deserve careful study. Those elements which interfere with the formation of magnetite and produce, instead, an internal film of protective goethite, are those which have proved useful in the low-alloy steels to be described later. de Miranda's results, based on extensive electrochemical measurements and interpreted with the aid of Pourbaix diagrams, should be read in the original report. de Miranda considers that his internal goethite layer may be identical with the $FeO_x(OH)_{3-2x}$ amorphous layer described by Misawa, and with the amorphous Fe_3O_4 layer described by Okada.

1. J. R. Duncan, *Werkstoffe u. Korr.* 1974, **25**, 420.
2. L. de Miranda, *Cebelcor Rapp. tech.* RT221 (1974), esp. pp. 28, 83, 84, 90, 94.

Quantitative study of rust production. The passage of current has been demonstrated in experiments with a model in which a weighed piece of iron or steel representing the anode is covered with filter paper soaked in $FeSO_4$ solution which in turn carries a piece of rusted steel (Fig. 26). The

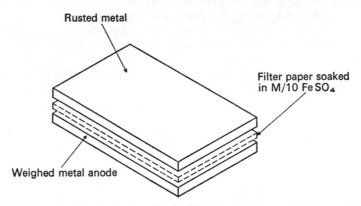

Fig. 26   Plate experiments designed to decide between two mechanisms. A weighed metal specimen is covered with filter paper soaked in M/10 $FeSO_4$, on which is laid a sheet of rusted iron; the rusted iron is connected to the lower specimen either directly or through a milliammeter. The specimen gains or loses weight in different cases. The results favour the Electrochemical Mechanism and not the Acid Regeneration Cycle (U. R. Evans and C. A. J. Taylor).

latter is joined to the anode through a milliammeter, and is turned upside down every 5 minutes. It is found (Fig. 27) that the current flowing sinks during the 5 minute period, but recovers when the upper electrode is inverted, because fresh ferric rust, previously produced by the action of air on

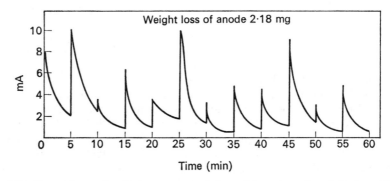

Fig. 27   Current passing between rusted iron (above) and weighed specimen (below). In this experiment the upper specimen was inverted every 5 minutes, so as to allow the magnetite formed by cathodic reduction of the ferric rust to be re-oxidized by air to the ferric state. The milliampere-hours generated are of the order of magnitude to be expected if the mechanism is electrochemical; slight divergences from the numbers predicted by means of Faraday's law are discussed in the text (U. R. Evans and C. A. J. Taylor).

magnetite, now becomes available. The weight loss of the lower specimen can be compared with that which would correspond to the total number of milliampere-hours actually observed, and is of the right order of magnitude. For pure iron it is slightly too small, probably because in addition to the main anodic reaction

$$Fe = Fe^{2+} + 2e$$

a secondary reaction

$$Fe^{2+} = Fe^{3+} + e$$

is proceeding; this is known to occur under conditions where the ferrous ions are prevented from moving away rapidly from points where they have been produced by the anodic reaction—as shown in early work by Thornhill.[1] In the case of steel the weight loss is slightly too high, probably because the corrosion eats its way around cementite particles which become dislodged and loosened. It seems impossible to explain these observations except on the electrochemical cycle indicated above.

Earlier work from Cambridge (1968 book p. 369) had shown that iron

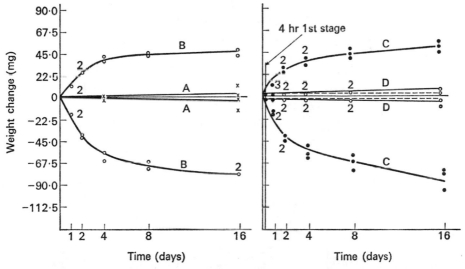

Fig. 28   Experiments showing influence of SO_2. Curves A relate to iron exposed to moist air without SO_2, and curves B to iron in moist air containing SO_2; both gain and loss of weight are much greater when SO_2 is present. Curves C represent two-stage experiments; after 4 h in moist air containing SO_2, the specimens were moved to moist air without SO_2; yet the weight change is similar to B, showing that when once $FeSO_4$ has been formed, SO_2 in the gas phase is not needed for corrosion. Curves D represent three-stage experiments; after 4 h in moist air with SO_2, the specimens were placed in water for 1 day and then in moist air without SO_2; the weight change is small because the $FeSO_4$ has been removed (U. R. Evans and C. A. J. Taylor).

1.   R. S. Thornhill and U. R. Evans, *J. Chem. Soc.* 1938, p. 614.

exposed to moist air containing SO_2 develops a barely visible film of moisture containing Fe^{2+} before ever any rust is observed. Later, however, ferric rust appears, and at that stage, it is possible to move the specimen to a vessel containing moist air free from SO_2 and yet the development of rust will continue. These qualitative observations were extended in quantitative fashion by Taylor, who after each experiment measured the gain in weight of the specimen due to conversion of iron to rust, and also the loss of weight observed after the removal of that rust in acid (under conditions chosen to avoid attack on the metal basis). If the corrosion product consisted solely of FeOOH (which could also be written $Fe_2O_3.H_2O$), the ratio of loss to gain should be [Fe]/[OOH] or 1·69. In fact, especially in the early stages, other substances (such as hydrated ferrous sulphate) will be present, and the ratio found (as an average of a number of experiments) was only 1·33 after 1 day at 25°C, rising to 1·52 after 4 days and to 1·68 after 16 days; the latter is close to the theoretical 1·69, but probably small amounts of substances other than FeOOH were still present, the effect of those giving a low ratio being balanced by those giving a high value. In the graphs of Figs. 28 to 32, the

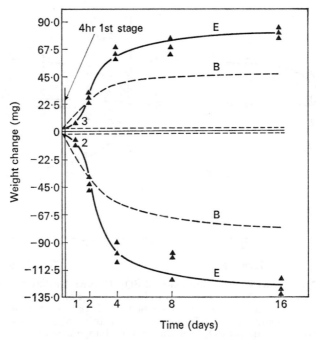

Fig. 29   Experiments with replenishment of SO_2. Curves E represent specimens exposed for 4 h to moist air with SO_2, and then moved to a tube containing fresh moist air with SO_2; thus they receive a double ration of SO_2. Curve B, in which the specimen receives only a single ration, is reproduced from Fig. 28 for comparison. The *final* effect of the double ration is nearly to double the weight change, although the initial effect of the extra SO_2 is a slight retardation (U. R. Evans and C. A. J. Taylor).

upper curves show weight gains due to rusting, and the lower curves losses after de-rusting.

Curve A in Fig. 28 shows iron exposed to moist air free from SO_2; the changes produced are small. Curve B shows iron exposed to air containing SO_2, and here the changes are at the outset much more rapid, but become slow after about 8 days. If a specimen is exposed for 4 hours (only) to moist air containing the same amount of SO_2 used as that used in B, and is then removed to air containing the same amount of moisture but no SO_2, the result is curve C of Fig. 28, which, it will be noticed, is practically the same as B. If, however, after the 4 hours' exposure to moist air with SO_2, the specimen is placed in water for 24 hours (which will remove any $FeSO_4$ and/or H_2SO_4 present) and then moved into moist air free from SO_2, the result is curve D— not very different from curve A. It would seem that, although SO_2 is needed for rapid rusting, its function is to produce $FeSO_4$; when once that has been formed, the presence of SO_2 in the gas phase is not necessary for rusting; if the $FeSO_4$ is removed by the immersion in water, rusting becomes slow.

Curve E (Fig. 29) shows the effect of exposing specimens for 4 hours to a moist air containing SO_2 and then transferring them to a vessel containing

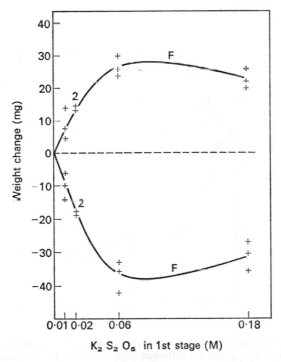

Fig. 30   Experiments with variation of the amount of SO_2. Curves F represent specimens exposed to moist air containing different quantities of SO_2 for 4 h, and then to moist air without SO_2 for 44 h. The final weight change increases with increasing SO_2 up to a point and then decreases (U. R. Evans and C. A. J. Taylor).

new quantity of moist air with the same amount of SO_2 as was originally present; the final amount of rusting after 16 days is nearly twice the amount produced when there was no replenishment of SO_2 (curve B, reproduced for comparison), presumably because twice the amount of $FeSO_4$ will have been rendered available. But immediately after the transfer, the attack seems to be slower on E than on B. The difference is not large and might be ascribed to error, but further experiments suggest that the effect is real. In Fif. 30, curve F shows the effect of varying the amount of SO_2 (produced by the action of H_2SO_4 on $K_2S_2O_5$, the amount of which could be varied) in the opening 4 hours, before removal for 44 hours to moist air free from SO_2. The amount of rust produced first increases with the amount of SO_2 provided and then appears to decrease; the probable reason is that, when too much SO_2 is present, an alternative anodic reaction becomes possible. If no SO_2 is present in the gas phase during the second stage, the cathodic reduction of $FeOOH$ to Fe_3O_4 must be balanced by the anodic attack on Fe (giving Fe^{2+}); if, however, SO_2 is present, a second anodic reaction (the oxidation of SO_3^{2-} to SO_4^{2-}) may account for part of the current flowing—explaining why

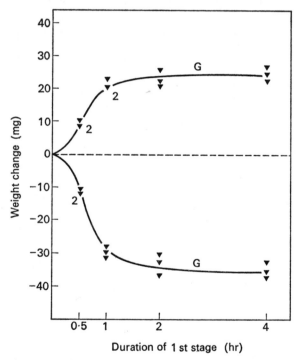

Fig. 31   Experiments with variation of time in moist air containing SO_2. Curves G represent specimens exposed for different times to moist air containing a standard quantity of SO_2 and then to moist air without SO_2 for a period making a total of 48 h. The weight change becomes constant when the first stage exceeds about 2 h—presumably the time needed to use up the SO_2 forming $FeSO_4$ (U. R. Evans and C. A. J. Taylor).

the attack on the iron is smaller.* In Fig. 31, curve G shows the effect of varying the duration of the first stage (exposure to moist air containing SO_2), the amount of SO_2 being now kept constant. It will be noticed that the amount of corrosion finally produced increases for the first 2 hours of

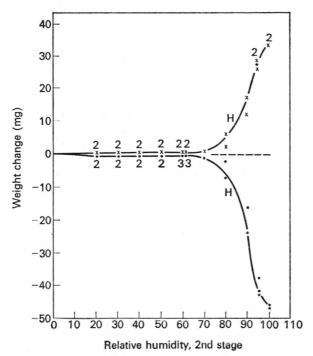

Fig. 32   Experiments with variation of relative humidity. Curves H represent specimens exposed to moist air with SO_2 for 4 h and then to air without SO_2, but at different humidities. The weight change increases abruptly about RH 70% (U. R. Evans and C. A. J. Taylor).

exposure to SO_2, but after that there is no further effect, presumably since the whole of the SO_2 has been used up in producing $FeSO_4$.

Curve H (Fig. 32) shows the effect of humidity. Specimens were exposed to moist air containing SO_2 for 4 hours and then transferred to air free from

* The fact that the corrosion produced first increases with the amount of SO_2 present but diminishes with still larger quantities has been confirmed by M. S. Al Ghali and D. Eurof Davies. Their careful chromatographic estimation of the SO_2 remaining, combined with continuous microgravimetric measurement of the weight increase, indicates that the presence of SO_2 diminishes the amount of O_2 available for the corrosion process. This seems the most probable explanation of the diminution of corrosion when the amount of SO_2 present in the atmosphere is large. Their work, which includes a study of the effect of (1) O_2 concentration, (2) SO_2 concentration, (3) humidity and (4) flow-rate, confirms the electrochemical mechanism. It will probably be published in *Corrosion Science*.

SO_2. If the RH in this second stage is less than about 70%, there is little attack, but above 70% the amount of rusting increases rapidly.

It is interesting to recall that the classical work of Vernon placed the critical humidity at about 70% although the experimental conditions were different. Czech work[1] suggests that for air containing 1 ppm SO_2 the critical humidity falls at 75%; it is somewhat lower (67%) when the amount is 10 or 100 ppm. As already stated, if the atmospheric contamination consists, not of SO_2, but of sea-salt, the critical value lies much lower.

Talbot[2] has found that an iron specimen placed in moist air containing 10 to 25 ppm SO_2 at a RH exceeding 64% develops rust, and when once hydrated $FeSO_4$ has been formed, the specimen can be moved to moist air free from SO_2, and yet rusting will continue; if, however, it is placed for 24 hours in water to remove the $FeSO_4$, little fresh corrosion is produced in moist air free from SO_2. This confirms Taylor's observations, but Talbot finds that if the SO_2 content is 400 ppm the situation is different, since now sulphite and sulphide are formed.

Franz[3] has found sulphide, sulphite, sulphate, thiosulphate and elementary sulphur in the rust formed on iron exposed to air containing SO_2 (apparently at high concentrations).

Chandler[4] has studied the behaviour of steel covered with synthetic rust containing a salt and exposed in a vessel containing air with controlled humidity. Changes were inappreciable at RH values below 42%, but at about 52% there was marked corrosion—which increased with the amount of salt present. In the range above 56% RH, 5% Cl^- caused more corrosion than 5% SO_4^{2-}, which in turn caused more than 0·5% SO_4^{2-}; 'pure rust' (free from salt) caused less rusting, and steel to which no rust had been applied suffered even less.

A recent research by Singhania and Sanyal[5] concerns the effect of previously formed rust on steel exposed outdoors, or to an atmosphere containing SO_2 in the laboratory; the pre-formed rust increases the rate of attack for the first 2–3 months but then there is a decline of corrosion rate. Rust contaminated with $FeSO_4$ or $FeCl_3$ enhances the rusting rate, whereas 'rust-converters' such as phosphates have beneficial effects. Experiments in which the pre-rusted specimens were first exposed to SO_2 and then to humid air free from SO_2 showed that a temperature (18°C) and humidity (50% RH) too low to initiate the formation of the primary product $FeSO_4.4H_2O$ on clean steel, allow sufficient formation of that compound to enable rusting to continue in moist SO_2 free air.

Distribution of ferrous sulphate in rust. Taylor's experiments, quoted above, refer to specimens in closed vessels exposed to constant

    1. K. Bartoň and Z. Martoňová, *Werkstoffe u. Korr.* 1969, **20**, 216; 1972, **21**, 85 (2 papers).
    2. J. Talbot, *Corrosion (Paris)* 1973, **21**, 97.
    3. E. Franz, *Werkstoffe u. Korr.* 1973, **24**, 598.
    4. K. A. Chandler, *Brit. Corr. J.* 1965, **1**, 264.
    5. G. K. Singhania and B. Sanyal, *Brit. Corr. J.* 1973, **8**, 224.

conditions. The special feature of atmospheric corrosion as met with in ordinary life is the variation of condition with time. Studies carried out on specimens exposed outdoors to the weather, or indoors under conditions of fluctuating humidity, are of great practical importance.

Ross and Callaghan,[1] using an electron probe microscope to study rust formed on mild steel exposed outdoors at Manchester, showed that for the first 4 months there was only slight fixation of sulphur; during the next 8 months there was a marked increase in formations containing sulphur, mainly in bands at the metal/rust interface. In the succeeding 4 months, the sulphur-containing bands moved outwards to form nests. Nests of ferrous sulphate have been observed by many investigators elsewhere, and are generally regarded of as great importance in connection with the breakdown of paint coats applied to the rusted surface.

Indoor rusting has been studied, also at Manchester, by Ali and Wood[2] in laboratory air with RH values held between 44% and 46% for most of the period (a smaller variation than would occur in an ordinary living room). The weight of each specimen remained almost constant for 20 days and then increased slowly; after 60 days the weight increase became rapid, with the formation of Fe_3O_4 and possibly $\gamma\text{-}Fe_2O_3$; later $\alpha\text{-}Fe_2O_3$ was found. Different crystal faces behaved differently; {100} rusted more quickly than {112}, whilst polycrystalline material was intermediate in behaviour; on the {100} faces the rusting rate became slower after 80 days.

Hydration State of Ferrous Sulphate in Rust. Shreir, with Fyfe and Shanahan,[3] has studied transformations between different hydrates of $FeSO_4$. The results show that $FeSO_4.7H_2O$ crystallizes from the solution formed during the rusting of steel in atmospheres containing SO_2. During periods of low humidity, this becomes converted to $FeSO_4.4H_2O$ which, when once formed, persists; it is not re-converted to $FeSO_4.7H_2O$ during periods when the humidity becomes high.

Distribution of Atmospheric Pollution

General. Most of the information today available regarding the distribution of oxides of sulphur and nitrogen in the air has been collected in connection with the problems of human health; much of it, however, has a bearing on the problems of metallic corrosion. McKay[4] states that $(NH_4)SO_4$ rather than H_2SO_4 is the commonest oxidation product of S in fuel. Stevenson[5] states that some of the sulphur in rain-water may be derived from the sea. Mason[6] discusses the oxidation of SO_2 to H_2SO_4 in the droplets of clouds; apparently this process is catalysed by $FeCl_2$. However, the content of SO_4^{2-} found in rain-water cannot be accounted for in that way, and Mason

1.   T. K. Ross and B. G. Callaghan, *Corr. Sci.* 1966, **6**, 337.
2.   S. T. Ali and G. C. Wood, *Brit. Corr. J.* 1969, **4**, 133.
3.   D. Fyfe, C. E. A. Shanahan and L. L. Shreir, *Corr. Sci.* 1968, **8**, 349.
4.   H. A. C. McKay, *Brit. Corr. J.* 1970, **5**, 235.
5.   C. M. Stevenson, *Quart. J. roy. meteorolog. Soc.* 1968, **94**, 56.
6.   B. J. Mason, 'Physics of Cloud' (Clarendon Press), 1971; esp. pp. 70–2, 75, 78.

attributes it to the presence of ammonia, which is found to catalyse the conversion of SO_2 to $SO_4{}^{2-}$. Further information regarding the oxidation of SO_2 to $SO_4{}^{2-}$ is provided by Papetti and Gilmore.[1]

Shaw[2] discusses the part played by oxides of sulphur and nitrogen in promoting corrosion. SO_2 is present in flue gases emitted from industrial plants in amounts up to 0.2% v/v. If conditions allow oxidation to SO_3, the dew-point of the stack gases will be raised, and unless their temperature is kept sufficiently high, there will be corrosion by H_2SO_4. He states that in general oxides of nitrogen do not directly cause corrosion to the plants in which they are produced, but they help to oxidize SO_2 to SO_3. It appears that most of the NO_2 is formed at hot spots, but the maintenance of low temperature combustion will not entirely prevent its formation; indeed NO_2 can be produced by oxidation of combined nitrogen present in the fuel itself.

A quantitative study of the major pollutants emitted from Diesel engines has been carried out at Peking.[3] Those most hazardous (to Man) are NO, NO_2, CO, formaldehyde and acrylaldehyde; SO_2 and SO_3 are also present.

Dispersal Policy. The (British) Central Electricity Generating Board have claimed that by dispersing flue gases from high chimneys, nuisance is avoided. This may be true of trouble close to the source of the pollution, but high chimneys merely transfer pollution from one point to another.

The manner in which SO_2 is dispersed by emission from high chimneys is shown in diagrams provided by Sutton.[4] The point where the concentration at ground level will be highest is generally some distance from the chimney.

A discussion on air pollution organized by Forrest and Mason[5] should receive attention. The paper by Billinge describes a method of removing SO_2 by a dry reaction which does not involve cooling the gas, and thus does not destroy the buoyancy of the plume at the chimney top. Part of the SO_2 is made to react with sodium aluminate, producing sulphate, which at a higher temperature is reduced to H_2S, re-forming the sodium aluminate. The H_2S reacts with another part of the SO_2 to give sulphur, which can be marketed.

Although the high chimney policy of the Electricity Generating Board seems to have been successful in controlling the SO_2 content of air at ground level close to the power station, it cannot solve nuisance due to emission from domestic chimneys or from small industrial plants. It has been suggested[6] that where low-sulphur fuels are available, they should not be used at the power stations but rather by the small consumer.

The corrosion of the chimneys themselves can present a problem. Sometimes these are clad with aluminium. There is a system which 'features a relatively light-gauge inner shell to give a quick warm-up above the acid dew-point

1.   R. A. Papetti and F. R. Gilmore, *Endeavour* 1971, **30**, 107.
2.   J. T. Shaw, *Chem. Ind.* 1969, p. 1365.
3.   *Chinese Med. J.* 1973, No. 1, p. 13 (English abstract).
4.   P. Sutton, *Process Engineering*, March 1971, p. 53.
5.   J. S. Forrest and B. J. Mason, *Phil. Trans.* 1969, **265**, 139–318. See especially papers by B. H. M. Billinge, A. C. Collins, J. Graham and H. G. Masterson, p. 309.
6.   J. Golden and T. R. Morgan, *Science* 1971, **171**, 381.

level to minimize corrosion. . . . The space between the shells is filled with mineral insulation.'[1]

Shaw[2] has provided a corrosion map of the British Isles, embodying information accumulated by personnel of various observatories, anemographic stations, rain-gauge sites and atmospheric pollution measurement stations.

Lawther[3] states that the Clean Air Act of 1956 has 'led to a spectacular reduction of pollution by smoke, and to a less extent by SO_2'. On the other hand, F. Taylor[4] utters the warning that the increased use of heating of large buildings such as hotels by heavy oil with sulphur contents of up to 4%, may lead to an increase in pollution in the near future. He states that, unlike Japan and certain continental countries, Britain had no legal provision to limit the sulphur content to the safe level of about 0·5%.

Progress in the avoidance of Pollution. Turner's article[5] entitled 'Corrosion can be fought and fought economically' deserves attention; its author's work for the Pure Air Movement, coupled with his contribution to the solution of practical corrosion problems, qualifies him to present a balanced assessment of the position. The article concerns all types of corrosion, but the section entitled 'Corrosion and Air Pollution' includes a table which

TABLE IV

POLLUTION IN DIFFERENT ATMOSPHERES (T. H. Turner)

	Smoke	Sulphur dioxide
Open Country	16 to 214	7 to 198
Industrial	114 to 1032	174 to 738
Residential	165 to 1335	206 to 1256

shows 'Industry to be cleaner than Housing'. His figures for smoke and SO_2 in January 1968, in 15 open country sites, 35 industrial sites and 42 densely populated residential sites, are reproduced in Table IV; the Units are $\mu g/m^2$.

It had long been believed that the products of combustion emitted from domestic chimneys can be responsible for more pollution than industrial fumes. It might well be the case that a change of fuel practice would have affected the situation. Turner's numbers, however, seem to show that in 1968, this statement was still broadly correct.

Turner also provides data about the air pollution testing programme organized in 1961 by the Warren Spring Laboratory, Stevenage; in 1972 daily tests for SO_2 were being carried out at 1194 sites in the UK and for smoke in 1227 sites. The reduction in smoke pollution, largely due to the Clean Air

1. *Process Engineering*, Aug. 1971, p. 7.
2. T. R. Shaw, Central Engineering Generating Board, 1972.
3. P. J. Lawther, *Proc. Roy. Inst. G.B.* 1971, **44**, 74; esp. Figs. 13, 14, pp. 90, 91.
4. F. Taylor, *The Times*, June 30, 1970.
5. T. H. Turner, *Corr. Sci.* 1969, **9**, 793; esp. table on p. 799; *Brit. Corr. J.* 1973, **8**, 49.

K

Acts of 1956 and 1963, has brought about a great improvement in the aspect of British industrial cities, and has doubtless helped in controlling corrosion also, since Vernon's early work showed that soot accelerates attack on steels.

Stanners[1] has discussed environmental factors in atmospheric attack, and considers the most important to be the presence of sulphur compounds and of chlorides, the pH of the rain, the metal temperature and the length of the periods during which it remains wet. He quotes studies of zinc in British Columbia where the duration of the wet period was found to be related to the variation of the RH value; there was a difference between the behaviour of the groundward and skyward surfaces of a specimen.

One possible method of removing SO_2 from flue gases, which would not only diminish pollution but also make possible recovery in useful form, depends on the use of adsorbent carbon. Indian work[2] shows that, at the optimum temperature (600°C), sugar charcoal or coconut charcoal removes 91·6–97·4% of the sulphur present. By heating the charcoal, after it has taken up the sulphur from the gas, in hydrogen at 800°C, the sulphur can be recovered as H_2S. A certain amount of the S from the fumes is converted to H_2SO_4, but this is largely avoided if the carbon has been freed from oxygen by outgassing at 750°C or 1000°C before being used. The overall performance of the charcoal remains about the same—whether or not there has been preliminary outgassing. Other Indian work[3] has shown that tests started in the months when the atmosphere is relatively non-aggressive reveal lower corrosion rates than those started in the aggressive months.

Another plan for the removal of SO_2 from stack gas involves adsorption on coal.[4]

An interesting method of ascertaining whether or not an atmosphere contains excessive amounts of SO_2 has been used in Holland.[5] It depends on the fact that certain lichens which normally grow in pure air become scanty when the SO_2 content reaches 25 $\mu g/m^3$ in summer and 50 $\mu g/m^3$ in winter (1 $\mu g = 0\cdot001$ mg). The attack on zinc is about 5 times as fast when the pH is 4·3 as when it is 6·1. Values are provided for the SO_2 contents found in eleven Dutch towns. Over 20% of the area of Holland, the SO_2 is now too high for the growth of any lichens, whilst over 50% only a single hardy type of lichen can survive. It is stated that Swedish rain is often acid, owing to SO_2 brought by winds blowing from Germany, Holland, Belgium and the UK.

A fresh description of the Delphi pillar has been provided by Wranglén.[6] He considers that, while the composition of the iron, particularly its low S

1.   J. F. Stanners, *Brit. Corr. J.* 1970, **5**, 117.
2.   B. R. Puri, D. P. Mahajan and S. S. Bhardwaj, *Chem. Ind.* 1973, p. 483.
3.   M. L. Projapat, G. F. Singhania and B. Sanyal, *Labdev. J. Sci. Tech.* (*Kanpur*) 1969, **7A**, 34.
4.   A. K. Kotb, *J. appl. Chem.* 1970, **20**, 147.
5.   J. F. H. van Eijnsbergen, *Tijdschr. oppervlakte tech. Metal* 1970, **13**, 334. See also *The Times*, Nov. 30, 1970, for Scandinavian complaints of acid rain from Western Europe.
6.   G. Wranglén, *Corr. Sci.* 1970, **10**, 761.

and high P contents, is conducive to a low corrosion rate over the part exposed above ground, the decisive factor is the dry and unpolluted air; the large mass of the pillar (causing rapid drying after rain) is a contributory factor. The part of the pillar which is below ground has become covered by a rust layer about 1 cm thick and shows deep pitting. Above ground, the pillar today carries a protective oxide-film 50 to 500 μm thick.

Low-Alloy Steels resistant to Atmospheric Corrosion

General. A review by Chandler and Kilcullen[1] summarizes the present position of the use of low-alloy steels. The alloy known as Cor–Ten apparently first found employment as an architectural material at Moline, Illinois about 1963; since then Cor–Ten has been used unpainted on many buildings in North America; it develops a protective rust, darker and more adherent than that formed on ordinary mild steel. Cor–Ten Grade A contains Cu, Cr and Ni, whilst Grade B has Cu, Cr, a little V and more Mn than Grade A. British conditions are more corrosive than those prevailing in the Middle West States of America, and it is still not certain whether such materials could safely be left unpainted in the UK.

The review mentioned seeks to assess the performance of low-alloy steels in the UK, and particularly discusses the effect of various elements, alone or in combination. P generally improves resistance to corrosion, especially if Cu is also present (a possible reason for the good effect of this combination is suggested on p. 268). The one element which is definitely harmful is S, and several authorities consider that the beneficial effect of Cu is due to its control of the adverse effect of S.

The corrosion rate of low-alloy steels falls off with time: tests at Rotherham showed that after 9 years the corrosion rate of such material had become extremely slow, whereas that of unalloyed steel was only slightly slower than during the first year. This is impressive, but the review shows that only limited information regarding British conditions is today available. American results suggest that, so far as corrosion performance is concerned, steels containing suitable amounts of copper and phosphorus may be almost as effective as steels of the Cor–Ten type. British data on such steels is lacking, but the question is under investigation. For Belgian tests, see p. 274.

Appearance of Alloying Elements in Rust. The distribution of the elements in the rust layer has been studied at Bethlehem, Pa., USA.[2] The concentrations of Cr, Cu and Ni increase slightly on passing from the air surface to the steel surface, whilst the contents of Si, P and Ca tend to decrease inwards. Large amounts of these elements are found in industrial regions; presumably they are derived from external dust. Other information about the distribution of the elements has been provided by Miranda.[3]

Recent work at Rome,[4] based on electron microprobe examination of rust

1. K. A. Chandler and M. B. Kilcullen, *Brit. Corr. J.* 1970, **5**, 25.
2. J. B. Horton, M. M. Goldberg and K. F. Watterson, *Amsterdam Cong.* 1969, p. 385.
3. L. de Miranda, *Cebelcor Rapp. tech.* **206**, addendum (1973).
4. R. Bruno, A. Tamba and G. Bombara, *Corrosion (Houston)* **29**, 95.

formed on low-alloy steels, reveals, beside an overall enrichment of the alloying elements, the presence of sharp concentration peaks in layers about 10 μm wide. The more resistant the steel is to atmospheric corrosion, the higher is the rust-to-steel concentration gradient, and the closer comes the concentration peaks to the steel surface. Measurement of corrosion potential and polarization studies on rusted specimens in $0.1M$ Na_2SO_4 shows that the interfacial enrichment enhances the tendency to anodic polarization and produces something approaching passivation; it is believed that potential sweep techniques on rusted specimens could be useful in evaluating new steels for resistance to the atmosphere.

Japanese work[1] has shown that the rust produced on low-alloy steel becomes enriched in Cr, Cu, P and S; there is much less enrichment of S in rust formed on ordinary carbon steel. The quantity of S in the inner layer of the rust is 20 times greater than that in the steel, indicating that it is derived from the SO_2 in the atmosphere.

However, Fyfe[2] using a scanning electron microscope has shown that when low-alloy steel suffers rusting, the alloying elements found in the rust are generally present in much the same proportion as they occurred originally in the metal. The exposure conditions may have been different.

The sequence of events in the rusting of low-alloy and unalloyed steel has been followed by radioactive tracer methods.[3] A specimen which had been exposed to the atmosphere for the desired time was immersed in a Na_2SO_4 solution containing radioactive S, and autoradiographs were then produced. Since the corrosion current generated during immersion in the sulphate solution consisted in the transport of SO_4^{2-} ions to the spots where anodic attack was proceeding, the dark spots produced on the photographic film indicated the distribution of anodic sites on the rusted surface. Autoradiographs obtained from low-alloy or unalloyed steels after 7 months indicated that the rust layer had not during that period become protective. After a year the number of anodic sites had decreased drastically on low-alloy steel; on the unalloyed steel the number of these sites had decreased, but their size was greater. After 4 years the low-alloy steels had practically no anodic sites, suggesting that the rust had become continuous and protective.

Comparison of different low-alloy steels. Hudson[4] has provided evidence of the different behaviour of the various steels in different atmospheres. Tests at Sheffield, UK, and at Bayonne, NJ, USA, showed that ordinary unalloyed steel rusted faster than Cu steel, whilst Cu–Cr steels rusted more slowly than either. At Lille, however, the behaviour of Cu and Cu–Cr steels was not very different. Hudson favours Copson's view that the resistance of copper steel is due to the insolubility of copper basic sulphate, which blocks the pores in the rust. As the SO_4^{2-} content of the rust becomes greater, so the corrosion rate declines; this is shown by Copson's data and

1. T. Riichi and K. Satoshi, *Corr. Control Abs.* 1970, No. 3, p. 22.
2. D. Fyfe, *Brit. Corr. J.* 1970, **5**, 103.
3. I. Matsushima and T. Ueno, *Corr. Sci.* 1971, **11**, 129.
4. J. C. Hudson, *Annales des Travaux publiques de Belgique*, No. 3 (1968–69).

confirmed by British work. It is pointed out, however, on p. 268, that, quite apart from the pore-plugging action, the removal of SO_4^{2-} in insoluble form would increase the resistance of the corrosion cells, and thus reduce the corrosion rate.

Japanese exposure tests[1] lasting one year only (carried out for the purpose of designing an accelerated test) led to the results shown in Table V. They may awaken surprise since they indicate that the effect of small amounts of alloying constituents at the four sites used is not so marked as might have been expected from some of the statements published elsewhere.

TABLE V

CORROSION IN DIFFERENT ATMOSPHERES (mdd)
(Y. Sakae, N. Kikuchi and K. Najima)

Material	Site			
	Coastal	Urban	Coastal and industrial	Rural
Mild steel	16·3	13·9	22·0	11·7
High tensile steel	11·9	11·4	16·3	8·3
'Atmospheric-corrosion resistant steel'	11·5	10·9	13·3	7·4
'Sea-water resistant steel'	11·1	10·7	11·3	7·5

French tests reported by Herzog[2] show the effect of geographical situation on the relative behaviour of different steels. On specimens exposed close to the coast, the weight losses of Cr–Al–Ni steels were much lower than those of 3·5% Cr steel, but the advantage of using the complex steel declined with growing distance from the sea.

The effect of geometry on the behaviour of low-alloy steels is clearly brought out by Japanese work.[3] It was found that inclined or horizontal roofings gave results similar to specimens tested on standard racks, with development of dark-brown protective rust, formed within 1 year. However, horizontal louvres of a design liable to retain water failed to produce stable rust even after 4 years. On a vertical surface, the higher portion became darker and developed an appearance suggesting that rust formation was ceasing in a shorter time. The difference was ascribed to water running down; on window frames the water tended to run along certain preferred paths, and at one site a uniform appearance was never obtained. Concrete bases, when present, quickly became rust-stained; staining was quickest during the early months, but continued after 5 years.

Tests carried out at Brancote, Notts, UK,[4] lasting 5 years showed the

1. Y. Sakae, N. Kikuchi and K. Nijima, *Tokyo Cong. Ext. Abs.* 1972, p. 407.
2. E. Herzog, *Tokyo Cong. Ext. Abs.* 1972, p. 311.
3. I. Matsushima, Y. Ishizu, T. Ueno, K. Kanazashi and H. Horikawa, *Tokyo Cong. Ext. Abs.* 1972, p. 321.
4. M. F. Taylor, P. J. Boden and E. Holmes, *Brit. Corr. J.* 1971, 6, 61.

beneficial effect of 1% alloying additions to stand in the same order after 1 and 5 years, namely

$$Cu > Mo > Si > Cr > Ni > Mn.$$

The authors reporting the results regard the adhesion of the rust as important.

Recent Explanations of Effect of Alloying Constituents. Anything which increases the resistance of either the electrolytic or electronic limb of a corrosion cell is likely to reduce the rate of atmospheric corrosion. The well established effect of small amounts of copper in steel in reducing the rate of atmospheric corrosion can thus be explained by an increase in the resistance of the electrolytic limb. The SO_4^{2-} present in the liquid filling the pores is precipitated as a sparingly soluble basic copper sulphate. Copson[1] considers that this substance plugs the pores, and it is clear that any solid matter produced will decrease the cross section available for the flow of ions and thus increase the resistance of the electrolytic limb. But the removal of SO_4^{2-} from the liquid will also decrease its specific conductivity, thus raising resistance and depressing the corrosion rate.

Shreir and his colleagues[2] have reviewed the explanations put forward for the good effect of copper in steel. They disagree with the views of Copson, who thought that basic copper sulphate plugs the pores, and quote Tomashov, who in 1948 advanced the opinion that the effect of Cu is electrochemical in nature, since the reprecipitated Cu provides effective cathodes and thus facilitates passivation of the iron surface. They describe new experiments which suggest that the good influence of Cu is to remove S as sparingly soluble copper sulphide.

Wranglén[3] also considers that the action of Cu in reducing the atmospheric corrosion rate of iron is connected with the low solubility of copper sulphide in acid moisture; this appears reasonable.

Another way in which the presence of the elements Mn, P, Si, Cu and Ni retards the atmospheric corrosion of low-alloy steels has been discovered by de Miranda (this volume, p. 265). He finds that whilst a solution containing $FeSO_4$ and $Fe(SO_4)_3$ treated with alkali normally produces a black, magnetic precipitate which X-ray analysis shows to be magnetite, no magnetite is formed if the elements mentioned above are present. He has also shown that, whereas γ-FeOOH (*lepidocrocite*) produces magnetite when treated with iron powder in presence of $FeSO_4$ solution, no production of magnetite occurs when α-FeOOH (*goethite*) is thus treated; 'green rust II' is formed in presence of the elements mentioned.

Since magnetite is needed for the electronic limb of the circuit, the retardation of corrosion receives a simple explanation.

The influence of relatively small amounts of chromium in low-alloy steels may be due to more than one cause. The spinel formed during oxidation may then contain both Fe and Cr, and will be a less good electronic conductor

1.　H. R. Copson, *Proc. Amer. Soc. Test. Mat.* 1945, **45**, 554.
2.　D. Fyfe, C. E. A. Shanahan and L. L. Shreir, *Amsterdam Cong.* 1969, p. 399; *Corr. Sci.* 1970, **10**, 817.
3.　G. Wranglén, *Corr. Sci.* 1969, **9**, 585.

than magnetite (on magnetite the easy exchange of an electron between an Fe^{2+} and an Fe^{3+} ion may be due to the fact that it leaves the energy situation unchanged; for a spinel containing Cr that would not be the case; that in itself might slow down the oxidation rate by increasing the resistance of the electronic limb). However, probably the main cause of the improved resistance to acid atmospheres is connected with the instability of the divalent state in chromium. This, as explained in chapter IX, is the cause of the resistance of stainless steel to dilute acid containing traces of dissolved oxygen. The alloys containing only a few per cent of chromium cannot truthfully be described as non-rusting, but (especially when other metals and also phosphorus are present) the rust is of a protective character.

Behaviour of Stainless Steels. Ferritic alloys containing 13% Cr or more, and austenitic alloys containing 18% Cr and, perhaps, 8% Ni, are correctly described as 'stainless', and under favourable conditions remain bright in the atmosphere. T. E. Evans[1] describes tests on various types of stainless steel exposed at Birmingham for 10 years. The results show that, under conditions of high pollution, some of these alloys, despite the high chromium content, undergo marked changes. He found that 13% chromium steel, with or without nickel, and also ferritic or martensitic steels containing 17% Cr, quickly developed rust which at the end of the test might cover 90% of the surface. On the other hand the 18/8 Cr/Ni austenitic steel produced rust which was much thinner and was confined to patches covering perhaps half of the total area; rust patches were somewhat more extensive on the 18/8 alloy containing 0·55% Ti than on a similar alloy with only 0·12% Ti; this may have been due to the presence of titanium carbide particles, which possibly act as nuclei for the development of rusting and pitting. A group of four austenitic Cr–Ni steels containing Mo remained substantially free from rust throughout the 10-year period. A steel containing 25% Cr and 20% Ni, without Mo, also remained almost free from rust—presumably owing to the higher Cr and Ni contents. Tests carried out at 14 different locations in 5 different countries showed considerable variation between the results. The nature of the pollution, the proximity of the exposure site to the sources of pollution, its location relative to the prevailing winds, and climatological features such as the duration of periods of high humidity, temperature and temperature variations, all interact in a complicated manner, making any simple generalization impossible.

Effect of Acid Fumes other than Sulphur Dioxide

Hydrogen Chloride. Swiss work[2] has shown that iron exposed over HCl solution or in dry HCl gas and then exposed to damp air at a controlled humidity suffers corrosion. The attack increases as the RH in this second stage is raised from 55% to 70%. There is a maximum about 70–75%; then, for some unexplained reason, the corrosion decreases to a minimum, rising again as the RH approaches 100%.

The corrosion product formed on steel after exposure in air containing

1. T. E. Evans, *Amsterdam Cong.* 1969, p. 408.
2. E. Längle, *Schweizer Archiv.* 1968, **34**, 109, 147 (2 papers).

HCl and then in damp air often assumes a 'horn-shaped' form. This *coniform corrosion* has been studied by Hill,[1] who has observed growth rates up to 3 mm per day. The corrosion product was identified as magnetite, and the results were explained by an electrochemical mechanism involving the intermediate production of $Fe(OH)_2$ and hydrated ferric oxide. The underlying metal became deeply pitted. Coniform corrosion may be important in the paint industry, being observed on paint containers exposed to the thermal decomposition products of chlorinated polymers.

Organic Acid Vapours. It is well known that many woods and some plastics give off acetic or formic acid, whilst H_2S is liberated by vulcanized rubber and HCl by chlorinated rubber. The problems involved have been investigated by Donovan and Stringer.[2] Other discussions of the matter come from Czech[3] and Indian[4] sources.

A paper by Sanyal and others[5] describes precautions necessary in India for avoiding deterioration in electrical equipment. Design should avoid bimetallic contacts, crevices and box sections. Packing cases should have a waterproof lining. Cushioning material should contain less than 0·02% chloride and 0·12% sulphate. Wood for cases should be seasoned; but even seasoned wood may contain 10–12% moisture and may establish 80% RH. Organic acid fumes, from paints or wooden cases—especially formic and acetic acid—often do damage. Kraft paper impregnated with volatile inhibitors can be extremely helpful.

Corrosion due to Perspiration. Lind[6] has reviewed the effect of human sweat on goods handled at a factory, pointing out that the corrosion produced can diminish the aesthetic appeal, and even cause dimensional changes; sometimes the damaged articles are rejected on inspection, owing to deterioration of surface finish; there can also be a loss of goodwill. The tendency to perspire varies from one individual to another, and even in a single person it fluctuates with physical and mental condition, dress, diet, and the supply of salt and water; it is affected by the RH and temperature of the working room. Perspiration is essentially a dilute aqueous solution, containing only about 0·5–1% of solids, of which 75% are inorganic and 25% organic (including organic acids and urea); the pH varies between 3·8 and 6·5.

The corrosion of metallic articles contaminated with sweat after handling is an example of atmospheric corrosion, and NaCl is the main corrosive agent; a critical value of the RH must be exceeded if corrosion is to occur rapidly— usually between 45% and 60%. It is mainly a summer problem. Lind recommends the provision of cooled drinking water. The use of gloves may some-

1.   G. V. G. Hill, *Brit. Corr. J.* 1973, **8**, 128.
2.   O. D. Donovan and J. Stringer, *Amsterdam Cong.* 1969, p. 537.
3.   D. Knotková-Cermáková and J. Vičoká, *Brit. Corr. J.* 1971, **6**, 17.
4.   G. K. Singhania, B. Sanyal and B. Lal, *Labdev. J. Sci. Tech.* (*Kanpur*) 1968, **6A**, 31; 1969, **7A**, 130 (2 papers).
5.   J. N. Nanda, P. N. Agarwal and B. Sanyal, *Labdev. J. Sci. Tech.* (*Kanpur*) 1970, **8A**, 105.
6.   S. E. Lind, *Corr. Sci.* 1972, **12**, 749.

times provide a solution, but in some industrial operations they are impossible; also the accumulation of sweat in gloves can lead to skin trouble.

The constituents of finger perspiration include various organic acids; of these, pyruvic acid ($CH_3.CO.COOH$) attacks zinc faster than lactic acid ($CH_3.CHOH.COOH$), producing a lower pH value. The problem has been discussed in connection with zinc-coated telephone equipment.[1] Passivation of the zinc produces a time-lag before corrosion starts. Various processes have been described; one of these (full passivation) leaves an excess of hexavalent chromium; this is used in the tropics, and a lacquer coat is applied. In the UK another process ('colourless passivation'), which leaves less chromate, is found to suffice, without a lacquer coat.

The idea that grease solvents can provide the answer to the problem is a mistake; degreasing of goods which have been handled does not remove the salt particles. (It may be added here that, when once the salt has started to cause rows of pits producing a 'thumbograph' pattern, washing with water may be ineffective, and indeed the EMF of the corrosion cells set up may cause the aggressive ions to move further inwards into the pits when a water film exists on the surface.) Oil-in-water emulsions and soap solution (especially if warm) may be helpful, and some of these leave a film on the metal which provides temporary protection. Supply waters containing appreciable amounts of salt should be avoided in making up these solutions.

Schweisheimer[2] mentions the case of a plant in California where 4 out of a group of 20 workers left sufficient perspiration on the goods handled to cause complaints from purchasers. An expert, consulted, advised that hand-washing with cake soap should be carried out at 2-hour intervals, followed by thorough rinsing to remove traces of soap. A word of caution was added about the use of corrosion inhibitors which, it was suggested, may cause dermatitis.

Temporary Protection during Storage. Useful information is provided by Rosenfeld and his colleagues[3] regarding inhibitors for different situations. They deal with a new class of volatile compounds, the nitrobenzoates of amines and imines; of these, hexamethylene-imine metanitrobenzoate has come into extensive use in Russia, but other nitrobenzoates show good inhibitive properties. There is a large variation of vapour pressures available, providing choice for meeting different situations; the vapour pressure of aminobenzoates range from 10^{-3} to 10^{-5} mm Hg at 41°C, whereas the nitrobenzoates give lower values (10^{-5} to 10^{-6}). The mechanism of inhibition varies in different cases; sometimes there is direct retardation of the anodic process, but in other cases there is an acceleration of the cathodic process (due to the NO_2 group) causing the potential to rise into the passive range.

1.   J. D. Underwood, K. Carvalho and A. McKinlay, *Trans. Inst. Met. Finishing* 1971, **49**, 123.
2.   H. Schweisheimer, *Anticorrosion, Methods and Materials* 1968, **15** (10) p. 19.
3.   I. L. Rosenfeld, V. P. Persiantseva and M. N. Polteva, *Amsterdam Cong.* 1969, p. 606.

Some of these inhibitors are very suitable for the impregnation of wrapping paper (in absence of inhibition, wrapping paper tends to set up corrosion at the point of contact, especially if the chloride content is high). In other cases, the volatility of the inhibitor provides protection at points which are not in contact with it. It is possible to use mixtures designed to deal with all situations. Thus in a mixture of sodium nitrite, ammonium phosphate and bicarbonate, the nitrites act mainly as contact inhibitors, whilst the ammonia evolved by hydrolysis gives protection at points which are not in direct contact.

For the wrapping of copper alloy articles, tissue impregnated with benzo-triazole has proved most useful.[1]

Atmospheric Behaviour of Non-Ferrous Metals

Zinc. McLeod and Rogers[2] have studied the behaviour of zinc in moving air containing moisture and SO_2. They find that the corrosion rate is high at first, but gradually decreases to a small constant value. The products are found to be zinc sulphate, zinc sulphite and sulphur.

N. and P. Dreulle[3] have also studied the atmospheric corrosion of zinc and favour an electrochemical mechanism. The anodic reaction ($Zn \rightarrow Zn^{2+}$) is balanced by the cathodic reaction ($O_2 \rightarrow OH^-$), and the products combine to produce a protective layer of $Zn(OH)_2$ or a basic carbonate; the latter is less soluble and more resistant; thus the presence of CO_2 is a favourable factor. Cl^- is not an adverse factor, since basic chlorides can be formed, but SO_2 is found to be very objectionable (doubtless forming H_2SO_4 which dissolves the oxide coat); increasing frequency of rain enhances the rate of attack. At a given place the rate of attack upon high purity zinc (99·99%) is not very different from that upon 99·5% zinc, with or without copper. Sea air causes attack only slightly faster than rural air, but the proximity of factories producing SO_2 or H_2SO_4 causes the attack to be very much faster. Corrosion is enhanced by differential aeration conditions which lead to the production of Zn^{2+} and OH^- at well separated parts of the surface; such conditions can be set up by the presence of inert objects resting against the zinc surface. Bimetallic contacts (especially Pb or Cu coats with disconti-nuities) also stimulate corrosion. Normally the zinc develops an adherent compact layer next to the metallic surface with a porous, less adherent layer further away from it.

Regarding the mechanism of the atmospheric corrosion of zinc, van Eijnsbergen[4] writes as follows:

'Reactions of SO_2 on zinc surfaces are comparable to those on steel. However, ZnO is, contrary to Fe_3O_4 (magnetite), not electroconductive. Therefore reactions proceed at a slower rate, in spite of the fact that zinc is

1.   T. B. Marsden, *Brit. Corr. J.* 1972, **7**, 251.
2.   W. McLeod and R. R. Rogers, *Corrosion (Houston)* 1969, **25**, 74.
3.   N. and P. Dreulle, *Corrosion (Paris)* 1970, **18**, 503.
4.   J. F. H. van Eijnsbergen, *Second South African Corrosion Conf.*, March 1972.

less noble than iron. From a relative humidity of 50%, zinc sulphate, zinc sulphite, zinc sulphide or hydrated zinc sulphate are formed:

$$SO_2 + O_2 + Zn \rightarrow ZnSO_4$$
$$SO_2 + 2O_2 + 2Zn \rightarrow 2ZnSO_3$$
$$SO_2 + 3Zn \rightarrow ZnS + 2ZnO$$
$$2ZnSO_4 + 2H_2O + O_2 + 2Zn \rightarrow 2ZnSO_4.Zn(OH)_2$$

An insoluble zinc patina may react with SO_2,

$$2Zn(HCO_3)_2 + 2SO_2 + O_2 \rightarrow 2ZnSO_4 + 4CO_2 + 2H_2O$$

forming the soluble $ZnSO_4$ which is washed out by rain water.'

A useful review of the behaviour of zinc has been produced by the Battelle Memorial Institute at Columbus, Ohio.[1] This includes a restatement of much early work by Vernon, Schikorr and others.

Aluminium. Schikorr[2] has studied the attack on aluminium by air containing SO_2, and finds that aluminium takes up less SO_2 than most other metals. In the case of Ni and Zn the corrosion produced appears to be equivalent to the SO_2 taken up; in the case of Fe, owing to the regeneration of acid, one molecule of SO_2 will attack many atoms of iron. In the case of aluminium, however, it seems that some other substance, e.g. dust, acts as an intermediary.

Arnold[3] has investigated the use of aluminium as a material of construction for swimming baths using chlorinated water. If the aluminium has been anodized, it appears to be suitable, provided that the design is such as to prevent direct contact between Al and heavy metal components. Chromium plated mild steel, and also painted steel, were found to suffer severe rusting under these conditions.

Filiform corrosion of anodized aluminium alloys has been studied by Sick.[4] Starting from a corrosion pit, the attack extends in a random direction, forming a snail's trace; two tracks do not cross one another; if they approach, they turn aside or cease to advance. The rate of advance can be 0·03 to 0·08 mm per day, but the maximum depth is only about 0·005 mm—so that this is really a surface phenomenon. Filiform corrosion is particularly noticeable on an alloy containing about 3% of Mg, and occurs at places exposed to air but not washed by rain.

The same filiform attack is discussed by Hinchliffe,[5] who attributes it to dust particles containing chlorides, and also to the presence of MgO, which enters the cavities generally present on an extruded surface; hygroscopic $MgCl_2$ sets up pitting or filiform attack.

The behaviour of aluminium windows is discussed by Pourbaix.[6] Window-frames, unanodized, will usually develop aesthetically objectionable changes. If there is anodizing followed by sealing, such changes can generally be

1. C. J. Slunder and W. K. Boyd, Report, 1971 (Battelle Mem. Inst., Columbus, Ohio).
2. G. Schikorr, *Aluminium* 1967, **43**, 108.
3. G. G. E. Arnold, *Anticorrosion* 1972, **19** (3) 5.
4. H. Sick, *Aluminium* 1968, **44**, 476.
5. J. C. W. Hinchliffe, *Australasian Corr. Engg* 1972, **16** (6), 7.
6. M. Pourbaix, *Cebelcor Rapp. tech.* **113** (1969).

avoided, provided that the thickness of the oxide-film is sufficient. He reports cases where alkaline water running off from 'artificial stone' has caused attack on aluminium, but states that this can be avoided if a temporary protective layer of grease or paraffin is applied during the erection of a building, and subsequently removed.

Bronze. An Italian paper[1] gives a description of the patina present on ancient Etruscan bronzes, and provides guidance regarding preservation of exhibits in museums. After mentioning the use of dicyclohexylamine as an inhibitor at Leningrad, the authors describe tests carried out in Italy to assess resistance towards traces of SO_2 or H_2S; benzotriazole and 2–mercaptobenzothiazole were found to yield the best results.

The 500 bronzes at the Fitzwilliam Museum, Cambridge, which were found to be suffering from bronze disease after the 1939–45 war and were treated about 1951 by Rayner, using the zinc-nib method (1960 book, p. 492), were reported[2] to be in good condition in 1972; there had been no recurrence of the trouble. It would seem that the method has successfully drawn out the aggressive anions from the pit bottoms. The method has received praise[3] as an example of 'the successful application of sound principles and simple methods to the solution of a vexatious corrosion problem'. It does not seem to have been used elsewhere, nor can any record be found of the application of the same principle to other cases of virulent localized attack.

Further References

Addendum (1975). Recent tests conducted at Liège by the Centre de Recherches Metallurgiques (CRM)[4] also show that low-alloy steels compare favourably with ordinary mild steel, although by no means immune from corrosion. After three years, three non-alloyed steels suffered weight losses between 11·9 and 13·5 gm/dm^2, whilst four 'patinable' steels suffered losses between 5·3 and 6·2 gm/dm^2—nearly half the amount. However, after 10 years, measurements showed that the alloying constituents were building up resistance. The weight losses of the same three non-alloyed steels fell between 26·6 and 32·1 gm/dm^2, whilst those of two 'patinable' steels were 8·5 and 8·8. Steels containing copper as sole alloying constituent showed a behaviour intermediate between those of the non-alloyed and 'patinable' classes.

Pourbaix and Miranda[5] attribute the good behaviour of the low-alloy steels to a layer of α-FeOOH (goethite), formed below the γ-FeOOH (lepidocrocite). It is only produced if the potential is sufficiently high for copper and other alloying constituents to pass into the rust and catalyse the conversion of γ-FeOOH into protective α-FeOOH. If the potential is too low for Cu and other elements to pass into the rust (or if alloying elements are absent from the steel), non-protective magnetite is formed, and the rust does not

1. P. A. Borea, G. Gilli, G. Trabanelli and F. Zucchi, *Ferrara Symp.* 1970, p. 893.
2. Letter from M. Rayner, Oct. 7, 1972.
3. S. C. Britton, *Brit. Corr. J.* 1971, **6**, 55.
4. *Cebelcor Rapp. tech.* **225** (1975).
5. M. Pourbaix and L. R. de Miranda, *Cebelcor Rapp. tech.* **227** (1975).

prevent corrosion. Misawa[1] had previously expressed a view that Cu and P can catalyse the formation of a protective compound. These two researches should be studied in the original papers, along with the important work of M. S. Al Ghali and D. Eurof Davies (this volume, p. 259, footnote).

Papers. 'Air pollution in the United Kingdom'.[2]

'Corrosion of Steel Structures'.[3]

'Atmospheric corrosion sites in Europe available for tests on steel protection'.[4]

'Macroscopic structure of the rust layer formed in the atmospheric corrosion of steel'.[5]

'Relationships between environmental pollution and the corrosion of zinc and galvanized steel'.[6]

'Schutz von Bauwerken gegen Verwitterungen'.[7]

'Effect of relative humidity on adsorption of sulphur dioxide on to metal surfaces'.[8]

'Discrete areas of SO_2 adsorption from the atmosphere on to iron'.[9]

'Pollution control by tall chimneys'.[10]

'A review of Filiform Corrosion'.[11]

'Crystalline and amorphous primary products during the rusting of metallic iron in SO_2 containing atmospheres at different moisture contents'.[12]

'Atmospheric corrosion of iron with periodic moistening'.[13]

'Influence of carbon particles on the corrosion of iron in a humid sulphide dioxide containing atmosphere'.[14]

'The Corrosivity of the Atmosphere'[15] (comparative figures for steel and zinc at 45 different places).

'Aspects scientifiques et relations entre la pollution et la corrosion métallique'.[16]

'Les relations entre la pollution de l'environment et la corrosion du zinc et de l'acier galvanisé'.[17]

1. T. Misawa (with T. K. Yuno, W. Suetara, S. Shimodaira, K. Asami and K. Hashimoto), *Corr. Sci.* 1971, **11**, 35; 1974, **14**, 279.
2. S. R. Crayford, *Chem. Ind.* 1973, p. 733.
3. R. Scimar, *Brit. Corr. J.* 1973, **8**, 154.
4. K. Bartoň and J. F. Stanners, *Brit. Corr. J.* 1973, **8**, 156.
5. J. Honzák, *Brit. Corr. J.* 1973, **8**, 162.
6. N. and P. Dreulle, *Corrosion (Paris)* 1973, **21**, 114.
7. E. B. Grunau, *Werkstoffe u. Korr.* 1973, **24**, 486.
8. J. R. Duncan and D. J. Spedding, *Corr. Sci.* 1973, **13**, 993.
9. J. R. Duncan, D. J. Spedding and E. E. Wheeler, *Corr. Sci.* 1973, **13**, 69.
10. D. Lucas, *New Scientist* 1974, **63**, 790.
11. G. M. Hoch, *Williamsburg Conf.* 1971, p. 134.
12. E. Franz, *Werkstoffe u. Korr.* 1973, **24**, 598.
13. D. Knotková-Čermáková, J. V. L. Crova and D. Kuchynaka, *Werkstoffe u. Korr.* 1973, **24**, 684.
14. R. Ericsson, B. Heimler and N. G. Vannersberg, *Werkstoffe u. Korr.* 1973, **24**, 207.
15. J. B. Mohler, *Plating*, Jan. 1974, p. 62.
16. J. Talbot, *Corrosion (Paris)* 1973, **21**, 97.
17. N. and P. Dreulle, *Corrosion (Paris)* 1973, **21**, 114.

CHAPTER XIV

Protection by Paints and Non-metallic Coatings

Preamble

Picture presented in the 1960 and 1968 volumes. Chapter XIV of the 1960 book opens with a short discussion of the economic advantages potentially obtainable from protective painting. The theoretical basis of protection by painting is then discussed; it is pointed out that the widely held belief that paint coats protect simply by the mechanical exclusion of corrosive substances does not explain the facts. In truth, the protection provided by paint may be due, in different cases, to at least four causes, including (1) sluggish ionic movement through the paint coating, (2) the presence of inhibitive pigments of appreciable, but not excessive, solubility in water, (3) the formation of inhibitive compounds (e.g. azelates) by the degradation of metallic soaps present in an oil-paint and (4) cathodic protection in the case of metal-pigmented paints—consideration of which is deferred to chapter XV. Hardly less important than the protective power of a fresh paint coat is the question of breakdown arising from a number of causes, including (1) decomposition of the vehicle by light, (2) peeling and softening due to alkali formed in the cathodic part of the corrosion process and (3) the pushing away of the paint coat due to the local expansion which occurs when even a small quantity of rust or other corrosion product is formed below the coat; the volume of corrosion product is generally greater than that of the metal destroyed. The breakdown of a paint coat depends very much on the state of the metallic surface existing when the paint is applied; a paint which gives admirable protection for a long period when applied to a clean surface may break down quickly if applied to a surface carrying rust and/or mill-scale. Thus methods of de-rusting and descaling are of great practical importance. But users of paints might reasonably plead with the paint industry that some of the effort now being put into endeavours to develop paints which are super-excellent when applied to a perfectly clean surface should be diverted to formulating paints capable of giving reasonably good protection when applied to a dirty surface; in the imperfect world of today, it cannot be assumed that proper cleaning will be carried out.

The choice of paint, like decisions regarding surface preparation, will vary greatly according to the environment to which the painted surface are to be exposed. Marine paints are different from those intended for inland protection, where drying oils with soap-forming properties can be valuable components; in the sea, however, an oil-vehicle is liable to destruction, loosening and/or softening by alkali formed in the cathodic part of the corrosion process.

The paint manufacturer has a large armoury of materials at his disposal. These include inhibitive pigments for the priming coat and flaky pigments for the outer coat; for the vehicle, apart from the vegetable drying oils, tars, bituminous substances and other products which have long been used, there are today available innumerable synthetic materials provided by the skill of the organic chemist, which not only possess valuable properties but are, in general, more reproducible in composition than materials derived from natural sources. Thus, although there is no one perfect paint suitable for all situations, it should be possible today to provide a good painting system (priming coat covered with 2 or 3 outer coats) suitable to any desired purpose, provided that the conditions of intended use are accurately known in advance; the information required includes, not only the external environment (liquid or gaseous), but also the nature of the metallic basis; the paint system suitable for coating galvanized iron or a light alloy is different from that which would be used on ordinary mild steel.

Developments during the period 1968–1975 and Plan of the new Chapter. The understanding of the passage of water and salts through paint and lacquer films has increased during the period under review, largely due to the work of Mayne and his colleagues. This is reviewed in the present chapter. A discussion of pre-treatment processes, including phosphating, follows. The electrodeposition of paint, already briefly mentioned in chapter I in connection with the painting of motor cars, receives further attention. The choice of paints, including the value of phosphate pigments, is reviewed. The question of paints suitable for application over rust or badly prepared surfaces is discussed—also the prevention of microbiological attack on paints. A section follows dealing with paints for special purposes, including pipe-lines, bridges, ships and aircraft: also paints suited for power-poles and for the coating of non-ferrous metals. The chapter ends with a discussion of testing methods for ascertaining the relative merits of different painting systems.

Fundamental Researches

Movement through Films. Mayne's important researches[1] into the mechanism of the passage of water and salts through lacquer films continue. He finds that in pentaerythritol alkyd varnish two types of film are met with. In one type (the direct or D-films) the conductivity increases as the conductivity of the solution with which it is in contact increases (in other words, conductivity decreases as the solution becomes more dilute). In the other type (inverse or I-films) the conductivity drops with increasing concentration of electrolyte in the solution, probably because the water-activity is lower in the more concentrated solution. In one series of experiments the percentage of I and D samples taken from different castings of the film varied, but on the average only 15% were D for the pentaerythritol alkyd varnish, 50% were D for phenol-formaldehyde tung-oil varnish, whilst 76% were D for the

1. A. J. Appleby and J. E. O. Mayne, *J. Oil Colour Chem. Assoc.* 1967, **50**, 897. E. M. Kinsela and J. E. O. Mayne, *J. Oil Colour Chem. Assoc.* 1968, **51**, 816. J. E. O. Mayne, *Brit. Corr. J.* 1970, 5, 106; *Farbe u. Lack* 1970, 76, 243. J. E. O. Mayne (with E. Kinsela and J. D. Scantlebury), *Brit. Polymer J.* 1969, **1**, 173; 1970, **2**, 240; 1971, **3**, 41, 237.

epoxy polyamide varnish. On a single piece of film there may be areas of D and I conductivity distributed as a mosaic over the surface. It is believed that when a film comes in contact with a solution of electrolyte, the more cross-linked portions take up both water molecules and ions and show D-type conductivity; this depends mainly on the movement of cations through the film. It appears that, in unpigmented lacquer films, cross-linking is very uneven, and films with greatly improved protective value would be produced if the cross-linking could be rendered more homogeneous by the use of a modified dryer system.

The two types of conduction are interrelated. A film which exhibits D-properties in a dilute salt solution acquires the characteristics of an I-film as the concentration of the solution is increased; on the other hand, an I-film can acquire D-properties after it has been exposed to a solution of an electrolyte and ion-exchange has taken place. The D-areas on a film are significantly softer than the I-areas, and this provides an easy method of distinguishing the two. The changes in resistance of an I-film depend only on the available water in the solution. The water-activity will be greater in more dilute solutions, but the nature of the solute is not important. An increase of resistance can be obtained by adding either KCl or sucrose to the solution; in both cases the water-activity is reduced, so that less water can enter the film.

Holmes-Walker[1] makes valuable comments on Mayne's work. 'In the more cross-linked ("I" film) only water is taken up, whereas in the less cross-linked areas ("D" film) ionic penetration is possible. This suggests two possible methods of increasing the barrier characteristics:—

(1) to provide films of greater uniformity with more highly cross-linked structures.

(2) as in the case of thin film, the incorporation of polar groups into the polymer structure to alter the ion-exchange characteristics.

'. . . There are, of course, other ways in which permeability can be reduced, namely

'(1) by the physical mixing of two dissimilar materials. For example, in the case of epoxy/coal-tar the permeability to oxygen and water is about half that of alkyd paints and a quarter that of epoxy polyamides. An ingenious technique recently reported uses a 90 : 10 blend of a conventional resin, such as an epoxy, with a material composed of fumed silica rendered hydrophobic by treatment with a silane. Instead of attempting to provide a continuous pinhole-free coating, this (silanox) coating, deposited by the normal electrostatic power-spray technique, functions by insulating the metal surface from water by means of an air layer.

'(2) by use of two-layer coatings where not only the barrier properties, but also the mechnical properties such as adhesion, flexibility and so one, reinforce one another. Also the base coat can be used as a vehicle for corrosion inhibition and the top coat to carry protection against ultraviolet degradation.'

Swedish workers[2] have used Mayne's work as the basis of their experimental

1. W. A. Holmes-Walker, *Met. Mat.* 1973, **7**, 225.
2. U. Ulfvarson and M. Khullar, *J. Oil Colour Chem. Assoc.* 1971, **54**, 604.

work and have measured the ion-exchange capacity of numerous pigmented paint films, suggesting that the exchange capacity may provide a method of characterizing anti-corrosive paints; measurements of the capacity at time intervals during the exposure of paint system may provide, they think, 'a useful tool for following the progressive deterioration'.

The passage of water through non-pigmented films has been studied by Perera and Heertjes.[1] For chlorinated rubber films, the mean diffusion coefficient appears to be independent of the mean water concentration, whereas in epoxy resin and alkyd films it increases with water concentration. They have studied diffusion through pigmented films. The presence of a pigment would be expected to decrease permeability, and for paints based on alkyd resins pigmented with titania this is found to be the case. If wetting is imperfect, air inclusions may be present, and sometimes cause an increase of permeability.

The effect of pigment particles has also been studied by Funke.[2] He finds that water is often taken up between the pigment particles and the vehicle; a water-filled cavity is formed around each pigment particle, so that water absorption increases. When the pigmentation becomes higher, so that the particles are very close together, the absorption falls off again.

Studies of the subject by Rosenfeld[3] suggest that in films based on alkyd resins or nitrocellulose, the penetration of electrolytes takes place through true pores; but for one copolymer it was found to occur through the paint substance itself.

Polarization studies by Epelboin[4] of iron immersed in dilute sulphuric acid show that the electrochemical reaction remains the same whether or not the iron is coated with epoxy resin; on the coated iron the reaction occurs only at the bottom of microscopic pores or crevasses, so that the area of metal taking part is greatly reduced.

Permeability to vapour and gas. Katz and Munk[5] find that in highly polar films, like nitrocellulose, permeability is influenced by the nature of the substrate. Thus nitrocellulose films cast on tin foil have a higher water vapour permeability than those cast on polyethylene; the oxide on the tin exerts a polar influence; the lower side of a film, facing the substrate, will contain more polar groups than the upper side facing the atmosphere. The solvent also influences the permeability of water vapour through nitrocellulose, chlorinated rubber and alkyd resin films. For instance, a chlorinated rubber film deposited from a solution in methyl isobutyl ketone is more pervious than one deposited from trichlorethylene.

1. D. Y. Perera and P. M. Heertjes, *J. Oil Colour Chem. Assoc.* 1971, **54**, 313, 395, 546, 589, 774 (5 papers).
2. W. Funke, *Werkstoffe u. Korr.* 1969, **20**, 12.
3. I. L. Rosenfeld, K. A. Zhigalova and Y. N. Bur'yanenko, Corrosion of Metals and Alloys. Collection of Russian papers No. 2. Vol. II; translator and editor, C. J. L. Booker and A. D. Mercer (Boston Spa), p. 263; I. L. Rosenfeld, E. K. Oshe and A. G. Akimov, p. 302.
4. I. Epelboin, M. Keddah and H. Takenouti, *J. electrochem. Soc.* 1973, **120**, 230C; abstracts Nos. 66 & 71.
5. R. Katz and B. F. Munk, *J. Oil Colour Chem. Assoc.* 1969, **52**, 418.

The penetration of SO_2 through alkyd resin films, clearly a matter of some importance, has been studied in Czech work.[1] Penetration often takes place easily, so that this factor affects protective properties of coatings intended for use in industrial atmospheres. The penetrating particles are SO_2 molecules, not H_2SO_3. Movement through alkyd films occurs easily, but a film based on chlorinated rubber and chlorinated paraffin is relatively impervious; probably the size of the SO_2 molecules is too great for the openings in such a film. If the mixture is pigmented with metallic lead, access of SO_2 to the metal is even better suppressed.

Permeability towards Cl^- ions is also of great importance. Passivating pigments used in paints must have limited solubility; otherwise they would soon be washed away. If, therefore, Cl^- ions arrive quickly at the metal–paint interface, passivity will break down. Murray[2] has studied three surface coatings, as films of a thickness related to practice. Cellulose acetate showed permeation by Cl^- at a rate constant with time; epoxy/polyamine in ratio 1 : 1 shows a rate increasing with time, but when the ratio is 2 : 1, the rate falls off with time. Davies[3] has worked out a method for the determination of Cl^- in rusted steel panels.

Pre-treatment Processes

Degreasing and Pickling. A useful collection of formulae has been provided from a French source.[4]

Sanyal[5] has studied the acid attack sometimes suffered when degreasing is carried out with trichlorethylene, which can be transformed by autoxidation to dichloracetyl chloride with liberation of HCl, CO and $COCl_2$; light, heat and moisture accelerate the change, and finely divided metal is thought to act as a catalyst. Acidity can also be derived from lubricant traces on the metallic surface, which may contain fatty acids.

Sequence of Treatments. Williams and Cronin[6] describe the continuous treatment of zinc-coated or uncoated cold-rolled steel which moves at controlled speed through a continuous plant, being first cleaned, then phosphated, then coated and finally cured. The metal receives in turn

(1) alkali degreasing at 70°C, lasting about 25 s,
(2) treatment with abrasive, applied on nylon brushes mounted on a rotating shaft,
(3) hot-water rinse,
(4) phosphating bath,

1. M. Svoboda, H. Klicova and B. Knapek, *J. Oil Colour Chem. Assoc.* 1969, **52**, 677; 1973, **56**, 171; also *Farbe u. Lack* 1969, **75**, 12.
2. J. D. Murray, *J. Oil Colour Chem. Assoc.* 1973, **56**, 210.
3. J. E. Davies, *J. Oil Colour Chem. Assoc.* 1971, **54**, 281.
4. M. Massard and P. Orlowski, *Corrosion (Paris)* 1970, **18**, 510; see formulae on pp. 513–14.
5. B. C. Srivastova and B. Sanyal, *Labdev. J. Sci. Tech. (Kanpur)* 1968, **6A**, 221.
6. E. L. C. Williams and J. A. Cronin, *Brit. Corr. J.* 1968, **3**, Supp. Issue, p. 21.

(5) cold-water rinse,

(6) rinse in water containing trivalent and hexavalent chromium,

(7) and (8) two coats of paint,

(9) oven-treatment,

(10) quench.

The primer may be a modified epoxy paint pigmented with strontium chromate. The finishing coat may be polyvinyl chloride, as plastisol or organosol, or it may be a solution paint with vinyl, alkyd or other resin; a coating based on polyvinylidene fluoride, which is distinctively expensive, has a reputation for long term life under outdoor weathering conditions.

A new process[1] which carries out degreasing, descaling, surface cleaning, etching and phosphating in a single operation may arouse interest. The medium, which can contain abrasive, glass beads or plastic polymer is contained in a sump which slopes towards the inlet of a large bore pump; this pulls the mixture from the sump and circulates it at high volume through the discharge gun positioned in the cabinet section above the sump. On impact the particles of solids are heavily cushioned from direct contact with the metallic surface by the water. It is claimed that this mechanical preparation produces a fine network of reaction points, and that 'as there is no surface contamination the chemical adherence of the coating is considerably stronger than that produced by any other pretreatment'; coatings have been produced exhibiting three times the adherence of conventional coatings.

The pre-treatment of metal before electropainting is discussed by Steinbrecher and Machu.[2] The corrosion resistance of steel is improved by a phosphate treatment; generally zinc phosphate coats give the best performance, although, with some electropaints, iron phosphate has been found suitable. A passivating rinse in a bath containing both hexavalent and trivalent Cr may improve performance if the surface is to enter the painting bath wet or only air-dried, but confers no advantage if the surface is dried in a conventional oven. Washing with deionized water is important, since any electrolyte (including chromate or chloride) will elevate the conductivity of the paint bath. Rinsing after paint application can generally be carried out in tap-water; but some paints are sensitive to tap-water, even if followed by a rinse in de-ionized water. In pre-treating aluminium carrying scrath-lines, an etching of the surface in strong alkali followed by a de-smutting operation is often desirable.

The pre-treatment of cars before electropainting is described by Quick and Shaw.[3] The cars pass through process tanks, their lower halves being immersed whilst the upper halves are subjected to spray; arrangements are made for heating and filtration. The schedule comprises (1) pre-rinse with sprays only, (2) alkali clean, (3) first rinse, (4) second rinse, (5) phosphate bath,

1. I. H. A. Yearsley and P. D. Darkins, *Trans. Inst. Met. Finishing* 1973, **51**, 173.
2. L. Steinbrecher and W. Machu, *Amsterdam Cong.* 1969, p. 731; *Industry Finishing* 1970, **46** (7) 50.
3. H. L. Quick and R. E. Shaw, *Trans. Inst. Met. Finishing* 1971, **49**, 93.

(6) third rinse, (7) fourth rinse, (8) rinse with deionized water (spraying only). Electropainting follows.

Removal of Rust from Steel. Two papers by Dasgupta and Ross[1] deserve attention. They found that, in preparing a rusty surface for painting by blasting, small abrasive particles are more effective than large ones, since they remove salts from the pits better. A combination of abrasive- and water-blasting has been found effective, especially in the USA, where sand and water is used. In the UK, sand is forbidden owing the the silicosis danger, but it has been suggested that wet-blasting with sand would minimize the amount of silica dust in the air and might render the method acceptable.

The best conditions for the blast-cleaning of steel have been studied by Remmelts,[2] who finds that the removal of rust is fastest if the blast is applied perpendicularly to the surface; abrasive of small grain size is advantageous.

Laboratory tests of processes designed to convert soluble ferrous salts into insoluble compounds have been carried out.[3] Pre-treatment with solutions of barium compounds, such as barium hypophosphite, gave better results than the incorporation of a barium compound into the paint. However, the paints containing barium hypophosphite or metaborate gave fairly good results; barium orthophosphate was not much inferior, despite the fact that it is less soluble. Cambridge work (1968 book, pp. 393, 411) in which barium ortho-phosphate was used in a paint containing metallic zinc and cadmium is mentioned in the Dasgupta–Ross paper with approval, but that mixture was not tested; probably paints containing cadmium would not be favoured today owing to its toxicity, but paints pigmented with metallic zinc and some barium orthophosphate added could probably be used. It is believed that with such a mixture, the EMF of the cell Fe/Zn would draw SO_4^{2-} ions out of the pits into the paint, where they would interact with barium phosphate to give $BaSO_4$ and an iron phosphate.

Mechanism of Phosphating. Canadian work[4] suggests that there are four steps during the phosphating of steel in a bath produced from ZnO, H_3PO_4 and HNO_3 at 95°C:

(1) Electrochemical attack in the acid medium leads to the production of crystals of the acid phosphate $M(H_2PO_4)_2$.
(2) Precipitation of amorphous phosphate $(M_3(PO_4)_2)$ occurs when the attack on the metal has modified the state of the solution sufficiently to produce supersaturation.
(3) Crystallization and growth occur later.
(4) Finally crystal reorganization sets in—a reaction probably taking place at the metal–solution interface; this modifies the porosity of the coating, decreasing the exposed metallic surface.

1.   D. Dasgupta and T. K. Ross, *Brit. Corr. J.* 1971, **6**, 237, 241 (2 papers).
2.   J. Remmelts, *Brit. Corr. J.* 1969, **4**, 199.
3.   D. Dasgupta and T. K. Ross, *Brit. Corr. J.* 1971, **6**, 237, 241 (2 papers).
4.   E. L. Ghali and R. J. A. Potvin, *Corr. Sci.* 1972, **12**, 583.

The reaction mechanism concerned in the production of conversion coatings in phosphate and oxalate baths has been discussed in detail by Machu,[1] with special reference to complexing and chelating.

A discussion of phosphating with contributions from several authors was published in 1971.[2] Machu in particular discussed a problem important in the motor industry, namely a phosphating bath suitable for composite systems comprising different metals. He stated that it is difficult to find a single bath suitable for all cases, but that certain zinc phosphate baths containing nitrates and fluorides (sometimes nitrites also) can be used for zinc coated steel and also for bare steel. Another discussion of the subject has been contributed by Cupr.[3]

Accelerator systems for zinc phosphate processes used before electroplating are discussed by James and Freeman.[4] Nitrates, nitrites, chlorates, perborates and H_2O_2 have all been included in the baths.

Work by Gebhardt[5] has shown that phosphate coatings are always textured; the quality of the phosphate crystals increases with the quality of the metal crystals, and the protection afforded depends on the nature of the layer crystallites, their texture, their growth and the adhesive bond to the metal substrate.

Work has been carried out on different methods of rinsing the surface between phosphating and painting.[6] A treatment in water containing Cr^{III} and C^{VI}) improved behaviour on subsequent salt-spray testing.

Painting Procedures

Electrodeposition of Paint. A review of electropainting has been provided by Milner,[7] who discusses the economics of the subject and the advantages and disadvantages of various resins and pigments. The polymer used must have a reasonably low molecular weight to provide water solubility and minimize viscosity; the acid groups are neutralized by a base to render the resin soluble as a salt, which nevertheless provides a water insoluble finish after deposition through the lowering of pH value accompanying the anodic reaction. The base chosen is usually monoethanolamine or triethylamine; NaOH or KOH have been used at times. In the early days the resins were alkyds or maleinized oils; today epoxydized alkyds, which provide better resistance to salt spray, are much used.

The reactions involved are discussed by Beck.[8] Where the paint contains polymeric anions, these are rendered insoluble on the surface to be painted, which is made the anode, by the H^+ formed in the anodic reaction. Probably metallic cations such as Fe^{2+} and Fe^{3+} also render the paint insoluble. In

1. W. Machu, *Werkstoffe u. Korr.* 1973, **24**, 361.
2. J. F. Andrew and P. D. Donovan, *Trans. Inst. Met. Finishing* 1971, **49**, 162. Cf. J. A. Scott, p. 170, W. Machu, pp. 170, 214.
3. V. Cupr and M. Pleva, *Metalloberfläche* 1971, **25** (3) 89.
4. D. James and D. B. Freeman, *Trans. Inst. Met. Finishing* 1971, **49**, 79.
5. M. Gebhardt, *J. electrochem. Soc.* 1973, **120**, 230C. Abstract No. 73.
6. A. Askiemazy and V. Ken, *Corrosion (Paris)* 1970, **18**, 56.
7. D. G. Milner, *Metal Finishing J.* (*London*) 1970, **16**, 150.
8. F. Beck, *Farbe u. Lack* 1966, **72**, 218.

a paint containing polymeric cations these migrate to the cathode and are rendered insoluble by the OH⁻ ions there formed.

Processes for curing electrodeposited coatings at room temperatures are described by North.[1] Such processes may be particularly useful in coating the inaccessible areas of large metallic structures which could not be stoved owing either to their size or to temperature limitations of the material. A pigmented epoxy resin is mixed with a polyamide epoxy adduct and dispersed in water. Satisfactory deposition can be obtained over a range of solid content varying from 20% down to 0·8%, although adjustment of voltage is necessary. The solid content of the bath can thus be reduced to a low level at the end of a days work. The curing is satisfactory and can occur under water.

The electropainting of cars has been discussed briefly in chapter I, but those desiring greater detail should study several of the papers delivered at the symposium of 1968.[2] Information on the procedure used at certain large plants is provided by Patton[3] and Petrocelli.[4] A review of pigments suitable for electropainting has been provided by Entwhistle.[5]

Electron-beam curing.   A curing process brought about by a high energy electron-beam, which produces solvent-free coatings in a short time at a low temperature, is already in operation at certain works for the internal parts of cars, and seems to have promise where the output is large. Vrancken[6] points to the rapidity of the process and relatively low cost per unit area as advantages, but the high capital cost is a disadvantage.

Choice of Paints.   Zinc phosphate as an inhibitive pigment has received strong recommendation from Harrison.[7] Its good inhibitive properties, which are not in doubt, are attributed by him to the stifling of the corrosive action of ammonium ions. It seems perhaps more likely[8] that its main function, in the average atmosphere, is to deal with the H_2SO_4 derived from sulphurous fuel; reaction with zinc phosphate will remove the H_2SO_4, giving $ZnSO_4$ and phosphoric acid—both of them relatively harmless. Harrison has stated that he finds 'nothing to disagree' with this suggestion.

Evidence for the anti-corrosive action of zinc phosphate is provided by tests carried out at Burry Port,[9] lasting $2\frac{1}{2}$ years. Various vehicles were tried, including chlorinated rubber, catalyzed epoxy resin, oil modified alkyd and urethane. It is stated that zinc phosphate paints can be used with advantage as first and second coats, and that the adhesion of subsequent coats has been

1.   A. G. North, *J. Oil Colour Chem. Assoc.* 1970, **53**, 353.
2.   *Inst. Mech. Engrs Proc.* 1967–68, **182**, Part 3J, esp. A. J. Bee, p. 6, B. M. Letzky, pp. 69, 72, R. J. Brown, p. 131.
3.   W. G. Patton, *Iron Age*, April 29, 1971, p. 52.
4.   A. G. Smith and J. V. Petrocelli, *J. Paint Tech.* April 1968, **40**, No. 519, p. 174.
5.   T. Entwhistle, *J. Oil Colour Chem. Assoc.* 1972, **55**, 480.
6.   A. Vrancken, quoted by H. W. Talen, *Verfkroniek* 1973, **46**, 106.
7.   J. B. Harrison, *Brit. Corr. J.* 1969, **4**, 55, 58; *Trans. Inst. Met. Finishing* 1970, **48**, 83.
8.   U. R. Evans, *Trans. Inst. Met. Finishing* 1970, **48**, 145.
9.   *Paint Tech.* 1970, **34**, No. 1.

found to be excellent. Blistering, often encountered when more soluble pigments are used, is not a serious problem. Excellent results have been obtained with zinc phosphate containing 20% (v/v) TiO_2 in polyvinyl chloride as vehicle and a top coat pigmented either with micaceous iron ore or with aluminium.

Another method of introducing phosphate into paint is to use a resin with P_2O_5 groups copolymerized. This possibility has been discussed by Lowe.[1]

The contrast between old-fashioned and modern paints is brought out by Hudson.[2] He points out that traditional paints such as red lead in linseed oil followed by various coats based on boiled oil are still serving well on railway bridges and the like, providing lives up to 5 years. Such paints are less susceptible to the effects of bad surface conditions than the more sophisticated modern paints, although in recent years they may have been surpassed by multi-pigment paints, and especially by zinc-rich paints. In general, modern paints require very careful surface preparation, and behave in a disappointing manner if the conditions of application have not been correct.

A review of painting developments over 50 years has been provided by Hoar;[3] amongst other innovations he mentions the incorporation of micaceous iron ore into epoxy vehicles to improve resistance to salt-spray.

Another useful component of paints is china clay, which is found to improve the brushing properties of zinc-rich and red lead primers.[4] In the paint as supplied in the pot, the particles, which are extremely small plates, are randomly oriented; on brushing, the platelets align themselves parallel to the surface, forming a number of slip-planes which offer less resistance to brushing than lumpy nodular particles. The objectionable setting in the pot sometimes experienced with paints containing heavy pigments is largely avoided if china clay is added as an extender. Under acid conditions the particles carry negative charges on the faces but are positively charged along the edges; the attraction between edges and faces produces a weak flocculant structure, and hard setting is avoided.

A relatively recent development is the production of cored pigments. The subject has been discussed by Hernelin.[5] Tests have shown that it is only the surface of a pigment particle which is effective in providing inhibitive material. Thus the interior can consist of an inactive substance, relatively cheap and light. The use of this principle in providing lead paints which are lighter than conventional lead paints, and, owing to their lower lead content, less toxic, is discussed on p. 287.

The painting of badly prepared surfaces. Chandler[6] stresses the need for careful surface preparation, pointing out that an inferior paint applied to a good surface under good conditions will in many circumstances perform better than one that is intrinsically superior. It might be added that a superior

1. P. Lowe, *Anticorrosion* 1967, **14** (12) 8.
2. J. C. Hudson, *Annales des Travaux publiques de Belgique* No. 3 (1968–9).
3. T. P. Hoar, *J. Oil Colour Chem. Assoc.* 1971, **54**, 203.
4. *Metal Finishing J.* (*London*) 1969, **15**, 316.
5. R. Hernelin, *Anticorrosion* 1972, **19** (1) 3.
6. K. A. Chandler, *Paint Tech.* 1969, **33** (4) 35.

paint capable of providing protection when applied to a surface which carries rust, scale and grease would be even more welcome.

There are, indeed, occasions when paint must be applied to a rusty surface. The results of doing so have been investigated by Schwarz[1] at Stuttgart and elsewhere. Comparisons of various paints applied over six types of rust were carried out, and in practically all cases, an oil paint containing red lead was the best. One coat of linseed oil paint pigmented with red lead gave better protection than two coats of a quick-drying paint based on a phthalate resin similarly pigmented; one coat of the latter gave better results than two coats of chloro-rubber paint pigmented with red lead.

Various explanations for the good behaviour of the oil paint are suggested by Schwarz, but the important work of Mayne and van Rooyen (1960 book, p. 547) is not discussed; these authors showed very clearly that lead linoleate, which is present in the oil paint but not in the other types tested, becomes degraded to lead azelate, which is an excellent inhibitor. It should be added, however, that the presence of free PbO in the red lead is needed for the best results, doubtless because it is required for the production of lead linoleate. Comparative tests reported in 1939 show that the addition of PbO to modern non-setting red lead improves performance.[2] Such a paint could not, of course, be supplied ready mixed, as it would 'set in the pot'; but it might be possible to provide a dispersion of litharge in an inert liquid (e.g. a hydrocarbon) and instruct the painter to add a small quantity, say one spoonful, to each potful of non-setting red lead paint just before application.[3]

Other German tests[4] at Sylt and Berlin confirm Schwarz's finding that classical red lead oil paints are superior to most of those recommended by by manufacturers today as being suitable for painting over rust. However, zinc dust paints were almost equal to the red lead paint, provided that they were applied to a sand-blasted surface providing good adhesion and covered with a top coat of a paint exhibiting good weather resistance.

Further views on the problem of painting upon a rusty surface are put forward by Fritz[5] and van Oeteren.[6] Tests carried out in India are described by Sanyal.[7]

The use of lead paints is forbidden in many countries on account of their toxicity. This is probably to be commended, although the excessive quantity of lead commonly found in patients at hospitals will not immediately be rectified by such a prohibition. A report from New York[8] shows that, although deaths attributed to lead declined between 1959 and 1969, 'cases' of lead poisoning discovered increased. In 1971 there were about 450 000 apartment units in New York City in such a bad state of repair that a child living in them would be exposed to the hazard of lead poisoning; apparently

1. H. Schwarz, *Werkstoffe u. Korr.* 1972, **23**, 648.
2. S. C. Britton and U. R. Evans, *J. Soc. chem. Ind.* 1939, **58**, 90.
3. U. R. Evans, *Brit. Corr. J.* 1972, **7**, 50.
4. J. Sickfeld and D. Wapler, *Werkstoffe u. Korr.* 1970, **21**, 77.
5. G. Fritz, *Farbe u. Lack* 1969, **75**, 937.
6. K. A. van Oerteren, *Fette, Seife, Anstrichsmittel* 1971, **73**, 102.
7. B. Gangali and B. Sanyal, *Labdev. J. Sci. Tech. (Kanpur)* 1968, **6A**, 136.
8. V. F. Guinee, *Trans. N.Y. Acad. Sci.* 1971, **33**, 539.

children often eat the chips of peeling paint. Another source of lead poisoning is petrol fumes.

Attempts have been made to provide paints containing a smaller amount of lead but nevertheless inhibitive. Basic lead silicochromate is recommended as pigment by Bates.[1] The grains of this pigment have silica cores covered with a thin shell of the lead compound. Mayne thinks that the inhibitive properties of this paint are due to the formation of a lead soap; he considers that practically no $CrO_4{}^{2-}$ ions would pass into solution.

Prevention of microbiological attack on paints. In absence of precautions, many paints are liable to suffer bacterial degradation; emulsion paints are specially susceptible, and gram-negative bacilli are a more serious menace than gram-positive species; fungi are more resistant to chemical disinfection than are bacteria. For many years the inclusion of mercury compounds in paint was relied on to deal with the problem; they were found to be highly efficient, but they are extremely toxic, and the discovery of large amounts of mercury in fish has led to restrictive legislation against such compounds. The subject has been discussed by Carter,[2] who distinguishes two main problems; the preservation of paints against bacteria in the can, and the protection of paint films on coated metal against fungi; he states that none of the available products seem to provide a satisfactory answer to both problems. As an in-can preservation, he states that 1 : 2-benz-isothiazolin-3-one is a particularly effective product. For paint films, ZnO has appreciable antifungal properties, but about 20% is required. Barium metaborate is claimed by some authorities to be effective owing to the release of Ba^{2+} as well as $(BO_2)^-$ ions; 2-(4-thiotolyl)-benzimidazole is stated to be satisfactory, but has no in-can preservative properties. For the moment, it would seem that a combination of additions may be needed to cope with both problems.

R.T. Ross[3] discusses the manner in which bacteria affect decorative properties, durability and adhesion. The particular organism most commonly isolated from defective paint coats is easily controlled by preservatives, but there are others more resistant. Destructive enzymes produced by organisms growing in a raw material before a preservative is added may continue to be active when incorporated into a paint—even though the organisms themselves have been killed by preservative added at that stage. Ross discusses tests for assessing various types of deterioration.

Paints for Special Purposes

Pipe-lines. The procedure followed by the (British) Gas Council[4] is as follows. The external surface of the pipe is pre-coated at the mill with a flood-coat of coal-tar enamel incorporating an inner fibre-glass reinforcement and an outer thermoglass wrap. The internal surface is coated with a single

1. R. P. Bates, *J. Oil Colour Chem. Assoc.* 1971, **54**, 945. Cf. J. E. O. Mayne, p. 956, and reply by R. P. Bates, p. 957.
2. G. Carter, *J. Oil Colour Chem. Assoc.* 1973, **56**, 302; *Verfkroniek* 1973, **46**, 254 (with G. Huddart).
3. R. T. Ross, *J. Paint Tech.* 1969, **41**, 266.
4. *Brit. Corr. J.* 1971, **6**, 6.

application of epoxy paint, not less than 2–3 thou thick. At the site the pipe-lengths are welded together, and the weld area, after an inspection by radio-graphy, is coated by the same flood method, and again inspected. The trench and back-fill must be free from loose stones and flints.

The protection of a pipe-line carrying crude oil from Louisiana to Illinois, crossing two major rivers besides swamps, clay soils and the like, is described by Layne.[1] A glass-lined pipe intended for the petroleum industry has been described.[2] The older glass-coated pipes, intended for the chemical industry, are too expensive for the oil industry, but the new pipe has a much thinner coating (20 mils instead of 50–80 mils) chemically bonded to the pipe.

Power Station Structures. Bayliss and Wall[3] discuss a problem confronting electric generating authorities—the provision of a 'holding primer' which will prevent rusting for a considerable period before the appli-cation of the permanent coats. There are a number of formulations which will hold blast-cleaned steel free from rust for six months in an aggressive environment; if the delay is much longer, rust generally appears. An etch-primer pigmented with zinc chromate is favoured, since this requires only one 'pack', presents no problem in regard to re-coating and has little effect on welding; it does not (like epoxy) harden with age, whilst the surface, if rust should appear, can easily be prepared for fresh coating with the wire-brush. The durability of a holding primer is proportional to its thickness, but a given thickness, applied as two coats, will last longer than a single coat.

Bridges. The painting of steel bridges has been described by Smith and Day,[4] who consider that a bridge in the UK should last 120 years and require maintenance painting 10 to 20 times. They consider that red lead is still almost unrivalled for a steel surface from which it is not possible to remove all rust; in such cases brushing is still the best method of application. Automatic methods of painting are much used today, and the capital cost of equipment is soon defrayed on a large bridge; the control of film thickness, and freedom from 'holidays', make for a better and more economic method. Metallic coats are sometimes applied; hot-dip galvanizing is preferable to spraying for low coating weights. Surface cleanliness is all-important. Piles driven into undisturbed waterlogged ground are said to suffer little damage after some initial pitting, even in the presence of sulphate reducing bacteria. However, porous ground through which water is moving, with renewal of dissolved oxygen, may cause trouble. Above ground, enclosure in concrete may be needed.

The painting of the Victoria Falls Bridge is described by Johansen.[5] The history over a period of 65 years is sketched. The bridge has had to resist unusually corrosive conditions in a tropical climate. In the early days oil paints were used, but at the last repainting in 1963 a three-coat system based

1.   H. B. Layne, *Mat. Prot.* 1969 (7) 23.
2.   F. W. Nelson, U. J. Arbter and S. L. Henry, *Mech. Engg* 1969, **91** (1) 14.
3.   D. A. Bayliss and D. C. Wall, *J. Oil Colour Chem. Assoc.* 1968, **51**, 792.
4.   D. W. Smith and K. J. Day, *Brit. Corr. J.* 1970, 5, 151.
5.   R. P. Johansen, *Farbe u. Lack* 1971, **77**, 149.

on an epoxy tar combination was adopted with a view to increasing the intervals between repaintings and protecting against H_2S as well as against mechanical damage; two coats of plain epoxy tar and one of epoxy tar containing aluminium were used. After 5 years a close inspection revealed no corrosion or mechanical damage.

Ships. A book on 'Recommended Practice for the Protection and Painting of Ships' contains authoritative information.[1] Bult[2] considers that the best results are obtained if a shop-primer is applied to give protection during the building period; he considers that a paint pigmented with red lead or aluminium based on vinyl resins, chlorinated rubber or epoxy tar is better than a zinc-rich primer, the behaviour of which is somewhat unpredictable. Some of the modern paints which give a thick layer in one operation are generally preferred, but they afford full protection only if the surface conditions are correct. Ellinger[3] refers to the revived interest in red lead paints, whilst acknowledging the objections connected with health hazards.

The paints used in the (British) Royal Navy are discussed by Smith.[4] The paint favoured in 1969 contained basic lead sulphate, non-leafing aluminium, barytes and iron oxide in a vehicle consisting of stand oil and tung-oil containing a modified phenol-formaldehyde resin. It is moderately resistant to alkali and can therefore be used with a sacrificial anode system during fitting out, but not with an impressed current sytem. If the latter is desired, an epoxy tar paint is used.

Anderton[5] describes blister formations on cathodically protected ships. He quotes Mayne's suggestion that water tends to congregate around ionogenic sites in a polymer. Anderton suggests that it may also concentrate in water-rich envelopes around hydrophilic pigment particles as well as on the hydrophilic metallic substrate. He thinks that the 'water-rich' layer at the steel–coating interface plays a part in destroying adhesion; it may provide a relatively conductive path for ionic current to flow between anodic and cathodic areas.

Other information about marine paints suitable for use under conditions of cathodic protection is provided in chapter VIII.

It is known that anti-fouling paints containing copper, whether as metal or oxide, can cause corrosion of steel. For this reason, the possibility of anti-fouling systems based on organotins will interest the corrosion specialist; they may also prove to be more efficient in preventing fouling than ordinary formulations containing cuprous oxide, which are stated, under sub-tropical conditions, to require renewal every two years. As part of a programme[6] aimed at finding improved systems to protect ships of the Royal Australian Navy, trials

1.   'Recommended Practice for the Protection and Painting of Ships' (British Ship Research Association, and Chamber of Shipping for the UK) 1973 (drafting by J. C. Hudson).
2.   R. Bult, *Australian Paint J*. Dec. 1969, p. 7; also *Verfkroniek* 1969, **42**, 217.
3.   M. L. Ellinger, *Paint Manufacture* 1970, **40** (5) 48.
4.   J. Smith, *The Engineer* 1969, **227**, 212. J. Smith and C. A. S. Palmer, *3rd Internaval Corr. Conf., London*, 1969.
5.   W. A. Anderton, *J. Oil Colour Chem. Assoc.* 1970, **53**, 181.
6.   *Tin and its Uses* 1973, No. 96, p. 7.

have been reported with rubber impregnated with tri-n-butyl tin compounds. Favourable results have been obtained in preliminary studies, and if the more advanced development work maintains the promise, these chemicals which are powerful biocides, may play a useful part in maintaining the efficiency of Australian vessels, In any formulation for an anti-fouling paint, however, it is necessary to control the leaching rate of the biocide; otherwise the period of efficiency will be short.

Aircraft. The subject of aircraft painting has been authoritatively reviewed by Hoey.[1] Resistance to cold oil constitutes the main requirement for an aircraft paint, but there are occasions when hot fluid can come into contact with paintwork; thus resistance at or above 70°C may be required. The degradation products of fuels, which include aggressive alcohols and acids, have to be taken into account; also hydraulic fluids may reach the paint, and although some of these, consisting of mineral oil, are innocuous to most paints, others like phosphate esters or castor oil, can cause severe attack. Inhibition of corrosion is considered in the same article. On steel, protection at gaps in a coating can generally be obtained by sacrificial action, but on aluminium this is difficult, and here there is reliance on chromates. Curves showing the amount leached after a different number of days are useful in predicting the behaviour of different chromates. A high leaching rate gives the best protection over a short period, but this will not continue after the chromate has become exhausted; also a high solubility can lead to blistering. Optimum protection is given by primers containing strontium chromate. calcium or zinc potassium chromate.

Scott[2] tells the story of protective methods at a large aircraft works. Up till about 1960 etch-primers were used, extensively supplemented in critical areas by alkyds or epoxy finishes. About that time a study of breakdown was carried out, which showed that the paint itself was being attacked by contaminants and that it was sometimes insufficiently flexible to withstand the constant movement of the air-frame. It was decided that a paint must be chosen which was both chemically resistant and flexible. An epoxy paint cured with a polyamide seemed to offer the best compromise. A primer pigmented with strontium chromate was used to provide inhibition. The colour of the finish was chosen to give good protection against light, combined with durability and aesthetic appeal. A clean surface for application of the paint was found to be essential; ordinary degreasing and pickling was not sufficient to meet the requirements for the application of an epoxy paint. In general, the priming coat, and sometimes the finishing coat, is now applied to each part before assembly, whilst the surface is still clean.

As stated above, aircraft manufacturers have experienced difficulty in obtaining good adhesion of epoxy paint finishes. Cleanliness is essential, but tests indicating the degree of cleanliness obtained have been unsatisfactory. Miller[3] considers that water-break tests (according to which the surface is

1.   C. E. Hoey, *Trans. Inst. Met. Finishing* 1968, **46**, 95.
2.   J. A. Scott, *J. Oil Colour Chem. Assoc.* 1969, **52**, 593.
3.   F. N. Miller, *Mat. Prot. Perf.* 1973, **12** (5) 31.

taken to be clean if the rinse water forms a continuous film) are unnecessarily severe. He recommends a method based on the diameter of a 5 ml drop, which he states to be rapid, accurate and versatile; it is regarded as suitable for surfaces to be electroplated, painted, adhesive-bonded, anodized or treated for conversion coatings.

Walker[1] has described coatings for space-craft, including satellites. He states that many inorganic pigments are unstable to ultraviolet light *in vacuo*; some of them show a reversible degradation. Treated ZnO is stable under the testing conditions. As a vehicle, silicone resin has good application properties and stability towards ultraviolet light; it also shows resistance to thermal shock, but the coatings are soft and the adhesion poor. Silicate paints often show surface cracking and crazing on cure; they do not adhere well to substrates other than Al, and their resistance to thermal shock leaves something to be desired. Thus much work remains to be done before the perfect coating is found.

Paints for Non-Ferrous Metals. Newton[2] has studied paints containing zinc dust and ZnO in the ratio 80/20 intended for use on galvanized steel. Numerous vehicles were tried. Chlorinated rubber and phenolic varnishes gave the best overall results.

The causes of failures of paint coats applied to galvanized steel is discussed by van Oerteren.[3] He considers that it may often be due to contamination picked up from the cooling bath. The galvanized surface, when it emerges from the bath of molten zinc, is free from oil, but it may become unclean as a result of passage through the water bath used for cooling; the water is sometimes dosed with an oil inhibitor for the prevention of white rust.

Another problem is the corrosion of zinc by vapours given off by polyester resins.[4] Attempts to prevent this by reducing agents designed to accelerate the decomposition of peroxide catalysts gave somewhat disappointing results, but the use of additives which will behave as acid acceptors, notably sodium bicarbonate, was effective. The influence of the resin structure is important; an almost non-corrosive resin has been devised, based on maleic anhydride, phthalic anhydride and propane-1,2-diol along with a small amount of $NaHCO_3$.

Lacquers suitable for zinc and copper have been studied by Christie and Carter.[5] The lacquer for zinc should contain rubeanic acid, whilst that designed for copper should contain bensotriazole. It is stated that the outdoor lives can be three years on zinc and five years on copper.

Methods of Testing

Early work and present needs. During the past 50 years, several comprehensive series of exposure tests carried out in different countries by

1. P. Walker, *Verfkroniek* 1970, **43**, 58.
2. D. S. Newton, *J. Oil Colour Chem. Assoc.* 1969, **52**, 133.
3. K. A. van Oerteren, *Werkstoffe u. Korr.* 1967, **18**, 184.
4. R. Cawthorne, W. Flavell, N. C. Ross and F. J. Pinchin, *Brit. Corr. J.* 1969, **4**, 35.
5. I. R. A. Christie and V. E. Carter, *Trans. Inst. Met. Finishing* 1972, **50**, 19.

qualified persons have provided knowledge regarding the relative suitability of paints of different types to protect steel against various climatic conditions —including rural, marine, urban and industrial atmospheres. Examples of exposure tests are provided in the 1960 book, pp. 576–85. Similar work has been carried out in a number of countries; there is no serious disagreement, and it might be felt that all the necessary information is today available regarding choice of painting systems. That is probably true, provided that one important condition is fulfilled—namely that the coatings remain *intact*. In practice, however, they may become discontinuous, either through abrasion by wind-borne grit or detachment resulting from bending of the basis metal or denting on accidental impact. Perhaps for that reason, more effort has been devoted in recent years to mechanical testing of painted surfaces than to further series of exposures. Generally the mechanical testing can be carried out in laboratory or workshop under strictly controlled conditions; but the choice of tests demands careful thinking to ensure that it really represents the sort of operation which will produce damage in service. For instance, the direction of the forces that cause detachment of a paint coat in practice may be parallel to the painted surface, and laboratory measurements of the force perpendicular to the surface needed to tear off the coating are not necessarily helpful in such cases. Again, even in detachment, chemical or electrochemical factors may be quite as important as purely mechanical factors; for instance, paints with vehicles containing saponifiable substances exposed to salt water may suffer detachment owing to softening and creepage by cathodically produced alkali. It seems to be thought detachment of paint by alkali can only occur when the paint contains a saponifiable vehicle. In early work, cases were noticed when a paint coat detached itself by the liquid creeping between paint and metal, without visible attack on the former. It was pointed out in 1929 that such detachment becomes possible from the energy standpoint if $\sigma_{MC} > \sigma_{ML} + \sigma_{LC}$, where σ is the interfacial energy between the phases suffixed, M being the metal, L the (alkaline) liquid and C the paint coat.[1]

Recent Work. Two tests recently put forward may be mentioned, one being a purely mechanical test, whilst the other involves electrochemical action.

An abrasion test can be delightfully simple. Boers[2] describes a test in in which grit is allowed to drop from a funnel on to a painted specimen inclined at a gentle angle to the horizontal. The damaged area is plotted against the height through which the grit falls. The influence of different types of grit is noted, and the power of different types of paint to withstand abrasion.

Stone[3] describes a test which may prove useful in selecting paints for motor cars; here it is clearly desirable to select a paint which will minimize detachment by the action of de-icing salt; a paint which would suffer alkaline peeling (1960 book, p. 552) would be clearly unsuitable. A scratch is made on

1.   U. R. Evans, *Trans. electrochem. Soc.* 1929, **55**, 243; esp. p. 246.
2.   M. N. M. Boers, *Verfkroniek* 1969, **42**, 251.
3.   J. Stone, *J. Paint Tech.* 1969, **41**, 661.

the surface of a painted steel specimen, using a standard scriber, the width being 0·1 mm and the length 9·52 mm. An adhesive gasket serves to confine 5 drops of 5% NaCl on the desired area. A cathodic current of 9 mA is applied for 15 minutes at 22°C, Pt wire being used as anode. After drying, adhesive tape is applied and pulled off. The width of the strip laid bare, less the original width of the scratch (0·1 mm), provides a measure of the detachment caused by the cathodically produced alkali.

Caution in drawing conclusions from tests. The two examples given above suggest that testing methods need not be elaborate in order to provide useful information. A word of warning, however, would seem necess-ary. A test which merely assesses resistance to abrasion or detachment will not in itself provide evidence that the painting system will inhibit corrosion. There has been a tendency to read into specifications guarantees which are not justified. An example is given on p. 175 of this book. B.S. Specification 4164 (1967) for hot-applied coating materials based on coal-tar is essentially a consideration of mechanical properties and contains no chemical tests.

Apart from under-rusting, there are several manners in which a coating itself may suffer deterioration as a result of factors which are not embodied in the generally accepted tests. For instance, bituminous coatings can suffer blistering under many conditions; the most familiar type of blister occurs when a coating has been applied to a porous substrate containing water and the surface is exposed to solar heat; this type is important on wood but not, as a rule, in connection with the protection of metals. A different sort of blister, however, has been described,[1] which is formed, not at the interface between the coating and the substrate, but within the coating material itself. This type of cavity may in time reach a size sufficient to cause perforation of the layer; the production of the cavity is stated to be the same, whether the substrate is steel, glass or plastic. Here the cavities responsible for breakdown do not arise through alteration of temperature but fluctuations of pressure; they are analogous to the bubbles produced when a bottle of soda water is opened, reducing the pressure, so that CO_2 can be formed as a gas.

Accelerated Testing. Since outdoor exposure tests must be continued for years before results emerge (sometimes results obtained over one year may be misleading if applied to the relative merits of different painting systems after, say, ten years), the desire for quick results obtainable by laboratory examination is understandable. However, many of the laboratory tests used today are of limited value, The idea generally is that by imposing severe conditions, quicker results can be obtained; that is no doubt true, but the order of merit of different painting systems as indicated by short exposure to severe conditions will not, in general, be the same as that emerging after long exposure to relatively mild conditions. Rather an attempt should be made to reproduce in the laboratory conditions similar to those which will have to be withstood in service, and then to examine the specimens after a relatively

1 P. Biloen, R. Bonn, P. D. Marys and W. A. Spoon, *J. appl. Chem.* 1972, **22**, 165.

short period; chemical indicators are today so sensitive that it should be possible to detect the start of corrosion long before any change is visible to the eye.

The specimens tested must represent the surface condition of the metal as it will in fact be painted in practice; it may be quite different from the condition as laid down in the set of instructions which will be found in the office filing cabinet; these probably prescribe the absence of rust. If the persons responsible for the testing have reason to believe that the instructions of the filed copy will not be carried out in the field, and that the paint will be in fact applied over rust, the specimens used in the test must be rusty.

If the surface condition of the steel as well as the composition of the paint and corrosive atmosphere to be withstood in service can be accurately reproduced in the testing laboratory, a number of tests carried out for different times should show up differences in the rate of formation of corrosion product, long before there is any visible change of the painted surface. The traces of corrosion product may be found after dissolving off the paint coat, either as insoluble matter adhering to the metal basis or suspended in the solvent used for the removal (perhaps sometimes dissolved in it). Given a sensitive reagent, it should be possible to put the paints in an order of merit which is more likely to correspond with true protective behaviour than observations upon painted specimens exposed to severe, unnatural conditions.

When once corrosion has started at a point below paint, it is likely to continue, because the bulky rust will push up the paint coat. Thus the recording of the initiation period which must elapse before any corrosion product can be detected is probably a better criterion of the protection provided by a paint system than any figure based upon (say) the proportion of the painted surface carrying visible rust after a much longer period; it is true that the number obtained for the period will depend on the sensitivity of the chemical test used, but if the chemical test is standardized, the order of merit provided is likely to have practical value.

It is possible that the initiation of corrosion might be detected without removing the paint coat, by electrochemical studies of the surfaces exposed for different periods. Wormwell[1] refers to research carried out at Teddington based on measurements of electrode potential, resistance and capacitance. He states that 'it was shown that breakdown of a paint could be assessed quite quickly . . . The method has been used by paint firms.' Further development on these lines would be welcome.

The condemnation of *artificially* accelerated tests need not be taken to imply condemnation of rapid tests, if the latter represent service conditions. The rotor test developed 25 years ago at Teddington[2] is sometimes described as an 'accelerated test' because it provides in 50 days an order of merit between paint systems which it would require at least a year to obtain on a raft test. It can, however, be regarded as a comparatively realistic test. The rotor speed generally employed (1500 rpm) represent 20 knots peripheral

1.   F. Wormwell, 'Corrosion of Metal Research 1924–1968' (Nat. Phys. Lab.).
2.   F. Wormwell, T. J. Nurse and H. C. K. Ison, *J. Iron Steel Inst.* 1948, **160**, 253.

velocity; if the breakdown of ship's paint is related to the ship's speed, such a test should provide more accurate prediction of behaviour under service conditions than the raft test—besides furnishing the results more quickly. If, however, breakdown in service is due to sulphur compounds found in polluted water, then a raft anchored in a polluted estuary may furnish results which are of greater practical value.

The fact that today laboratory tests and field tests do not always agree is not surprising. Johansson[1] suggests why a salt-spray test sometimes gives different indications from an outdoor exposure test. In the latter type of test, erosion caused by sand and water probably plays an important role during the early stages in the deterioration of the protective film. In the salt-spray chamber, on the other hand, the high temperature (35°C) may cause rust-proofing material to flow off the surface and thus remove the film. Evidence sometimes adduced for a close correlation between the two tests must be ascribed to the play of chance.

Success in paint testing depends mainly on the judgement of the person who decides which of all the many tests available is most suitable for ascertaining which of many available paint systems will best withstand the service conditions prevailing in the case under consideration. Probably a visit to the place where the painted metal will be exposed and a study of the manner in which paint coats have broken down at that place in the past will be necessary. If the inspection reveals that these have failed owing to abrasion by wind-blown particles, an abrasion test is called for; but this will be utterly useless for situations where alkaline softening is the cause of failure. If the correct decision has been made on the type of destruction to be feared, the testing apparatus need not be elaborate or expensive, nor need the personnel employed possess a high degree of dexterity. If the wrong decision has been made, no apparatus, however expensive, can save the situation.

An interesting method of assessment has been described by Boers[2] in a research designed to study the influence of climatic conditions on the extension of rusting on painted steel where the coating has been interrupted by a scratch line; the painted panels were exposed at Delft, and results in an industrial situation might be very different. The panels of cold-rolled steel carried a very porous ground coat, covered with a richly pigmented outer coat. A specimen was exposed on the first day of every month and exposed for exactly one month; the under-rusted area was found to be uniform in breadth, which could be measured with an accuracy of 0·5 mm. The spread of the rusting was much quicker in winter than in the summer months, being 0·5 mm in June and August (1972) but rising to 3·0 mm in November. The difference was correlated with variations in relative humidity (RH) and chlorine content in the rain-water; the breadth reached in a month was equal to 0·0085 (RH − 70) [Cl]). Taken literally, this suggests that there would be no rust spreading at all if the RH were below 70% or if Cl^- was absent.

If this test is to be applied to compare the merits of different painting systems, it may be well to combine it with a scratch test. A good paint system

1. U. Ulfvarson and K. Johansson, *Mat. Prot.* 1969, **8** (6) 43.
2. M. N. M. Boers, *Verfkroniek* 1973, **46**, 157.

should have the outer coat sufficiently resistant to scratching to ensure that the steel basis only becomes exposed at infrequent points, whilst the innermost coat should be so chosen that at places where the steel starts to be attacked, the rusting spreads only slowly over the surface.

Further References

Addendum.   Important studies by Mayne and Mills on mild steel coated with three different varnishes and exposed to 3·5M KCl show that at points where corrosion sets in the film is found to be of the D type and to have a low resistance; where corrosion is absent, the film is of the I type and has a high resistance. The resistances of D areas on platinum are much higher than on mild steel; the resistance of I films is independent of the presence and nature of substrate. When zinc chromate is used as a pigment, the steel substrate is passivated and behaves in a manner similar to platinum. It is believed that the protective value of coatings (under immersed conditions and without ultraviolet radiation) can be predicted from resistance measurements.

Codes of Practice and general guides.   'Protection of Iron and Steel Structures from Corrosion' (British Standards Code of Practice, CP *2008* (1966)).

'Painting of Buildings' (*British Standards Codes of Practice*, CP *231* (1966)).

'Cleaning and Preparation of Metal Surfaces' (*British Standards Institution*, CP *3012* (1972)).

'Painting: Iron and Steel' (*Building Research Establishment Digest 70* (1973)).

Papers.   'Metal Preparation before painting'.[1]

'Electrochemical methods of assessing the corrosion of painted metal'.[2]

'A study of the formation and the protective properties of zinc phosphate coatings'.[3]

'Corrosion Protection of steel highway bridges; full-scale trials of paint systems'.[4]

See also lists of French specification for protection by metallic coats and paints.[5]

'Corrosion under Organic Coatings'[6]

1.   J. E. O. Mayne and D. J. Mills, *J. Oil Colour Chem. Assoc.* 1975, **58**, 155.
2.   J. W. Holme, *Corr. Sci.* 1973, **13**, 521.
3.   C. Kosarev, *Werkstoffe u. Korr.* 1974, **25**, 327.
4.   R. R. Bishop, *Brit. Corr. J.* 1974, **9**, 149.
5.   *Corrosion (Paris)* 1973, **24**, 193, 255 (2 articles).
6.   E. L. Koehler, *Williamsburg Conf.* 1971, p. 117.

Metallic Coatings and Metal-pigmented Paints

Preamble

Picture presented in the 1960 and 1968 volumes. Chapter XV of the 1960 book opens by presenting the procedures available for applying metallic coatings, which are more diverse than those used in painting. There are *dry processes*, including the procedures based on (1) dipping into molten metal, (2) heating in metallic powder, (3) deposition from vapour, (4) spraying with molten globules which flatten and coalesce on impact and (5) cladding with a veneer of the metal chosen; there are also *wet processes* including (6) electroplating under an external EMF and (7) chemical deposition without any applied EMF. To an even greater extent than in painting, a proper surface condition of the basis metal is important for success; the right choice of pretreatment process is essential.

It might perhaps be thought that, provided that the metal chosen for the covering layer can resist the liquid or atmosphere with which the coated article is to be in contact, further discussion is unnecessary. In practice, however, coatings are often discontinuous, and methods of measuring the number of pores per unit area, and also the fraction of the surface which is left bare, deserve attention. The number of pores in a plated surface diminishes as the thickness increases; but if the plating carries internal stresses, the risk of spontaneous cracking may increase with thickness; thus in some systems it may be impossible to prescribe a thickness which will ensure freedom from porosity and at the same time immunity from cracking. Furthermore, certain plating methods contribute hydrogen to the basis metal, and this may greatly enhance cracking when the plated article is subjected to tensile stress in a corrosive atmosphere—a matter discussed in chapters XI and XVI of the present volume.

Since in practice (owing to porosity, cracking or presence of small areas which have escaped coating), the basis metal and the coating metal may both come into contact with a corrosive liquid, their relative polarity becomes important. If the metal used for the coating is relatively noble, we shall get the dangerous combination of large cathode and small anode, and pitting of the basis may occur at the pores; in some cases, application of a discontinuous coating consisting of a noble metal may lead to more corrosion than if the basis had been left uncoated. If the coating is anodic, it will be preferentially attacked and the basis metal will receive, for a time, cathodic protection; however the destruction of the coating will cause the gap to extend and, unless the system is one which the cathodic reaction deposits a coating on the bare area (e.g. $CaCO_3$ from a hard water containing $Ca(HCO_3)_2$, or a basic salt containing the metal forming the coating), attack on the basis metal will set in sooner or later. This type of protection is generally known as

298 METALIC COATINGS AND METAL-PIGMENTED PAINTS

'sacrificial protection'; but, except in special cases, it is not economically sound to protect a cheap material, such as steel, by sacrificing an expensive one; in most cases where this type of protection at gaps is economically acceptable, it is 'sacrificial' only in the opening period.

Just as a painting system almost always consists of several layers of different paints, so a commercial plating system often comprises layers of different metals. Early in the century, it was considered sufficient to provide steel with a coat of nickel to prevent rusting; the rusting was indeed prevented, but in a moist atmosphere containing sulphur dioxide, the nickel itself underwent a change known as 'fogging', and lost its reflectivity owing to a film of basic nickel sulphate. It was then found that a much thinner layer of chromium deposited on the nickel largely protected it against fogging, although internal stresses in the chromium were found to cause cracking; in some cases, it has been found advantageous to encourage this cracking, since a network of fine cracks provides a larger anodic area (and hence less intense corrosion) than just a few isolated cracks. Endeavours by the plating industry to avoid the cost of polishing by using baths capable of depositing bright layers of nickel raised fresh problems, since the bright deposits were, in effect, less noble than the old-fashioned matt or semi-matt plating. The challenge has been met in different ways; in one relatively simple system, a semi-matt layer of nickel is first deposited, covered with the bright layer and finally the thin outer coat of chromium. The Ni/Cr cell set up at breaks in the chromium coat causes anodic attack on the bright nickel layer; this, however, does not quickly penetrate to the steel basis, but tends to turn sideways, because the lowest coat of non-bright nickel is noble in relation to the bright layer. In some cases, more complicated systems (sometimes involving copper as an intermediate layer) are used, but sufficient has been said to indicate the principles involved.

Developments during the period 1968–1975 and Plan of the new Chapter. No very spectacular developments have been noted in the period under review, but a large amount of detailed work has been carried out which, if properly applied, should improve protection. The Chapter starts with a main section dealing with coatings of aluminium and zinc on steel. Sprayed coatings are first considered, and then zinc coatings obtained by hot-dipping, which in terms of area covered represent the most important protective method today. Sherardizing, galvanealing, electrodeposition and vapour phase deposition receive brief treatment; zinc-rich paints are discussed at greater length. Tin coatings, as used in connection with canning, are next discussed, with special attention to efforts now being made to improve the quality of tin-plate; there is brief mention of protection by coatings of intermetallic compounds containing tin. The subjects of nickel and chromium plating are necessarily treated together, and attention is directed to the possibilities of chromium deposition from the trivalent state. Two forms of micro-discontinuous chromium, the use of which reduces the penetration of pits into the underlying nickel, receive attention. Finally there is a discussion of the deposition of precious metals, which are used considerably for electrical contacts.

Coatings of Aluminium and Zinc on Steel

Sprayed Coatings. The use of aluminium spraying for the protection of steel has been discussed by Scott.[1] The best protection is obtained when the purity of the aluminium used is 99·5%. No consistent improvement is obtained by adding 1% or 5% Zn or 5% Mg. It is stated that powder pistols and wire pistols give similar results. No marked advantage has been found through conducting the spraying in a non-oxidizing atmosphere. Long term protection—up to 15 years in a marine atmosphere—has been provided to steel by a thin Al-spray coating (0·003 inch or 0·076 mm). In the case of Al-sprayed steels, the early appearance of slight rust-staining does not necessarily mean that serious deterioration is imminent. (With Zn coatings, the situation is different; when once rust appears on a Zn coating, subsequent deterioration is likely to be progressive.)

Bailey,[2] discussing Scott's results, states that they confirm those of Hudson and Stanners and the conclusions of the American Welding Society; the latter show that coatings of Al 0·003 inch thick are unaffected after 12 years' exposure to severe marine and severe industrial exposure (the same thickness of zinc permits rust to form after 12 years). If the sprayed Al coat is sealed with vinyl lacquer or covered with paint, protection may last as long as 20 to 30 years. An example is quoted of a severe atmosphere within a steelworks at Sheffield producing no corrosion after 34 years. Blistering may occur with high purity Al coatings; but with 99·5% or 99·0%, blistering is avoided.

Sprayed Al coatings have been considered for protecting the faying surfaces of bolted structures. The situation has been discussed by Bullett.[3] In the past the practice has been to leave such surfaces unpainted in the case of high strength material, because it is essential that the coefficient of friction between the surfaces shall be high. Doubts have been expressed as to whether this practice is wise, since, if the fit is not tight, there is a possibility of corrosion. Steel joints made 4 to 10 years ago, recently examined, show that unprotected surfaces had become brown with rust, although the attack was serious only if the joint had been a bad fit. Al coatings improved the situation; tests designed to compare seven methods of treatment showed that sprayed Al metallization gave the best results; sprayed Zn was only slightly inferior.

Tests carried out by Stanners[4] at Lighthouse Beach, Lagos, on specimens sloping at 45° and facing south, showed that sprayed Al coatings successfully prevented corrosion for $8\frac{3}{4}$ years. Zn applied by hot-dipping (50 μm thick) was fairly effective, but sprayed zinc (48 μm thick) was less successful. Specimens coated with Pb started to develop rust-stains within a year. Stanners[5] has also described tests at other sites; these included an extensive set of metal-sprayed steel specimens, some carrying an outer paint coat and some unpainted. At Shoreham, where conditions are marine and industrial, sprayed Al behaved generally better than sprayed Zn; spraying with Zn–Al

1. D. J. Scott, *Trans. Inst. Met. Finishing* 1971, **49**, 111.
2. J. C. Bailey, *Trans. Inst. Met. Finishing* 1971, **49**, 174.
3. T. R. Bullett, *Paint Technology* 1970, **34** (12) 12.
4. J. F. Stanners, *Brit. Corr. J.* 1971, **6**, 211.
5. J. F. Stanners and K. O. Watkins, *Brit. Corr. J.* 1969, **4**, 7.

alloys gave promising results. Among the paints used, micaceous iron ore performed better than titania (but the former was apparently applied as a thicker coat). At Stratford the exposure station was situated in a railway triangle, where conditions had become less severe owing to electrification of the railways; here all metals used for spraying seemed to behave rather similarly, lasting 5 years in the unpainted condition. At Osu (a tropical surf station) all the Zn coated specimens failed within $4\frac{1}{2}$ years, whilst those sprayed with Al performed well, especially when finished with thick micaceous iron ore paint. Exposure to the weather after priming usually had an adverse effect.

Further information regarding the behaviour of sprayed coatings on bridges, chimneys and gas cylinders has been provided by Sheppard.[1]

Twelve-year tests on metallized steel are described by Fenton.[2] In mild or moderately salt air, Zn has behaved extremely well; under severe marine conditions, it has corroded rather badly. The air-borne contaminants of urban and industrial atmospheres have promoted corrosion when retained by the porous surface; this has occurred with unsealed panels, and even with sealed panels in cases where ultraviolet light has degraded the sealer. In comparative tests with equivalent amount of sealer, Zn has remained cleaner and brighter than Al. Unsealed Al at Kure Beach (80 ft from the sea) has developed blisters, but they have not increased in number in recent years, nor do they interfere with the protection of the steel. All types of Al coatings tested appear to provide good protection at each of the sites. Under conditions of complete immersion in sea-water, sealed Al generally provides excellent protection, but sprayed Zn is not suitable for immersed conditions in sea-water.

In exposures up to 2 years Bonner[3] has drawn tentative conclusions regarding the filling up of pores in sprayed coatings by corrosion product. He states that the mechanism of protection is a combination of pore-filling by corrosion product and, in varying degrees, cathodic or sacrificial protection. The protection afforded by electrodeposited Zn–Fe alloy coatings is apparently connected with the formation of an adherent layer of corrosion product. A premature failure of hot-dipped Al coatings was noted on some specimens; this is attributed to the heterogeneous character of the outermost layer.

Hot-dipped Coatings.   It is well known that Al, up to 0·2%, is added to galvanizing baths to increase fluidity and suppress the formation of Zn–Fe alloy below the main Zn coating, so that bending without crack formation becomes possible. Preliminary tests by N. and P. Dreulle[4] suggest that such addition causes susceptibility to intercrystalline attack and that Pb, if present, increases the tendency.

However, the character and performance of the coating depend not merely on the composition of the Zn bath but also that of the steel basis. Information about the effect of constituents on adherence to wire has been made available.

1. J. A. Sheppard, *Anticorrosion* 1972, **19** (7) 8.
2. E. A. Fenton, *Amer. Welding Soc. Report*, AWS Feb. 11, 1967.
3. P. E. Bonner, *Amsterdam Cong.* 1961, p. 751.
4. N. and P. Dreulle, *Rev. Met.* 1968, **65**, 515. *Corrosion (Paris)* 1970, **18**, 503.

Gladman[1] states that C and P accelerate the alloying reaction generally, but render the brittle layer (described as the Γ phase) discontinuous—which results in an improved adherence. Si decreases the thickness of the alloy layer formed after short periods of immersion; this also improves adherence on low-carbon steels. Vyse and Jones[2] state that galvanized coats on tubes of rimming steel containing less than 0·2% P generally failed the British Standard adhesion tests, but that larger amounts of P render the brittle γ-layer either absent or discontinuous, so that the adhesion became good. On killed steel, it seems likely the Si may play the same role as P.

Jackson[3] has compared the five methods of applying zinc coats. Of these, hot-galvanizing is the most important; sherardizing is useful where dimensional tolerance is small, e.g. in threaded work; electrogalvanizing is suitable for many articles used in kitchen and bathroom; spraying is applicable to components too big to enter a treatment vessel, whilst zinc-rich paints are useful for on-site work. A table of Dutch origin provides comparative costs of galvanizing and three different painting systems (the costs of pickling or grit-blasting are included). The figures suggest that painting can be more costly than galvanizing and provides protection for a shorter period. However, all such cost calculations must vary with the time and location. In extremely severe industrial conditions, e.g. in Sheffield, galvanizing gives protection for only 5 years; here Jackson suggests galvanizing followed by painting; preparation of the galvanized surface before the application of paints is needed. It is stated that the Dutch and Swiss railways have used galvanized steel for decades, and that the German railways are now replacing painted steel by galvanized material.

In Michigan, which has 11 000 miles of major highways, the authorities, after making studies of cost, started in 1963 to call in all the painted guard and bridge rails (previously ungalvanized), sending them to a galvanizing plant and re-installing them in the zinc coated condition; the cost was almost the same as repainting on the site would have been. A similar change from painting to galvanizing has been carried out in New York State and elsewhere. On a bridge situated between Montreal and Quebec, it was decided, after cost studies, to galvanize the entire steel-work, including the faying surfaces. This was carried out in 1963, and it is hoped that the galvanized surface will last at least 30 years without maintenance. It appears that in this case a zinc coat of $2\frac{1}{2}$ to 3 oz/ft² was applied, and it must be recognized that such a thickness is not easily obtainable. The Swedish State Power Board find that by using silicon-killed steel, an increased thickness, which is necessary for tower footings and advisable for inaccessible upper sections, can be obtained. The silicon-killed steel is now used for the construction of the complete tower. By careful choice of steel composition, of time and of immersion temperature, the thickness of the coat can be trebled.

Jackson provides details of the design of treatment baths, and gives the

1.   T. Gladman, B. Holmes and F. B. Pickering, *J. Iron Steel Inst.* 1973, **211**, 765.
2.   R. E. Vyse and D. Jones, *J. Iron Steel Inst.* 1973, **211**, 693.
3.   D. J. Jackson, *Metal Finishing J. (London)* 1970, **16**, 145.

sequence of operations. Porter[1] furnishes information about the different processes, which agrees well with that of Jackson. The life of a coat is roughly proportional to its thickness, and the following figures convey typical results:

Sprayed Zinc	$100\mu m$	20 years
Galvanized Structures	$100~\mu m$	20 years
Galvanized Sheet	$20~\mu m$	4 years
Zinc-rich Paint	$20~\mu m$	4 years
Zinc Plated	$12\mu m$	$2\frac{1}{2}$ years
Thinly Zinc Plated Sheet	$3~\mu m$	$\frac{1}{2}$ year
Sherardized	$25~\mu m$	6 years

Porter also discusses the painting of galvanized surfaces. If the paint is to be applied to a new unweathered surface, it is necessary to obtain a bond by means of phosphating, chromating or an etch-primer. Many types of paint provide adhesion in the initial stage, but this is not maintained, owing to the formation of a friable zinc soap at the interface. Formic acid, produced by some paints, can attack zinc, and crystals of zinc formate are often found in situations where paint has become detached; this occurs especially with linseed oil, alkyds and epoxy esters as vehicles.

Useful information about Zn coatings is provided by van Eijnsbergen.[2] Hot-galvanizing is the method most used; dipping in molten zinc at 445–465°C (or at 525–545°C if ceramic pots are available) is recommended. Continuously galvanized coils and sheets take up 50–80% less zinc than 'job-galvanized' steel; silicon-killed steels acquire thicker coatings than unkilled steel, whilst surfaces blasted with sand or grit acquire heavier coats than pickled surfaces. Since the period of protection conferred by the coat is proportional to thickness, these differences are important. In rural climates, a highly protective hydrated zinc carbonate containing 13–16% carbonic acid is formed in about 6–18 months, provided that condensed water evaporates quickly; at recesses where evaporation does not easily occur, voluminous products which are more soluble are found; these contain only 1–4% carbonic acid, and corrosion of the outer coating of unalloyed zinc proceeds at 20 to 30 times the normal rate. Under favourable conditions, however, the corrosion rate of galvanized steel may be only 1/30 or 1/40 that of ungalvanized steel. In marine climates, the corrosion rate increases with the amount of Cl^- present and also with the number of days per year on which the humidity exceeds 90%. Under conditions where Zn alone would not provide the desired life, a procedure is recommended, in which the steel is first galvanized and then receives 1 to 3 coats of paint. In one system, a modified epoxy resin paint is applied immediately after galvanizing to the still hot surface. It is stated that articles coated by this method can be transported, stacked outdoors or assembled, without major damage to the coat.

1. F. C. Porter, *Brit. Corr. J.* 1969, **4**, 179.
2. J. F. H. van Eijnsbergen, *Second South African Corrosion Conf.*, March 1972.

Exposure tests at Heligoland[1] provide some information about the protection to be obtained from Zn coatings produced by spraying or hot-dipping, with or without a covering coat of paint; specimens were exposed for about nine years under (1) fully immersed conditions, (2) alternate immersion, and (3) in the 'splash-zone'. Steel specimens carrying a 100 μm coating of sprayed zinc started to show rust spots after 2 years in the splash-zone; after 4 years about half the area was rusty. In the other two zones protection lasted longer, but the conclusion was reached that even a coating of 150 to 180 μm thickness would provide a life of only about 6 years. It is recommended that inaccessible parts of a structure protected with sprayed Zn should have an outer coat of paint. Hot-dipped specimens carrying a coat 85μm thick developed a dark grey layer after 3 years in the splash-zone which in the following year started to blister locally. Numerous rust spots appeared after 7 years. Methods of preparing the surface before painting were studied; application of a wash-primer gave results less favourable than those obtained with degreasing in trichlorethylene; chromate treatment produced some improvement, but the best results were obtained when a ground-coat of zinc dust epoxy resin was followed by a tar-epoxy covering coat.

Advice regarding choice of procedure to give economical protection is provided in a recent publication.[2] An important use of galvanizing is for protection of steel against supply waters. It is well known that galvanized pipes resist many types of hard water, forming a protective deposit, but that waters from chalk soils which contain large amounts of free CO_2 give rise to complaints of loose sandy deposits. Campbell[3] states that the sandy deposit consists of zinc carbonate (smithsonite), whereas the adherent protective layer is a basic carbonate. Doubtless the sandy carbonate is produced out of contact with the metal, and thus could not protect, whatever its structure; but it is likely that the structure of the basic carbonate is specially favourable to protection.

Sherardizing and Galvannealing Processes. Price[4] provides an interesting history of the Sherardizing process, originally introduced in 1901, but steadily improved by modification of furnace design and control methods. The need for a clean dry surface free from scale and rust is emphasized. Sherardizing provides the hardest coat of all the processes available and is well suited to parts liable to abrasion.

The formation of an alloy layer by galvannealing is discussed by Smith and Batz.[5] The time required to alloy the coating completely in the temperature range between 450° and 650°C increases with the amount of Al in the molten Zn bath. Alloying time is longer on rimming steel than on steel killed with Al. This is probably due to the larger amount of 'free nitrogen' in the rimming

1. E. Brauns and W. Schwenk, *Stahl u. Eisen* 1967, **87**, 713.
2. 'Galvanizing Guide', published jointly by the Zinc Development Association and the Galvanizers Association. See Appendix 5; also remarks in *Brit. Corr. J.* 1972, **7**, 54.
3. H. S. Campbell, *Water Treatment and Examination* 1971, **20**, 11, esp. p. 16.
4. G. C. Price, *Engineering Materials Design* 1972, **15** (5) 435.
5. H. Smith and W. Batz, *J. Iron Steel Inst.* 1972, **210**, 893.

steel; if the nitrogen is removed by treatment in dry hydrogen for 72 hours at 677°C, which reduces the N-content from 0·005% to 0·001%, alloy formation becomes more rapid and the time required is comparable to that needed in the case of Al-killed steel.

Electroplating. Bright Zn plating from a pyrophosphate bath is described by Domnikov;[1] beside sodium pyrophosphate the bath contains glue, sodium silicate and thiourea. Tests in 3% NaCl show that the corrosion resistance of the bright coatings is 3 to 4 times that of matt coatings, also obtained from a pyrophosphate bath. Anodic testing in Na_2SO_4 containing $K_3Fe(CN)_6$ shows that pores are absent from coats obtained from a pyrophosphate bath, but are present in those deposited from a zincate or acid bath.

The comparative value of zinc and cadmium plating has been considered by White.[2] The idea that Cd is always superior is regarded as mistaken; Zn generally gives a better performance than Cd in urban and industrial atmospheres. Cd will perform better than Zn in very humid tropical conditions. Its toxicity is a reason for avoiding it wherever possible. Tin provides coatings which are themselves resistant, but do not provide good protection against the atmosphere owing to porosity and the fact that Sn is generally cathodic to steel. The deposition of Sn on the top of Cd improves appearance, solderability and corrosion resistance in some slightly acid natural waters.

Interest is being shown in the deposition of magnesium and beryllium from organic baths. Brenner and Sligh[3] point out that whereas the deposition of Al from a hydride–ether system require the expensive LiAl hydride, Mg can be plated from a cheap Grignard reagent. It is thought that at some future date coatings of Mg may replace Zn. Methods of depositing Be are discussed by the same authors.

Deposition from the Vapour Phase. 'The technique of vapour phase reduction or decomposition as a means of producing coatings continues to find increasing application. . . . Methods have now been established for the satisfactory deposition of molybdenum and rhenium coatings.'[4] Cr is often deposited from the vapour phase. Wakefield[5] describes its deposition on a niobium alloy. A mixture of HCl and argon passes through a bed of heated Cr and takes up $CrCl_2$ by means of the reaction

$$Cr + 2HCl \leftrightharpoons CrCl_2 + H_2$$

proceeding in the left-to-right direction. Further hydrogen is then added and the mixture produced is passed at a higher temperature over the substance to be coated. Under the changed conditions, the reaction proceeds in the right-to-left direction, and Cr is deposited.

Zinc-rich Paints. Drisko[6] has compared the use of inorganic and organic paints rich in zinc. The inorganic coatings based on sodium silicate

1. L. Domnikov, *Metal Finishing (Westwood, N.J.)* 1969, **67** (5) 70.
2. P. E. White, *Met. Finishing J. (London)* 1968, **14**, 192.
3. A. Brenner and J. L. Sligh, *Trans. Inst. Met. Finishing* 1971, **49**, 71.
4. *Fulmer Research Inst. Newsletter*, No. 10, June 1970.
5. C. F. Wakefield, *J. electrochem. Soc.* 1969, **116**, 5.
6. R. W. Drisko, *Mat. Prot.* 1970, **9** (3) 11.

provide less than 2 years' protection under conditions prevailing in the US Navy, unless a top coat is applied; it is necessary to remove the last traces of the curing agent (generally an acid) before applying the top coat. Inorganic zinc-rich paints carrying a top coat are extensively used in the atmospheric zone on steel off-shore platforms.

Organic zinc-rich coatings are usually tough and have reasonably good resistance to abrasion—although this is somewhat less good than that of inorganic coatings. The application of a top coat is usually easier on an organic than on an inorganic paint.

Zinc-rich coats are often applied by spraying; a pot-agitator is needed to keep the zinc dust in suspension.

Much information about zinc-pigmented paints was provided in an international conference held in 1969.[1] Further data regarding the varieties of zinc dust then available can be found in a brochure.[2]

Another comparison between the inorganic and organic types of paint is provided by Nicholas.[3] He states that a 200-mile pipe-line in Australia coated in 1935 with inorganic zinc-rich paint and cured by heat was still in good condition in 1969; he recognizes, however, that there are objections to heat-curing. Organic zinc-rich coatings have gained much favour through their less critical requirements for surface preparation and their ability to adhere to organic films which may be covering adjacent sections. The application of top coats is easy, and they can be applied by brush—which allows for touch-up and repairing procedures. Against this the organic type is less resistant to high temperatures; the inorganic type can be used at temperatures approaching the melting-point of zinc.

Regarding the relative merits of the two types, Dahlke[4] states that in a marine atmosphere coats of inorganic primers have a longer life than those of organic; far from the sea there is little difference. For submersion in fresh water some or the organic types have longer lives. In general, where no top coat is to be used, an inorganic paint may be preferred, but if a solvent-based top coat is to be applied, an organic zinc-rich primer may sometimes be better.

Further information regarding zinc-rich primers is provided by Bayliss and Wall,[5] who emphasize that the behaviour depends on the zinc content. If this falls below 92%, rust appears more quickly. At 92·5% Zn, the two-pack epoxy types are superior to single-pack types.

It might be mentioned, however, that, although where the paint consists solely of metallic Zn and vehicle, 92% of Zn is needed, this is probably due to the fact that a smaller amount of Zn would imply an excessive amount of vehicle. If a third ingredient is permitted, it should be possible to reduce the zinc content well below 92% by weight. It is unfortunate that some of those

1.   International Conference on Protecting Steel with Zinc-rich Paints, 1969. Papers by T. R. Bullett, C. O. Munger, S. Sakemi, J. N. Grove, J. C. Moore, N. Lamme and H. Spellrink.

2.   'Technical Notes on Zinc Dust', published by Zinc Development Assoc. 1969.

3.   L. J. Nicholas, *Mat. Prot.* 1969, **8** (11) 11.

4.   C. A. Dahlke, *Mat. Prot.* 1971, **10** (9) 25.

5.   D. A. Bayliss and D. C. Wall, *J. Oil Colour Chem. Assoc.* 1968, **51**, 792.

who have drawn up specifications have chosen to prescribe the *minimum* quantity of *zinc* by *weight*, whereas it would have been more logical to prescribe the *maximum* quantity of *vehicle* (left after evaporation of the volatile solvent or thinner) by *volume*. The aim should always be to ensure that a zinc particle is not separated from its neighbours by a thin film of vehicle, which would prevent the mixture as a whole from being an electrical conductor and thus interfere with electrochemical protection at a scratch-line or other gap where the steel basis is exposed. Calculations based on probability principles suggest that a considerable proportion of the zinc present in classical zinc-rich paint could be replaced by a non-conducting substance, and this suggests the possibility of developing a paint which would be cheaper, lighter and perhaps more easily applicable than the classical paint and yet retain the power of gap protection. The first attempts to replace zinc were based on the belief that an electrically conducting substance was needed. Good gap protection was obtained with mixtures containing Cd (1968 book, pp. 393–425), but Cd is more costly than Zn and is moreover highly toxic; it is unlikely that paints containing Cd will come to be widely used. Later attempts (partly unpublished) have introduced non-conducting substances; gap protection is indeed achieved—as predicted by theory; but the presence of such substances as silica and alumina reduce weldibility. It has been shown by work at Niagra Falls,[1] that considerable quantities of Fe_2P can be introduced into metallic-zinc paints and yet preserve protection at a score-line and also avoid welding difficulties. It should be noted that Fe_2P is an electronic conductor and some authorities may attribute the good gap protection to that fact. It is, however, hardly lighter than metallic zinc, although presumably it would be cheaper if produced on a large scale. Time will show whether a paint of this sort fills a need.

Unpublished work by Taylor shows that it is possible to obtain protection at gaps with a paint containing a large amount of non-conducting pigment material, so that the content of metallic zinc per unit volume of paint is only about half that of classical zinc-rich paint. However, the formation of rust, although prevented for a period, starts sooner than occurs when the classical paint is used; probably rusting starts when the metallic zinc becomes locally exhausted, and this will occur quicker with paint in which the zinc content is halved, so that the zinc per unit area is halved. If two coats of this paint are applied, the results obtained appear to be comparable to those obtained with one coat of the classical paint; but there is then no overall economy of metallic zinc. Probably the most hopeful way of obtaining such economy will be to reduce the rate of attack on the metallic zinc to the lowest value consistent with protection of the steel exposed at a gap. Such a possibility is now being explored.

The choice of top coat to cover a primer pigmented with metallic Zn is all-important. Keane[2] points out that, in the absence of a top coat, a zinc-rich primer cannot be used in environments with pH below 6 or above 11. The open structure of a zinc primer allows a considerable choice of top coats.

1. V. P. Simpson and F. A. Simko, *J. Oil Colour Chem. Assoc.* 1973, **56,** 491.
2. J. D. Keane, *Mat. Prot.* 1969, **8** (3) 31.

However, in the case of organic zinc paints, the top coat should have a composition related to that of the primer; over inorganic zinc-rich paints, numerous covering coats based on vinyl, epoxy, chlorinated rubber, acrylic or coal-tar vehicles have been used.

It has been found possible to deposit zinc-pigmented paints by the electro-painting system. Since the coat of metallic Zn thus obtained is itself an electrical conductor, a second coat of non-conducting paint can be applied upon it. Thus a complete painting system can be carried out electrically.[1]

Alexander[2] states that the use of zinc dust in paints doubled between 1957 and 1962, and more than trebled in 1967. The main use is now in the zinc-rich type, and much of the demand is connected with the protection of the under-bodies of cars from de-icing salt.

Tin Coatings

Difficulties of tin-plating. Early in the century tin-plate was made by passing steel sheet through molten tin, but the shortage of tin during the second world war led to the general adoption of electrodeposition, which allowed thinner coatings to be obtained (1960 book, p. 646). Deposition from a potassium stannate bath, although somewhat slow, produces a better coating than is obtained from a stannous sulphate bath. Conditions must be maintained under which the tin anode dissolves solely as Sn^{4+}; if any Sn^{2+} is formed, there may be disproportionation to $Sn + Sn^{4+}$, leading to a spongy, non-adherent deposit. Another trouble is the danger of the anode becoming passive if the CD becomes excessive. One way of obtaining quick deposition without risk of passivity is to use a very large anodic area, but this is obviously inconvenient. Lowenheim[3] many years ago discovered that the presence of Al in the Sn anode allows a much higher CD to be used—apparently by modifying the character of the anodic film. Recent work at Sheffield under Gabe[4] shows that the optimal Al content is 0·8% for potassium stannate (1·0% for sodium stannate); the CD needed for passivity, which is only 0·9 mA/cm^2 with pure tin, can be raised to 20 mA/cm^2.

Corrosion of Tin Cans. An important factor affecting the life of cans is the presence of nitrates. These have come to be used as curing agents for bacon.[5] However, certain vegetables also introduce nitrates. Rapid detinning by carrots and green beans, due to a high nitrate content, has been reported[6] (in the case of spinach, oxalates may be more important in the attack upon the tin). The pH is critical in determining the behaviour of nitrates; during the storage of green beans the pH may drop, finally reaching a level favourable to corrosion. This explains the apparently capricious action of nitrates.

1. Circular No. 68/3023 (Zinc Development Association and Lead Development Association).
2. J. R. Alexander, *Amer. Paint Journal*, Oct. 22, 1968.
3. F. A. Lowenheim, *J. electrochem. Soc.* 1949, **96**, 214.
4. D. R. Gabe and P. Sripatr, *Trans. Inst. Met. Finishing* 1973, **51**, 141.
5. A. C. Hersom and E. D. Hulland, 'Canned Food' (Churchill), 1969, p. 211.
6. *Tin and its Uses* 1970, No. 83, p. 11.

It is stated that two different varieties of beans differ in their take-up of nitrates; but the time of harvest, and the nature of the crops preceding the beans, also exert their influence.[1]

Sherlock and Britton[2] find that anionic surface-active agents create a negative shift of potential and inhibit the corrosion produced by nitrates, whilst cationic agents have the opposite effect; sodium lauryl sulphate produces strong inhibition at 5 ppm, and almost suppresses the effect of nitrate at 30 ppm.

The attack upon tin by fruit juices depends very largely on the formation of complexes. Sherlock and Britton measured the stability constants of various complexes, using an electrochemical method. The constant is highest for oxalic, less high for citric and malic, and rather lower for tartaric acid. The corrosion rates of tin exposed to an oxygen free solution of the acid fall in the same order as the stability constants.

Studies of complex formation have also been carried out by Willey,[3] whose work has covered the behaviour of Sn, Fe and Sn–Fe alloys in various acids and in food media. In order to be protective, Sn must be anodic to Fe; this condition is fulfilled in most—but not all—canned foods under anaerobic conditions. In fruit acids, complexing anions can influence the relative positions of Sn, Fe and the Sn–Fe alloy—and thus the corrosion mechanism.

Pitting of the steel by acid food can be brought about by the presence of certain fungicides which decompose to produce CS_2; this lowers the hydrogen overvoltage of Sn and thus reduces cathodic protection.[4]

An attempt has been made to predict pitting on tin-plate by means of capacitance measurements.[5] The shape of the curve connecting capacitance with potential is correlated with the type of corrosion, which sometimes leads to pitting, and sometimes to de-tinning.

Improvement of Tin-plate quality. In the early stages of corrosion of a can interior by foodstuff, the reduction of oxygen is the main cathodic reaction, and the compositon of the steel basis may have little influence on the corrosion rate. When oxygen is nearly exhausted and hydrogen evolution becomes important, the quality of the steel exposed at pores and scratches affects behaviour, since tin is a poor cathode for hydrogen evolution. Hence the study of the exposed steel area, and a comparison of the cathodic polarization behaviour of a given sample of tin-plate with that of solid tin, can provide useful information. This point, emphasized by Britton, has been investigated in detail; the results should be studied in his paper.[6]

A considerable improvement in the continuity of the alloy layer on electrolytic tin-plate can be obtained by the use of an oxidizing pickle before tintlating; 5% (w/v) HNO_3 is recommended.[7] The increased continuity of the

1. *Tin and its Uses* 1971, No. 90, p. 13.
2. J. C. Sherlock and S. C. Britton, *Brit. Corr. J.* 1972, **7**, 180; 1973, **8**, 210.
3. A. R. Willey, *Brit. Corr. J.* 1972, **7**, 29.
4. P. W. Board, R. V. Holland and O. Britz, *Brit. Corr. J.* 1968, **3**, 238.
5. P. W. Board, R. V. Holland and R. J. Steele, *Brit. Corr. J.* 1972, **7**, 87.
6. J. C. Sherlock, J. H. Hancox and S. C. Britton, *Brit. Corr. J.* 1972, **7**, 222.
7. L. R. Beard, D. R. Gabe and L. M. Warner, *Trans. Inst. Met. Finishing* 1971, **49**, 63.

$FeSn_2$ layer leads to improved corrosion resistance, as shown by the 'ATC test'. This test, which has achieved general acceptance as a measure of alloy continuity, depends on the measurement, after a stabilization period, of the current passing between pure tin and tin-plate which has been de-tinned so as to expose the alloy layer; the liquid employed is a prepared grapefruit juice.

An improvement in the intermediate alloy layer on tin-plate enables thinner coatings to be used. The electron microscope allows the alloy crystallites to be studied, using a shadowing technique. The pattern is different on the {100}, {110} and {111} faces. The crystals formed on the {100} faces grow outwards from the steel surface instead of lying flat upon it, and these naturally provide the poorest coverage of the steel. Research at the Tin Research Institute[1] is being directed to attempts to modify tin-plating conditions so as to produce better coverage of the steel by the alloy layer.

Chromate Passivation of Tin-plate. Treatment of electrolytic tin-plate in dichromate solution, with or without cathodic polarization, improves resistance to corrosion or staining by sulphur products. Britton,[2] examining the surface layer by an X-ray fluorescence method (which measures the amount of Cr irrespective of its form) showed that there are two main constituents, respectively soluble in acid and alkali. The acid-soluble material appears to exist as Cr^0—i.e. in the metallic state. He produced his layers by cathodic treatment between pH2 and 6.

More complicated methods have been used. One Japanese procedure[3] has been described in which tin-plate receives cathodic treatment in a solution which deposits chromium as a compound of tetragonal structure.

Coatings of Intermetallic Compounds. A tin-nickel electroplate possesses several practical advantages.[4] The coating does not tarnish, and, unlike many ductile materials, grows no whiskers. Brass coated with Sn–Ni alloy shows practically no change when exposed to vinegar, fruit juice or eggs. Owing to absence of ductility, it is not suited for flexible surfaces, except when very thin. It should be noted that the NiSn recrystallizes above 300°C to give $Ni_3Sn + Ni_3Sn_4$. Recent work in Hoar's laboratory[5] based on Auger spectroscopy suggest that the oxide-film which confers the greatest corrosion resistance on Ni-Sn alloys (which resistance is best in neutral and moderately acid environments) is a glassy nickel polystannate containing water and OH^-.

Clarke[6] draws attention to the fact that several intermetallic compounds develop passivity which is stable over a wide pH range. These include NiSn, CoSn, $FeSn_2$, Cu_6Sn, Cu_6Sn_5 and SbSn; the last-named is used on Babbitt bearings and its passivity may be beneficial in service. Experiments carried out on bright CoSn deposited from an acid fluoride bath show that the coating is hard and fine-grained; it is completely passive in aqueous media between

1. B. T. K. Barry and C. A. Mackay, *Tin and its Uses* 1971, No. 91, p. 8.
2. S. C. Britton, *Brit. Corr. J.* 1975, **10**, 85.
3. S. Teramae, K. Yamada and Y. Miyoshi, *Corr. Abs. (Houston)* 1968, p. 123.
4. *Tin and its Uses* 1968, No. 78, p. 5.
5. T. P. Hoar, M. Talerman and E. Trad, *Nature Phys. Sci.* 1973, **244**, 41.
6. M. Clarke, R. G. Elbourne and G. A. Mackay, *Trans. Inst. Met. Finishing* 1972, **50**, 160.

pH 1·4 and pH 14, but dissolves in concentrated HCl; it tarnishes at 350°C. Its properties are similar to those of NiSn, but it is probably slightly less protective. Further study would seem to be worthwhile.

A Japanese paper[1] states that SnCo is more ductile than SnNi. Anodic polarization curves obtained on the electrodeposited alloy provide more favourable results than those obtained on the cast alloy.

Nickel and Chromium Plating

Early History. Isserlis[2] has provided interesting information of the development of nickel–chromium plating. He writes: 'In the early days of nickel–chromium plating, the articles to be coated had to be polished by hand. In the case of steel articles a thick initial layer of copper was often deposited first because, being a soft metal, copper was easier to polish to a mirror-bright finish than steel. This was followed by a layer of nickel from a Watts bath. which again had to be mechanically polished to a mirror-like finish. Finally a very thin layer of chromium was deposited, which then also had to be finished by buffing. The whole procedure was lengthy, tedious and therefore costly. Early innovations were the development of various semi-bright nickel solutions, which very much later reached a high standard of sophistication.

'In the late 1930s, semi-bright plating processes were developed in the USA and the UK by Weisberg and Hinrichsen respectively, by the co-deposition of a nickel–cobalt layer. The process was fully commercialized and satisfactorily operated for a long time. This was followed in the 1940s by the development of fully bright nickel coatings by the incorporation of specific sulphur bearing organic compounds into the bath. The resultant deposits were bright but very brittle. This brittleness was overcome by the addition of other organic compounds which were capable of counteracting this deleterious effect without diminishing the degree of brightness of the coating. It was also found that the corrosion resistance of bright nickel coatings was reduced by the incorporated sulphur, and ways and means had to be found of overcoming this. The answer was found in the duplex nickel processes, in which a relatively thick, corrosion resistant semi-bright coating is applied first, followed by a thinner fully bright one, and a very thin superstrate of chromium. The corrosion is brought to a virtual halt at the interface between the bright and semi-bright nickel coats, since the bright nickel is anodic to both semi-bright nickel and chromium. In addition, the early bright nickel baths had a tendency to produce pitted deposits. This was overcome by the incorporation into the baths of anti-pitting agents. Finally, the necessity to start with a mirror-bright substrate surface was overcome by the development of yet other organic compounds with levelling characteristics. . . .

'Considerable developments have also taken place in improving the performance of chromium deposits, which culminated in the early 1960s in the introduction of micro-cracked and micro-porous chromium. These coatings

1. Y. Tsuji and M. Ichikawa, *Corrosion (Houston)* 1971, **27**, 168.
2. G. Isserlis, *Trans. Inst. Met. Finishing* 1973, **51**, 1.

ensure a longer life for the duplex undercoat by more evenly spreading the corrosion of the bright nickel, thereby avoiding deeper local penetration.'

Recent Developments. Much information is provided by Dennis and Such.[1]

The effect of thiourea on the corrosion resistance of nickel plating has been studied by Roffey and Shreir.[2] Brighteners and levellers containing sulphur provide a deposit incorporating that element and showing lowered corrosion resistance. New experiments indicate that at an optimum concentration of 10^{-4}M, thiourea produces a minimum value for the anodic over-potential and the same concentration has a maximum effect on the cathodic plating. It is thought that undecomposed S compounds inhibit dissolution, presumably by providing a chemisorbed layer, but the decomposition products may behave differently. Sulphide anions appear to facilitate the escape of Ni^{2+} ions. Tafel curves show that H_2S reduces the activation energy needed for anodic attack; thiourea, if undecomposed and present in sufficient quantity, appears to nullify the effect of H_2S.

Micro-levelling has been studied by Zak,[3] who suggests explanations which the reader should consider.

The corrosion aspects of certain electroless nickel coatings is discussed by Andrew and Heron.[4] The phosphorus content was found to range between 9·7% and 12·5%. The method of preliminary cleaning can affect the properties of the coating, especially the porosity. There seems to be little to choose between three proprietary processes, if the same post-coating heat-treatment, is applied. The heat-treatment, however, greatly affects the protective behaviour. Application of coatings 4 μm thick can prevent corrosion and staining of steel; the salt droplet test suggests that 3 hours at 175°C provides good protection, whilst 1 hour at 400°C allows rusting. Properly applied electroless coatings prevent damage from finger markings during warm damp weather. Outdoor exposure tests and tests with salt droplets show that the protection of steel is increased as the coating thickness is increased up to 50 μm; beyond that thickness improvement is small. There seems to be some difference of opinion as to whether electroless deposits give better or worse protection than those obtained from a Watts plating bath.

Chromium deposition from the trivalent state. The ordinary method of chromium plating uses a bath containing hexavalent Cr; early attempts to start from the trivalent state produced only coarse or otherwise unsuitable deposits. More recently a bath with a trivalent Cr compound dissolved in a 10% v/v mixture of dimethylformamide (DMF) and water has produced a good deposit, which is micro-porous at 0·25 μm and becomes fully micro-cracked at 1·25 μm thickness. Preliminary static exposure tests showed that in industrial, marine and rural atmospheres, DMF chromium was superior to

1. J. K. Dennis and T. E. Such, 'Nickel and Chromium Plating' (Butterworth), 1972.
2. C. G. W. Roffey and L. L. Shreir, *Trans. Inst. Met. Finishing* 1967, **45**, 206.
3. T. Zak, *Trans. Inst. Met. Finishing* 1971, **49**, 220.
4. J. F. Andrew and J. T. Heron, *Trans. Inst. Met. Finishing* 1971, **49**, 105.

regular chromium; it was possible to reduce the nickel thickness, and yet obtain good protection, although dulling was rather rapid on static exposure in severe industrial atmospheres; tests on vehicles showed that there was no dulling under mobile conditions. New outdoor exposure tests, conducted by Carter and Christie[1] at industrial, urban and marine sites, as well as on operating cars, have led to the conclusion that 'it is possible to reduce the nickel thickness when over-plating with DMF chromium, whilst still maintaining the same degree of protection as is achieved under normal nickel thicknesses under regular chromium'. The tests were carried out on zinc based die-castings plated in a bath containing $CrCl_3$, NH_4Cl, boric acid in 40% DMF (pH 1·0) at 25°C. An acceptable life of 9 to 18 months was obtained at industrial sites and over three years at other sites. This compares favourably with the 5 to 15 months obtained from regular chromium, but performance is inferior to the figure (14 to 24 months) obtained with micro-porous chromium deposited from a hexavalent bath when they are both exposed under static conditions at an industrial site. At other sites, performance of the two types is similar; both are superior to regular chromium.

Another method of depositing Cr or Cr–Ni alloys, starting with a trivalent Cr compound, is described by Levy and Momyer.[2] They use hexa-ammine-Cr(III) formate in a 30% (mol) acetamide–formamide mixture. Cathodic efficiencies up to 50% are obtained. By adding suitable amounts of hexa-ammine-Ni(II) formate alloys with Ni contents varying between 0 and 100% can be obtained at will. If the water in the bath exceeds 400 ppm, aquation of the Cr complex occurs and plating efficiency declines; NaSCN improves efficiency.

Alternatives to Nickel. Chisholm[3] has studied the deposition of nickel–chromium alloys, obtaining compositions varying over a wide range. He uses a bath containing $NiSO_4$ and CrO_3 and obtains good adhesion and useful thickness. Co-deposition is possible also from a fluoborate bath.

Another development is the substitution of cobalt for nickel in bright-plating baths. Mathieson and Sedghi[4] state that the system copper/cobalt/crack-free chromium gives better protection to steel than copper/nickel/crack-free chromium.

An interesting project is protection by Ni–Mo alloys.[5] Although there is little prospect of depositing unalloyed Mo from aqueous solution, it is possible to obtain Ni–Mo alloys from baths containing sodium citrate, nickel sulphate and sodium molybdate, buffered to pH 10·5 with NH_3.

Micro-discontinuous chromium. Carter[6] has discussed the effect of thickness on the corrosion resistance of (1) micro-cracked Cr and (2) micro-

1.   V. E. Carter and I. R. A. Christie, *Trans. Inst. Met. Finishing* 1973, **51**, 41.
2.   D. J. Levy and W. R. Momyer, *J. electrochem. Soc.* 1971, **118**, 1563.
3.   C. U. Chisholm, *Trans. Inst. Met. Finishing* 1968, **46**, 147.
4.   R. T. Mathieson and M. Sedghi, *Trans. Inst. Met. Finishing* 1972, **50**, 152.
5.   R. A. E. Hooper, D. R. Gabe and J. M. West, *Trans. Inst. Met. Finishing* 1970, **48**, 182.
6.   V. S. Carter, *Trans. Inst. Met. Finishing* 1970, **48**, 16, 19 (2 papers).

porous Cr. Where a micro-cracked deposit is being used, great thickness is needed to produce high crack density (the object is not to provide greater protection); a minimum of 1.0 μm is suggested. The increased crack density reduces the size of particles detached by attack on the nickel below, and indeed reduces the tendency to become detached at all; it causes small blisters, but on the whole increased thickness causes improved appearance and adds to the effective protection.

Where micro-porous Cr is desired, increased thickness reduces the loss of brightness on static exposure to severe environments; Cu undercoats improve protection.

Another review of these discontinuous Cr plating processes may be welcomed.[1] It shows how micro-cracked and micro-porous coatings can be distinguished, but states that tests have shown no great superiority for one class over the other. The advantage of having many discontinuities is to spread the corrosion current over a large area of exposed Ni, so that the penetration of pits is delayed. In some processes the chromium develops its own crack pattern, e.g. where selenium is present; in others the Cr is deposited on a specially stressed Ni basis—which causes the Cr to develop cracks. The 'Post nickel strike system'[2] shows special promise, here a thin, highly stressed layer of Ni is deposited from a $NiCl_2$ bath containing additives before the Cr plating, which makes it possible to obtain an adequate crack density when the Cr thickness is only 0.3 μm, the thickness usually favoured for decorative purposes; with other procedures, a thickness of 0.8 μm is needed to develop the desired density of cracking. Tests by Carter[3] show that the corrosion resistance proved when the Cr thickness is only 0.3 μm is at least as good as that provided by 0.8 μm produced by other methods.

Turner and Miller[4] discuss the influence of copper either as a layer below the nickel or a layer separating two layers of bright nickel. Cu has little influence when regular Cr is the outer coat, but a basis of Cu below the Ni is beneficial with micro-cracked or micro-porous Cr as the outer layer.

Deposition of Precious Metals

Uses. Precious metals are rarely used for protection against severe conditions, since at any pore the high cathode–anode ratio would cause intensified attack. Although porosity could doubtless be avoided by thick coatings, that would involve great expense. However, these coatings are used extensively in precision electronics for covering contacts. Where switch devices operate at low voltages, a low contact resistance is essential, and gold, in particular, provides this. Recent experimental study of precious coatings has largely been directed towards means of reducing porosity.

Gold. Coatings on reed relays which have failed in service have often been found to be black owing to the presence of carbon; experiments using

1. *Met. Finishing* 1970, **6**, 15.
2. G. C. Postins and J. E. Longland, *Trans. Inst. Met. Finishing* 1971, **49**, 84.
3. V. E. Carter, *Trans. Inst. Met. Finishing* 1974, **52**, 25.
4. P. F. Turner and A. G. B. Miller, *Trans. Inst. Met. Finishing* 1969, **47**, 50.

radioactive tracers show that the carbon has been derived from the cyanide in the plating bath. Work by Holt and Stanyer[1] shows that baths containing a cobalt brightener produce more carbon than those without brighteners. Low bath temperatures favour the uptake of carbon.

Inclusions present on the metallic basis can also lead to porosity. This matter has been studied by Leeds,[2] who points out that if they exactly fitted the pore bottoms, they should prevent corrosion; but in practice this does not occur.

The structure of gold deposits varies with the bath used; under some conditions microscopic spikes are produced.[3] Tests for porosity of gold deposits are discussed authoritatively by Clarke.[4]

The unpredictable character of porosity is emphasized by Campbell.[5] Generally there is less than one pore per cm^2, but occasionally an area will be found where there are more than 10 pores per cm^2. He describes two types of pores:

(1) Small pores of irregular shape which are numerous when the coating is only $0.1-0.2$ μm thick, but disappear as the thickness increases.

(2) Larger pores of roughly triangular shape which are often associated with foreign particles containing silicon—probably derived from buffing compounds; some pores of this type may be associated with pits or other defects in the substrate.

Scanning electron microscopy reveals the nature of the particles, and sometimes discloses the presence of matter containing carbon—perhaps derived by polymerization from cyanide.

Clarke[6] discusses the effect of pre-treatment of a copper substrate on the porosity of gold electrodeposits—particularly the effect of mechanical abrasion, mechanical polishing, scratching, wire-brushing, electropolishing, etching, bright-dipping and chemical polishing. Gold plating is excellent for electronic circuits provided that the gold is free from pores; otherwise sulphides appear on the surface and contact is poor. Smoothing the surface to remove *major* asperities reduces porosity, but brightening, designed to remove *minor* asperities, has little effect. Annealing can reduce porosity by increasing grain-size. In contrast, polishing can increase porosity as the surface becomes smoother.

Leeds and Clarke[7] measure porosity by ascertaining the voltage needed to pass some desired current when the plated specimen is made the anode in a bath in which gold is unattacked, so that the current passes only from the exposed substrate via the pores. The gradient dV/dI represents the electro-

1.   K. Holt and J. Stanyer, *Trans. Inst. Met. Finishing* 1972, **50**, 24.
2.   J. M. Leeds, *Trans. Inst. Met. Finishing* 1969, **47**, 222.
3.   H. Y. Cheh and R. Sard, *J. electrochem. Soc.* 1971, **118**, 1737.
4.   M. Clarke and A. J. Sansum, *Trans. Inst. Met. Finishing* 1972, **50**, 211.
5.   G. L. Cooksey and H. S. Campbell, *Trans. Inst. Met. Finishing* 1970, **48**, 93.
6.   M. Clarke and A. M. Chakrabarty, *Trans. Inst. Met. Finishing* 1970, **48**, 99, 124 (2 papers).
7.   J. M. Leeds and M. Clarke, *Trans. Inst. Met. Finishing* 1968, **46**, 1.

lytic resistance of the pore channels. dV/dI is a porosity index which varies inversely with porosity; a high value indicates low porosity. The test should be carried out on an area of 20 cm². Confirmation of the validity of this test was obtained by comparing the results with a test in a chamber containing a humid atmosphere with 10% SO_2, generated by the action of H_2SO_4 on $Na_2S_2O_3$. Pores are shown up as sharp brown or black spots with very little spreading stain, unless the coating is extremely porous. It may be advisable to carry out both tests, since the electrical test by itself does not distinguish a specimen with many small pores from one with a few big ones.

Electrodeposition of the Platinum Metals. An excellent review of the deposition of the six platinum metals has been provided by Reid.[1] Owing to their limited availability, the platinum metals can present no serious challenge to the pre-eminence of gold, but they offer advantages in respect of wear resistance, stability at high temperatures and resistance to erosion.

Ruthenium is cheaper than the other noble metals used for electrical contacts, and for that reason arouses interest. Plating has not always been easy, but a method[2] depending upon a compound generally written RuNC appears suitable. The composition of RuNC is $(NH_4)_3[Ru_2NCl_8(H_2O)_2]$. It is used in an acidified solution containing either HCl, H_2SO_4 or sulphamic acid, with ammonium sulphamate added to prevent the formation of NCl_3.

Economy in Plating with Precious Metals. The relative advantages of using a proprietary solution or a 'home brew' provide a subject for discussion among platers. Stubley[3] emphasizes the fact that reliability is the criterion. If the choice of the home brew necessitates even a small increase of thickness to obtain this, it may be cheaper in the end to purchase a proprietary solution.

Testing Methods

Thickness. Guidance regarding methods and instrumentation needed in measuring the thickness of coating will be found in a book by Plog and Crosby.[4] Information about methods based on ultrasonic gauges is provided by Sidwell,[5] who emphasizes their advantages over other methods. Studies of the distribution of thickness of sprayed deposit as produced by pistols of different designs are available from a Dutch source.[6]

Porosity. Information regarding testing for porosity has been provided on p. 314 of this volume: also in the 1960 book, pp. 625–32, and the 1968 book, p. 234.

Internal Stresses. Methods for the measurement of internal stresses in electrodeposits are described by Walker.[7]

1. F. H. Reid, *Trans. Inst. Met. Finishing* 1970, **48**, 115.
2. G. S. Reddy and P. Taimsalu, *Trans. Inst. Met. Finishing* 1969, **47**, 187.
3. J. Stubley, *Trans. Inst. Met. Finishing* 1972, **50**, 130.
4. H. Plog and C. E. Crosby, 'Coating Thickness Measurement' (R. Draper), 1971.
5. R. J. Sidwell, *Anticorrosion* 1972, **19** (6) 4.
6. G. H. Douna, A. Th. A. Hartong and P. Selier, *Verfkroniek* 1972, **45**, 167.
7. R. Walker, 'Internal Stresses in Electrodeposits' (*Met. Finishing J.*), 1969.

Further References

Books. A second edition of West's book[1] has appeared, whilst a new volume by Gaidi[2] is intended for 'Electroplaters, Chemists and Students'. A translation of Pavlova's[3] historical survey of electroplating has been prepared.

Papers. 'Protective Metallic Coatings—a Review'.[4]

'Tin diffusion coatings on Steel'.[5]

'Sherardized coatings on iron and steel articles'.[6]

'Why zinc-rich coatings have maintained the lead as anti-corrosive champions'.[7]

'Passivity of electrodeposited Sn–Co alloy'.[8]

'Rapid method for determining the degree of cleanliness of metal surfaces'.[9]

'Corrosion resistant heavy chromium plating'.[10]

'Organic zinc-rich primers effectively prevent under film corrosion'.[11]

'Spontaneous growth of whiskers on tin coatings: 20 years of observation'.[12]

'The rust-preventing mechanism of Zn-dust paints'.[13]

'Tin-plate corrosion in nitrate containing media: the role of nitrite'.[14]

1. J. M. West, 'Electrodeposition and Corrosion Processes' (Van Nostrand Reinhold Co.), 1972.

2. B. Gaidi, 'Electroplating Science' (R. Draper), 1971.

3. O. I. Pavlova, 'Electrodeposition of Metals: a historical review' (Israel Program for scientific translation: US Dept. Commerce), 1968.

4. M. Clarke, *Brit. Corr. J.* 1971, **6**, 197.

5. C. J. Thwaites and E. A. Speight, *J. Iron Steel Inst.* 1973, **211**, 476.

6. *British Standard Institute*, BS 4921 (1973).

7. H. Kredentser, *Canadian Paint Finishing* 1971, **45** (8) 16.

8. Y. Tsuji and M. Ichikawa, *Corrosion (Houston)* 1971, **27**, 168.

9. P. N. Miller, *Mat. Prot. Perf.* 1973, **12** (5) 31.

10. R. W. Ludwig, *Trans. Inst. Met. Finishing* 1974, **52**, 19.

11. W. W. Page, *Met. Finishing*, April 1974, p. 38.

12. S. C. Britton, *Trans. Inst. Met. Finishing* 1974, **52**, 95.

13. T. Theiler, *Corr. Sci.* 1974, **14**, 405.

14. D. Britz and H. Luft, *Werkstoffe u. Korr.* 1973, **24**, 290.

CHAPTER XVI

Intergranular Corrosion and Stress Corrosion Cracking

Preamble

Picture presented in the 1960 and 1968 volumes. It is common knowledge that alloys which consist of a single phase at a high temperature, but of two phases at low temperatures, tend on cooling to deposit the minor phase preferentially at grain-boundaries, where the loose structure favours nucleation; this may cause the part of the major phase close to the grain-boundaries to have a composition different from that of the grain-interiors. In 'unstabilized' stainless steel, the precipitation of particles rich in Cr at the grain-boundaries leaves an adjacent band denuded in Cr; similarly, an Al alloy containing Cu, depositing an intermetallic compound along the grain-boundaries, leaves a band depleted in Cu. In some liquids, the denuded layer may be anodic relatively to the grain body, and in such a case it is to be expected that corrosion will eat its way along the depleted material and finally, perhaps, loosen the grains, so that they fall out virtually unattacked. If the precipitate chances to be more noble than the major phase, this may enhance the attack on the denuded band appreciably, since, even though the addition to the cathodic area is small, the EMF available will be increased and the precipitate is very close to the denuded area which is acting as anode; it may provide a very active basis for cathodic reactions. In any case, the advance of the intergranular attack is likely at first to be rapid, since the dangerous combination of large cathodic area and small anodic area has been established. In general, however, the attack will tend to slow down as it advances along the grain-boundaries, since the resistance of the path joining anodic and cathodic areas steadily grows (the main cathodic region will be the external face of the specimen). A situation relatively favourable to the continuation of attack will exist when, owing to previous rolling, the structure of the alloy consists of thin flattened grains arranged parallel to the surface; in such a material the corrosion will turn sideways and loosen the flattened grains, producing a situation recalling the structure of flaky pastry, with paths left between the flakes wide enough for the passage of current. This is known as *Layer Corrosion*. The more general case where there is penetration between the grains of equiaxed material is known as *Intergranular Corrosion*. If it should happen that the minor phase is itself anodic relatively to the major phase in the corrosive liquid, loosening of the grains becomes possible; this is most likely to be met with if the minor phase forms a continuous network around the grains instead of being precipitated as discrete particles.

If a material susceptible to intergranular attack is held under tensile stress while immersed in the corrosive liquid, the penetration along grain-boundaries may *in some cases* be enhanced. If, for instance, the anodic path (the denuded band, or, in some cases, the minor phase) is discontinuous, so that the advance of anodic attack would be periodically held up, the stress, becoming concentrated on the bridges which are resisting attack, may rupture them mechanically; then the advance can be resumed. In such a case, the advance, plotted against time, should produce a jerky graph. Jerky patterns have indeed been observed by experimenters, and clicks have been detected aurally, which some experimenters have attributed to the breaking of bridges. That interpretation has been disputed; it has been suggested that the jerks are merely due to the sticking of unsatisfactory recording instruments, and that the clicks heard are really due to the bursting of hydrogen bubbles. Such criticism may have been in some cases justified, but it seems unlikely that all experimenters have been deceived in those ways. If two materials are tested carefully on the same machine and under identical conditions, and if one material consistently provides jerks and the other none, it would seem likely that the jerks are a genuine phenomenon. Moreover, the mechanical jerks are often accompanied by electrochemical jerks (sudden fluctuations of potential—presumably caused by the exposure of fresh film-free material whenever a bridge is broken); sudden potential movements were observed in early work (1968 book, p. 261) which deserves more study than it receives today; it is unlikely that they were due to unsuitable instrumentation.

In cases where genuine jerks occur, we must envisage two sets of phenomena occurring during alternate periods:

(1) slow attack on the anodic phase until removal of material has been sufficient to make possible
(2) the mechanical failure of the phase which is not susceptible to anodic attack and which would otherwise hold up the advance of the fissure.

Clearly this alternation of phenomenon will only be possible in materials where there is a favourable geometrical distribution of the various phases (precipitate, denuded zone and main grain body). *Only a few* of the materials which show slow intergranular corrosion in the absence of stress suffer rapid cracking when stress is applied in the corrosive environment.

There are other cases, however, where the advance appears to be steady, without jerks, and is best explained by assuming that the stress intensification at the tip of the advancing crack increases the distance between atoms, so that the energy situation is altered in a sense favourable to the escape of metal as ions; the activation polarization opposing anodic attack will then be reduced, and corrosion at the crack-tip in preference to attack elsewhere will be favoured. Such reduction of activation polarization has, indeed, been demonstrated experimentally in Hoar's laboratory, where also electrical measurements have pointed to a current density consistent with the rate of steady advance of the crack which has been measured geometrically (1968 book, p. 254). This appears to be a satisfactory demonstration of the situation, but difficulty has been felt by some who consider that, if the increased interatomic distance produced by local strain were sufficient to produce localized attack at the tip, pure metals should show stress corrosion cracking. Such an objection has largely been removed by the demonstration (1968 book, p. 265) that pure copper is susceptible to stress corrosion cracking in a liquid containing a high concentration of $Cu(NH_3)_5^{2+}$; it does not occur in a liquid free from copper containing ions. A little thought will show the reason for this distinction; if the liquid is initially free from copper, the production of Cu^+ (or Cu^{2+}) at the crack-tip will produce sufficient concentration polarization to counterbalance the result of the reduced activation polarization due to the greater interatomic distance; thus the effect of applied stress will be negligible. If, however, the liquid already contains a high concentration of $Cu(NH_3)^{2+}$—alike in the crack and on the face outside—the concentration effect can be disregarded, and the reduction of activation energy will show itself.

What is clear is the fact that stress corrosion cracking is a case of conjoint action; it is connected with a number of different mechanisms, and the effects of the corrosion and the mechanical destruction are *not additive*; each helps the other. Stress may remove activation polarization by increasing the interatomic distance, as just suggested; this is a case of mechanical action aiding corrosion. However, a corrosion trench produced along a steeply aligned grain-boundary or a set of pits set on a straight line, can cause stress intensification, and thus allow mechanical breakage to occur at a stress too low to cause damage on an uncorroded surface; that would be a case of corrosion aiding mechanical breakage. Other instances of conjoint action could be suggested.

The situation is further complicated by the fact that often a system which produces stress corrosion cracking is capable of evolving hydrogen. There is little doubt that hydrogen does play a part in many cases of cracking under stress. It can act in at least two ways. Atomic hydrogen in the metal phase will change the mechanical properties, whilst if atomic hydrogen produces molecular hydrogen at a high pressure in internal cavities (and as explained in chapter XI, very high pressures are theoretically possible, and indeed have been demonstrated) this will superimpose an additional stress system on the tensile stress applied from an external source.

Hitherto it has been assumed that the cracking occurs along grain-boundaries. Often in practice it runs across the grains, following planes along which, in absence of corrosion but under high stress, slip would occur. The distinction between intergranular and transgranular corrosion may not always be sharp. Work in Nutting's laboratory (1968 book, p. 264) has shown that attack on brass which appears, with insensitive methods of observation, to be a case of intergranular corrosion, is really directed against dislocations piled up on slip-planes close to the grain-boundaries; this is probably not true of all cases of intergranular corrosion.

Cases of definite transgranular cracking following planes which would be slip-planes if stress was applied in absence of chemical action can arise in different ways. Pits produced at the original surface may deepen into tunnels which penetrate deeply into the metal; probably these tunnels follow paths which are anodic to the main material. If a series of tunnels spaced closely together lie in the same plane, they may merge together to produce a fissure. If, however, the tunnels keep changing course, producing irregular tangles not confined to a single plane, continuous cracks are unlikely to develop. Regarding the nature of the susceptible material providing the paths for the tunnelling, there is no definite knowledge; but it is at least possible that the principle generally accepted for intergranular corrosion cracking could be applied to transgranular cracking also. According to circumstances, stainless steel can suffer either intergranular or transgranular cracking with stress in a corrosive environment. The former is explained by the fact that carbon atoms tend to collect at the grain-boundaries, since their presence at such places will minimize discontinuity of structure and thus keep low the energy of the system; carbides rich in chromium are there deposited, leaving bands depleted in chromium adjacent to the boundaries, which become the paths of intergranular attack. May not the same thing be occurring at dislocations on slip-planes? The association of foreign atoms with dislocations is no new idea. Although there is no direct evidence of chromium depletion along dislocations within the planes (whether dissociated to produce stacking faults or not), such an explanation of the formation of tunnels (which later merge into cracks) involves no *ad hoc* assumption, and appears less speculative than some of the other explanations advanced.

Whether the cracking is intergranular or transgranular, a factor favouring the advance of a crack appears to be the existence of a large unstressed area next to a small stressed one. It has been shown (1960 book, p. 389) that if two electrodes are placed in the same liquid, one being under tensile stress and the other unstressed, a current will flow, the stressed electrode being the anode. In practical cases of stress corrosion cracking, the tip of an advancing crack forms a small anode and the face of the material outside the crack a large cathode. This is a sufficiently dangerous situation, but if, in addition, the stressed portion of the member is joined to a much larger area of unstressed material, facilities for the cathodic reaction (often the reduction of oxygen) are greatly improved; the current available for anodic corrosion at the tip is likely to be increased and the crack will advance more rapidly. In designing testing apparatus, more attention might well be given to the provision of an almost unstressed area on the specimen, bearing a reproducible relationship to the stressed area.

Developments during the period 1968–1975 and Plan of the new Chapter.   The period under review has witnessed a very large amount of work on stress corrosion cracking; this is due to the fact that it is materials showing great strength in the absence of corrosion, which most readily develop cracks when subjected to high stress under corrosive conditions. Considerable increase in the understanding of intergranular attack under stress has been gained through the work of Hoar, who has obtained evidence of the electrochemical mechanism by correlating the crack propagation rate calculated from electrochemical measurements with that observed directly. This theoretical argument cannot be applied to cases where periods of electrochemical attack alternate with periods of mechanical cracking; in such cases the crack propagation may occur at a faster rate than electro-chemical measurements would predict.

Another feature of the period has been the application of the principles of fracture analysis to the problems of stress corrosion cracking, and here the work of B. F. Brown has been outstanding. He has also contributed to knowledge by introducing an experimental method based on the use of liquid nitrogen which enables the surprisingly low pH values existing in cracks to be determined with accuracy.

The argument of the chapter can perhaps best be indicated by stating the titles of the headings. These include 'Stress Corrosion as a Borderline Effect', 'Conditions for Crack Propagation', 'Electrochemical Basis of the Rate of Crack Advance', 'Acid production in the fissure Fracture', 'Hydrogen as a Factor in Cracking', 'The Spectrum of Phenomena', based on an exposition by Parkins, 'Early Work on Stress Intensification', 'Fracture Mechanics', 'Testing Methods', 'Stress Corrosion Cracking in various Materials'; the materials discussed include mild steel, stainless steels, high strength steels, aluminium alloys, titanium alloys, copper alloys, zinc alloys and the alloys of noble metals.

General

Stress Corrosion as a Borderline Effect. In early chapters it was pointed out that in service the most serious cases of corrosion occur at the borderline between corrosive and non-corrosive conditions. Thus on iron in a salt solution the attack is generally spread over a broad area, so that the loss of thickness is small even after a long period. If a large amount of inhibitor is added to the liquid, attack may be prevented altogether. If the amount added is sufficient to prevent corrosion on the main area, but still allows it at certain small susceptible spots, the dangerous combination of large cathode and small anode may lead to rapid penetration. The susceptible spots may be places where stresses cause local breakage of any protective film, or perhaps inclusions where such films will be discontinuous; the local corrosion may then take the form of pitting. In other cases, crevices, where the inhibitor will not readily be replenished, provide the small anodic area; a meniscus—a special type of crevice—may constitute the point of attack, and perforation may occur along a water-line whilst the immersed part of a specimen remains unattacked. Stress is a factor which may greatly inten-sify attack, producing a crack where otherwise a pit or rounded trench would develop. The continued cracking of the film at the tip of a crack and the fact that an increased interatomic distance must favour escape of ions from the metal into the liquid, will promote anodic attack at the crack tips, and the distribution of potential will be such as to reduce the chance of attack on the walls of the fissure, so that rapid crack propagation becomes almost inevitable. The production of acidity by the anodic reaction—which has long been recognized as a factor in pitting—is particularly important in stress corrosion cracking.

Conditions for Crack Propagation. The important work of Scully[1] has indicated the conditions likely to produce cracking. If as in Fig. 33(A) a metal specimen carrying an oxide-film is subjected to stress so that gliding takes place (B), an area free from oxide is produced on the slip-plane. If (C) the conditions are such that healing takes place over the greater part of the plane, leaving the tip bare, anodic attack on this small area will lead to the formation of a fissure extending into the metal. If healing of the film takes place very rapidly, we may get no attack at all. If healing takes place too slowly, we shall, instead of a crack, obtain a pit, or even general corrosion. When once a fissure has started to form, cracking and localized anodic attack is more likely to take place at the tip than anywhere else. In order

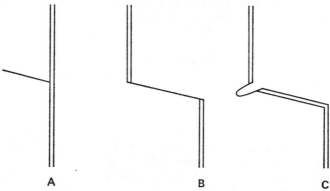

A   B   C

Fig. 33   Schematic diagram suggesting the nature of the attack that occurs on the emergent slip-step surface in alloys that develop wide slip-steps when exposed to a passivating environment. (A) Alloy surface. (B) Same surface after plastic deformation and the production of a wide slip-step. (C) Subsequent re-passivation of nearly all the surface; only a small area remains active (J. C. Scully).

to confine cracking to the tip, the walls must remain shielded with oxide; under favourable conditions these walls, as well as the face outside, can act as cathodes. Events can repeat themselves, and Fig. 34, due to Staehle,[2] suggests the course of events as the fissure propagates into the metal. It is evident that stress corrosion cracking can only take place under a limited range of conditions; its occurrence depends on the relation between the rate of corrosion on the bare surface and the rate at which a protective film is formed. Mere measurement of the current passing at a bare area does not provide this relation, since part of that current is connected with the formation of a fresh oxide-film and the rest with anodic dissolution of metal. A new technique, known as *tribo-ellipsometry*, has been developed by Ambrose and Kruger[3] to decide how much is due to film-formation. An apparatus has been designed in which a rotating polishing wheel abrades off the oxide-

1.   J. C. Scully, *Ericeira Conf.* 1971, pp. 1, 127; also *Brit. Corr. J.* 1966, **1**, 355; *Corr. Sci.* 1969, **8**, 513; *Anticorrosion* 1972, **19** (9) 5.

2.   R. W. Staehle, *Ericeira Conf.* 1971, p. 223.

3.   J. R. Ambrose and J. Kruger, *Corrosion* (*Houston*) 1972, **28**, 30. Also *National Bureau of Standards Report* 10865 (1972).

film and then is quickly removed. The transient current connected with anodic attack on the bared surface is measured and simultaneously optical measurements are taken showing the re-growth of the film. Since the initial phase of re-passivation is completed in less than 50 milliseconds, ellipso-metric measurements capable of rapid response are necessary. The combination of the electrical and optical methods makes possible a determination of the re-passivation rate and reveals how much of the transient current is devoted to anodic attack on the metal and how much to film-growth. Measurements are made of the total charge passed during re-passivation

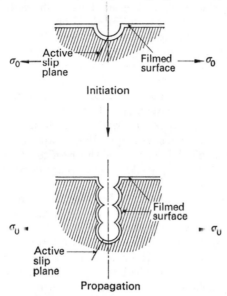

Fig. 34 Schematic aspects of crack propagation by successive step emergence and dissolution (R. W. Staehle).

(1) of a ternary Al–Zn–Mg alloy and (2) of an alloy possessing the composition of the precipitate-free zone which results during the heat-treatment of such an alloy. The results show that the precipitate-free zone re-passivates itself more slowly than the matrix in an N NaCl solution; this seems to explain the susceptibility of such an alloy to intergranular attack on film-rupture. The technique can be used to compare the re-passivation rate for a low-carbon steel in a nitrate solution with that obtained in a nitrite solution. The re-passivation rate is found to be slower in the nitrate than it is in the nitrite, and a larger proportion of the current is involved in metal dissolution in the nitrate. This may explain why there is susceptibility to stress corrosion cracking in nitrate but not in nitrite.

A model which should enable stress corrosion cracking to be placed on a mathematical basis—at least in a simplified case—has been proposed by Bignold.[1]

1. G. J. Bignold, *Corrosion* (*Houston*) 1972, **28**, 307.

Electrochemical Basis of the Rate of Crack Advance. Hoar[1] with several colleagues, has correlated the rate of advance of the crack with the current flowing from a cracking electrode, maintained as anode at constant potential. In order to explain the observed rates of crack propagation, high current densities at the crack advancing edge must be postulated, up to 1 amp/cm² for stainless steel in hot chloride solution. Electrochemical measurements by Hines showed that, with advancing edges of 10 to 100 Å in width, sufficient current could be generated by a corrosion cell containing hot $MgCl_2$ solution and electrodes of stainless steel to explain the observed rate of progress of the crack. Subsequent work showed that an anodic CD of up to 0·2–0·3 amp/cm² can be maintained on *yielding* stainless steel wire anodes, which, when static, will pass only about 10^{-5} amp/cm² at the same potential. Very high anodic activity can be obtained on continuously scraped stainless steel, under conditions where any oxide-film formed is continuously removed and the metal immediately below it is continuously deformed. Galvele studied mild steel in hot nitrate solution which shows anodically accelerated stress corrosion cracking; when such a specimen is subjected to yielding under *constant strain rate* conditions, the film cracks and the yielding bared metal dissolves at rates corresponding to about 2 amp/cm². The correlation shown in Table VI, based on the results of seven experimenters working in Hoar's laboratory, would seem to place beyond doubt the electrochemical basis of stress corrosion cracking.

In cases where there is equivalence between the rate of crack propagation and the current density, it is possible to suggest means of improving resistance to cracking; in a paper with Lees and Ford, Hoar[2] presents some 'guide lines'. Improved resistance can be achieved (1) by decreasing the rate of oxide rupture or (2) increasing the rate of passivation on the surface of the yielding metal. Most methods used today aim at an alteration of microstructure, generally achieved by a modification of the fabrication procedure. But it might also be brought about by alteration in tha composition of the alloy, and perhaps in the future this may prove the most hopeful method. In special circumstances, alteration of the environment may be possible,

Another research[3] indicates a relationship between the CD on the straining electrode and the rate of passivation found on cessation of plastic deformation. The shape of the current–time transient confirms the view that the change in reaction rate upon straining is controlled by the rate of oxide rupture (caused by slip-step emergence) and by the subsequent passivation rate of the newly bared metal.

The equivalence of crack propagation rate with current density may not be valid in all cases. Engell[4] has pointed out that the maximum CD observed in ordinary studies of anodic dissolution range from 10 amp/cm² for aluminium to 100 amp/cm² for iron. In the absence of complicating factors,

1. T. P. Hoar, *Ericeira Cong.* 1971, p. 105; also (with J. R. Galvele), *Corr. Sci.* 1970, **10**, 211.
2. D. J. Lees, F. P. Ford and T. P. Hoar, *Met. Mat.* 1973, **7**, 231.
3. T. P. Hoar and F. P. Ford, *J. electrochem. Soc.* 1973, **120**, 1013.
4. H. J. Engell, *Ericeira Conf.* 1971, p. 86.

this would account for rates of crack propagation up to 5×10^{-3} cm/s. For precipitation-hardening Al–Zn–Mg alloys, the measured propagation rates can reach 10^{-2} cm/s. Engell concludes that 'the limitation mentioned above shows that any explanation of these crack velocities cannot be given by any reasonable electrochemical hypothesis alone'; the final word 'alone' would seem to be important.

TABLE VI

COMPARISON OF CRACK PENETRATION RATES[1]

(a) Estimated from maximum current density on bared metal and
(b) Measured microscopically (T. P. Hoar)

Material	Solution	Temperature	Potential	Bared metal CD	(a) Crack pen. rate (from CD)	(b) Crack pen. rate (micro-scopic)	Ref.
18/8 Stainless steel	42% $MgCl_2$	154°C	V(SHE) −0·14 −0·14 0·065 0·15	A/cm² 0·3 0·16 0·6 0·3	mm/h 0·4⎱ 0·2⎰ 0·7 0·4⎰	mm/h 0·5–1·0	West[2] Scully[3] Lees[4]
Carbon steel	4M $NaNO_3$ pH 4·8	Boiling	−0·22 −0·18 −0·14	0·25 0·8 1·5	0·3 1·1 2·0	0·2⎱ 0·9⎰ 1·5⎰	Galvele[5]
	4M $NaNO_3$ pH 3·2	Boiling	−0·05 0·00 +0·75	0·8 2·2 5·0	1·1 2·6 6·2	1·1⎱ 1·6⎰ 6·9⎰	Lees[6]
Carbon steel	10M NaOH pH 14·5	121°C	−0·90 −0·80 −0·75 −0·70 −0·65 −0·6 −0·55	0 0·15 0·12 0·09 0·07 0·08 0·03	0 0·20 0·16 0·12 0·10 0·10 0·04	0 0·14 0·12 0·13 0·12 0·10 0·09	Jones[7]
α-brass	M NH_4Cl, 0·05M $CuSO_4$ + NH_3 to give pH 7·3	25°C	+0·25 +0·26	0·05 0·20	0·07⎱ 0·3 ⎰	0 3	Podesta[8] Lees[9]
Al–7% Mg	M NaCl pH 2–5·5	25°C	−0·80 −0·75	0·1 up to 80	0·11 up to 90	0·25 very high	Ford[10]

1. T. P. Hoar, Lecture at Firminy (France).
2. T. P. Hoar and J. M. West, *Proc. Roy. Soc. (A)* 1962, **268**, 304.
3. T. P. Hoar and J. C. Scully, *J. electrochem. Soc.* 1964, **111**, 348.
4. D. J. Lees and T. P. Hoar (in preparation).
5. T. P. Hoar and J. R. Galvele, *Corr. Sci.* 1970, **10**, 211.
6. D. J. Lees and T. P. Hoar (in preparation).
7. T. P. Hoar and R. W. Jones, *Corr. Sci.* 1973, **13**, 723.
8. T. P. Hoar, J. J. Podesta and G. P. Rothwell, *Corr. Sci.* 1971, **11**, 241.
9. D. J. Lees and T. P. Hoar (in preparation).
10. T. P. Hoar and F. P. Ford, *J. electrochem. Soc.* 1973, **120**, 1013.

The cause of the discrepancy may be found in the fact that Hoar's basis of calculation is only applicable to cases of continuous advance. Where there is jerky advance, no correlation in the sense of Faraday's law can be expected. In such cases the function of the current is to eat its way around places where the material is not susceptible to preferential anodic attack, until the stress intensification becomes sufficiently high for these resistant bridges to be ruptured mechanically. Even where there is continuous advance, another factor requires consideration. West,[1] working in Hoar's laboratory, showed that, under yielding conditions, activation polarization can practically disappear. Whether this is attributable to the fact that during plastic deformation atoms sometimes find themselves momentarily in unstable positions of high energy, or whether it is due to the fact that at certain places the yielding produces an abnormally high value for the interatomic distance, need not here be considered. The observation would seem to suggest that at the crack tip the CD might sometimes reach values higher than those commonly noted in electrochemical studies carried out on uncracked specimens.

There is considerable evidence from other sources that—at least in cases where there is no intermittent mechanical rupture—an explanation based on anodic dissolution is acceptable. Parkins[2] has shown that steel in a nitrate solution can maintain a CD on a film free surface over an order of magnitude higher than in a hydroxide solution. This would suggest that the crack propagation rate in NH_4NO_3 should be faster than that in $NaOH$ by a similar factor. The measured rates are about 2·5–3·5 mm/hour in NH_4NO_3 and only about 0·04–0·10 mm/hour in $NaOH$; this is in reasonable agreement with the prediction and supports the suggestion that crack propagation occurs by a process of anodic dissolution.

Wilde,[3] after a reasoned consideration of alternative mechanisms, reaches the conclusion that the Hoar–Hines model is the one most generally applicable.

The average velocity of propagation of a crack in a copper alloy containing 1·8% Be in aqueous ammonia at pH 12·5, studied by Sparkes and Scully,[4] is found to be roughly a rectilinear function of the initial stress intensity factor. At low values of the stress intensity factor the fracture is intergranular, but becomes partly transgranular at higher values. Propagation is thought to occur by slip-assisted rupture of the tarnish film.

A mathematical calculation of crack growth due to electrochemical processes on the assumption of crack branching after a certain stress intensity factor has been attained, has been provided by Russian scientists.[5]

A recent Japanese paper[6] describes the measurement of the anodic dissolution current accompanying stress corrosion crack propagation by connecting the stressed specimen (coated with teflon except for a 0·2 mm slit)

1.   T. P. Hoar and J. M. West, *Proc. roy. Soc. (A)* 1962, **268**, 304.
2.   R. N. Parkins, *Ericeira Conf.* 1971, p. 167, esp. p. 180.
3.   B. E. Wilde, *J. electrochem. Soc.* 1971, **118**, 1717.
4.   G. M. Sparkes and J. C. Scully, *Corr. Sci.* 1971, **11**, 641.
5.   G. P. Cherepanov, L. V. Ershov and G. G. Kuzmin, *Corrosion (Houston)* 1973, **29**, 100.
6.   T. Suzuki, M. Yamabe and Y. Kitamura, *Corrosion (Houston)* 1973, **29**, 70.

as anode with an unstressed cathode through an external circuit; potentio-static control prevents the potential difference from falling during the pro-pagation. The method is designed to 'distinguish the initiation and propagation processes and clarify the material and environmental factors affecting SCC'.

Acid Production in the Fissure. The drop of pH inside pits has already been discussed (this volume, p. 72). This appears to be an important factor in stress corrosion cracking, and an exact method for determining pH in a cavity has been worked out by B. F. Brown;[1] it depends on the use of liquid nitrogen to freeze the liquid present in the fissure and thus allow examination at leisure after the specimen has been broken open. The proce-dure has varied according to the problem under examination. The method adapted for use in connection with the metal-wedge test may be men-tioned. A rectangular specimen is used, having a saw-cut into which a metal wedge is pressed, using a vice to stress the root of the saw-cut in tension. The specimen is then placed in either distilled water or 3·5% NaCl solution (pH 6·5), with the wedge either paraffin-coated or kept above the water-line to prevent galvanic effects. After the corrosion crack has pene-trated about 1 cm, the specimen is immersed in liquid nitrogen and broken apart. Paper carrying a pH indicator (or $K_3Fe(CN)_6$ to detect Fe^{2+}) is pressed along the walls of the crack. (Agreement between values obtained from the two opposing walls of the same crack shows that the test is valid.) The tests show that acidity is high in the cracks and also that metallic ions are present. In the case of an Al alloy and also 0·45% C steel, pH values of 3·5 and 3·8 respectively are reached near the advancing crack tip; probably the true value is slightly lower. In the case of a titanium alloy containing Al, Mo and V, the pH value falls to 1·7.

Although the acidity of the liquid in the crack is today generally recog-nized, the high concentration of metallic salts attained receives less atten-tion; it is clearly important, since the acidity itself depends on the concen-tration of metallic ions, which will also affect the prospect of continued corrosion in the crack in various ways. The matter has been studied by Bianchi and his colleagues,[2] who state that a solution typical of localized corrosion on iron is nearly saturated $FeCl_2$ with traces of $FeCl_3$; saturation may be attained in pits, but is not necessarily reached in a stress corrosion crack, where the volume keeps increasing as the crack develops. They cal-culate from hydrolysis considerations that the pH value in such a solution should reach 3·7 at 25°C, and this value corresponds fairly well with experi-mental determinations. The high concentration is reached because the trans-port number of the highly hydrated cations formed by the anodic attack is virtually zero. A high cation concentration will shift the potential in a noble

1. B. F. Brown, 'Stress-Corrosion Cracking in High Strength Steels and in Titanium and Aluminium Alloys' (Naval Research Laboratory, Washington), 1972. *Ericeira Conf.* 1971, p. 186. Also *Met. Reviews* **129** (1968). *J. electrochem. Soc.* 1969, **116**, 218 (with C. T. Fujii and E. P. Dahlberg). *Corrosion (Houston)* 1970, **26**, 249. *Brit. Corr. J.* 1969, **4**, 284. *Amer. Soc. Test. Mat.* 1969; *Corr. Sci.* 1970, **10**, 839 (with G. Sandoz and C. T. Fujii).
2. G. Faita, F. Mazza and G. Bianchi, *Williamsburg Conf.* 1971, p. 34.

M

direction, but the low free-water concentration, the acidity arising from hydrolysis and the halide concentration cooperate to make passivation more difficult.

In considering the effect of the high concentrations, a warning sounded by Pourbaix[1] should be borne in mind. Commenting on another paper, he remarks, 'these concentrated 4·2M solutions may not . . . be considered as "perfect", and the calculations relating to dilute solutions are not valid as such. There is an urgent need for a better knowledge of the thermodynamic equilibria in such concentrated solutions.'

Hydrogen as a Factor in Cracking.   Brown emphasizes the point that, on plotting the combination of pH and potential observed during cracking on a Pourbaix diagram, the points always fall in that region where hydrogen evolution is possible. He states that 'there does not appear to be any valid reason to require a mechanism other than hydrogen embrittlement to account for the cracking of high-strength steels in salt water'. Wilde and Kim[2] also believe that in high strength martensitic steel the cracking can be explained by hydrogen embrittlement, but in the case of austenitic stainless steel they show that hydrogen embrittlement plays little or no part in the crack propagation process, which is attributed to strain assisted anodic dissolution. Although hydrogen is evolved at the free corrosion potential, none is absorbed into the steel during cracking. Sodium arsenite, which would be expected to help hydrogen absorption, increases the time to failure—the opposite of what would be expected if a hydrogen absorption mechanism was accepted. In unbuffered solution, there is little dependence on the initial or final pH value—perhaps because the acidity is soon used up. Rhodes,[3] however, discussing austenitic stainless steel in chloride solution, states that the results point to the importance of hydrogen evolution at the crack tip, and do *not* support the idea of stress assisted anodic dissolution. Although hydrogen diffusion in austenite is slow, that in ferrite is quicker, whilst martensite platelets, if they exist, may carry hydrogen ahead of the crack tip. It is postulated that hydrogen absorption facilitates the formation of those platelets.

For low-alloy steels, without applied potential, Snape[4] attributes failure mainly due to hydrogen embrittlement. Baker and Singleterry[5] quote a case where in strongly buffered solution the time to failure increases 100 times on passing from pH 4 to pH 7; this is thought to support the idea of hydrogen embrittlement.

Some writers have rejected or minimized the role of hydrogen in cracking, because it occurs at potentials where the liberation of hydrogen would be impossible at the pH value of the bulk liquid; Wilde,[6] however, points out that the now well established formation of acid in pre-cracks, pits and

1.   M. Pourbaix, *Williamsburg Conf.* 1971, p. 149.
2.   B. E. Wilde and C. D. Kim, *Corrosion (Houston)* 1972, **28,** 350.
3.   P. R. Rhodes, *Corrosion (Houston)* 1969, **25,** 462.
4.   E. Snape, *Brit. Corr. J.* 1969, **4,** 253.
5.   H. R. Baker and C. R. Singleterry, *Corrosion (Houston)* 1972, **28,** 340.
6.   B. E. Wilde, *Corrosion (Houston)* 1971, **27,** 326.

crevices may invalidate that sort of argument. He holds that under sufficient anodic polarization, corrosion following an active path may be produced, shortening the life of a stressed specimen by crack propagation; under sufficient cathodic polarization, hydrogen embrittlement may occur, also shortening the life. However, the corrosion potential prevailing when no polarization is applied may not be the value at which active-path corrosion gives place to hydrogen embrittlement. Thus for certain substances, a small amount of cathodic polarization may lengthen the life, suppressing active-path corrosion without causing hydrogen embrittlement; for other substances, a small amount of anodic polarization may lengthen the life, suppressing hydrogen embrittlement without starting active-path corrosion.

Townsend[1] has studied the stress corrosion cracking of high strength steel wire in H_2S solutions, which occurs at room temperatures under a stress less than 15% of the ultimate tensile strength; cracks usually form at 45° to the direction of the tensile stress; the time needed for failure is increased by anodic polarization and is decreased by cathodic polarization—as would be expected if cracking is due to hydrogen. Susceptibility to cracking increases with rising temperature until a maximum is reached near 25°C and then falls. Experiments on the build-up of molecular hydrogen inside a hollow steel cylinder, the outside of which was exposed to H_2S solution, showed that the pressure attained increases with rising temperature over the same range in which cracking susceptibility abates. The conclusion is reached that the cracking is caused by hydrogen *atoms* diffusing through the lattice and becoming adsorbed on the surfaces of embryonic cracks (dislocation pile-ups), thereby reducing the energy needed for creating new surfaces when a pile-up is transformed into a true crack; this is the mechanism proposed by Petch. It is considered inconsistent with the view, generally attributed to Zappfe, that the hydrogen atoms combine in internal voids to form molecular hydrogen, the pressure of which pushes the metal apart from within.

Kennedy and Whittaker[2] emphasize the practical importance of distinguishing between the cracking due to hydrogen and that due to anodic attack at the crack tip, since such information is necessary in deciding on a protective treatment. If the mechanism of fracture can be shown to be of the second type, then a coating consisting of a metal anodic to the basis metal should provide cathodic protection at gaps (at least for a time), whereas if the mechanism of failure is connected with hydrogen, the choice of an anodic metal is likely to be disastrous; phosphating and painting may be more suitable. They discuss methods of distinguishing between the two types of failure, reaching the conclusion that 'various techniques show promise' but that 'absolute diagnosis is not yet possible'.

A comprehensive report, published by W. Beck and two colleagues,[3]

1. H. E. Townsend, *Corrosion (Houston)* 1972, **28**, 39.
2. J. W. Kennedy and J. A. Whittaker, *Corr. Sci.* 1968, **8**, 359.
3. W. Beck, E. J. Jankowsky and P. Fischer, 'Hydrogen Stress Cracking of High Strength Steels', Naval Air Development Center, Warminster, Pa., NADC-MA-7140.

deserves study. It is largely confined to cases of stress corrosion cracking in which hydrogen is considered to play a part, but the authors evidently consider that this includes many of the important types; however, they quote—apparently with approval—an opinion that 'caustic cracking is not caused by embrittlement by hydrogen'. The cases considered include embrittlement resulting from pickling and chemical milling, as well as failures arising in electroplated steel aircraft components, weldments, the petroleum industry, pressurized water reactors, pressurized hydrogenation units and parts cathodically protected. Methods for the determination of hydrogen in steel, and its effect on fatigue strength, are considered in some detail.

Results from various sources are quoted showing that maraging steel appears to be relatively non-susceptible to plating embrittlement. Tests on an alloyed martensitic stainless steel containing Mo and V carried out in 3% NaCl, with NaOH or HCl added to adjust the pH value, show that at pH 6·5 either a cathodic CD or an anodic CD can greatly reduce the time to fracture. At pH 12·5 an anodic current will greatly reduce the time, but a cathodic current has much less effect. At pH 1, the time to failure is very short even in the absence of applied current. It would appear that hydrogen is playing a part in the rapid failure under cathodic polarization, but that failure under anodic conditions is due to selective attack upon the crack tips. The authors quote views emphasizing the difficulty of distinguishing between the two types of failure by inspection of the fracture; they point out that hydrogen embrittlement is predominatingly nucleated from subsurface sites, and often shows dimples and marked hair-line indications in the intergranular regions. In contrast, true stress corrosion cracking is associated with predominatingly surface nucleation of intergranular fracture and regions of pronounced secondary cracking or deep crevices; hair-line indications on the intergranular surfaces are less pronounced than in the case of hydrogen embrittlement.

In the cases met with in the oil and gas industries, sulphide corrosion often plays an important part.

The introduction of hydrogen during plating is discussed in the final chapters of the report, which should not be overlooked. It is emphasized that a cadmium plating bath based on cadmium ammonium fluoborate and brighteners produces far less embrittlement than the ordinary cyanide bath. This may seem surprising, since the fluoborate bath is slightly acid (pH 3·8), whereas the cyanide bath is alkaline; an acid reaction would seem likely to favour the production of hydrogen. It must, however, be remembered that it is not the hydrogen evolved as gas which causes damage, but the hydrogen entering the metal in the atomic state. According to the views tentatively put forward on p. 216 of this volume, a discharged hydrogen ion will generally escape as an H_2 molecule if another atom is being discharged on an adjacent site, and this is most likely to happen if the liquid is rich in H^+ ions; it is less likely to happen if most of the sites are blocked by adsorbed CN^- or other ions.

The explanation favoured by Beck appears to be a little different. He would seem to picture, not prevention of escape as H_2, but some positive assistance to the entry of atomic hydrogen into the metal provided by strongly adsorbed groups, possibly by their influence on the relevant bond-strengths. He points out that S^{2-} ions and CN^- ions, both of which are also very strongly adsorbed, are powerful promoters of hydrogen embrittlement, whereas the BF_4^- ion, which is weakly adsorbed, causes little hydrogen embrittlement.

The current efficiency is also mentioned. This is high for the fluoborate bath and low for the cyanide bath. Obviously, for reasons of economy a high current efficiency is to be welcomed, and if 100% efficiency could be reached, the production of hydrogen would be impossible. However, if the efficiency falls below 100%, the question to be answered is what part of the waste current is devoted to the production of atomic hydrogen which can enter the metal, and what part is used on relatively harmless reactions. The reasons for the higher efficiency of the acid bath is probably that plenty of

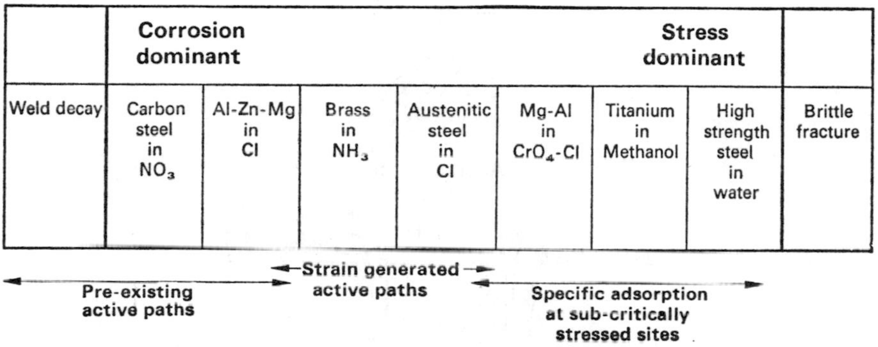

	Corrosion dominant				Stress dominant			
Weld decay	Carbon steel in NO_3	Al-Zn-Mg in Cl	Brass in NH_3	Austenitic steel in Cl	Mg-Al in CrO_4-Cl	Titanium in Methanol	High strength steel in water	Brittle fracture

←————— Pre-existing active paths —————→ ←— Strain generated active paths —→ ←— Specific adsorption at sub-critically stressed sites —→

Fig. 35   The stress corrosion spectrum (R. N. Parkins).

Cd^{2+} is present, whereas in the cyanide bath most of the cadmium is locked up as complex anions. The Cd deposition potential is thus less negative in the acid bath and less hydrogen is postulated.

Townsend[1] has studied the effect of stress on the entry and permeation of hydrogen in iron. He attributes the increase in hydrogen permeation through a membrane on the application of tensile stress to an expansion of the lattice, leading to an increase of the exchange current for the reaction

$$2H^+ + 2e \leftrightarrows H_2$$

The Spectrum of Phenomena.   A lecture delivered at Harrogate by Parkins[2] displays the various stress corrosion phenomena as a continuous spectrum, varying between weld decay (which occurs in the absence of applied stress) at one end and brittle fracture (which occurs without corrosive influences) at the other. The spectrum is shown in Fig. 35; there is a continuous passage between corrosion-dominant and stress-dominant cases.

Often the character of the phenomenon varies with conditions. Mild steel

1.   H. E. Townsend, *Corrosion (Houston)* 1970, **26**, 361.
2.   R. N. Parkins, *Brit. Corr. J.* 1972, **7**, 15.

in nitrate solution suffers intercrystalline corrosion, but this does not penetrate to a greater depth than 10^{-2} mm, unless either stress or anodic polarization is applied; the latter can cause virtual disintegration even in the absence of stress. The stress corrosion cracking of carbon steels is *not* due to differences between the free energy of iron atoms situated on differently oriented crystal faces and/or grain-boundaries. Pure iron in bulk is not susceptible to stress corrosion, indicating that the pre-existing active path is likely to be defined in terms of chemical segregation. However, the state in which the carbon exists is important. As the carbon content increases, so the proportion of pearlite in the steel increases, and the number of carbide

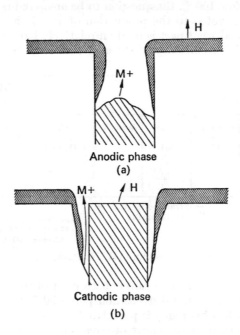

Fig. 36   Galvanic cell mechanism of intergranular attack (R. N. Parkins).

particles in the boundaries decreases. Thus the threshold stress needed to produce cracking in annealed mild steel exposed to boiling 4N NH_4NO_3 increases with C content. Parkins's schematic representation of a corrosion fissure developing at a grain-boundary in the two cases where the precipitated phase is either (a) anodic or (b) cathodic is shown in Fig. 36. His version of the penetration connected with alternate deformation with rupture and fresh oxidation is reproduced in Fig. 37. In the alloys Al–Mg and Al–Zn–Mg, the precipitates are preferentially corroded; but in the alloy Al–Cu, the precipitation of $CuAl_2$ leaves the adjacent regions denuded of Cu and these regions suffer preferential attack. The fact that cracking can occur when there are gaps between discrete particles of precipitate has directed attention to the 'precipitate-free zone', but experiments designed

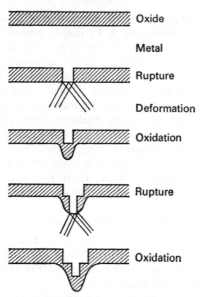

Fig. 37   Oxide-film rupture mechanism of stress corrosion cracking (R. N. Parkins).

to determine the exact location of the crack path indicate that this coincides with the grain-boundaries and not with the precipitate-free zones.

Regarding the idea that bursts of mechanical crack propagation are interspersed between periods of electrochemical propagation, Parkins states that the evidence (discontinuous elongation and acoustic emission) can often be interpreted in other ways, and that 'for the softer ductile alloys, there is no unequivocal evidence suggesting that crack extension occurs by cleavage. On the other hand, it is possible that in the stronger, notch-sensitive materials, such as some of the high-strength steels, mechanical propagation may contribute significantly to the overall rate of crack propagation.'

Environmental studies have perhaps favoured the idea of specific adsorption as an important factor. 'Uhlig supports this contention with the arguments that specificity of environments causing cracking . . . are more consistent with an adsorption mechanism than any alternative.' Parkins comments that Uhlig's preference for an adsorption mechanism appears to be due to chemisorption being specific, whilst electrochemical reactions are not, because the similar conductivities of different ionic solutions would require all such environments to promote cracking. Parkins observes 'why the many other variable properties of ionic solutions should be insignificant in comparison with electrical conductivity is not clear, but there are no inconsistencies between an electrochemical mechanism of cracking and solution specificity, as earlier defined, or between this mechanism and the effect of extraneous anions or cathodic protection'.

Regarding a hydrogen embrittlement mechanism, Parkins states, 'It has not infrequently been argued that the formation of hydrogen bubbles in the vicinity of cracks is evidence of a hydrogen embrittlement mechanism; however, it would appear at least as logical to suggest that hydrogen entry into the metal is likely when that element is not escaping as a gas.'

The last remark is very pertinent, but it might be added that if the over-potential is sufficient for the saturation of the solution with hydrogen and the consequent evolution of molecular hydrogen in bubbles, it will probably be sufficient for the entry of atomic hydrogen into the metal, which may become converted to molecular hydrogen at considerable pressure at internal cavities. The pressure of molecular hydrogen calculated to be in equilibrium with the concentration of atomic hydrogen indicated by the measured over-potential is generally extremely high; there is considerable evidence that these high pressures can be produced (1960 book, pp. 403, 404).

Stress Intensification

Early Work.[1] In 1913 Inglis[2] showed that at an elliptical hole in material the factor representing the local stress intensification is $1 + 2\sqrt{L/\rho}$, where L is the half-length and ρ the radius at the tip; thus, for a circular hole where $L = \rho$, the factor becomes 3.

Griffith[3] suggested that the low strength exhibited by glass was due to sub-microscopic cracks producing stress intensification. He imagined these cracks to be situated in the interior, but Andrade and others have shown that they exist at the surface. Glass fibres can be made which exhibit surprisingly high strength—presumably because cracks are absent; also fibres which are weak can be strengthened if the surface layers are removed chemically. If a crack is to propagate, Griffith considered that two conditions must be fulfilled:

(1) Propagation must involve a drop in the energy of the system; this means that the reduction in strain energy must exceed the surface energy connected with the creation of the two new surfaces at the sides of the crack.
(2) There must be a molecular mechanism by which the energy transformation can take place.

Griffith used a brittle material (glass) for his work, where elasticity theory can be used and plasticity phenomena can be ignored. The materials sensitive to stress corrosion cracking are mainly high strength alloys which are distinctly notch-sensitive. However, the application of the energy principle to relatively ductile metals has been found possible.

Fracture Mechanics. If a metal plate subjected to a uniform tensile stress carries a short crack, the elastic stress field at the leading edge of the

1.   Those requiring an elementary discussion of these matters should consult J. E. Gordon, 'New Science of strong metals' (Pelican), 1968, pp. 60, 78–80, 102–5.
2.   C. E. Inglis, *Trans. Inst. Naval Architects* 1913, **55**, Part 1, 219.
3.   A. A. Griffith, *Phil. Trans. Roy. Soc. (A)* 1921, **221**, 163.

crack is characterized by a single parameter K proportional to the product of the nominal stress σ and to $\alpha^{1/2}$, where α is the crack size. Relationships between K and $\sigma\alpha^{1/2}$ are available for different geometrical situations. K increases with increase of stress and of crack size, and when it reaches a critical value, K_c, unstable fast fracture will set in.

Readers who experience difficulties on consulting papers on fracture mechanics may care to use a report[1] 'intended to assist non-specialists'. Further information is supplied in a book by Liebowitz.[2]

Application to Stress Corrosion Cracking.   Attempts have been made to extend the principles of fracture mechanics—generally accepted today for non-corrosive conditions—to the case of stress corrosion cracking. The equations indicating the size of crack needed to give fast fracture under different situations are complicated. Readers should consult an excellent review by B. F. Brown.[3] In the particular case where the length of a surface crack is at least 10 times its depth, the appropriate equation is considered to take the simple form

$$\alpha_{cr} = 0 \cdot 2 \left(\frac{K_{ISCC}}{\sigma_Y} \right)^2$$

where α_{cr} is the depth of the shallowest crack expected to propagate a stress corrosion crack, and σ_Y the yield stress. Here K_{ISCC} is the minimal initial value of K at which an environmental crack growth occurs, so that the value of K grows until it finally reaches K_c—at which stage unstable fast fracture will set in.[4]

Evidence has been collected to support the suggestion that a threshold value of K, written K_{ISCC}, below which cracking in a corrosive environment will not occur, really does exist, at least in some materials. The practical importance of this point is obvious. In his book published in 1972 B. F. Brown wrote: 'There appears to be a threshold value of K below which SCC stops altogether for some metals. This is almost certainly true of Ti alloys; it appears to be true of high-strength steels, but it does not appear to be true of Al alloys.'

Views regarding the true existence of a threshold value are somewhat conflicting. Speidel[5] states: 'It is sometimes reported that there is a threshold stress intensity below which SC crack growth does not occur. . . . However, an analysis of the experimental evidence shows that such a threshold has not yet been established experimentally for aluminium alloys. . . . In view of the extension of region I crack growth to very small levels, and the

1.   'Fracture Mechanics—what it is, what it does'. N.E.L. Report No. 465 (East Kilbride, Glasgow), 1970.
2.   H. Liebowitz, 'Fracture: an advanced treatise' (Academic Press), 1971.
3.   B. F. Brown, *Met. Reviews* **129** (1968). Also *J. Metals* (*A.S.T.M.*) 1970, **5**, 786. 'Stress-Corrosion Cracking in High-Strength Steels and in Titanium and Aluminium' (Naval Research Laboratory, Washington), 1972. Also *Ericeira Conf.* 1971.
4.   On this point a paper by R. N. Parkins, F. Mazza, J. J. Royuela and J. C. Scully, *Brit. Corr. J.* 1972, **7**, 154, may be consulted.
5.   M. O. Speidel, *Ericeira Conf.* pp. 289, 345.

extremely small velocities observed there, the measurement of a "threshold" stress intensity for SCC of aluminium alloys with such arbitrary experimental cut-off times as 10, 100 or 1000 hours could lead to mistakes when data are used in actual design. Moreover, such data would almost certainly be useless for fundamental investigations.'

The types of curves useful in presenting information are shown in Fig. 38

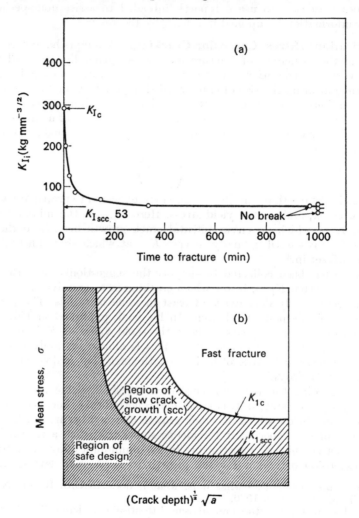

Fig. 38 Dependence of stress corrosion cracking on conditions. (a) Relation between stress intensity and time to fracture for steel of type 4340 (0·38–0·43% C, 1·65–2·80% Ni, 0·70–0·90% Cr, 0·20–0·30% Mo, 0·60–0·80% Mn and 0·20–0·35 Si) exposed to 3·5% NaCl; note the positions of K_{IC} and K_{ISCC}.
(b) Schematic indication of loci of mean stress plotted against crack depths for K_{IC} and K_{ISCC}, showing regions of safe design and regions to be avoided (M. G. Fontana).

provided in a paper by Fontana.[1] If K_{ISCC} is a reality, the curve in (a) should become asymptotic at high values of the time.

Brauns and Ternes[2] have studied the behaviour of austenitic stainless steel in hot chloride solution at different stresses. They reach the conclusion that down to 2 kg/mm² there is no discernible threshold below which corrosion does not occur in the absence of applied current (they were measuring actual values of applied stress, and not an intensification factor due to a pre-existing fissure).

Divergent views are indeed held regarding the utility of Fracture Mechanics generally. Wells,[3] who provides equations for K, considers that the K_{ISCC} concept is useful even when local plasticity arises, provided that the size of the yielded zone does not approach the other dimensions of the loaded body. Scully thinks that fracture mechanics, although practically useful as a screening test in selecting alloys, has limited usefulness in determining mechanisms. Fielding, speaking as a designer concerned to obtain a low weight structure without risk of catastrophe, welcomes the fracture-mechanical approach of B. F. Brown, because it enables the behaviour of a large structure to be predicted from that of a small specimen. Simple tests are possible, he thinks, giving results in days rather than in years. He reports good correlation between results in different laboratories.

The methods of presenting results are indicated in a review by T. R. Beck and his colleagues.[4] These refer specially to titanium alloys. The method used for smooth surfaces is shown in Fig. 39(a), whilst two methods suitable for notched and for pre-cracked specimens are shown in (b) and (c). Typical experimental curves obtained from titanium alloys are shown in (d). The method shown in (b) was initially used by Brown and others; time to failure is plotted against stress intensity. Such curves can provide evidence for a threshold stress intensity K_{ISCC}, for the maximum time of exposure, which is usually obtained after about 6 hours in the case of titanium alloys. In the method shown in (c), the logarithm of the crack velocity is plotted against the stress intensity. This shows three regions with region II horizontal. The real experimental curves of (d) often include a transition region denoted as IIa. If two materials are to be compared, comparison should be made between the same regions of crack growth. A possible interpretation of the three regions has recently been offered.[5]

Testing Methods. Simple methods used in early work were mentioned in the 1960 book (pp. 693–9), but many other procedures have been used.

Orman[6] has described a rapid test designed specially to evaluate the effect

1. M. G. Fontana, *Corrosion (Houston)* 1971, **27**, 129; esp. Fig. 7, p. 139.
2. E. Brauns and H. Ternes, *Werkstoffe u. Korr.* 1968, **19**, 1.
3. *Met. Mat.* 1969, **3**, 173. Contributions from A. A. Wells, p. 173, J. C. Scully, p. 174, A. H. Priest, p. 175, J. Fielding, p. 177.
4. 'Advances in Corrosion Science and Technology', M. G. Fontana and R. W. Staehle (Editors), Vol. **3** (1973); chapter by M. J. Blackburn, J. A. Feeney and T. R. Beck: see p. 123.
5. J. G. Williams and G. P. Marshall, *Proc. Roy. Soc. (A)* 1975, **342**, 55.
6. S. Orman, *Corr. Sci.* 1969, **9**, 849.

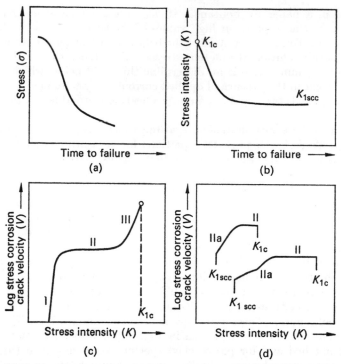

Fig. 39 Methods of presenting SCC data. (a) smooth specimens and (b and c) notched (and pre-cracked) specimens; also (d) typical curves for α or $\alpha + \beta$ titanium alloys tested in aqueous solutions (M. J. Blackburn, J. A. Feeney and T. R. Beck).

of various heat-treatments on a variety of alloys. It is essentially a tensile test carried out at constant strain rate in a controlled atmosphere; the results are said to be in agreement with those of longer term tests.

German work[1] suggests that direct observation of the surface of a specimen under tension in boiling $MgCl_2$ solution can provide more information than the analysis of a time–extension curve. The setting in of gas evolution reveals the start of the surface attack upon the metal, and this accords satisfactorily with the inflexion point on the time–extension curve.

Grubitsch[2] has studied potential profiles on various alloys in a state where two or more phases are present. In several cases where normally only small variations of potential are obtained on traversing the surface along a line, large variations are obtained if the specimen is subjected to an anodic current. Potential variation, due to local variation of composition, is obtained on crossing a weld-line on ferritic stainless steel with 17% Cr, also on traversing the site of a pit on 18/8 Cr/Ni austenitic steel, and finally

1. H. U. Borgstedt, I. Michael, S. Müller and G. Wittig, *Werkstoffe u. Korr.*, 1971, **22**, 121.
2. H. Grubitsch and A. Zirkl, *Werkstoffe u. Korr.* 1972, **23**, 565.

at the grain-boundaries of an aluminium alloy containing 4% Cu after a treatment which produces a copper depleted zone; approval is expressed for the early work of Dix and Akimov (1960 book, p. 671).

For ascertaining rapidly whether certain minor constituents can improve or diminish resistance to cracking, accelerated procedures have been adopted. Stainless steel is often tested in boiling $MgCl_2$ at different concentrations, but this has been criticized as unrealistic and difficult to control, since solutions of different concentrations have different boiling-points. Streicher[1] points out that small losses of water vapour from the testing apparatus can easily lead to increases in the boiling temperature; he describes an apparatus designed to minimize such losses. Ito[2] has found that the fracture-time of a given alloy becomes minimal at a certain boiling-point, but that this varies from one composition of stainless steel to another. There is often difficulty in obtaining reproducibility as a result of variation in surface condition. Staehle[3] states that manual preparation produces greater scatter than chemical treatment; chemical polishing leaves a state in which fracture is obtained fairly rapidly, and with only a small scatter between the results of duplicate experiments.

In the case of copper free Al–Zn–Mg alloys, the so-called ASC test (anodic salt chromate) has been used at Kingston, Ontario,[4] and the results stand in reasonably good agreement with those obtained in other laboratories, and with both marine and industrial exposures. A CD of 0.047 mA/cm^2 is applied to specimens immersed in 2% NaCl containing 0·5% Na_2CrO_4.

Accelerated tests for estimating the resistance of aluminium alloys to exfoliation have been discussed by Ketcham,[5] who favours an acidified salt-spray test but holds out hopes of replacing this by an immersion test and ultimately by electrical methods. The test in the form described has been used to establish the correct duration of the second stage in a two-stage heat-treatment designed to minimize exfoliation. The solution used is based on the principle of producing borderline conditions (between corrosion and passivity) by adding a critical amount of inhibitor to a chloride solution, and thus obtaining intense localized attack.

In research on the scientific mechanism of stress corrosion cracking, it is generally convenient to use a specimen with a smooth surface. In such cases there is often an incubation period before a perceptible crack is formed, followed by a period of crack growth. For practical work intended to show whether it is safe to use a given material in some machine, vehicle or structure, tests with smooth specimens are often unsatisfactory, since there may already be a flaw or microscopic crack at some vital part of the structure, producing serious stress intensification. Various tests based on the use of pre-cracked specimens have been put forward. In some of these, a bolt is used to produce

1. M. A. Streicher and A. J. Sweet, *Corrosion (Houston)* 1969, **25**, 1.
2. N. Ito and M. Yoshino, *Tokyo Cong. Ext. Abs.* 1972, p. 163.
3. R. W. Cockran and R. W. Staehle, *Corrosion (Houston)* 1968, **24**, 369.
4. P. W. Jeffrey, T. E. Wright and H. P. Goddard, *Amsterdam Cong.* 1969, p. 133.
5. S. J. Ketcham and I. S. Shaffer, *Amsterdam Cong.* 1969, p. 803.

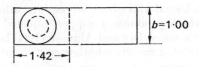

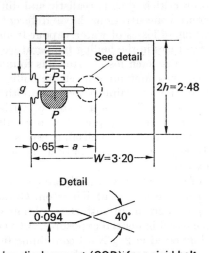

V^*=crack opening displacement (COD) for a rigid bolt

Fig. 40   Modified fracture specimen used to determine K_{ISCC} for steel alloys (H. V. Hyatt).

stress in the specimen without the need for a tension testing machine. The design of the fracture specimen favoured by Hyatt[1] is shown in Fig. 40. The procedure used by Speidel[2] is shown in Fig. 41. Here again the specimen is loaded by a bolt and mechanically pre-cracked until the crack front reaches a certain length (α_0). At this stage the corrosive liquid or atmosphere is admitted, and stress corrosion cracking may start. As the crack length increases, the stress intensity at the crack tip decreases according to the equation

$$K = \frac{E \; \delta h[3h(a + 0 \cdot 6h)^2 + h^3]}{4(a + 0 \cdot 6h)^3 + h^2 a)}$$

where E is the modulus of elasticity, δ is the deflection at the load-line, a the length of the crack and $2h$ the breadth of the specimen. The rate of crack growth decreases with time. Finally the specimen is broken open, allowing the crack face and the crack front to be inspected. The area of the stress corrosion crack shows up, being different in appearance (generally brighter) than the area of the mechanically produced pre-crack and the mechanical fracture made after the test.

1.   H. V. Hyatt, *Corrosion (Houston)* 1970, **26**, 487.
2.   M. O. Speidel, *Ericeira Conf.* 1971, p. 289; esp. p. 295.

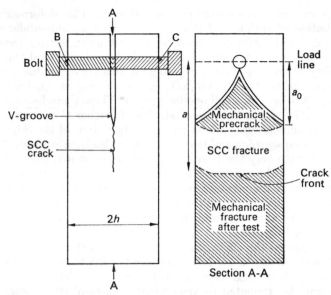

Fig. 41 Double cantilever-beam specimen, and stress intensity calibration (M. O. Speidel).

Stress Corrosion Cracking in Various Materials

Iron and Mild Steel. General. In many salt solutions (e.g. sodium chloride) iron suffers well distributed attack without localization or cracking, even if stress be applied; in other cases (e.g. sodium nitrite) there is immunity. Intergranular attack, which may become cracking if stress is applied, is met with in presence of nitrates (especially nitrates of weak bases, which yield acid solutions). Recently it has been studied in carbonate solutions. Caustic cracking in boilers has long been known and feared. Cracking produced in boiling 35% NaOH under conditions of slow deformation is the basis of a French test. These cases will be studied in turn.

Mild Steel in Nitrate Solution. Work in Parkins's laboratory[1] has helped to clear up apparently conflicting results regarding the effect of cold-work and composition on the tendency towards cracking in a nitrate solution. In cases where the steel used has initially been highly susceptible, cold-work has usually reduced that susceptibility. Where the steel has a high C content and is inherently more resistant, plastic deformation has served to induce cracking. Tests in $4N$ NH_4NO_3 show that 0.017% C steel becomes less susceptible to cracking as the degree of cold-work is increased; in contrast a 0.13% steel, which is not susceptible in the annealed condition, becomes susceptible after 10% cold-work. Thin-film electron microscopy shows that cold-work reduces the size and number of ridged grain-boundaries

1. M. J. Humphries and R. N. Parkins, *Amsterdam Cong.* 1969, p. 151. R. N. Parkins, P. W. Slattery, W. R. Middleton and M. J. Humphries, *Tokyo Cong. Ext. Abs.* 1972, p. 173; *Brit. Corr. J.* 1973, **8**, 117.

(the ridges are due to segregated carbon). It seems that deformation alters the distribution of carbon and that carbon-ridged grain-boundaries act as local cathodes and facilitate attack on the adjacent ferrite. Deformation reduces the amount of C segregated at grain-boundaries, and thus reduces susceptibility to stress corrosion cracking; it also greatly reduces the depth of penetration in boiling $4N$ NH_4NO_3 in the absence of applied stress but with the application of a small anodic current. These remarks apply to steel with only 0.017% C; in the case of steel with higher C content, deformation increases susceptibility whilst reducing the extent of the ridging.

The effects of quenching and tempering have also been studied. Quenching renders steels with very low C more resistant to cracking, whereas on steels with over 0.1% C, the opposite effect is observed. Subsequent tempering at a sufficiently high temperature increases the resistance of the steel with higher C content, but decreases the resistance of those containing very little C. These changes are attributed mainly to the nature and distribution of sites for localized corrosion.

Flis[1] has assessed the effect of carbon on the depth of intergranular corrosion (without stress) in $5N$ NH_4NO_3 and $Ca(NO_3)_2$. Small amounts of C increase susceptibility and large amounts decrease it. The adverse effect of small amounts is attributed to 'depassivating action' (the lowering of the protective properties of the film), whilst the beneficial effect of large amounts is attributed to 'impeding action' (increased deposition of magnetite and improvement of its protective properties).

Herzog[2] discusses the effect of purity on intergranular corrosion in a hot concentrated solution of a nitrate or of caustic soda. A protective film of magnetite is formed preferably on the ferrite faces, but the grain-boundaries adsorb NO_3^- or OH^- ions and remain active; selective attack occurs here, especially if small amounts of pearlite are present at the boundaries. On pure iron there is no grain-boundary attack, but iron with $0.01-0.08\%$ C is rapidly attacked under tension equal to 0.3 to 0.4 of the apparent elastic limit; attack becomes slow when the pearlite content exceeds 30%. Above 50% pearlite, it is necessary to deform the steel strongly to produce intergranular cracking.

Other experiments have been carried out by Hoar and Galvele[3] on mild steel in a hot nitrate solution under yielding conditions. The yielding of oxide-covered steel is accompanied by a large increase in anodic CD. The increase is almost independent of strain rate and is a rectilinear function of the total strain. The anodic CD on the yielding metal bared by rupture of the oxide can exceed 2 amp/cm^2. The rate of crack propagation was estimated microscopically and sometimes reached 0.15 cm/hour, which is equivalent to anodic dissolution at the advancing edge of the crack equivalent to 2 amp/cm^2. The sides of the crack are protected, probably because they are oxide-covered, so that the attack concentrated on the tip produces a narrow fissure.

1.   J. Flis, *Corrosion (Houston)* 1973, **29**, 37.
2.   H. Herzog, *Corrosion (Paris)* 1972, **20**, 187.
3.   T. P. Hoar and J. R. Galvele, *Corr. Sci.* 1970, **10**, 211.

Lees and Lockington[1] find that NH_4NO_3 produces intergranular trenching in the absence of stress, and this contributes to destruction if stress is subsequently applied. $Ca(NO_3)_2$ and $NaNO_3$, in the absence of stress, produce only grooves unless an external EMF is applied. Pre-immersion in $Ca(NO_3)_2$ has no marked shortening effect on subsequent stress corrosion life.

Since cracking in hot nitrate can follow grain-boundaries even in the absence of stress, it might be thought that stresses play only a minor part in promoting failure. However, another research by Flis[2] in which specimens were first subjected to 'pre-corrosion' (without stress) in a NH_4NO_3 or a $Ca(NO_3)_3$ solution, and then to stress in the same liquid, showed that time to failure decreases as the pre-corrosion period is increased—up to a certain limiting value. Conjoint action of corrosion and stress is necessary for the occurrence of rapid crack propagation leading to failure; this is true of both the nitrates used.

Tests on heat-treated low-alloy steels in aqueous $Cu(NO_3)_2$ or NH_4NO_3 at 100°C carried out by Cocks and Bradspies,[3] show that, even on highly polished surfaces, at least half the time required to produce failure is occupied in a process which is not accelerated by the application of a tensile stress. Their observations suggest that the good effect of peening in preventing cracking is connected with the production of a highly deformed surface layer within which the preferred cracking paths (generally grain-boundaries) have been disrupted; residual compressive stress may help.

Steel in Alkali. Stress corrosion cracking in carbonate solution has been studied by Parkins,[4] who considers that it may be responsible for several industrial failures, reported but hitherto unexplained. He has carried out potential sweeps in 2N $(NH_4)_2CO_3$ at 75°C under potentiodynamic conditions, and finds that curves obtained at a fast change of potential (100 mV/min) are sometimes different from those obtained at a slow change (20 mV/min); this difference is found over the potential range between −200 and −600 mV, but not elsewhere; he attributes it to the fact that the slow change allows time for a protective film to be established, except at places (like crack tips) where it will be continually broken. It is just over the potential range mentioned that stress corrosion cracking occurs. Parkins believes that a method for discovering solutions liable to produce cracking could be based on carrying out potentiodynamic sweeps, first under fast and then under slow conditions, with observations as to whether the attack is dependent on structure.

Such a method of prediction has, in fact, been applied to cracking in a phosphate system by Middleton,[5] who obtained polarization curves using both rapid and slow sweeps; the two curves were compared. The argument is based on the belief that cracking will occur, not at the potential where

1. D. J. Lees and N. A. Lockington, *Brit. Corr. J.* 1970, **5**, 167; 1971, **6**, 95.
2. J. Flis, *Brit. Corr. J.* 1973, **8**, 57.
3. F. H. Cocks and J. Bradspies, *Corrosion (Houston)* 1972, **28**, 192.
4. J. M. Sutcliffe, R. R. Fessler, W. K. Boyd and R. N. Parkins, *Corrosion (Houston)* 1972, **28**, 313. R. N. Parkins, *Brit. Corr. J.* 1972, **7**, 15.
5. W. R. Middleton, *Brit. Corr. J.* 1973, **8**, 62.

the divergence between the two curves is greatest, but rather at a potential slightly more noble than the active peak as obtained on the slow-sweep curve. The conclusion reached is that 'it is possible to predict from the polarization curves an approximate potential range where stress corrosion cracking will occur . . . in 1M $NaH_2PO_4.2H_2O$ at both room temperature and 100°C'. There appear, however, to be some unexpected features of the situation, not yet fully explained.

The mechanism of caustic cracking has been studied by Hoar and Jones.[1] The anodic behaviour of 0·1% carbon steel wire has been studied in 10M NaOH at 121°C at potentials between −0·60 and −0·85 V (standard hydrogen scale); if the wire is unstrained the CD falls to about 0·2 mA/cm^2 after 40 minutes, since a fairly coherent, adherent film of magnetite is formed; under strain, the CD at the parts bared by the cracking of the film can rise to over 8 mA/cm^2. The results support the theory of crack propagation based on the bared metal at the tip of the advancing crack dissolving much faster than the walls, which are protected by the film.

Herzog and his colleagues[2] have studied the behaviour of steel in boiling 35% NaOH under conditions of slow deformation. Their tests enable different steels (or the same steel after different treatments) to be classified as regards susceptibility to stress corrosion cracking; the aim has been to reproduce in the laboratory conditions typical of industrial practice. They find that oxygen favours the initiation of attack at grain-boundaries. Slow deformation propagates attack, leading finally to brittle fracture.

Parker,[3] whose work is concerned not only with NaOH but also nitrate, ammonium acetate and 'turbine condensate', considers that the initiation of stress corrosion, and also that of general corrosion, is associated with certain non-metallic inclusions rather than with grain-boundaries. He obtained pitting in all the liquids mentioned, and, at least in the case of NaOH, the stress corrosion cracking appears to be closely associated with the pits. The nature of the inclusions which cause pitting and ultimately cracking, is not stated, but in view of the results of other investigators, it would seem likely that sulphides are dangerous (this volume, p. 66).

Stainless Steel. The now generally accepted idea that intergranular attack, when it occurs, is generally due to chromium depletion has received confirmation from Frankenthal[4] and others;[5] it is believed in some quarters that the depleted zone is situated in spaces between dendrites of carbide. However, during attack by HNO_3, there is attack on the carbide particles themselves—which by some (but not all) authorities is held to be important. A research by Herbsleb and Schwabe[6] points to two types of intergranular attack, (1) typical attack resulting from Cr depletion (2) dissolution of carbo-nitrides segregated at grain-boundaries. The first occurs in steel which

1. T. P. Hoar and R. W. Jones, *Corr. Sci.* 1973, **13**, 725.
2. E. Herzog, M. Hugo and J. Bellot, *Corrosion (Paris)* 1970, **18**, 287.
3. J. G. Parker, *Brit. Corr. J.* 1973, **8**, 124.
4. R. P. Frankenthal and H. W. Pickering, *J. electrochem. Soc.* 1973, **120**, 23.
5. F. C. Wilson and F. B. Pickering, *J. Iron Steel Inst.* 1972, **210**, 37.
6. G. Herbsleb and P. Schwabe, *Werkstoffe u. Korr.* 1968, **19**, 484.

has become sensitized through inappropriate heat-treatment, and the second in steels which have been 'stabilized' by addition of Ti or Nb; the difference between the behaviour of steels of various compositions is attributed to different dissolution rates of the various carbo-nitrides produced. Tests by Bond and Lizlovs[1] indicate that both Nb and Ti are effective stabilizers in preventing the susceptibility as shown up in the Strauss test (using $CuSO_4$–H_2SO_4), but that steels containing Ti are sensitive to 65% HNO_3 whereas those containing Nb are not. Good results with Nb stabilized steels are reported by Spähn.[2]

Work in Vermilyea's laboratory[3] also confirms the essential correctness of the chromium depletion explanation for the susceptibility of austenitic stainless steel which has been treated in the range 550–800°C. Calculations have been made of the Cr concentration to be expected at different distances from the $Cr_{23}C_6$ precipitated at the grain-boundaries (also, in cases where the precipitate is discontinuous, at different distances from a particle measured along the grain-boundary). In these calculations an equilibrium concentration of Cr just outside a carbide particle has been assumed (many previous investigators have assumed this to be so low that it can be neglected). Grain-boundary attack is shown, by means of the electron microscope, to be continuous in some materials even when the particles do not constitute a continuous film around the grains, but in other cases the attack is patchy. Experimental studies of the rate of grain-boundary attack as a function of temperature and composition have confirmed the predictions of theory—especially as regards the interdependence between susceptibility to intergranular corrosion and the carbide particle spacing.

A study of austenitic stainless steel by Joshi and Stein[4] provides additional evidence of the importance of chromium depletion. After treatment at 1050°C for 2 hours followed by water-quenching, impurity elements like S are found to have segregated at the grain-boundaries, but the concentrations of Cr and Ni at the boundaries are not different from those prevailing in the grain-interiors. When, however, there is subsequent treatment in the range 600–850°C, Cr depletion and Ni enrichment at the boundaries occur. When the material is tested for intergranular susceptibility in the $CuSO_4$–H_2SO_4 solution specified for the Strauss test, there is good correlation with Cr depletion. If, however, it is tested in a HNO_3–$K_2Cr_2O_7$ solution, there is correlation between the corrosion properties and the S segregation; Cr depletion, although present, is not in that case the factor deciding behaviour. The peculiarity of the Huey test (involving boiling 65% HNO_3) is attributed by Hersleb[5] to the 'autocatalytic acceleration of corrosion in crevices as a consequence of the formation of Cr^{VI} and the depolarization of the HNO_3 reduction by hexavalent chromium'.

1.   A. P. Bond and E. A. Lizlovs, *J. electrochem. Soc.* 1969, **116**, 1305.
2.   H. Spähn and U. Steinhoff, *Werkstoffe u. Korr.* 1969, **20**, 733.
3.   C. S. Tedmon, D. A. Vermilyea and J. H. Rosolowski, *J. electrochem. Soc.* 1971, **118**, 192.
4.   A. Joshi and D. F. Stein, *Corrosion (Houston)* 1972, **28**, 321.
5.   G. Hersleb and W. Schwenk, *Amsterdam Cong.* 1969, p. 463.

Čihal[1] also regards Cr depletion in the neighbourhood of the carbide particles as the primary cause of intergranular attack. He says 'Extreme intercrystalline corrosion appears at such annealing temperatures and times at which the majority of the grain-boundary is occupied by a more or less continuous network of carbide particles.' There is, however, a combination of temperature and time at which precipitation of Cr carbide occurs without causing intergranular attack, because in this combination the rate of Cr diffusion is sufficient to prevent Cr depletion in the neighbourhood of the carbide particles. Truman[2] states that high Ni contents improve resistance to chloride cracking but have little effect on caustic cracking. A ferritic structure is more resistant to cracking than an austenitic structure, but shows a tendency to pitting.

Staehle[3] has provided extensive information about the effect of composition on the stress corrosion cracking of stainless steels in boiling 45% $MgCl_2$ at 154°C. He states that an alloy of iron containing 10–15% Cr and 10–15% Ni is highly resistant and can be improved by adding Al, Be and C. The alloy with 20% Cr and 15% Ni is resistant if it contains 0·1–2·0% Be, 0·1–2·0% Al and 0·05–0·2% C. Further addition of platinum elements, N, P, As, Sb, Bi and Mo are deleterious.

Japanese work[4] shows that, in steels containing less than 15% ferrite, impurities such as P and C increase the susceptibility to stress corrosion cracking; if the ferrite content falls between 15% and 30% they do not cause cracking. Another Japanese research[5] on austenitic stainless steel indicates that P, N and Mo are detrimental in boiling $MgCl_2$ at 154°C; if P and N are controlled, a 18/10 Cr/Ni steel can become immune to cracking. A third investigation[6] shows that Mo, although beneficial in reducing active chloride corrosion, increases susceptibility to cracking under stress. A fourth paper[7] states that P, N and Mo increase the susceptibility of austenitic steel in boiling 42% $MgCl_2$ (especially those rich in Ni and containing Si). A fifth paper[8] shows that immunity to cracking in boiling $MgCl_2$ at 154°C could be obtained by control of P and N; Mo and As are detrimental.

Loginow and Bates[9] find that C and Ni improve the resistance of annealed austenitic material, whilst Ni and Si improve cold-worked material; Mn, Cu, Cr, S and Al have little effect. N decreases the resistance of annealed material, whilst P and Mo decrease that of cold-worked material. A new steel with 18% Cr, 18% Ni, 2% Si and 0·06% C, with low contents of P and Mo, is recommended. Engell[10] suggests a reason for the bad effect of Mo in the

1.   V. Čihal and I. Kašova, *Corr. Sci.* 1970, **10**, 875.
2.   J. E. Truman, *Met. Mat.* 1968, **2**, 208.
3.   R. W. Staehle, J. J. Royuela, T. L. Raredon, E. Serrate, C. R. Morin and R. V. Farrar, *Corrosion (Houston)* 1970, **26**, 451.
4.   G. Ito, T. Ishiharo and Y. Shimizu, *Amsterdam Cong.* 1969, p. 75.
5.   M. Kowaka and H. Fujikawa, *Corr. Abs. (Houston)* 1971, p. 362.
6.   H. Okada and Y. Hosoi, *Tokyo Cong. Ext. Abs.* 1972, p. 165.
7.   M. Ueda, E. Sunami, H. Abo and T. Muta, *Tokyo Cong. Ext. Abs.* 1972, p. 159.
8.   M. Kowaka and H. Fujikawa, *Williamsburg Conf.* 1971, p. 437.
9.   A. W. Loginow and J. F. Bates, *Corrosion (Houston)* 1965, **25**, 15.
10.   F. Schreiber and H. J. Engell, *Werkstoffe u. Korr.* 1972, **23**, 175.

19/10 Cr/Ni alloy; it suppresses the formation of α-martensite, which, when present, is responsible for good resistance to cracking. However, commercial and pure materials are found to behave differently. Armijo[1] states that the intergranular corrosion of non-sensitized austenitic stainless steel is promoted by the presence of Si and P; elimination of these elements from austenitic steels of commercial purity decreases the susceptibility by a factor of 6 to 8; additions of C, N, O, Mn and S have little effect.

Another study by Herbsleb[2] of stainless steel containing Nb in 42% $MgCl_2$ boiling at 145°C shows that at any particular tensile stress there is a critical potential necessary for stress corrosion cracking, whereas at any particular potential there is a critical tensile stress needed for cracking. Re-examination of the results of earlier tests shows no incompatibility with the new findings.

The addition of Ni to ferritic steels containing 17–25% Cr and 0–5% Mo was found to have an unfavourable effect by Bond and Dundas;[3] an alloy free from Ni and Cu did not crack in $MgCl_2$ boiling at 140°C, but there was transgranular cracking if 1% Ni or 0·5% Cu was present. Uhlig[4] found that the addition of 2% Ni to a ferritic steel containing 18% Cr produces cracking in $MgCl_2$ boiling at 130°C or 154°C, but not in $Ca(NO_3)_2 + NH_4NO_3$ boiling at 110°C. Uhlig[5] has also discussed the effect of nickel in such an alloy, and of various salts ($NaNO_3$, NaI, sodium acetate and sodium benzoate) in the hot $MgCl_2$ solution used in cracking tests, on the basis of a shift of critical potential for cracking. The influence of nitrogen on cracking in austenitic stainless steel under corrosive conditions—a matter of some interest in view of the proposals to use nitrogen as a cheap austenite-stabilizer—has been studied by West.[6] Holzworth and Symonds[7] describe a method of preventing stress corrosion cracking by pre-treatment of stainless steel in 25% lithium silicate at 60–70°C.

Much of the work quoted above has been carried out on commercial materials. Important research on stainless steel made from high purity materials has been described by Montuelle and others.[8] A very pure material was obtained by melting the constituents in a plasma induction furnace and avoiding a refractory crucible. Such alloys (containing 18/14 Cr/Ni) were found to be completely insensitive to stress corrosion cracking in boiling $MgCl_2$. A duplex material with sensitive 18/10 alloy covering high purity 18/14, suffered cracking in the outer layer, but when the crack reached the boundary of the pure 18/14 material, its advance was arrested. It is con-

1. J. S. Armijo, *Corrosion (Houston)* 1968, **24**, 24.
2. G. Herbsleb and W. Schwenk, *Werkstoffe u. Korr.* 1970, **21**, 1.
3. A. P. Bond and H. J. Dundas, *Corrosion (Houston)* 1968, **24**, 344.
4. R. T. Newberg and H. H. Uhlig, *J. electrochem. Soc.* 1972, **119**, 981.
5. H. H. Uhlig, *J. electrochem. Soc.* 1969, **116**, 173 (with E. W. Cook); 1973, **120**, 1629 (with R. T. Newberg).
6. R. V. Maskell and J. M. West, *Brit. Corr. J.* 1971, **6**, 10.
7. M. L. Holzworth and A. E. Symonds, *Corrosion (Houston)* 1969, **25**, 287.
8. J. Montuelle (with F. Bourelier, M. da C. Belo, G. Chaudron and D. Colin), *Comptes rend. (Série C)* 1970, **270**, 903; 1971, **272**, 1098; 1972, **274**, 477.

cluded that stainless steels of sufficiently high purity can be proof against stress corrosion cracking.

Of the elements tested, carbon added to the very pure 18/14 alloy had no effect on stress corrosion cracking—as is generally agreed. Nitrogen (generally considered to be objectionable) did not provoke cracking. Mn alone did not modify behaviour, but Mn and N together rendered the material seriously sensitive to stress corrosion cracking—an effect commonly attributed to nitrogen alone. Pt greatly promoted stress corrosion cracking, rendering the attack intergranular on the very pure material with 14% Ni, but transgranular on that containing 10–12% Ni and on the commercial alloys.

A study of 26/1 Cr/Mo ferritic steel made from materials refined in the electron beam process has been carried out at Murray Hill, New Jersey.[1] Some of the materials thus produced are entirely resistant to stress corrosion cracking in boiling $MgCl_2$ and there is also good general corrosion resistance in organic acids, oxidizing acids, phosphoric acid and concentrated NaOH.

A useful review of the intergranular corrosion of Fe–Ni–Cr alloys has been provided by Cowan and Tedmon.[2] As regards the austenitic stainless steels, they state that recent results have provided strong evidence in favour of the Cr depletion theory. However, for nickel based alloys, the situation is more complicated. There appear to be at least two mechanisms which will cause them to be susceptible to intergranular attack. Much more research is required before the situation becomes clear, and the same is true of the duplex stainless steels which are increasingly used today. A literature survey of work on intergranular corrosion of austenitic stainless steel has been provided by Wilson.[3]

Other Alloy Steels.   B. F. Brown[4] has emphasized the concern felt about stress corrosion cracking in high strength steels owing to (1) the use of such steels at increasingly high stress levels, (2) the fact that high strength steels are more susceptible than low strength steels, (3) the consequences of failure —which may be catastrophic brittle fracture or rapid fatigue crack propagation. His detailed discussion of the whole subject should be studied in the original book and papers.

It might be remarked here that the great susceptibility of strong materials is not surprising. Such materials are strained to an extent that the increase of interatomic distance is far greater than that occurring in weaker materials; the escape of atoms will presumably become easier. Moreover, unless the

1.   R. J. Hodges, C. D. Schwartz and E. Gregory, *Brit. Corr. J.* 1972, **7**, 69.
2.   'Advances in Corrosion Science and Technology', M. G. Fontana and R. W. Staehle (Editors), Vol. 3 (1973); chapter by R. L. Cowan II and C. S. Tedmon Jr.: see p. 394.
3.   F. G. Wilson, *Brit. Corr. J.* 1971, **6**, 100.
4.   B. F. Brown, 'Stress-Corrosion Cracking in High Strength Steels and in Titanium and Aluminium Alloys' (Naval Research Laboratory, Washington), 1972. *Ericeira Conf.* 1971, p. 186. Also *Met. Reviews* 129 (1968). *J. electrochem. Soc.* 1969, **116**, 218 (with C. T. Fujii and E. P. Dahlberg), *Corrosion (Houston)* 1970, **26**, 249. *Amer. Soc. Test. Mat.* 1969; *Corr. Sci.* 1970, **10**, 839 (with G. Sandoz and C. T. Fujii).

strength of the oxide layer is increased by the alloying additions to an extent commensurate with the increased strength of the metallic phase, film-cracking will certainly take place readily on such material.

Green and Haney[1] have studied the behaviour of a maraging steel with 19/5/9 Ni/Mo/Co in NaCl–NaOH solutions. Failure becomes rapid when the amount of OH^- added is just insufficient to produce passivity. This is easily understood. The OH^- addition needed for passivity depends, of course, on the Cl^- content, and the amount can be ascertained by electrode potential measurements.

The transgranular cracking of austenitic steels containing Mn but no Cr has been studied by Bäumel.[2] This occurs more slowly than the intergranular cracking of Cr steels; another difference is that on the Cr steel intergranular attack can occur in the absence of stress, although it is accelerated when the material is in tension; for the Mn steels, stress is needed. Engell[3] states that certain steels containing Cr and C suffer intergranular attack in the precipitation annealed condition but transgranular attack after solution annealing. Certain low-carbon steels containing Cr up to 4% are susceptible to transgranular attack, but increase of the Cr content to 8% changes the form of attack from cracking to pitting.

The effect of various constituents and also heat-treatment on the behaviour of low-alloy steels in $Ca(NO_3)_2$ has been studied by Rädeker.[4] P has little effect, whilst Cr (even when present up to 2%) does not improve resistance. Mn and Mo generally produce improvement, but V is beneficial only in certain cases. Overheating produces sensitivity, annealing at 750°C produces relatively good resistance, whilst normalizing is intermediate in effect. The results suggest that the situation is not simple; those wishing to apply them in a practical case would do well to study Rädeker's diagrams.

Eutectoid steel wire is used in civil engineering in a high strength condition obtained either by cold drawing or by isothermal quenching or (again) by oil-quenching followed by tempering. Gilchrist[5] has examined its behaviour under conditions chosen to represent engineering applications. The time to failure depends on the stress applied and also on the potential. Cathodic polarization greatly shortens the time to failure—evidently owing to hydrogen; a small movement of the potential in an anodic direction also reduces the time to failure—evidently owing to anodic attack—but in some solutions the time starts to increase again above a certain potential value. Both effects have been obtained after certain heat-treatments in solutions containing $Fe_2(SO_4)_3$, NH_4NO_3 or H_2S (at pH 5·3), but the behaviour depends both on the previous heat-treatment and on the temperature of the liquid used to produce cracking.

Aluminium Alloys. Certain alloys of aluminium, suitably heat-treated, attain high strength; since they are also light, they are extensively used in

1. J. A. S. Green and E. G. Haney, *Corrosion (Houston)* 1967, **23**, 5.
2. A. Bäumel, *Werkstoffe u. Korr.* 1969, **20**, 387.
3. W. Prause and H. J. Engell, *Werkstoffe u. Korr.* 1971, **22**, 421.
4. W. Rädeker and B. N. Miskro, *Werkstoffe u. Korr.* 1970, **21**, 69.
5. J. D. Gilchrist and R. Narayan, *Corr. Sci.* 1971, **11**, 281.

aircraft. Unfortunately they are often susceptible to cracking in certain environments (including moist air) at stresses far below those required for gross yielding. This is to be expected. Aluminium is an extremely reactive substance (the free energy drop associated with its oxidation greatly exceeds that attending the oxidation of heavy metals), and the excellent corrosion resistance of aluminium in absence of stress is connected with the ready formation of a highly protective oxide-film, which is also a very poor electron conductor. On unalloyed aluminium, local breakage of the film need not necessarily provoke a very serious situation, since the film-covered region around the break, although providing a large cathodic area, will also be inefficient for the cathodic reaction owing to the low electron conductivity. If, however, alloying elements have been added to improve the strength, the electronic conductivity of the oxide may well be increased, and the cracking of the film may then produce a situation more favourable to attack at the tip of an advancing crack. If, in addition, the heat-treatment designed to enhance strength has led to the presence of three or more phases near the grain-boundaries, and if one of these (either the precipitated phase, or the depleted zone left by the precipitation) suffers anodic attack at a potential where the other two cannot suffer it, there should be no surprise if the fissure is found to advance rapidly into the interior of the metal. However, the situation is not always simple and further factors may have to be taken into account, as suggested below.

Galvele[1] has studied the intergranular corrosion of an Al alloy containing 4% Cu without applied stress, after heat-treatment designed to precipitate particles of the θ-phase (generally regarded as Al_2Cu) along the grain-boundaries, leaving a depleted zone which, if equilibrium is achieved, should contain 0·2% Cu; the interior of the grains is assumed still to contain 4% Cu. It is recalled that various authors have demonstrated that in the corroding alloy a potential difference of 100 mV exists between grain-boundary and grain body. Nevertheless it is argued that 'a mechanism of anodic and cathodic zones does not explain why the presence of Cl⁻ ions is necessary to get intergranular corrosion. . . . The purpose of this work was to determine the function of the Cl⁻ ions.' It is shown by iV curves with pure Al in 0·1M NaCl that at a certain 'critical potential', the current increases very suddenly, and at that point pitting is detected; this supports, at least for the alloy in question, the existence of a critical pitting potential—about which doubt had been expressed. The value of the critical potential varies with the Cl⁻ concentration and is different for the solid solutions containing 4% and 0·2% Cu respectively; the former shows values similar to those obtained in experiments on specimens composed of Al_2Cu, whilst the latter shows values similar to 'pure' Al (99·99 or 99·9999%). It is considered that the intergranular corrosion of the 4% Cu alloy is due to a difference in the breakdown potential of the two phases. It will occur only if

(1) there is a solute depleted zone along the grain-boundaries,
(2) the medium contains anions which break down passivity,

1.   J. R. Galvele and S. M. de Micheli, *Amsterdam Cong.* 1969, p. 439; also *Corr. Sci.* 1970, **10**, 795.

(3) the breakdown potential of the depleted zone is lower than that of the grain bodies,

(4) the natural corrosion potential of the alloy lies above the breakdown potential of the depleted zone and below that of the grain bodies.

The research just mentioned has been discussed by Kaesche,[1] who points out that the situation is not so simple as is sometimes imagined, but ends with the words 'The Dix concept of intergranular corrosion eventually has been shown to be correct.'

Cocks[2] has studied the aluminium alloy containing 5·5% Zn, 2·2% Mg, 1·2% Cu and 0·18% Cr. This is generally reputed to be highly sensitive to stress corrosion, but the new work shows that in the opening stages corrosion can sometimes be equally damaging in the absence of stress. For instance, machining of a surface may have produced a disrupted layer in which the grain-boundaries have been microscopically fragmented; in such a condition, the corrosion is not, at the outset, accelerated by stress. However, when the disrupted layer has been penetrated, true stress corrosion cracking can be developed.

Watkinson and Scully[3] have studied the behaviour of an alloy containing 6% Zn and 3% Mg, in environments which cause little general corrosion; the moisture in the laboratory atmosphere was found to set up cracking under stress on material which had been treated for 1 hour at 460°C, water-quenched and then heated for 24 hours at 120°C. Relatively long initiation times and the infrequency of secondary cracks suggested that the starting of a crack was a difficult process. The discontinuous curves obtained with sensitive apparatus indicate that bursts of fast fracture do occur, but the movements observed as beam deflections cannot distinguish between an increase in crack length and the effects of plastic deformation which would allow wider opening of a crack.

An important research at Swansea[4] concerns the alloy with 4·3% Zn and 2·4% Mg. This alloy (along with certain other alloys containing Zn and Mg) is peculiar in being susceptible to cracking under stress, although it does not normally show intergranular corrosion in the absence of stress—an observation that may be difficult to reconcile with the widely held view that stress corrosion, when it follows grain-boundaries, is just a stress-intensified form of intergranular attack. The new work, in which potentiometric control and various refinements have been introduced, shows that, if an anodic potential of −921 mV (saturated calomel electrode) is applied, intergranular corrosion is, in fact, obtained without applied stress. Structural changes, obtained by suitable heat-treatment, which eliminate intergranular corrosion, impart high resistance to stress corrosion. This would seem to clear up the anomaly. The best procedure for removing sensitivity is a double heat-treatment; after

1. H. Kaesche, *Williamsburg Conf.* 1971, p. 516; esp. p. 524.
2. F. H. Cocks (with J. F. Russo and S. B. Brummer), *Corrosion (Houston)* 1969, **25**, 345; 1970, **26**, 157.
3. F. E. Watkinson and J. C. Scully, *Amsterdam Cong.* 1969, p. 140.
4. D. Eurof Davies, J. P. Dennison and M. L. Mehta, *Williamsburg Conf.* 1971, p. 608; also *Corrosion (Houston)* 1971, **27**, 371.

solution treatment and *air-cooling*, the material is left for 72 hours at ambient temperature, then 24 hours at 90°C followed by a second treatment for 24 hours at 160°C. It was found that this produces a microstructure with a few widely spaced particles precipitated on the grain-boundary and a wide (7000 Å) precipitate free zone; in contrast, material which had been *water-quenched*, followed by the same after-treatment, had a narrower zone (700 Å) and closely spaced grain-boundary precipitate. The results would seem to suggest that the grain-boundary precipitate acts as the anodic path; this is also indicated by the results of Middleton and Parkins[1] and also those of Sedriks, Green and Novak.[2]

The exfoliation (layer corrosion) of an alloy with 5% Zn and 1% Mg has been studied in Swedish work.[3] Naturally aged alloys suffer exfoliation, but artificially aged alloys do not. In the case of welded specimens, artificial ageing after welding does not greatly reduce susceptibility in the precipitate free zones produced at a distance from the weld, but it does reduce it in the zone adjacent to the weld-band. An electrochemical interpretation is provided; particles of the α-Al(Fe,Me)Si phase constitute cathodes, whilst in naturally aged alloys the anodic attack falls on adjacent areas which have become enriched in Zn and Mg; in artificially aged material, the $MgZn_2$ provides anodes, but the particles are evenly distributed, so that there is no exfoliation and the corrosion is fairly general.

Titanium Alloys. The introduction of titanium as a relatively cheap material appeared at one time to be a red letter day in the fight against corrosion; later the discovery that titanium alloys were subject to stress corrosion cracking came as a serious disappointment. If, however, it is a fact that in titanium alloys the threshold value of K_{ISCC} is something to be relied upon—which is not the case with some other light alloys—the special value of titanium alloys may once more come to be accepted.

An account of the behaviour of titanium alloys under conditions of stress and corrosion has been provided by T. R. Beck.[4] In the case of the alloy containing 8% Al, 1% Mo and 1% V, tested in 0·6M KCl and in 12M HCl, he was able to suggest a model representing the electrochemical basis of a propagating stress corrosion crack, which accounts for his observations in a semi-quantitative manner. He rejects explanations based on wedging of the crack opening with voluminous oxide and also those connected with brittle cracking of the oxide, on the grounds that the crack actually observed progresses so quickly that it outruns the oxide. Beck finds that both in water and in methanol, acceleration is produced by Cl^-, Br^- and I^-, whilst inhibition is obtained with SO_4^{2-} and NO_3^-; cathodic protection also tends to inhibit cracking.

1.   W. R. Middleton and R. N. Parkins, *Corrosion (Houston)* 1972, **28**, 88.
2.   A. J. Sedriks, J. A. S. Green and D. L. Novak, *Williamsburg Conf.* 1971, p. 569.
3.   E. Mattsson, L. O. Gullman, L. Knutsson, R. Sundberg and B. Thundal, *Brit. Corr. J.* 1971, **6**, 73.
4.   T. R. Beck, *Ericeira Conf.* 1971, p. 64; *J. electrochem. Soc.* 1968, **115**, 890.

A discussion of titanium alloys in aqueous solution has been provided by Feeney and Blackburn.[1] They find that all elements which promote stress corrosion cracking are those which strengthen a solid solution. In some systems (e.g. Ti–Al) additional strengthening occurs by the formation of ordered particles with the same crystal structure as that of the matrix. If sufficient alloying element is present, specimens will fail in a brittle manner when fractured by impact loading or in inert environments; for example, the transgranular cleavage of Al alloys and the intergranular separation of an alloy containing 11·5% Mo, 6% Zr and 4·5% Sn is of the same type as stress corrosion cracking. Glide processes occur by movement of dislocations in planar arrays that produce localized displacements. In those alloys which fail by transgranular separation, the metallurgical parameters that influence cleavage failure (e.g. grain-size) have a similar effect on stress corrosion cracking.

The behaviour of titanium in a solution of iodine in methanol has also received study.[2] In the case of pure unalloyed titanium, the mechanism appears to be anodic dissolution at the crack tip accelerated by stress, whereas in an alloy containing 5% Al there is mechanical transgranular failure.

A useful survey of the practical situation has been provided by Cotton.[3] He states that titanium can crack in some circumstances in contact with hot dry NaCl and in aqueous saline solutions at ambient temperature. Cracking can take place in certain chlorinated hydrocarbons used as grease removers, but this can in general be prevented by the addition of suitable stabilizers and inhibitors. In alloys free from Al, trouble can generally be avoided. When Al is present, there is a hazard, and this is accentuated at welds. In Ti alloys containing Al, a crack started by fatigue can be propagated further by static stress in salt solution, even at ambient temperature, if the applied stress exceeds a certain limit. When an aircraft component is in contact with wet salt, which later becomes dried by heat from the engine, this may be expected to produce hot-salt cracking. No example from service appears to be on record, but the effect has been produced and investigated in the laboratory.

The mechanism of the cracking in alloys containing α-titanium has been studied by Scully and Powell,[4] who find some evidence for the effect of hydrogen embrittlement.

Bretle[5] points out that most Ti alloys have an $\alpha + \beta$ structure. Elements which stabilize the β-phase (Nb, Mo, V and Ta) generally reduce or eliminate the susceptibility to cracking in aqueous chloride solution. Al, if present above 5% by weight, increases the susceptibility, apparently by producing Ti_3Al as an ordered phase. Cl^-, Br^-, I^- and ClO_3^- favour cracking, but in methanol solutions, cracking can occur in their absence. The results quoted

1.   J. A. Feeney and M. J. Blackburn, *Ericeira Conf.* 1971, p. 355.
2.   A. J. Sendricks, J. A. S. Green and P. W. Slattery, *Corr. Abs.* (*Houston*) 1969, p. 103.
3.   J. B. Cotton, Private Comm. Aug. 30, 1972.
4.   J. C. Scully and D. T. Powell, *Corr. Sci.* 1970, **10**, 719.
5.   J. Bretle, *Met. Mat.* 1972, **6**, 442.

provide considerable evidence that hydrogen and hydride formation are important in stress corrosion cracking.

Copper and Copper Alloys.   The fact that unalloyed copper appears to be capable of exhibiting stress corrosion cracking was mentioned on p. 319. Uhlig[1] expresses doubts as to whether this is really stress corrosion cracking, but Kruger[2] has produced cracking in copper acetate solution—which appears to be genuine; it was not, however, obtained in a copper sulphate solution. Japanese work[3] also provides evidence of stress corrosion cracking in unalloyed copper. It was found that hard-drawn wire can suffer cracking under a bending stress in a moist ammoniacal atmosphere with temperatures alternating between 70°C and room temperature. This might, perhaps, be criticized by some on the grounds that the fluctuating temperature brings the treatment into the category of corrosion fatigue. However, it was found that transgranular cracks can be produced on the surface of annealed copper uniaxially stressed in dilute ammonia at 70°C, and this is best ascribed to preferential anodic attack at the tip of the crack.

The development of cracking in brass has some historical interest. In the 19th century, it was found that brass cartridge cases stored in India developed cracking at a certain season of the year, when the air contained ammonia and moisture combined with a high temperature; it came, therefore, to be known as 'season cracking'. This led to Moore's classical work which showed that the cracking, although due to ammonia, was only serious if the brass contained internal stresses; suitable heat-treatment designed to relieve these stresses largely solved the problem.

Season cracking is today a less serious problem, but under service conditions where applied stresses are necessarily present, the cracking of brass, especially in presence of ammonia, is still a cause of trouble. It can be restrained by using brass of suitable composition. Parkins[4] finds that As and Sn increase the resistance of 80/20 brass to stress corrosion cracking in NH_3 at pH 7·3 and 11·3, and he attributes the improvement to the influence of these elements on film-forming properties rather than to their effect on the mechanical properties of the alloy—e.g. the stacking fault energy. Indian measurements[5] based on the electrical resistance of α-brass during stress corrosion cracking indicate that an incubation period occurs on annealed material, but not on cold-worked brass.

Japanese work,[6] based on the solution containing $CuSO_4$ and $(NH_4)_2SO_4$ used by Mattsson and by Hoar (1968 book, p. 264), shows that the stress and strain required for intergranular cracking is greater than that needed for transgranular cracking in NH_3 vapour. In the first case, a thick black film is produced which is believed to retard the movement of dislocations at and near the surface, arresting the development of slip-steps; at high angle

1.   H. H. Uhlig and D. J. Duquette, *Corr. Sci.* 1969, **9**, 557.
2.   E. Escalante and J. Kruger, *J. electrochem. Soc.* 1971, **118**, 1062.
3.   Y. Suzuki and T. Yamaguchi, *Tokyo Cong. Ext. Abs.* 1972, p. 137.
4.   B. C. Syrett and R. N. Parkins, *Corr. Sci.* 1970, **10**, 197.
5.   A. K. Lahiri and S. P. Nayak, *Brit. Corr. J.* 1971, **6**, 84.
6.   M. Takano and S. Shimodaira, *Trans. Japan Inst. Met.* 1967, **8**, 239.

grain-boundaries, the oxide-film is imperfect and slip occurs, leading to intergranular cracking. When the arresting action is weak, the initial micro-cracks grow along slip-planes and transgranular cracking is set up.

Work at New Haven[1] has determined the best annealing temperature for obtaining the maximal resistance to stress corrosion cracking in the case of three new commercial alloys. The tests were carried out on U-bend specimens in moist air containing NH_3 vapour and also in Mattsson's solution brought to pH 7·2 by addition of NH_3.

Indian work[2] shows that in annealed samples tested in Mattsson's solution, anodic polarization produces a minimum of cracking time, accompanied by a transition in mode from intergranular to transgranular. In some cold-worked samples which normally crack in a transgranular fashion, cathodic polarization produces a transition to the intergranular mode. Stress corrosion cracking is met with in solutions free from Cu^+ or Cu^{2+}, but not in solutions free from NH_3.

Pugh[3] considers that two distinct mechanisms can operate in the cracking of a α-brass in presence of ammonia. The first occurs in solutions which produce a tarnish film; here failure occurs by the rupture of that film. Intergranular penetration occurs in unstressed specimens, and it is suggested that the role of stress, when present, is primarily to accelerate penetration by continually rupturing the film and thus eliminating the necessity of transport over long distances through the film. The tarnish-rupture mechanism is also thought to be responsible for the failure of α-brass in certain citrate and tartrate solutions, and that of pure copper in aqueous cupric acetate—a case mentioned above. The industrially important type of failure, season cracking, is considered to be due to this mechanism. Pugh recommends that efforts to combat the practical problem should be focused on understanding and controlling the tarnishing process. In contrast, the cracking produced in non-tarnishing ammoniacal solutions is probably due to preferential anodic dissolution at the crack tip. This would seem to have less practical importance.

Measurements of films produced after immersion of 30/70 brass in ammoniacal $CuSO_4$ solution have been made by Booker and Salim[4] using gravimetric, electrometric and colorimetric methods; fairly good agreement between the three sets of numbers was obtained. Weight gain measurements are meaningless in this case, since the metal is passing into solution, and the change measured represents a difference between a gain and a loss; however, if the film is dissolved away in aqueous ethanolamine (50%), which hardly attacks the substrate (a correction based on behaviour in a blank experiment can be applied), numbers are obtained which possess significance. The potential–time curve obtained during cathodic reduction at constant current shows a short limb at a potential which falls with time, indicating the reduction of CuO, and a long limb with potential constant over the long period of

1.   J. M. Popplewell, R. P. M. Procter and J. A. Ford, *Corr. Sci.* 1972, **12**, 193.
2.   S. C. Sircar, U. K. Chatterjee, M. Zamin and H. G. Vijayendra, *Corr. Sci.* 1972, **12**, 217.
3.   E. N. Pugh, *Ericeira Conf.* 1971, p. 418.
4.   C. J. L. Booker and M. Salim, *Nature Phys. Sci.* 1972, **239**, 62.

time, indicating reduction of Cu_2O. It should be noted that the film, although largely Cu_2O, is black.

Although stress corrosion cracking in brass is commonly attributed to NH_3 in the atmosphere, it seems likely that SO_2 may in fact often be the cause. This view appears to have been held in Russia for some time,[1] and has been confirmed by British experiments carried out by Cotton[2] on looped specimens of 70/30 brass with air containing SO_2. When the SO_2 concentration was less than 0·01%, there was no cracking; at about 0·05% one or two deep cracks developed, but when the concentration reached 1% there was overall attack and possibly a little shallow cracking.

Zinc Alloys. About 30 years ago, a 'superplastic' zinc alloy containing 22% (wt) Al was introduced, but was found to be extremely susceptible to intergranular corrosion. Recent research[3] into the effect of ternary additions has shown that Cd has an adverse effect, that Ti and Si leave properties almost unchanged, but that Cu, Mg and Ag are beneficial. Of these Cu seems the best to adopt, since it practically abolishes the tendency to intergranular corrosion when 1% (atomic) is added; this addition actually enhances the superplastic properties. Mg is slightly more effective than Cu in combating intergranular corrosion, but reduces the superplasticity. Ag is more effective than Cu at very low levels, but less effective at high levels; it is, of course, more expensive.

Alloys of Noble Metals. New information, supported by electron microscopic evidence, has been presented by Swann.[4] He has worked mainly on alloys containing gold, particularly Ni containing 16·5% Au, and also the intermetallic compound Cu_3Au. How far the mechanism suggested would apply to commercial alloys is uncertain. Noble metal alloys are exceptional in their ability to become embrittled even in the unstressed condition. This type of embrittlement is very severe in Ni–Au alloys, which gradually disintegrate on exposure to $FeCl_3$ solutions. The process can be observed in the electron microscope; specimens develop cracks growing from the centres of colonies of tunnels; the colonies possess a sponge-like morphology. The nucleation of such fractures is ascribed to radial tensile stresses set up by the change in lattice parameter of the sponge during dissolution of the less noble component. The fracture takes place entirely within the sponge, indicating that the rate of cracking is controlled by the corrosion process. The tunnels can be formed with great rapidity; an alloy of composition Cu_3Au exposed for 30 seconds to a solution of $FeCl_3$ produces colonies often roughly circular in cross section; the diameter of an individual tunnel may be only about 75 atoms, but the diameter of a colony of such tunnels may be 1000–2000 Å.

A corrosion tunnel is formed when the dissolution rate along an axis is n times greater than the dissolution rate in all directions perpendicular to the

1. N. D. Tomashov, 'Theory of Corrosion and Protection of Metals' (Macmillan), p. 601, quoting A. V. Bobivev.
2. J. B. Cotton, Priv. Comm. Sept. 15, 1974.
3. D. L. Dollar, J. A. Clum and R. E. Miller, *Corrosion (Houston)* 1972, **28**, 296.
4. P. R. Swann, *Ericeira Conf.* 1971, p. 113, *Williamsburg Conf.* 1971, p. 104.

axis; the value of n may be anything between 5 and 100, according to the material and environment. If n is much less than 5, the feature is generally described as a pit. In order to form a perfectly cylindrical tunnel, n must be infinity; however, in practice the tunnel is not visibly tapered unless n is less than about 50.

The crack propagates through the colonies of corrosion tunnels, which tend to grow in groups along directions corresponding to slip traces. In Swann's work no cracks were observed to propagate beyond the corroded regions, and he believes that the stress corrosion of the alloys studied is controlled by the electrochemical process which leads to tunnel formation.

Further work on single-phase alloys containing noble metals is provided by Graf.[1] He considers the noble alloys to be particularly suitable to clarify the fundamental causes of stress corrosion cracking, because

(1) their reactivity is smaller than that of alloys of base metals;
(2) there is complete miscibility of the components;
(3) there is no influence of heat-treatments or impurity;
(4) in Ag–Au alloys there is no change of stacking fault energy;
(5) up to a certain Au content, corresponding to the so-called 'parting limit', there is no formation of protective layers; general surface attack and also stress corrosion cracking can both take place. If the parting limit is exceeded, neither surface attack nor cracking will occur.

Graf believes that a distinct electrochemical process is at work, with increased corrosion of the grain-boundaries and of disturbed areas on the surfaces (e.g. slip bands) even in the absence of tensile stress. As a result of this attack, notches are formed which, after an appropriate time, reach so deeply into the metal that the roots of the notches start flowing if tensile stresses are present. The time which elapses between the onset of corrosion and the start of the flowing process at the roots may represent the induction or incubation period. When the roots start to flow, the reactivity of the flowing areas greatly increases. The notches rapidly develop into cracks, and as soon as the tensile strength is reached in the ever-diminishing transverse section, the specimen ruptures with a ductile fracture extending over the unattacked part of the specimen.

Further References

Papers. 'Stress corrosion crack paths in Al–Zn–Mg alloys'.[2]
'Natural conditions for caustic cracking of a mild steel'.[3]
'Chemical analysis of the liquid within a propagating stress corrosion crack in 70/30 brass immersed in concentrated ammonium hydroxide'.[4]
'Electrochemical aspects of stress corrosion of steels in alkaline solutions'.[5]

1. L. Graf, *Ericeira Conf.* 1971, p. 399.
2. W. R. Middleton and R. N. Parkins, *Corrosion (Houston)* 1972, **28**, 88.
3. J. E. Reinoehl and W. E. Berry, *Corrosion (Houston)* 1972, **28**, 151.
4. H. Leidheiser and R. Kissinger, *Corrosion (Houston)* 1972, **28**, 218.
5. G. J. Bignold, *Corrosion (Houston)* 1972, **28**, 307.

'Stress corrosion cracking of carbon steel in carbonate solutions'.[1]

'The role of hydrogen in the mechanism of stress corrosion cracking of austenitic stainless steel in hot chloride media'.[2]

'A correlation between acoustic emission during stress corrosion cracking and the fractography of cracking of the zircaloys in various media'.[3]

'Cathodic protection and hydrogen in stress corrosion cracking'.[4]

'Stress corrosion cracking of α-brass in ammonium sulphate solution'.[5]

'Effect of iron on the stress corrosion resistance of 90/10 Cu/Ni in ammoniacal environments'.[6]

'Grain-boundary segregation'.[7]

'On the chemistry of the solution at the tips of SCC cracks'.[8]

'Resistance à la corrosion sous tension des aciers inoxydable de pureté élevée'.[9]

'Use of microelectrodes for study of stress-corrosion of aluminium alloys'.[10]

'Fracture Mechanics: how it can help Engineers'.[11]

'Aqueous corrosion of $\langle 001 \rangle$ and $\langle 011 \rangle$ tilt boundaries on aluminium bicrystals'.[12]

'Influence of surface conditions on the resistance of stainless austenitic Cr–Ni steels to transcrystalline stress corrosion'.[13]

'Problems of intercrystalline corrosion susceptibility of an inadequately stabilized steel'.[14]

'The role of hydrogen in the stress-corrosion cracking of titanium alloys'.[15]

'Some observations on the films formed on α-brasses tarnished in an ammoniacal solution of copper sulphate'.[16]

'Protection of aluminium alloys against stress corrosion cracking in saline waters by properly oriented anodic coatings'.[17]

1.   J. M. Sutcliffe, R. R. Fessler, W. K. Boys and R. N. Parkins, *Corrosion (Houston)* 1972, **28**, 313.
2.   B. E. Wilde and C. D. Kim, *Corrosion (Houston)* 1972, **28**, 350.
3.   B. Cox, *Corrosion (Houston)* 1972, **28**, 473.
4.   C. F. Barth and A. R. Troiano, *Corrosion (Houston)* 1972, **28**, 259.
5.   U. C. Bharta and A. K. Lahiri, *Brit. Corr. J.* 1973, **8**, 182.
6.   J. M. Popplewell, *Corr. Sci.* 1973, **13**, 593.
7.   M. P. Seah and E. D. Hondros, *Proc. Roy. Soc. (A)* 1973, **335**, 196.
8.   A. J. Sedriks, J. A. S. Green and D. L. Novak, *Corrosion (Houston)* 1971, **27**, 198.
9.   M. C. Belo and J. Montuelle, *Corrosion (Paris)* 1972, **20**, 105.
10.   J. A. Davis, *Williamsburg Conf.* 1971, p. 168.
11.   L. P. Pook, *Trans. N.E. Coast Instn Engrs Shipbuilders* 1974, **90**, 77.
12.   J. V. Boos and C. Goux, *Williamsburg Conf.* 1971, p. 556.
13.   G. Herbsleb, *Werkstoffe u. Korr.* 1973, **24**, 867.
14.   V. Masařík and V. Čihal, *Werkstoffe u. Korr.* 1974, **25**, 330.
15.   S. Orman and G. Picton, *Corr. Sci.* 1974, **14**, 451.
16.   E. F. I. Roberts, C. J. L. Booker, P. Osborne and M. Salim, *Corr. Sci.* 1974, **14**, 307.
17.   Th. Shoulikidis and A. Karageorgos, *Brit. Corr. J.* 1975, **10**, 17.

CHAPTER XVII

Corrosion Fatigue

Preamble

Picture presented in the 1960 and 1968 volumes. It has long been known that a piece of metal, subjected to alternating stress much lower than the stress which would be needed to produce fracture if applied continuously, can set up a crack which gradually extends through the cross section, ultimately causing breakage. Under non-corrosive conditions, there is a stress range below which, it is generally believed, no cracking will develop, and the life will, in theory, be infinite; even if this is an over-simplification, it is possible, for some practical purposes, to assume safety over a reasonable period if the stress range does not exceed a certain 'fatigue limit'. If, however, corrosion is present, there is probably no 'fatigue limit', and fracture will always occur, although the time required is lengthened if the stress range is kept low.

Our knowledge of the facts of corrosion fatigue is still largely due to the classical work of Haigh and McAdam—which deserves study today, even though doubts may have been expressed regarding some of their conclusions. In some of McAdam's experiments it was found that when a material was subjected to alternating stress in a corrosive environment, pits were first formed; some of these, causing local stress intensification, developed into cracks. Most of McAdam's experiments were two-stage tests; during the first stage the material was subjected to corrosion fatigue, but during the second stage there was no corrosion and the material was subjected to dry fatigue; the period which could be withstood during this second stage before fracture occurred can be regarded as a measure of the strength remaining after the corrosion fatigue of the first stage; it varies, of course, with the stress range and with the duration of the first stage, besides depending on the material under examination.

McAdam's one-stage tests are also worthy of study. He plotted the logarithm of the life to breakage against the stress range applied, but no simple relationship was obtained. Later Simnad (1960 book, p. 716), performing experiments in triplicate on steel (0·19% C) subjected to alternating stress in N/10 KCl, obtained good agreement between the various sets of values; the points fell well on a straight line when the stress range was plotted against the logarithm of the life; it is not known to what extent this simple law would apply to other materials and solutions; it certainly does not seem to be universally obeyed, and the subject deserves further examination.

An important research by Forsyth (1968 book, p. 270) on an Al alloy containing 7·5% Zn and 2·5% Mg established two modes of cracking:

(1) a *shear mode*, with the fracture generally following a plane inclined at approximately 45° to the applied tensile stress (such a plane will receive maximum resolved shear stress).

(2) a *tensile mode* with the fracture running roughly at right angles to the direction of the applied tensile stress.

In the fully aged alloy, the initial cracking occurs along slip-planes inclined at an angle close to that corresponding to the maximum resolved shear stress; but later the fracture becomes approximately normal. In the absence of corrosion, above the fatigue limit, the failure is thought to occur by one of the mechanisms generally accepted by physicists and engineers for dry fatigue. Below the fatigue limit, a few thousands of cycles of stressing produces fine slip (not always easy to observe), often following two sets of planes, so that dislocations become locked at intersections, forming sessile groups; material thus becomes hardened and no further deformation can occur in absence of corrosion. When, however, corrosion is present, the destruction of the points where locking has occurred radically alters the situation, and cracking can develop.

Work at the Fulmer Institute (1968 book, p. 271) shows that when the initial mode is at 45°, cathodic protection, if applied, is effective in the prevention of cracking; if the mode is at 90°, cathodic prevention is ineffective. This rather suggests that the 45° mode is connected with anodic corrosion, whilst the 90° cracking is the result of the hydrogen produced by the cathodic reaction.

Many methods have been tried for the prevention of corrosion fatigue—not without success. These include protective coatings, water-treatment, peening, nitriding and cathodic protection. The choice must, however, depend on the material and service conditions.

Developments during the period 1968–1975 and Plan of the new Chapter. Surprisingly little experimental work on corrosion fatigue has been published during the period under review. Considerable attention has been directed to the mechanism of fatigue under dry conditions, but the influence of corrosion has often been neglected by those writing on the subject; this is a somewhat disturbing feature of the present situation.

The contents of the short chapter which follows can best be indicated by section headings. These are 'Relation between Dry Fatigue and Corrosion Fatigue', 'Relation between fatigue risks and tensile strength', 'Relation between stress-range and fatigue life', 'Relation between Stress Corrosion Cracking and Corrosion Fatigue', 'Relation between Corrosion Fatigue and Frettage', 'Fatigue at Elevated Temperatures', 'Electrochemical Aspects of Corrosion Fatigue', 'Protective Measures', 'Fatigue of Offshore Structures'.

General

Relation between Dry Fatigue and Corrosion Fatigue. Several reviews of knowledge regarding fatigue are today available.[1, 2, 3] The portions dealing with dry fatigue should receive attention even from readers

1.   'Metal Fatigue: theory and design'. A. Madayag (Editor), 1969 (Wiley): review by K. A. Ridal, *J. Iron and Steel Inst.* 1970, **208**, 93.
2.   P. J. E. Forsyth, 'Physical Basis of Metal Fatigue' (Blackie), 1969.
3.   W. J. Plumbridge and D. A. Ryder, *Met. Reviews* 1969, **14**, 119.

whose main interest is corrosion fatigue. The fact that some of these authoritative accounts of the subject barely mention the corrosion aspect must, however, cause uneasiness among those who know the hazards connected with corrosive influences. Perhaps the best reference for those seeking information about 'the physical basis of metal fatigue' is Forsyth's book of that name, since, although the book does not pretend to deal with the chemical aspect in detail, the author has carried out distinguished work on corrosion fatigue, and does not underrate the importance of chemical influences.

It should be borne in mind that whilst, in the case of dry fatigue, mathematical calculations are possible to ensure that no part of a machine, vehicle or structure is subjected to a stress range exceeding the safe value (the 'fatigue limit'), no such safeguard is possible in the case of corrosion fatigue, if, as is generally believed, there is no 'corrosion fatigue limit'.

There has been discussion of the question as to whether chemical action influences the time needed for the initiation of cracking, the propagation rate when once a crack has started or both. Rollins, Arnold and Lardner[1] have carried out tests on a steel containing 1% C and 1% Cr in chloride media. They state that the corrosion fatigue process consists of five stages:

(1) Preliminary pitting;
(2) Anodically controlled dissolution at the base of the pit, in which the dissolution is aided by yield occurring during the tensile part of each cycle;
(3) Resistance controlled dissolution;
(4) Mechanical fatigue;
(5) Brittle fracture.

Sometimes fatigue cracks were found to advance more quickly in air than in the solution in which they were formed, or in distilled water; the distilled water, it is suggested, served to exclude oxygen. The results are held to throw doubt on the interpretation offered in 1947 for some two-stage tests conducted at Cambridge (1960 book, p. 710).

The research quoted assigns an important role to pitting. On the other hand, Uhlig[2] expresses the view that pit formation is not an essential prerequisite for fatigue damage. There is no reason to doubt the accuracy of the classical work of McAdam, which includes description of cases where pits have developed into corrosion cracks, but it is fair to remember that those who have reached different opinions on corrosion fatigue may have been working on different materials or under different conditions.

Hoeppner[3] describes work designed to disclose the relative roles of 'initiation' and 'propagation' in the corrosion fatigue of engineering materials, using unnotched specimens kept under observation in 'movies'. 3·5% NaCl was found to be the most deleterious of the solutions studied, and the reduction of the fatigue life becomes greater as the maximum cyclic stress decreases; this is held to indicate that the role of the corrosive substance is

1. V. Rollins, B. Arnold and E. Lardner, *Brit. Corr. J.* 1970, **5**, 33.
2. H. H. Uhlig, *Tokyo Cong. Ext. Abs.* 1972, p. 187; see also *Connecticut Symp.* (*Corr. Abs.* (*Houston*) 1971, p. 231).
3. D. W. Hoeppner, *Corrosion* (*Houston*) 1972, **28**, 469.

greatest during the initiation stage or the early propagation stage; corrosive attack at preferential sites leads to the development of a crack; in some cases, but not all, the corrosive attack is related to pitting. The 'movies' show bubble formation during all stages of the corrosion fatigue process.

Wanhill[1] has studied crack propagation into a titanium alloy containing 4% Al, 4% Mo, 2% Sn and 0·5% Si, interpreting his results in the light of fracture mechanics (this volume, p. 334); the propagation rate is 2 to 5 times faster in salt water than in air, under comparable conditions.

Just as in stress corrosion cracking (chap. XVI), it would, no doubt, be possible to arrange the phenomena to present a continuous spectrum extending from cases where mechanical action is all-important to those where chemical action is predominant.

The corrosion fatigue of high strength Al–Zn–Mg alloys has been studied by Stoltz and Pelloux.[2] The growth rate of the crack in NaCl solution was 10 times that in dry air and about 100 times that in argon. A brittle fatigue striation morphology was obtained in NaCl solution and also in distilled water, but ductile striations were obtained in dry air; addition of $NaNO_3$ to a NaCl solution replaced brittle by ductile striations.

Relation between fatigue risks and tensile strength. It is well known that certain steels which can withstand extremely high steady stresses, are liable to fatigue failure; probably many failures ascribed to 'metal fatigue' have really been due to corrosion fatigue. A conference organized by the West of Scotland Iron and Steel Institute[3] gave opportunity for the discussion of whether, in the interests of public safety, such materials ought to be used at all. Reference was made to the catastrophic explosion of a tanker carrying liquid ammonia which, in the view of one speaker, would not have occurred if thicker steel plates made of an old-fashioned (but weaker) type of steel had been used.

The causes of corrosion fatigue in these exceptionally strong steels has been the subject of much argument. In a discussion regarding the failure of rock drills made of high-carbon steel,[4] Parkins has suggested that hydrogen embrittlement may be involved, but others have stated that the metallographic evidence is on the whole against such a view. Cole considers that hydrogen does not embrittle at high cycling rates, but may do so at low cycling rates. Barker suggests that residual compressive stresses left in the drill might enhance resistance to corrosion fatigue.

de Kazinczy[5] attributes the reduction in fatigue strength to surface defects; pores are stated to have a larger 'notch effect' than micro-shrinkage cavities of the same size. No special reference is made to corrosion fatigue.

Relation between stress range and fatigue life. The rectilinear relationship between the stress range and log N, where N is the number of

1. R. J. H. Wanhill, *Brit. Corr. J.* 1973, **8**, 217.
2. R. E. Stoltz and R. M. Pelloux, *Corrosion (Houston)* 1973, **29**, 13.
3. *Steel Times*, Oct. 1970, p. 721.
4. See various views expressed by R. N. Parkins, H. G. Cole and W. Barker, *Brit. Corr. J.* 1971, **6**, 96.
5. F. de Kazinczy, *J. Iron Steel Inst.* 1970, **208**, 851.

cycles withstood, appears to be fairly general for corrosion fatigue. Waterhouse[1] summarizes the situation: 'I have found that with materials such as stainless steel or aluminium, the curve in corrosion fatigue is more or less a straight line. However with certain corrosion resistant materials such as 18/8 stainless steel and certain titanium alloys, particularly the Ti/25% Cu alloy, the curve deviates quite considerably from a straight line . . ., even these materials, when one has fretting plus corrosion fatigue, are again linear.'

Pook[2] has used the concepts of fracture mechanics (p. 334) in connection with the threshold effect in crack growth. Whether or not a crack will grow can be predicted from ΔK, the range of stress intensity factor during the fatigue cycle; a critical value ΔK_c is needed for crack growth; the value in general depends on the ratio of mean to alternating stress. His report provides threshold data for various materials including mild steel, a low-alloy steel containing Ni, Cr, Mn and V, and grey cast iron—also for steel weldments. His work refers to dry fatigue conditions, and his arguments elucidates the strange fact that the sloping line connecting stress range and $\log N$ suddenly gives place to a horizontal at a certain critical stress range. This assumes that no other factor than fatigue is present; under corrosive conditions, the horizontal limb is not obtained.

Relation between Stress Corrosion Cracking and Corrosion Fatigue. A continuous spectrum of phenomena extends from cases where a steady tensile stress is applied to the material, and those in which there is stress alternating, symmetrically or otherwise, about a mean which is zero. In practice, fatigue cracking is most dangerous in situations where the mean is not zero but definitely tensile. Such cases can be regarded as an alternating stress with a constant tensile stress superimposed.

A study of the complete situation with different amounts of tensile stress superimposed on alternating stress is long overdue; different forms of stress wave should be included. In fact, however, tests carried out for practical purposes do sometimes have a tensile stress superimposed (i.e. the cyclic stress fluctuates around a mean which is not zero), but relatively little has been published on the subject. Gräfen[3] brings out certain analogies and differences between corrosion fatigue and stress corrosion cracking. He points out that stress corrosion cracking depends very much on the composition of the metallic phase, although it can be produced in almost any medium; corrosion fatigue is possible only in certain media, but is not related to any particular metal composition.

It should be noted, however, that the mere fact that fatigue cracking is caused by the presence of a liquid does not necessarily mean that that liquid is causing chemical or electrochemical attack. There are many cases recorded of fatigue failure in cylinders subjected to pulsating internal oil pressures. The physical effect of pressurized oil penetrating into micro-cracks has been

1. R. B. Waterhouse, letter June 13, 1973. Also (with D. E. Taylor and M. H. Wharton) *Wear* 1970, **15**, 449; 1972, **20**, 401; 1973, **23**, 251.
2. L. P. Pook and A. F. Greenan, *N.E.L. Report* No. **571** (1974).
3. H. Gräfen, *Werkstoffe u. Korr.* 1969, **20**, 209.

examined mathematically.[1] There seems no reason to think that chemical attack is involved.

Stanley[2] has supplied information regarding stress distribution in a specimen undergoing corrosion fatigue. The experimental material was, however, an epoxy resin; it is not certain how far the distribution would be the same in a metal.

In studying changes due to alternating stress, frequency as well as stress range has to be taken into account. The subject is reviewed in a German paper.[3] Even in media which would produce stress corrosion cracking the fatigue component dominates the situation at about 50 Hz, and is noticeable at frequencies as low as 30 Hz; a typical feature of such failures is intercrystalline fracture. With decreasing load, the fatigue component gradually disappears and crack formation is initiated by localized attack from the surrounding medium; experiments are described on Cr–Ni steel in 35% $MgCl_2$ at various temperatures.

Relation between Corrosion Fatigue and Frettage. It has long been realized that one of the greatest dangers of frettage is that it may set up fatigue cracking which otherwise would not occur. The matter is discussed in chapter XVIII.

Fatigue at Elevated Temperatures

General. An important symposium was held at the University of Connecticut in 1972 dealing with fatigue at elevated temperatures. This included cases of cracking promoted simply by physical influences, but many of the papers gave adequate attention to corrosive influences. On the chemical side Rama Char[4] reviewed work carried out by many different methods, and reached the conclusion that 'electrochemical factors play a significant role in corrosion fatigue failure'. Several of the papers usefully introduced the concepts of fracture mechanics (this volume, p. 334). Gallagher and Wei[5] stated that there are two major categories of corrosion fatigue, operating above and below K_{ISCC} respectively. Barson[6] pointed out that the form of the cyclic stress wave is important for corrosion fatigue. Endo and Komai[7] also stated that the corrosion fatigue strength can depend on the form of the stress wave. At low frequencies the susceptibility of materials to stress corrosion cracking is an important factor. In high strength steels fairly insensitive to stress corrosion cracking, the fractures are largely due to the propagation of fatigue cracks with blunted ends.

1. *N.E.L. Report* Nos. **381, 444** (1969–70), East Kilbride, Glasgow.
2. P. Stanley, *Brit. Corr. J.* 1969, **4**, 39.
3. F. W. Hirth, O. Michel and H. Speckhardt, *Werkstoffe u. Korr.* 1972, **23**, 356.
4. T. L. Rama Char, *Connecticut Symp.*, see *Corr. Abs. (Houston)* 1971, p. 232.
5. J. P. Gallagher and R. P. Wei, *Connecticut Symp.*; see *Corr. Abs. (Houston)* 1971, p. 233.
6. J. M. Barson, *Connecticut Symp.*, see *Corr. Abs. (Houston)* 1971, p. 234.
7. K. Endo and K. Komai, *Connecticut Symp.*, see *Corr. Abs. (Houston)* 1971, p. 233.

Bearings. The part played by chemical influences in the fatigue of rolling-element bearings, gears and other machinery is brought out by Schatzberg.[1] The fatigue life is reduced by the presence of a small amount of water dissolved in a hydrocarbon lubricant, or by salt water emulsified in it. It is believed that water, although a minor constituent of the lubricant, may become concentrated in fatigue micro-cracks; a cell is then formed, the advancing tip being anodic and the rest of the crack cathodic. The propagation rate is increased by anodic dissolution of iron at the crack tip and also by hydrogen embrittlement. The influence of water is stated to be eliminated if a small amount of an amino-alcohol is introduced.

Engines. Coffin[2] attributes fatigue failures in aircraft engines, gas turbines and nuclear reactors to excessive oxidation from the air. He points out that at high temperatures the fatigue life depends largely on the duration of each cycle (the 'hold-time'), whereas at ambient temperature fatigue life depends usually on the number of cycles—regardless of the duration of each. Experiments conducted in high vacuo at high temperatures reveal the fact that the duration of each cycle (the hold-time) does not play a part in the fatigue life. This, and the fact that excessive oxidation has been observed in fatigue cracks, suggests that oxidation from exposure to the air during high temperature cycling is the probable cause of the 'hold-time effect', rather than creep, as has sometimes been suggested.

Electrochemical Aspects of Corrosion Fatigue

Polarization Curves. Endo and Komai[3] have studied the effect of dynamic loading on the slopes of polarization curves. They find that the Tafel gradient of the anodic curve is strongly influenced, whereas the cathodic line is hardly affected.

Doruk[4] has studied the effect of alternating stress on the polarization of mild steel in acid solution. He has determined the dissolution rate as shown by the polarization slope, which first increases with time and then becomes constant. This constancy is attained when the crack depth is such that the cathodic reaction no longer occurs on the face outside the crack but on the crack walls.

Russian work[5] has produced curves relating the corrosion fatigue rate to the polarizing CD. In the cathodic region low CD has an adverse effect, reducing the time needed for destruction, whilst a higher CD is beneficial, causing a smooth increase in the time to destruction. In the anodic region low CD causes an important reduction in the time to destruction, whilst a higher CD has a less marked effect.

1. P. Schatzberg, *Connecticut Symp.*, see *Corr. Abs.* (*Houston*) 1971, p. 231.
2. L. F. Coffin, *J. Iron Steel Inst.*, Sept. 1971, **209**; *Industrial News*, Sept., p. 30.
3. K. Endo and K. Komai, *Metalloberfläche* 1968, **22**, 378.
4. M. Doruk, *Corr. Sci.* 1968, **8**, 317.
5. S. V. Pushkina and V. V. Romanov, *Corr. Control Abs.* 1970, No. 2, p. 14.

Protective Measures. Rama Char,[1] reviewing the electrochemical aspects of corrosion fatigue, states that cathodic protection can be effective, making the corrosion curve for steel obtained in a corrosive environment similar to that obtained in air, provided that the optimum CD is applied. Above the optimum CD, the protection diminishes. Soluble inhibitors have some effect. In the case of magnesium alloys, nitrates are more beneficial than benzoates; neither, however, provides complete protection, and the concentrations required are higher than those demanded in stress corrosion cracking. Differential aeration cells are more important in the corrosion fatigue of magnesium alloys than in stress corrosion cracking.

Experimental work by Hirth and Speckhardt[2] shows that cathodic protection (best applied under galvanostatic rather than potentiostatic conditions) is effective in retarding the early stages of corrosion fatigue; but when once a crack has attained a certain depth, corrosion fatigue proceeds in the normal way.

Fatigue of Offshore Structures. Corrosion fatigue is causing anxiety to those concerned with the maintenance of offshore rigs, where welded structures may have a fatigue life only half of that expected under dry conditions. Welded structures contain crack-like defects, and virtually the whole fatigue life is occupied by crack growth from such defects. Results of tests made on unwelded specimens cannot be applied to welded structures, and a fracture mechanics approach is indicated; such an approach has proved most successful for interpreting the results of tests carrried out at East Kilbride.[3]

At stresses above the air-fatigue limit, application of salt-spray during fatigue testing did not affect the strength. Below that limit, the cracks grew at very low stress levels; at wave frequencies typical of offshore structures, the life was found to be halved. As a result of the information obtained, critical dimensions of cracks or defects needed for propagation can be determined for specific components under known conditions. Unfortunately, the effect of corrosion on fatigue at low frequencies is still imperfectly understood. When welded joints are subjected to corrosive conditions—in the North Sea—predictions must be confirmed by tests on representative test pieces under realistic loadings.

Addendum (Dec. 1975). A study of current transients on a 18/8 Cr/Ni stainless steel anode in 3·7M H_2SO_4 subjected alternately to tension and compression under square-wave conditions has been carried out by Rollins and Pyle.[4] Current flows immediately after tension or compression has been applied, but quickly dies away; the value is greater under tension than under compression. At potentials close to the transition between the active and passive conditions, the magnitude of the current increases with the cycle number, increasing perhaps 10 times after 6 or 7 cycles. They consider that 'if

1. T. L. Rama Char, *Corr. Prev. and Control*, 1972, **19** (5) 8.
2. F. W. Hirth and H. Speckhardt, *Werkstoffe u. Korr.* 1973, **24**, 774.
3. *N.E.L. Research Summary* 119 (1975).
4. V. Rollins and T. Pyle, *Nature* 1975, **254**, 322.

in corrosion fatigue stress concentration effects are sufficiently high to induce local plastic deformation on the surface of the metal, then the observed dissolution behaviour is likely to occur and could assist crack initiation by a combination of reversed slip and dissolution . . . For stainless steel in H_2SO_4, the maximum contribution of dissolution to the initiation of cracking should occur at the active–passive transition.'

A one-page summary[1] of a discussion on metal fatigue held in London in 1974 shows the attitude adopted today by many of those interested in the physical aspects of fatigue. The word 'corrosion' does not seem to occur, although there is occasional allusion to the effect of the environment. Ryder alluded to the 'torsional fatigue of an Al–Zn–Mg alloy in which a supposedly neutral environment gave the shortest fatigue life'. Knott, describing the fracture-mechanical approach to the subject, recognized three regions: (1) the threshold, strongly dependent on microstructure, mean stress and environment, (2) a region normally insensitive to metallurgical variables and (3) a region involving brittle or fibrous fracture.

Further References

Book. An authoritative treatment of the subject by Waterhouse[2] is awaited with interest.

Papers. 'Two theoretical models of fatigue crack propagation'.[3]
'Corrosion Fatigue initiated by Ocean Waves'.[4]

1. *Met. Mat.* May 1975, p. 45. See especially remarks by D. A. Ryder and J. F. Knott.
2. R. B. Waterhouse, 'Corrosion Fatigue' (Pergamon), in preparation.
3. I. N. M. Cartney and B. Gale, *Proc. Roy. Soc. (A)* 1973, **333**, 337.
4. J. Larsen-Basse, R. D. Ireland and F. M. Casciano, *Mat. Performance* 1974, **13** (1) 10.

CHAPTER XVIII

Other Types of Conjoint Action

Preamble

Picture provided in the 1960 and 1968 volumes. In previous chapters, especially those concerned with direct oxidation and soluble inhibitors, it has been emphasized that many reactions, under static conditions, stifle themselves by the formation of a protective film. If, however, this film is continually being broken or rubbed away by mechanical action, the reaction will not be halted and serious destruction will continue unabated. The most serious situation occurs when the action is localized on a small area.

It is common knowledge that wear occurs where there is relative movement of two surfaces. In general, this is spaced out over a large area, and the change of dimensions produced may not be appreciable—at least over short periods. A more serious situation arises if two surfaces which are supposed to be stationary and are in contact only over a microscopic area, are exposed to vibration which sets them into oscillatory movement relative to one another with an extremely small amplitude; here the destruction, concentrated on a very small area, can be intense and rapid. One case of attack occurs if the highly localized friction removes an oxide-film present at the point of contact, and then, assuming that air has access to the point in question, further oxide is rapidly formed—only to be removed in its turn. This mode of destruction requires the presence of oxygen or some oxidizing agent, and can be regarded as a form of localized corrosion. It may, however, happen that mechanical grinding continues after the metal is free from oxide, producing metallic powder, which may be converted to oxide at some point accessible to air, possibly situated at a distance from the point of grinding. At first sight, this would seem to be something which does not concern the corrosion specialist, since the mechanical destruction of metal requires neither oxygen nor corrosive agent. However, on many metals, the oxide produced by the action of air occupies a larger volume, and may be harder, than the metal destroyed in producing it. Consequently, if air can reach, or approach, the point where debris is being formed, the situation is worse than if oxygen were completely absent. If the oxide formed consists of hard grains, these may act as abrasive particles in the narrow gap where oscillatory motion is proceeding; in that case destruction may proceed more rapidly than if air had been excluded (for then the debris would have consisted of particles of relatively ductile metal). Thus the nature of the environment is important. Both these types of 'frettage', as it is called, are the proper concern of the corrosion specialist.

Another case of localized damage is produced by the impingement of bubbles on a metallic surface. These may be bubbles of air at roughly atmospheric pressure, in which case they will probably bounce off, perhaps dragging off a protective oxide-film and leaving the metal bare so that further oxide can be formed—only to be removed in its turn. Or they may be vacuum cavities produced by alteration of pressure where water passes rapidly through a channel of varying breadth; these cavities may collapse on impact, producing a very intense blow where the surface of the cavity originally remote from the metallic surface ultimately strikes it. In all these cases, there can be serious damage in the absence of oxygen and in liquid which would normally be non-corrosive. However, in some circumstances, the destruction is made much worse through the removal of a film which under static conditions would be protective. How far the damage to the film is due to the blow delivered on impact, and how far it should be attributed to the adhesion of the film to the surface of a bubble departing from the metal surface (and such adhesion must be expected if it represents the condition of minimal interfacial energy), is a matter on which there is incomplete agreement. Some corrosion specialists tend to overlook the possibility of film-removal through its adhesion to a departing bubble, whilst some physicists and engineers tend to overlook the chemical aspect of the matter altogether.

The effects produced by impingement of small liquid particles, or of a jet, on metal surrounded by air present some analogies. Here also there could be purely mechanical damage even if the gas phase contained no oxygen and the liquid was non-corrosive. But the alternate formation and removal of a film should not be overlooked as a possible cause of increased damage.

Developments during the period 1968–1975 and Plan of the new Chapter. Whilst no startling developments have been reported during the period under review, careful work has contributed to the lessening of risks of serious damage due to conjoint action.

The chapter which follows is divided into four main sections, dealing with wear, frettage, cavitation and erosion corrosion respectively. The section on wear is divided into sub-sections carrying the titles 'Metallographical Aspects', 'Roller Bearings' and 'Economic Factors'. The section on frettage has sub-sections entitled 'Spectrum of Phenomena' and 'Practical Cases of Fretting Damage'. The section on cavitation has sub-sections entitled 'Relation to other forms of Deterioration', 'Part played by Mechanical and Chemical Factors', 'Electrochemical Mechanism' and 'Cases in Service'. The section dealing with erosion corrosion has sub-sections entitled 'Relation to other forms of Destruction', 'Corrosion set up by Scouring' and 'Cases in Service'.

Wear

Metallographical Aspects. Certain aspects of wear are discussed by Eyre.[1] A white layer found on machine-gun barrels and wire ropes after usage is believed to be largely nitride or nitrogen-rich martensite. Hard, white-etching regions have been found on diesel engine liners and on roller bearings; the hardness far exceeds that induced by conventional hardening

1.   T. S. Eyre (with F. Wilson and A. Baxter), *Met. Mat.* 1969, **3**, 86; 1972, **6**, 435.

methods. Oxygen does not seem to be necessary for the formation of the white layer. It is generally agreed that its properties are due to a fine dispersion of second-phase particles, and that the structure is produced by a combination of surface temperature flashes and extensive cyclic deformation. It has been suggested that commercial advantage might well be taken of these friction hardened surfaces, if only the conditions for producing them could be standardized.

Roller Bearings.   The problem of overlay bearings has been discussed by R. W. Wilson and Shone.[1] Engine bearings today are generally coated with a precision plated overlay of Pb–Sn or Pb–In alloy, 20–40 Å thick, electrodeposited on a Cu–Pb or Pb–bronze substrate which is itself bonded to a steel backing. The use of unalloyed lead is not possible, since it is quickly attacked by the organic acids formed in mineral oil by oxidation; alloying with In or Sn prevents corrosion. Pb containing 4% In or 8% Sn is generally used, but the concentration at the surface may be reduced by diffusion into the substrate, or (in the case of In) by internal oxidation. If the Sn or In content is too low, yellow deposits are formed; these were once thought to be oxide or di-lead-tetra-alkyl oxide, but are now known to be lead soaps of naphthenic and straight chain monocarboxylic acids.

Pooley and Tabor[2] point out that static and kinetic friction may be very different. Polytetrafluorethylene (PTFE) or high density polyethylene have a static frictional coefficient of perhaps 0·2. However, when once sliding has commenced and the slider has acquired a preferred orientation, the coefficient of friction may fall to a value below 0·1. This behaviour does not depend on the degree of crystallinity or the crystalline texture of the polymer. The low friction number and the small amount of transfer which occurs during sliding seem to be connected with the smooth molecular profiles of the polymer. If bulky sidegroups are present, as in low density polyethylene, the kinetic frictional value remains the same as the static value.

Scott[3] describes recent work at East Kilbride on lubricants and coatings for roller bearings. In some applications, conventional liquid and semi-solid grease lubricants are unsuitable owing to temperature limitations; they may also be appreciably volatile and become lost by evaporation. Technological progress has introduced arduous conditions for rolling mechanisms, and this has necessitated the adoption of new lubrication policies. Lamellar solid-film lubricants can be effective under conditions where the conventional lubricants are unsatisfactory; the ultimate objective is lifelong lubrication. Today lubricants such as PTFE, as well as materials such as MoS_2 and graphite, which are built up of layers, are much used. New tests at East Kilbride show that of the solids tested, MoS_2 generally gives the best results, but that the method of bonding to the surface is important. Some interesting synergistic effects have been noted. A combination of 21% carbon fibre with 14% MoS_2

1.   R. W. Wilson and E. B. Shone, *Amsterdam Cong.* 1969, p. 788.
2.   C. M. Pooley and D. Tabor, *Proc. Roy. Soc. (A)* 1972, **329**, 251.
3.   D. Scott, *Wear* 1972, **21**, 155; also (with P. J. McCullagh) *Electrodeposition and Surface Treatment* 1972–3, p. 21.

and 65% PTFE keeps the weight loss very small under conditions where other combinations permit considerable losses; this combination is effective in reducing wear.

Despite stringent control, the scatter observed between the lives of roller bearings is a matter for concern in such applications as aero-engines, space-craft and atomic reactors; sometimes failure may occur quickly, for no obvious reason. Since the highly localized stresses associated with rolling contact occur at or just below the surface, tests have been carried out at East Kilbride on materials carrying an electrodeposited coating which should be free from deleterious inclusions. Since the coatings can be thick enough to contain the rolling contact stresses, it was hoped to use cheaper and more easily produced material as substrate. Tests have indeed shown that the contact fatigue life of hardened electrodeposited iron carrying Ni as a coating compares favourably with the performance of conventional roller bearing materials; it was obtained by deposition on carbon steel (or on suitable alloy steels previously etched anodically) from a bath containing $NiSO_4$, $FeCl_2$, sodium tartrate and an addition agent.

Black and Dunster[1] have compared the effectiveness of MoS_2 particles with solutions of molybdenum dialkyl-dithiophosphate and zinc dialkyl-dithiophosphate; they determined the 'load-carrying capacity' by increasing the load in steps until there was seizure or sudden increase in frequency torque; on that criterion suspensions of MoS_2 were found to be superior to the two solutions. However, at lower loads the molybdenum dialkyl-dithiophosphate showed the greater resistance and suffered the lowest rate of wear; it becomes less effective at elevated temperatures.

Economic Factors. Finniston[2] has stated that a saving of £20M per year would be possible in the metal industry by avoiding replacement and maintenance connected with unnecessary wear. He emphasizes the 'growing interdisciplinary nature of the phenomena of friction, wear and lubrication'. In this area, he thinks, the physicist, the chemist, the metallurgist and the engineer should work in consort.

Frettage

Spectrum of Phenomena. As in the cases of stress corrosion cracking and corrosion fatigue there is, in frettage, a continuous passage from wastage which is mainly mechanical aided by chemical change to wastage which is mainly oxidation but continues without interruption owing to the mechanical removal of a film which would otherwise provide protection. However, three basic processes are involved and are brought out clearly in Waterhouse's book.[3] These are:

(1) continuous removal of oxide-films from the metallic surface, which then quickly produces more oxide,

1. A. L. Black and R. W. Dunster, *Wear* 1969, **13**, 119.
2. H. M. Finniston, *Met. Mat.* 1969, **3**, 421.
3. R. B. Waterhouse, 'Fretting Corrosion' (Pergamon Press), 1972.

(2) mechanical removal of metal without oxidation, and

(3) abrasive action by oxide debris formed in (1) or by subsequent oxidation of metallic debris formed by (2).

Clearly in absence of oxygen only (2) is possible, but it is fairly certain that the damage rate is much slower if oxygen is excluded; consequently (1) and (3) must be taken seriously. Feng and Uhlig (1960 book, p. 743) found that the damage to steel in nitrogen is only about one sixth that in oxygen; in the former case the debris was metallic iron, whereas in the presence of air oxide was formed. Waterhouse deals with an early argument based on a study of a steel ball fretting on a glass plate, where the debris at its centre was observed to be black and was assumed to be metal. He points out that black oxides of iron exist, and these are just the compounds the formation of which would be expected at the points where the oxygen supply is small. There is little doubt today that oxygen plays an important part in frettage.

Waterhouse states that frettage produces debris, and the effect on an originally smooth surface depends largely on whether the geometry permits this debris to escape. If it cannot escape, deep holes are produced by abrasive action. If it can escape, there will usually be shallow dish-like depressions. The production of holes is also discussed by Funk.[1] An electrochemical study of fretting corrosion on 18/8 Cr/Ni stainless steel in NaCl has been carried out by Waterhouse and his colleagues.[2] In the region of the corrosion potential, the air-formed oxide-film is disrupted by the fretting and is not re-formed. The major part of the current increase on fretting can be attributed to plastic deformation of the metal surface resulting in the emergence of active slip bands. The corrosion current is a rectilinear function of the fretting amplitude and (at a given amplitude) of the frequency; it is much higher at low pH than at high pH.

An electrochemical study by Taylor and Waterhouse[3] provides evidence of the importance of the disruption of protective films; on certain metals, film-breakage would seem to be a more important factor than increased activity due to deformation of the metal basis. On Al, Cr and Ta they found substantial falls of potential produced when frettage was applied; these are really base metals which owe their corrosion resistance to the presence of a protective film. In contrast, the truly noble metals, Ag and Cu, exhibited no change of potential when subjected to fretting, whilst Ni showed only a small fall in potential, with much smoother fluctuation about the mean value than was observed with base metals.

Practical Cases of Frettage Damage. The main danger of frettage is that it sets up fatigue cracking; ordinary wear does not generally lead to fatigue, and indeed it may sometimes improve fatigue properties, possibly by rubbing away crack nuclei.[4] Several other serious effects of frettage, however, are mentioned in Waterhouse's book. In steel ropes, where damage

1.   W. Funk, *Metalloberfläche* 1969, **23**, 193.
2.   M. P. Sherwin, D. E. Taylor and R. B. Waterhouse, *Corr. Sci.* 1961, **11**, 419.
3.   D. E. Taylor and R. B. Waterhouse, *Corr. Sci.* 1974, **14**, 111.
4.   R. B. Waterhouse, Private Comm. Oct. 27, 1968.

occurs between contacting strands, they may be forced apart, since the oxide occupies a greater volume than the metal destroyed; the bulge formed on the exterior of the rope may have undesirable consequences. Again, debris may cause seizure—a serious matter in the case of a safety valve or governor. In cases where the debris escapes, there may be loss of fit, or reduction in clamping pressure, which in turn may lead to increased vibration. In machinery, escaping debris may fall or be carried on to other moving parts and cause fresh wear problems. The formation of fretting debris in electrical contacts, particularly those which have to transmit small currents (such as relays in telephone exchanges) causes change in contact resistance and results in distorted signals. The black powder formed by the frettage of aluminium surfaces can be pyrophoric—with alarming possibilities if aluminium containers have been used for live ammunition in ships' holds.

Preventive methods are very diverse and include the use of liquid lubricants, greases and various films, metallic or non-metallic. Sometimes small geometrical changes can be beneficial. On press-fitted railway axles, the provision of a groove at a suitable place has prevented the formation of fatigue cracks and has reduced frettage to a minimum; before the introduction of grooving, fatigue cracks usually appeared on London Transport axles after 160 000 miles travel—that is, after 8×10^7 wheel revolutions. Some rather unexpected facts are quoted in regard to the effect of geometry. In pinned connections, pins with a close fit were found to provide lower fatigue strength than those where some clearance had intentionally been provided.

Precautions against fretting corrosion are discussed by Paolo.[1] Vibration should be eliminated as far as possible and loads on the surface reduced to the minimum. Cyaniding, cementation and coating with Cu and Ni increase resistance of steel. Low viscosity oils give good results. Of solid lubricants, MoS_2 and graphite are the best.

Strochi[2] has studied fretting corrosion on various stainless steels. Sheet samples which had been transported for distances up to 500 kilometres became covered with brown spots due to the formation of corrosion products during friction between sheets; similar samples which had not been transported showed no change. Protective methods to be observed during the transport of aluminium are discussed by Scott and Skerrey,[3] who point out that today much aluminium is stored and transported with no more packaging than is required for mechanical stability.

White[4] discusses deterioration at places where austenitic stainless steel nuts have been used for bolting an alloy steel; the combined effects of high temperature operation, marine atmosphere, steam impingement and chemical contaminations cause heavy rusting and seizure of the bolting.

Nieth[5] has found that the iron–zinc alloy layer present on galvanized iron resists abrasion better than the outer layer of relatively pure zinc; in an

1. L. Paolo, *Corr. Control Abs.* 1970, No. 3, p. 24.
2. P. M. Strochi, *Electrochim. Metallorum* 1967, **2**, No. 1, p. 117.
3. D. J. Scott and E. W. Skerrey, *Brit. Corr. J.* 1970, **5**, 239.
4. P. E. White, *Process Engg*, May 1971, p. 103.
5. F. Nieth, *Metalloberfläche* 1968, **22**, 175.

abrasion test the amount of 'pure' zinc removed was 4·57 times that of the alloy layer, if measured by weight (or 4·64, if measured by volume). He suggests that steps should be taken to provide a thicker alloy layer.

Cavitation

Relation to other forms of deterioration. Orlowski[1] has stressed the importance of cavitation occurring during irregular water-flow through pipes—also types of corrosion promoted by vibration—and regrets that these receive so much less attention than intergranular and transgranular corrosion, or again pitting. Tichler and Scott[2] have obtained a correlation between resistance to cavitation and to rolling contact fatigue.

Part played by Mechanical and Chemical Factors. Schulmeister[3] states that cavitation damage is primarily mechanical, being initiated by repetition of the compressional stress arising from the collapse of bubbles. However, even small stresses due to cavitation can produce severe damage at the onset of corrosion due to the crushing of the normally protective film and the removal of corrosion products. New work on ferrous materials vibrated magnetically shows that the damage is influenced by the water composition when the material is one susceptible to corrosion; the stresses required to produce damage are small compared to the strength of the material. Attack is greatest when Cl^- and SO_4^{2-} are high, pH low, and the concentration of O_2 high; destruction can be reduced by use of a relatively inert atmosphere, such as N_2 or CO_2. In the cooling circuits of internal combustion systems, inhibitors have been effective.

Electrochemical Mechanism. Yahalom[4] has compared polarization curves obtained in presence of vibration with similar ones without vibration. The current passing in the passive range (i.e. the horizontal limbs in Fig. 9, p. 134) is very much greater when vibration is applied. The importance of this observation is obvious.

Cases in Service. Mølgaard[5] describes severe corrosion trouble on a cast iron centrifugal impellor in a pump used for chlorinated sea-water at a dock. Tests lasting 380 hours have shown that during *slow* rotation of specimens, such as causes no cavitation, chlorination actually decreased the weight loss observed, although the scale found at the end of a run was heavier. If the rotation was sufficiently rapid to cause cavitation, chlorination again increased the weight of the scale, but now there was an increased weight loss. Evidently cavitation can remove scale which, if it had remained adherent, would have been protective.

A super-cavitating pump has been designed at East Kilbride[6] in which the cavities collapse harmlessly at points clear of the blades. The main objectives

1. P. Orlowski, *Corrosion (Paris)* 1970, **18**, 357.
2. J. W. Tichler and D. Scott, *Wear* 1972, **16**, 229.
3. R. Schulmeister, *Metalloberfläche* 1967, **21**, 68.
4. G. Engelberg and J. Yahalom, *Corr. Sci.* 1972, **12**, 469.
5. J. Mølgaard, *Tokyo Cong. Ext. Abs.* 1972, p. 335.
6. *N.E.L. Progress Reports* 1968–70 (East Kilbride, Glasgow).

relate to problems in which corrosion plays no part (such a pump should be valuable in pumping near-boiling liquids), but the principle may be of value in situations where corrosion is involved. Similarly a discussion of cavitation in hydraulic equipment[1] may be read with advantage, although corrosion damage is not discussed.

Erosion Corrosion

Relation to other forms of destruction. Heitz and Loss[2] draw a distinction between the erosion corrosion product by liquids at high flow velocities from the cavitation produced by the sudden collapse of bubbles. In the experiments which they describe, erosion by solid particles was excluded. Their apparatus consists of two co-axial cylinders, an inner one of diameter 12 cm being fixed inside the outer one, which rotates so as to give any desired relative velocity up to 50 m/s (which means an averaged velocity of 25 m/s); the annular gap was 0·4 cm. If the flow velocity exceeds a certain value, erosion corrosion starts after a certain incubation period (often about 50 hours). During the whole run the overall corrosion current passing between the cylinders is found to remain constant, despite the fact that corrosion is uniform in the early stages and localized later; anodic areas develop at surface regions exposed to the strongest turbulence. After 50 and 100 hours, three characteristic regions can be distinguished:

(1) at low velocities (up to about 5 m/s), the current increases rapidly with flow velocity;
(2) at intermediate velocities, the current becomes nearly independent of flow velocity;
(3) above about 25 m/s, the current again rises sharply with flow velocity.

The observations are best attributed to corrosive attack caused by increased diffusion through pores in the layers of corrosion product, especially at high flow-rates and preferentially at sites of high turbulence.

Corrosion set up by Scouring. Work in Tomashov's laboratory[3] has been devoted to the effect of scouring on various electrochemical reactions. Scouring greatly reduces hydrogen overvoltage on metals like Fe, Ni and Pd, which have strong H-adsorbing capacity, but does not appreciably affect the overvoltage on Pb and Sn. A considerable decrease of the overvoltage of the cathodic reduction of oxygen is brought about by continual renewal of the surface of a Pd cathode. The overvoltage of anodic ionization of Fe and Ni is diminished by surface renewal; on Ti, Cr and Ni (but not on Fe), the CD needed to produce passivity is increased by some orders of magnitude— as also is the current passing in the passive state. The inhibiting effect of adsorption of halogen ions (especially I$^-$ and Br$^-$) on certain electrode reactions (including the cathodic evolution of hydrogen and anodic ionization of Fe and Ni in N H$_2$SO$_4$) can be eliminated by vigorous surface renewal.

1. J. M. Hobbs and D. Molley, *Met. Mat.* 1972, **6**, 27.
2. E. Heitz and C. Loss, *Tokyo Cong. Ext. Abs.* 1972, p. 201.
3. N. D. Tomashov and L. P. Vershinina, *Electrochim. Acta* 1970, **15**, 501.

Work on titanium by T. R. Beck[1] shows that anodic current can pass if the surface is renewed; the potential is maintained by means of a potentiostat at a level where there would be no current if the surface was not renewed; the counter-electrode is Pt. The renewal of the surface is brought about by scratching the end of a rotating Ti tube or through the sudden breakage of the Ti brought about by the drop of a heavy weight on to a lever arm; the current is strong immediately after breakage but soon dies away.

Cases in Service.   Damage caused by the erosion of Al brass at power stations is described by Tanabe.[2] It is largely prevented by a cathodic current if the flow velocity is less than 2 m/s, because the process is then largely electrochemical; it is not prevented at 4 m/s because of the mechanical characteristics of sand erosion under severe flow conditions. Suk[3] has compared the resistance of different materials to running well-water, and finds aluminium brass to be less resistant than unalloyed copper, although ternary alloys containing Cu, Ni and Fe are better.

Mattsson[4] has examined erosion corrosion caused by water flowing through copper tubes; these generally show good resistance, unless the flow velocity is too high, or unless flow is disturbed at bends or side-branches; no difference is found between the resistance of annealed and hard-drawn tubes. Higher flow-rates can be tolerated if the water is flowing only a quarter of the time or less, than if it is continuous; with continuous flow a velocity of 3 m/s can be accepted at 65°C and 6 m/s at 30°C for tubes of small diameter (6–4 mm).

The special destruction at places where flow is disturbed at bends or branchings (doubtless producing adjacent areas of high and low velocity) may be an example of the moto-electric effect—investigated in classical work by W. J. Müller (1960 book, p. 130).

Corrosive wear of steel and chromium plated guides in the textile industry has been discussed by Ijzermans.[5] It occurred with an emulsion carrying a cathodic surfactant, but was much less serious with one containing an anionic emulsifier, which apparently acted as an inhibitor—as also did a non-ionic emulsifier.

Erosion corrosion in a re-forming plant at a gas-works due to impact of water droplets in an atmosphere containing CO_2 at a bend in a steel pipe is described by Olden and Cameron.[6] It can be avoided by the use of stainless steel.

Further References

Conference Report.   The manner in which frettage sets up fatigue cracking was discussed at a meeting held at Munich.[7] Harris emphasized the im-

1.   T. R. Beck, *Ericeira Conf.* 1971, p. 68.
2.   Z. Tanabe, *Corr. Abs.* (*Houston*) 1969, p. 355.
3.   H. Suk, *Werkstoffe u. Korr.* 1972, **23**, 12.
4.   L. Knutsson, E. Mattsson and B. E. Ramberg, *Brit. Corr. J.* 1972, **7**, 208.
5.   A. B. Ijzermans, *Amsterdam Cong.* 1969, p. 812.
6.   M. J. F. Olden and R. W. N. Cameron, *Brit. Corr. J.* 1968, **3**, 271.
7.   AGARD Conference Proceedings No. 161 (München) 1974 (NATO). Especially W. J. Harris p. 7 (1) and R. B. Waterhouse p. 8 (1).

portance of residual tensile stresses, which can nucleate the fatigue either just inside or just outside the 'scarred region'; often the nucleation is subcutaneous. Waterhouse stated that often the oscillatory tangential relative movement of two surfaces in contact only occurs on part of the contact area, and that fatigue cracks can originate at the boundary between the slip and non-slip regions.

Reports. Metallurgical aspects of materials for tribological situations are discussed in various reports from East Kilbride;[1] attention is directed to the N. E. L. Rolling Four-ball Machine for testing lubricants and bearing materials.[2]

Papers. 'Zum Mechanismus der Erosionskorrosion in schnell strömenden Flüssigkeiten'.[3]

'An electrochemical investigation of fretting corrosion of a number of pure metals in 0·5M sodium chloride'.[4]

'Mechanism of erosion corrosion in fast flowing liquids'.[5]

'Flow dependence of the pitting of a CrNi steel in NaCl solution'.[6]

'The effect of applied potential and surface dissolution on the creep behaviour of copper'.[7]

1. *N.E.L. Reports* Nos. **458** (1970); **536, 538, 539** (1973) (East Kilbride, Glasgow).

2. *N.E.L. Report* No. **557** (1974) (East Kilbride, Glasgow).

3. C. Loss and E. Heitz, *Werkstoffe u. Korr.* 1973, **24**, 38.

4. D. E. Taylor and R. B. Waterhouse, *Corr. Sci.* 1974, **14**, 111.

5. C. Loss and E. Heitz, *Werkstoffe u. Korr.* 1973, **24**, 38.

6. F. Franz, E. Heitz, G. Herbsleb and W. Schwenk, *Werkstoffe u. Korr* 1973, **24**, 97.

7. R. W. Revie and H. H. Uhlig, *Acta Met.* 1974, **22**, 619.

CHAPTER XIX

Recent Research on Interference Tints

Object of the Work. Interference tints are familiar to all who work on oxidation and corrosion; they appear on metal heated gently in air, and are often observed on specimens partly immersed in salt solution, generally on the zone just below the water-line. The corrosion worker is accustomed to draw qualitative conclusions from his observations. If, for instance, he notices the colour on a specimen slowly changing from brown to mauve and then later to blue, he rightly draws the conclusion that the film is becoming thicker with passage of time. If a specimen which is not changing colour with time is brown on the left, mauve in the centre and blue on the right, he may draw the conclusion that the film is thinnest on the left and thickest on the right. Here he is on less sure ground. Vernon[1] has shown that the colour produced is not necessarily a simple function of the oxygen uptake per unit area. This has made the corrosion worker wary of drawing conclusions, and sometimes suspicious about the whole subject. More than one reason could be suggested for the fact that the colour is not a simple function of the average film thickness; the question is discussed later, but it may be here stated that the complexity of the situation is largely due to the fact that the films are generally produced at the expense of the metal to which they are attached. For instance, when metal is heated gently in air, oxide is produced on its surface; under certain conditions (but not all) oxygen penetrates inwards, perhaps along grain-boundaries, and the internally produced oxide will contribute little or nothing to the colour; thus a given uptake of oxygen per unit area will produce a different colour according as the oxidation is confined to the surface or partly internal. What appears at first sight to be a very simple case is really a complicated one.

A more suitable case for experimental study would be a situation where the film producing the colour is provided from an external source; this eliminates at least one complicating factor. Such a method was used in the research to be described. In interpreting the new results, a crude optical picture of the interference phenomena has been used, which may be welcomed by those who have not had time to acquire a knowledge of metal optics. This theoretical treatment will be condemned by some specialists as being an over-simplification. Undoubtedly it is an over-simplification, as indeed are all the chapters in the present volume, and probably most of the chapters of the books on which the specialists themselves rely for enlightenment on matters outside the subject of their specialization. Whether a greater dis-service is performed by providing an over-simplified picture than by providing an unintelligible account of the subject—that must be left for the reader himself to decide.

1.   W. H. J. Vernon, *Trans. Faraday Soc.* 1935, **31**, 1674.

Summary of early work. Temper colours must have been known since the days of the earliest steel implements; until the introduction of the thermocouple, they were used to control the temperature and time of tempering of tools. Colours of greater diversity are produced on heating non-ferrous metals in air (or even exposing them unheated to air containing hydrogen sulphide).

At the start of the century, such colours were ascribed to interference between light reflected from the two surfaces of the film. As the film thickens, the wavelength at which maximal interference occurs increases. When it is in the ultra violet (Fig. 42, Stage A) there is no visible change. When it enters

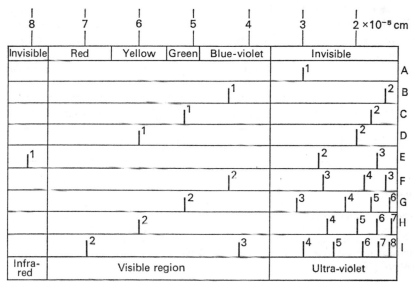

Fig. 42 Production of interference colours by films of gradual increasing thickness (A to I). A carries a very thin (invisible) film giving an interference band in the ultraviolet region—hence no colours. In B, C, D, the first band advances across the visible spectrum as the film thickens, producing first order colours. It leaves the visible region before the second band enters it—explaining the silvery hiatus (E). The second band then produces second order colours (F, G, H), but is more closely followed by the third band, explaining the production of a second order green (I); there is no green at the end of the first order.

the blue part of the spectrum (stage B), the light reflected, being deficient in blue, appears yellow. When, with increase of film thickness, the interference band reaches the green, the light, deficient in green but not in red or blue, appears mauve or purple (stage C); when it reaches the yellow, the appearance will be blue (stage D). This seems to explain the sequence of tints actually observed.

In 1918, however, objections were raised. If, when a film thickens the colours change in the order yellow → mauve → blue, they should change in the order blue → mauve → yellow when the film is made thinner—assuming the interference explanation to be correct. An attempt to reduce the thickness of the blue film by polishing failed to obtain the change to mauve and yellow; it was suggested[1] that the colours were due to some structural pattern in the film substance producing 'damped molecular periods'—not to interference.

It is, however, no easy task to obtain uniform thinning by mechanical polishing. A more suitable method would seem to be cathodic treatment in dilute acid, where the thinning would be controlled by the current density, which can easily be kept uniform. In 1925 this method was adopted at Cambridge.[2] (The thinning was produced by reductive dissolution—a phenomenon afterwards examined in greater detail,[3] when it was found that *reductive* dissolution of Fe_2O_3 in acid to give Fe^{2+} ions proceeded much more easily than *direct* dissolution to give Fe^{3+}.) In the research of 1925 a strip of iron was heated at one end to give the sequence of colours, and the lowest part only was subjected to cathodic treatment in $0.02M$ HCl (Fig. 43). The various colours moved relatively to their positions in the upper, unwetted

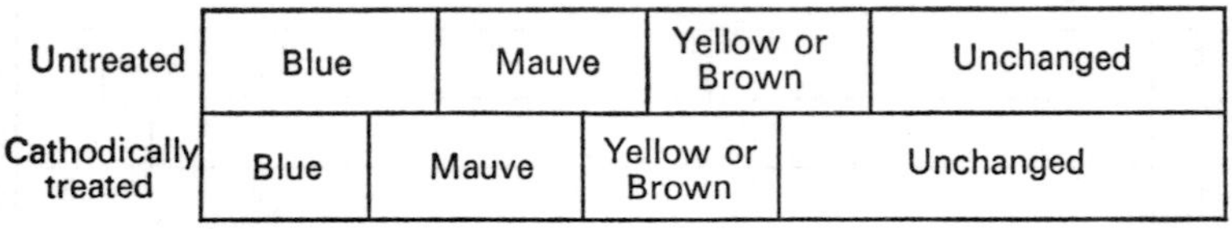

Untreated	Blue	Mauve	Yellow or Brown	Unchanged
Cathodically treated	Blue	Mauve	Yellow or Brown	Unchanged

Fig. 43 Experiment showing that colour depends on film thickness. When an iron strip, previously heated at one end, so as to produce a sequence of colours, is subjected to cathodic treatment in acid (a treatment which reduces the thickness of the film by reductive dissolution), the colours shift in the direction which would be expected if the tint is dependent on thickness. The specimen shown above has received cathodic treatment on the lower part whilst the upper part, which has remained unwetted, represents the original position of the various colours (U. R. Evans).

part of the strip, in the manner to be expected if the colour is determined by the film thickness.

Whilst this work was in progress at Cambridge, other methods were being used elsewhere. Mason[4] produced the colour change by immersion in dilute nitric acid without current—which was probably a case of direct dissolution as Fe^{3+} ions. Gale[5] used polishing and succeeded in obtaining the colour change where a previous investigator had failed. Thus in three different laboratories the change of colour predicted by the interference explanation was really obtained, and objection to that explanation was completely removed.

It was found possible[6] to remove the oxide-films formed on molten lead

1.   A. Mallock, *Proc. Roy. Soc.* (A) 1918, **94**, 506.
2.   U. R. Evans, *Proc. Roy. Soc.* (A) 1925, **107**, 228.
3.   M. J. Pryor and U. R. Evans, *J. Chem. Soc.* 1949, p. 3330; 1950, pp. 1259, 1266, 1274.
4.   C. W. Mason, *J. phys. Chem.* 1924, **28**, 1233.
5.   R. C. Gale, *J. Soc. chem. Ind.* 1924, **43**, 349T.
6.   U. R. Evans, 'Corrosion and Oxidation of Metals' (Arnold), 1960, p. 20.

to glass; oxide-films were transferred from nickel to plastic; in each case, the film after transfer presented a colour roughly complementary to that produced when it had been attached to the metal; an oxide-film which had appeared red when attached to nickel became green after transfer to plastic. This would be expected on the interference explanation, but not if the colour was due to some internal structure.

Complicating Factors. Later work suggested that the situation was not always simple. As stated above, Vernon showed that the colour was not necessarily a simple function of the oxygen uptake; a bright purple was produced on heating iron for 3 hours at 225°C which gave a weight increase of 0·30 mg/cm², whereas at 195°C purple started to appear only after 69 hours, the weight increase being then 0·66 mg/cm². A similar discrepancy was obtained by Miley.[1] It could be explained in more than one way. If heating at one temperature was to give an oxide-film of uniform thickness, whilst at another temperature the oxidation tended to spread inward along paths of weakness (e.g. grain-boundaries), such internally formed oxide sheaths would contribute nothing to the colour but would involve weight increase; thus a larger weight increase would be needed to produce a given colour than if the film were of uniform thickness. Conversely, if oxidation were to start at certain points on the surface and spread laterally, a smaller weight increase would be needed than under conditions producing uniform thickness. On the other hand, it might well be the case that oxide formed slowly at low temperatures really did possess different optical properties to that formed more quickly at higher temperatures

Proposed use of electrodeposited films. One way of deciding whether the colour is decided essentially by the thickness or whether the conditions of formation can modify colour might be provided by the study of colours produced by electrodeposited films, since here penetration inwards along grain-boundaries or other favoured paths would be unlikely. The film substance can be deposited slowly at low current density or quickly at high current density, and the colours compared.

It was found possible to deposit on a metal surface a film of molybdenum dioxide by cathodic treatment in a solution of ammonium molybdate; the composition seemed to depart from the formula MoO_2, as shown in Appendix II, but the term 'dioxide' will be used for convenience. The colours produced altered with film increase in the same way as when a film was formed by heating the metal in air. It is possible to vary the CD without altering the colour sequence.

There is another advantage in using electrodeposition films for the study of colours; it is possible to compare the appearance of the *same* film laid down on different metals; where a film has been formed by heating the metal in air, the comparison is between two *different* films. If, for instance, the film of nickel oxide produced by heating nickel in air is found to differ in appearance from the film of iron oxide produced by heating iron, it is impossible to know whether this difference is due to the substrate being different in the two cases

1. H. A. Miley and U. R. Evans, *J. Chem. Soc.* 1937, p. 1298.

or to the film substance being different. No such doubts arise if the film substance is the same.

Exploratory Experiments

Sequence of Colours on different substrates.   In early work on colours formed by heating in air, the whole sequence could be obtained on a single specimen simply by heating it at one end, so as to maintain a temperature gradient. When the films are to be deposited electrically, arrangements must be made either for the time of plating to increase progressively on passing from one end to the other, or for the CD to increase progressively. The *Emergence Method* described in Appendix I combines these two principles— a plan which obtains suitable spread of the colours. In most cases, the sequence is identical with that obtained previously for iodide-films on silver[1] and for oxide-films on nickel,[2] except that on many specimens the 5th or 6th order colours are not obtained, the appearance becoming grey prematurely. This is not surprising; to obtain the late-order colours requires very uniform deposition. The complete sequence is given in Table VII.

TABLE VII
SEQUENCE OF INTERFERENCE TINTS

Yellow, Orange, Mauve, Blue	1st order
Silvery Hiatus	
Yellow, Red, Blue, Green	2nd order
Yellow, Red, Green	3rd order
Red, Green	4th order
Red, Greenish grey	5th order
Reddish grey	6th order

This colour sequence was obtained, with only slight modification, by deposition of MoO_2 on the following substrates:

Rolled Mild Steel,* as rolled
Rolled Mild Steel, abraded

1.   U. R. Evans and L. C. Bannister, *Proc. Roy. Soc.* (*A*) 1929, **125**, 370.
2.   U. R. Evans, *Iron Steel Inst. Corr. Comm.*, 5th rep. (1938), p. 225.
  * The steel contained 0·065% C, 0·032% S, 0·015% P, 0·34% Mn, 0·08% Ni and 0·08% Cu.

Rolled Mild Steel, abraded and then electroplated with Nickel*
Rolled Nickel, abraded
Rolled Nickel, electroplated with Nickel
Rolled Nickel, plated alternately with Copper and Nickel (4 layers of each)
Steel, electroplated with Nickel and then with Copper
Rolled Copper, acid-cleaned but not abraded
Rolled Copper, plated with Nickel
Rolled Copper, plated with Mercury (early tints weak or absent, for reasons explained later)
Stainless Steel, abraded

Alloy AZ31, abraded [aluminium 3%, zinc 1%, rest magnesium].

One example of the 'slight modifications' mentioned above may be quoted; a difficulty was experienced in discerning first order tints on copper, owing to the colour of the metal; the second order colours produced by thicker films which obscure the metal were more conspicuous.

Interpretation of Exploratory Experiments. It will be noticed that the later part of the sequence shown in Table VII is not a strict repetition of the early colours. The reason has long been known, but a *résumé* of the argument may be convenient.

In the absence of complications, interference may be expected between the waves reflected at the outer and inner surface when the film thickness (y) is $\lambda/4n$, $3\lambda/4n$, $5\lambda/4n$, $7\lambda/4n$ etc. where λ is the wavelength, and n the refractive index. Thus a single thickness, y, will produce interference of light with wavelengths $4ny$, $4/3ny$, $4/5ny$ etc. Evidently the values representing the 1st, 2nd, 3rd and 4th interference bands are not spread out at equal intervals. As the films thicken, the first band will pass out of the visible spectrum (roughly an octave in width) at the red end before the 2nd enters at the violet end (Fig. 42, stage E). Thus a silvery hiatus is to be expected between the 1st and 2nd orders of colour, and is indeed obtained. However, the 3rd band follows the 2nd more closely, and will enter the violet before the 2nd has passed out at the red end (stage I). Thus a green, absent at the end of the first order, is conspicuously present at the end of the second order.

Interference cannot destroy light energy. There is a phase change at the internal reflection (CDG in Fig. 44) so that, if the thickness of the film is such as to cause the light following the path ABCDE to be *exactly* out of phase at D with that which would follow the path FDE, so that the light emerging along DE will be deficient in the wavelength in question, the internally reflected light DG will acquire an excess intensity of the same amount. This

* The nickel was deposited from a bath quoted by W. Machu in his 'Moderne Galvanotechnik', 1954, p. 241 (Verlag Chemie); it contains 150 g/l $NiSO_4.7H_2O$, 22 g/l $MgSO_4.7H_2O$, 10 g/l H_3BO_3 and 6 g/l NaCl; the content of metallic impurities was kept low by the use of 'Analar' nickel sulphate supplied to a specification demanding as maximal limits 0·003% Cl, 0·0005% Co, 0·002% Fe, 0·0025% Zn, 0·04% Pb (and other heavy metals) and 0·2% alkali metals as sulphates.

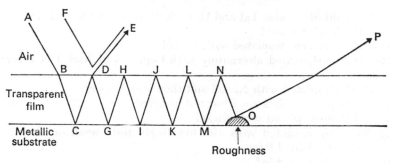

Fig. 44   Diagram showing why a thin film will only produce interference tints if there is some imperfection of a kind which will get rid of the imprisoned wavelengths. These may be either (1) imperfect transparency of the film substance, (2) imperfect reflectivity of the metallic base or (3) imperfect smoothness of the surface. Of these (1) and (2) will convert the imprisoned light energy into heat, whilst (3) will reflect it at an irregular angle, so that the colour as seen by scattered light is roughly complementary to that observed by regularly reflected light.

light will be reflected back along DG, and then along GH; it will meet at H more impinging light. If the wavelength is *exactly* that suited to cause maximal interference between FDE and CDE, the excess light will not escape at other points (D, H, J, L etc.) along DG, HF, JK and LM. If, however, the wavelength is *slightly less* or *slightly greater* than the value suited for maximal interference, a small amount will escape (along DE or some parallel course); if the film substance were to be absolutely transparent and the surfaces absolutely smooth, the *whole* of the light would ultimately escape; the number of reflections needed for (say) half of the light to escape would decrease as the wavelength diverged more and more from the interference value. Now in practice the film substance will *not* be perfectly transparent, and some of the light will be converted to heat; the amount thus transformed will be small if the wavelength is far from the interference value, but great when it is close to it, because the number of reflections needed for escape is greater; in other words, we shall get an interference band of finite width. However, the film thickness is also important. If the film is thick, then, for any given wavelength, there will be more conversion into heat, than if it is thin, because the path length between any two internal reflections is greater. It follows that the conversion into heat, leading to a deficiency of certain wavelengths in the reflected light, with consequent production of a colour sensation, may not occur to an important extent for the early (thin) films. In such cases the main mechanism for getting rid of the interfering wavelength may be reflection at an irregular angle if the inner interface of the film is rough. In Fig. 44, this is shown as occurring at O; the imprisoned light is escaping along OP. If the film substance is highly transparent, and if the surface is smooth and reflecting (humps like that at O being absent), there may be no way of preventing the light from emerging, except for wavelengths very close to the interference value; in other words, the interference band will be very narrow, and practically no colour sensation will be produced—doubtless the reason why the *early* colours, corresponding to very thin films, are very weak for films deposited on mercury, where the surface will be smooth. This is found to be the case.

It becomes evident that if the transparency of the film, the reflectivity of the metal and the smoothness of the surface are all nearly perfect, the interference band will be narrow (Fig. 45(A)) and the colour weak or absent. Any

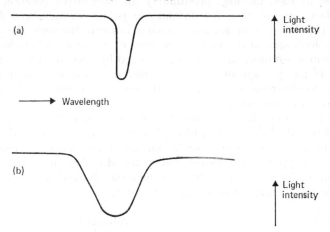

Fig. 45   Variation of breadth of interference band will conditions. (A) Narrow interference band obtained when the film substance is highly transparent, basis metal highly reflective and surface very smooth. (B) Broad interference band obtained when either transparency, reflectivity or smoothness is less perfect

lack of transparency of the film-substance, reflectivity of the metal and smoothness of the surface will broaden the band (Fig. 45(B)), and lead to more intense colour; this again is found to be the case.

Fig. 44 also explains why wavelengths missing from the regularly reflected light are often found in the scattered light (OP). Wood[1] showed that a collodion film formed on a bright silver surface which was of such a thickness as to produce a purple sensation by regularly reflected light appeared green by scattered light. His films of collodion on silver, which in some cases produced bright colours, in other cases produced no colour at all. This was due to the fact that the colour depended on a microscopic frilling of the collodion, providing a possibility for escape of the imprisoned wavelengths at an irregular angle. Frilling generally occurred in cases where the ether (used to dilute the collodion) had contained a little alcohol (water also helped frilling). If pure, dry ether, obtained by fresh distillation, had been used, there was no frilling and consequently no colour.

Study of Anomalous Cases

Failure to produce Colours on Aluminium.   Strips of aluminium in the as-rolled condition, subjected to the *Emergence Method*, failed to produce

1. R. W. Wood, *Phil. Mag.* 1904, **7**, 376, esp. p. 385; also 'Physical Optics' (Macmillan), 1934, p. 206.

any colour; very weak colours were obtained with abraded aluminium. The reason seems to be that the oxide-film on aluminium is a bad electronic conductor. Electrons are needed for the cathodic reduction of $MoO_4{}^{2-}$

$$MoO_4{}^{2-} + 4H^+ + 2e = MoO_2 + 2H_2O$$

Now when the aluminium is being treated cathodically in a molybdate bath, current is, in fact, flowing; presumably an alternative reaction must be taking place—except at a few favourable points on an abraded surface. Possibly this consists of protons moving inwards through the film, and producing hydrogen in the metal. The idea of a proton space charge in an alumina film is not new, having been discussed by several authors quoted by Tajima.[1] If the passage of current is related to cations moving inwards instead of to electrons moving outwards, no regular film will be produced, and consequently no colour.

Attempts to deposit colour films of molybdenum dioxide on as-rolled nickel led to only feeble tints, although satisfactory tints were obtained on the same nickel in the abraded condition, and also on nickel electrodeposited on a steel or copper basis; possibly the as-rolled nickel carries a thin film of some oxide which is a bad electronic conductor, derived from a minor constituent in the metal; magnesia suggests itself.

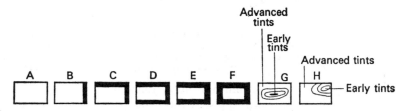

Fig. 46   Appearance of advanced colours along the edges of the exposed area on masked specimens. When the film is thin (A), the colour is usually uniform, but with gradually increasing thickness, breakdown along the edges of the area leads to abnormally advanced colours along 1, 2, 3 or 4 edges (B to F). If the abnormal colours move inwards uniformly from all four edges (starting at the same time), the final pattern of colours will be symmetrical (G), but often the movement is not uniform and an asymmetric pattern (H) is produced.

Attempts to produce uniformly coloured specimens. Using the method described in Appendix I, masked specimens with an area left bare (generally 1×1 cm, but sometimes 2×1 cm) were coated at a definite current for different times, and acquired colours in the sequence shown in Table VII; short times produced yellow, rather longer times mauve, still longer times blue, then silver followed by second order yellow and red. The specimens coated for short times usually became uniformly coloured, but before the first order blue was reached a tendency was noticed for a more advanced colour to appear along an edge than elsewhere; thus

1.   S. Tajima, chapter in 'Advances in Corrosion Science and Technology' edited M. G. Fontana and R. W. Staehle (Plenum Press), Vol. I (1970), p. 229, esp. p. 319.

when the main area was still mauve, a narrow blue line might be seen along an edge. At first, this generally happened only along one edge (Fig. 46(B)), but on longer treatment (giving a thicker film) two edges (either adjacent or opposite) developed abnormally advanced colours (Fig. 46(C) and (D)); with still longer treatment (Fig. 46(E) and (F)) three or four edges might develop advanced colours, so that occasionally the square area showed an advanced colour all round the edge, with a less advanced strip inside it and a rude square of the least advanced colour in the centre (Fig. 46(G)). Sometimes the arrangement of the colours might be asymmetric (Fig. 46(H)), probably because the development of advanced colours had started on the four edges at different times. Occasionally patches of advanced colour appeared in the centre, but that was unusual.

Causes of lack of Uniformity. It may appear surprising that uniform tinting was more easily obtained after *short* exposure, when irregularities accompanying the introduction of the specimen into the bath might be expected to produce greater disturbance, than after *long* treatment. The most obvious explanation for the more advanced colour along an edge would be based on exhaustion of molybdate in the centre; replenishment would be expected to be most thorough at the edges. If the cause was exhaustion, uniformity of tinting should be improved if the molybdate concentration was increased; two sets of experiments carried out at concentrations in ratio 5 : 1 did not show the expected improvement. Similarly, if the cause was exhaustion, an increase of CD should make matters worse; experiments did not show this. Clearly the non-uniformity observed cannot arise from local exhaustion.

The true explanation appears to be a *breakaway* due to stress in the film substance. This will occur preferentially at the edges, favouring increased deposition here, and consequently (if the total current flow is fixed by the approximately galvanostatic conditions) unexpectedly slow deposition at the centre. This would explain all the results observed, and involves no *ad hoc* assumption, since a breakaway of a film carrying internal stresses may be expected on simple energy principles and has indeed been observed often in practice. If, for instance, a film carries strain energy, this will be diminished if it becomes detached (except at a few points) so that it can assume an approximately unconstrained condition. If the strain energy per unit volume is ε, that per unit area will be $y\varepsilon$, where y is the film thickness. Until $y\varepsilon$ exceeds W, the adhesional work per unit area, spontaneous detachment is impossible. When the thickness y exceeds the critical value W/ε, detachment in the absence of applied stress becomes possible without conflicting with energy principles; in practice, however, it is unlikely to start until W/ε has been exceeded by a certain margin representing nucleation energy related to the interfacial energy of the first stable sub-microscopic cavity between film and metal. The amount of interfacial energy involved will be smaller at the edge than in the centre (if the conditions called for a spherical cavity at the outset in the centre, this would be a hemisphere at the edge; although in fact the shape will be different, it is fairly certain to be such as to require less nucleation energy at the edge than in the centre). It is thus easy to explain the more advanced colour at the edge.

The fact that breakaway, in absence of applied stress, occurs when the thickness has reached a certain value, is well known from work on the oxidation of zirconium by Wanklyn and Jones,[1] also by Coriou.[2] The film produced by high temperature water suffers breakaway on reaching a certain thickness; the time needed to produce that thickness varies with the purity of the material and the experimental temperature, but the breakaway thickness is much the same in all cases.

There is no doubt that internal stresses do exist in the film, which at great thicknesses corresponding to the end of the colour sequence often peels off altogether—such peeling will allow nearly complete release of the strain energy; this detachment would be expected to occur most readily if there is either gas or liquid under the film; in fact it is met with where the substrate is either magnesium alloy, which evolves hydrogen, or alternatively mercury; these cases are considered later. No *ad hoc* assumption, therefore, is made in advancing this explanation for the appearance of advanced colours around the edge.

The lack of uniformity within each specimen is accompanied by lack of reproducibility between the colour pattern produced on different specimens after apparently identical treatment. The breakdown along points on the edge is a probability effect, denoted by the expected number of breakdown points per unit length of edge; the expectation will increase with the mean film thickness. If, at a certain thickness, it is 0·5 per cm, the 'expected' number on 4 cm (the total length of the 4 sides on a 1 cm square) will be 2. This does not mean that on all the specimens two edges (only) will show advanced colours. The breakdown points on a long edge would not be spaced out regularly at 2 cm intervals, but distributed more or less randomly. If breakdowns were events without influence on one another, so that a Poisson distribution could be expected, then 27% of the specimens would show two breakdown points, but 14% would show none, and 9% would show four breakdowns (which, however, does not always imply advanced colours on all four edges, since sometimes two of the four might be on the same edge). In fact, however, mutual influence cannot be excluded, for a number of reasons; for instance, breakdown on one edge might spread round the corner to the next. The numbers quoted above would need correction for that and other reasons, but it is certain that behaviour will differ from one specimen to another, even though the treatment has been identical; this is found to be the case.

The situation is unfortunate, since it means that only the early part of the colour sequence can be used to answer the question posed at the start of this chapter. Furthermore the use of the thinner films (obtained by short times of treatment) introduces experimental error. As explained later, the adoption of an orange tint as standard represented the best compromise between requirements; films of that thickness are sufficiently thick to avoid excessive

1.   J. N. Wanklyn and P. J. Jones, *J. nuclear Mat.* 1962, **6**, 291.
2.   H. Coriou, *Corr. Sci.* 1961, **1**, 132.

errors associated with short experiments, but insufficiently thick for serious risk of breakaway.

General Conclusions

Relation between Thickness and Colour. The irreproducibility just discussed prevents exact tabulation of the colours corresponding to different thicknesses such as has been published for oxide-films and iodide-films, but a rough indication can be given. In the discussion which follows, the 'units' of thickness represent millicoulombs per cm^2, but these could be converted approximately into micrograms per cm^2 by means of the relationship provided in Appendix II.

When the substrate was mild steel sheet in the as-rolled condition (degreased in xylene but otherwise untreated), yellow persisted up to about 12 units, then becoming orange, but at 15 units the film was definitely mauve; the red tint disappeared with increasing thickness and at 26 units the colour was blue, becoming paler with increasing thickness. Around 50 units the effect was very pale blue or silver, but about 53 units a yellow line appeared along one or more edges, the interior being silver or very pale blue. In this range, uniformity and reproducibility became poor. About 105 units specimens showed green edges around a mauve centre, but this centre appeared green by scattered light; the green at the centre is probably the effect of missing wavelengths escaping by irregular reflection, as in Wood's experiments with frilled collodion films. The green at the edges, produced by thicker films viewed at the angle suitable for regular reflection, arises from a different cause—namely, breakaway. Third order yellow and sometimes red is seen round the edges in specimens receiving 110 units, with the centre second order green.

If the mild steel had been abraded with 1F emery before being coated with MoO_2, the changes from one colour (or pattern of colour) to another were observed at thicknesses differing little from those mentioned above for an as-rolled surface. It seemed that a few more units were needed to obtain a given tint. When, however, the steel was plated with nickel after abrasion and then coated with molybdenum dioxide, a definitely larger number of units were required for any given colour; seven specimens coated at 25 units were still orange. Once a standard had been chosen, it became possible, for each substrate, to decide the number of units needed to give a colour match. It was found that mild steel, abraded and plated with nickel, and then treated in ammonium molybdate for 25 units, produced a bright orange tint which was generally reproducible; occasionally a freak specimen gave a different tint, but such occasional breaks were to be expected on probability principles and were disregarded. Most of the specimens studied which had received identical treatment gave the same colour; moreover specimens in which the thickness of the nickel was different or specimens in which the CD used was different matched one another closely—provided that the millicoulombs per cm^2 was the same. Such specimens provided a suitable standard tint with which other specimens, based on different substrates, could be matched. Table VIII shows the number of units needed to produce the standard tint.

TABLE VIII

UNITS NEEDED FOR STANDARD TINT ON DIFFERENT SUBSTRATES
(U. R. Evans)

	Units (millicoulombs per cm²)
Mild Steel, abraded and plated with Nickel (standard)	[25]
Mild Steel, as rolled	14
Mild Steel, abraded	18
Mild Steel, plated with Iron	18
Copper (acid-washed, but not abraded), plated with Nickel	20
Nickel (not abraded), plated with Nickel	20
Stainless Steel, abraded	19

No great significance should be attached to these numbers as exact values Doubtless a mild steel from a different source and tested in the as-rolled condition would have yielded a different value. Abrasion under different conditions would also have affected results. Even among the samples cut from the same material, there was incomplete agreement. In some cases matching was rendered difficult by slight differences in the character of the colour. But a study of a large number of specimens (altogether about 800 were prepared for the research) showed that without doubt the number of units needed to match the standard depended on the substrate. The abraded steel carrying no nickel definitely required a smaller number of units than the same surface plated with nickel; the difference observed cannot be explained away by arguments based on poor reproducibility.

Other experiments showed that the colour was changed not only when the substrate was altered but also when the external phase (normally air) was replaced by some other transparent medium. When the specimen was wetted with a thin film of water, the colour was slightly different from that observed when the specimen was dry. On applying a layer of polystyrene lacquer outside the colour-film, there was a marked change, although the lacquer was in itself colourless and perfectly transparent; the original colour returned when the lacquer was removed by washing in xylene; in certain experiments, one half of the colour square was coated with lacquer, but on subsequently washing with xylene and drying it was impossible to detect any difference between the two halves (which had respectively been coated and left uncoated).

Causes of the Variation of Colour. The fact that the colour depended on the adjacent phase (whether substrate or external phase) is probably connected with the 'specific phase change' produced at a 'surface of reflection'— a matter studied by Winterbottom[1] and also, on anodized tantalum, by Charlesby and Polling.[2] The fact that abrasion alters the number of units required to match the standard is not surprising. At first sight it may seem surprising that the effect is not greater; it might perhaps have been thought

1.   A. B. Winterbottom, *Trans. Faraday Soc.* 1946, **42**, 487.
2.   A. Charlesby and J. J. Polling, *Proc. Roy. Soc.* (*A*) 1955, **227**, 434.

that, if the true area were (say) doubled by roughening, it would halve the difference between the paths of light reflected at the outer and inner surface of the film, and that the amount of material deposited would have to be doubled to obtain a colour match. This, however, neglects the fact that, on a rough surface, much of the light will impinge at an oblique angle even if the illumination is arranged to be normal to the general plane of the surface; where oblique impingement occurs, the path difference is greatly increased, especially when the refraction index is not too high.* Exact calculation would need knowledge about the profile of the rough surface, and would be a formidable task; but a little thought will show that the increase in the material needed to match the colour produced on a smooth surface will be much *less* than proportional to the increase of surface produced by roughening.

Effect of Gas or Liquid below the film. The study of colours on the magnesium alloy, AZ31, introduced fresh factors. When the CD used was of the order previously found to produce bright colours on iron and nickel were tried (say, 1 mA/cm^2), no bright colours were obtained, merely a faint indefinite yellowish grey, darkening to brown and then to sepia as the time of treatment was increased. Probably local action was proceeding; at some points, the magnesium was suffering anodic attack whilst at others molybdenum dioxide was being deposited by cathodic reduction—but unevenly so that no uniform film was produced and a definite colour sensation was not achieved. Local anodic attack might be smothered by a high CD, and it was found that very bright colours could be obtained with reasonable reproducibility at 8 or 9 mA/cm^2; if this was applied for sufficient time to give 60 units, a rose colour could always be obtained, which was remarkably uniform immediately after formation. However, the current was found to fluctuate wildly to an extent which the galvanostatic arrangements failed to prevent. Evidently, hydrogen was being evolved, and, since part of the current was being used up in hydrogen production, a relationship between millicoulombs and colour—such as had been worked out successfully for steel—could mean nothing in this case.

The tendency for the alloy to evolve hydrogen had other consequences; the thicker films peeled completely, to an extent not noted on iron or nickel, whilst the specimens carrying thinner films underwent spontaneous colour changes which implied no change of thickness but merely cessation of optical contact between film and substrate. For instance, the rose colour produced at 60 units generally remained rose, so long as the specimen was covered with a thin layer of water. If dried without precaution, it sometimes remained rose, but more often, after a few seconds, bright green appeared at certain points, and sometimes spread so that the whole area became green; sometimes the final effect was a patchwork of rose and green, often producing a dull orange sensation when viewed by the naked eye. The green almost certainly represented parts where the film and substrate were separated by a

---

* It can be shown that light impinging at a glancing angle will experience a path difference $n/\sqrt{(n^2-1)}$ times that corresponds to normal impingement; the multiplier becomes infinite when n, the refractive index, becomes 1·0.

o

thin layer of hydrogen (doubtless later replaced by air). Endeavours were made to avoid the colour change by conducting the washing operation in ice-cold water; these were generally successful, and the rose was preserved, although on some specimens there was a thin green line along an edge, where detachment would be easiest; sometimes green patches appeared even in the cold water. The irreproducibility of hydrogen evolution on small specimens owing to its preferential occurrence on highly dispersed sensitive spots is well known.[1] When warmed water (30–50°C) was used, green invariably appeared; generally the final result was a patchwork of red and green.

A liquid phase below the film would also be expected to favour detachment, although less effectively than gas; this was found to be the case. Copper specimens coated with mercury (which would contain about 0·002% copper in solution, assuming equilibrium to be attained) developed bright colours when treated cathodically in molybdate solution; here again the thicker films (produced after long treatment) peeled completely. In a very few cases, spontaneous colour changes after drying were noticed, but these could not be repeated at will. In one case, a specimen was coated to give a rose colour, and washed with water. From one half of a 1 × 1 cm square, the water was removed by absorption into a strip of filter paper applied normally to the surface; this left the other half still covered with a puddle of water which was left to evaporate. When the whole was dry, the first half was found to have become green, whilst the second half remained rose, but had developed a network of cracks revealing the silvery substrate. The explanation is simple; the suction of water from the first half had lifted the film just out of contact with the mercury, so that contraction and strain relief became possible without cracking, but the colour became green (for the same reason that a red nickel oxide film becomes green when transferred from nickel to celluloid[2]). On the other part, the strain had been relieved by the formation of cracks, optical contact being maintained, so that the rose colour survived.

Effect of CD on Colour.   In order to answer the question posed early in this chapter—as to whether a given quantity of film matter produces the same colour independently of the rate of deposition—experiments were carried out using 25 units deposited on nickel plated abraded steel; as stated above, the colour produced on a number of such specimens coated at the same CD generally showed good matching, although occasionally a freak was met with. Having established this fact for small areas (ideally 1 × 1 cm, coated for 25 seconds at 1·0 mA), other specimens with ideally 2 × 1 cm area exposed were coated at the same current for twice the time (in practice a small time correction was applied to compensate for the fact that the areas, as measured, differed slightly from the ideal 1·0 × 1·0 or 2·0 × 1·0 cm). If correctly adjusted, both experiments should provide 25 units (25 milli-coulombs/cm²). Comparison between the two sets showed a satisfactory colour match.

1.   For summary and references, see U. R. Evans and C. A. J. Taylor, *Corr. Sci.* 1972, **12**, 227, esp. Appendix on p. 243.

2.   U. R. Evans, *Iron Steel Inst. Corr. Comm.*, 5th rep. (1938), Table 124, p. 232.

The method just adopted has the advantage of being independent of the accuracy of the milliammeter. Other experiments were carried out with the same bare area in the two sets, but with different currents. Similar conclusions were reached. Work was also carried out on other substrates; the lack of reproducibility rendered conclusions rather less certain, and statistical studies upon a very large number of specimens would be needed before it could be stated with certainty that the CD had no effect; but it is certain that the effect is too small to be interesting for the purpose of the present discussion. The marked differences in coloration noted between films formed by heating iron to a certain weight increase above and below 200°C—observed by Vernon, as already mentioned—are not met with when a film is produced by electrodeposition.

Use of colours in following film thickening. Colour changes have long been used in studies of oxidation and corrosion. Qualitatively, the method can give useful indications, although even here caution is needed in interpreting the observations. If, for instance, the cathodic area on a corroding specimen is observed to develop colours in the sequence yellow, mauve, blue, it is fairly certain that a film is being formed which is progressively thickening. But even in such cases, there are possibilities of misleading conclusions. When, for instance, a set of specimens has been coated with grease or lacquer for purposes of preservation, colour changes have been observed which clearly did not indicate any alteration of thickness. Again, when films have lost optical contact with the metal basis, sharp changes of colour have been observed which indicated no change of thickness.

Measurements of thickness based on colour have been used at times. The pioneer work of Tammann[1] on the laws of film-growth was based on the belief that the thickness of an oxide- or iodide-film on a metallic basis could be assessed by matching it with the colour given by an air-film between two glass surfaces close together (as in the observation of Newton rings); by dividing the air-thickness needed for the corresponding colour (of the appropriate order) by the refractive index of the film substance, a value for the thickness of the oxide- or iodide-film could—it was thought—be obtained. Even in the absence of a specific phase change at the interface, results thus obtained would be slightly in error, since the division would give merely the thickness at which the *centre* of the absorption bands of the two films would fall at the same wavelength. But in fact the colour depends not only on the position of the centre of the band but also on the proportion of light lost at wavelengths situated on each side of that wavelength of maximal interference; since the band produced by the air-film will be much wider than that produced by a film on metal, the values obtained by this method will not be precise. However, a greater discrepancy is introduced by the existence of the specific phase jump, which Winterbottom states to be approximately equivalent to 220 Å for a Cu/Cu_2O interface, 150 Å for Fe/Fe_2O_3 and 160 Å for Al/Al_2O_3.

A better method was used by Bannister (see p. 381) in his study of the

1. G. Tammann, *Z. anorg. Chem.* 1920, **111**, 78.

growth of iodide-films on silver. Specimens of silver were exposed to a solution of iodine in an organic solvent for different times, and the thicknesses measured by four different methods: (1) the weight increase obtained on a micro-balance, (2) the number of millicoulombs needed to reduce the silver iodide to silver, (3) the amount of iodine estimated by nephelometry, (4) Tammann's method, described above. The fourth method, as was to be expected, gave results at variance with those obtained by the first three methods, which yielded values in good agreement with one another. It was concluded, therefore, that these represented the correct thicknesses, and accordingly a set of standards was prepared representing known thicknesses. In the experiments designed to establish the growth law, specimens were immersed in the iodine solution for different times, and their colours compared with those of the set of standards; in that way, the thicknesses of films produced after different times were obtained much more conveniently than if every specimen had to be weighed before and after exposure. The thickness measurement, of course, depended ultimately on the micro-balance, which even at the time of that early research was a reliable instrument (it has since been improved). It is believed that, for the purpose of the research, the method adopted was appropriate and accurate.

However, limitations exist to the use of methods based on colour matching with previously prepared standards. There are ranges where the variation of colour with film thickness is small; here the method will be insensitive. Very thin films (such as are produced on metals exposed to air at room temperature—a case of special interest) produce no colour at all, the interference band being situated beyond the violet end of the visible spectrum; here, obviously, the method is inapplicable.

Alternative Methods.   For both the ranges mentioned, other methods are available. For relatively thick films (in the range where colour is produced), accurate measurements of the wavelengths at which the light reflected from the film-covered surface is minimal can be made by means of a spectrophotometer. This method was used by Charlesby and Polling (p. 389), whose observations of the bright colours on anodized tantalum showed that the phase jump on reflexion at the metal surface is important and varies greatly with wavelength, being 0·2 radian at 2700 Å but rising to 0·8 radian at 8000 Å; by the aid of these constants, it was possible to calculate the absolute thickness of the film obtained at any voltage.

For thin invisible films the method developed by Tronstad[1] and Winterbottom (p. 389), based on polarized light and now known as 'ellipsometry', is probably unequalled. It has many applications, some of them still unexplored. The investigation of films responsible for passivity carried out by Kruger and others,[2] in which ellipsometric methods have played a large part, deserves special mention. For the early stages of the formation of invisible films on metals exposed to oxygen at ordinary temperatures, which proceeds very rapidly, methods capable of providing measurements of thicknesses

1.   L. Tronstad, *Trans. Faraday Soc.* 1935, **31**, 1151.
2.   C. L. M. Bee and J. Kruger, *Nature Phys. Sci.* 1971, **230**, 194.

developed in very short times possess interest; several laboratories are active in producing suitable instruments, and possibly some of the work still unpublished may best provide what is required. Reference may be made, however, to the procedure evolved in Bockris's Laboratory[1] giving measurements down to 0·01 s.

Use of Interference Tints for producing Decorative Finishes

General. Methods for colouring films on titanium and stainless steel have been worked out and are in industrial use; there are aesthetic possibilities.

Tints on Titanium. The anodizing of titanium produces films of thickness which depends on the EMF; thus by controlling the voltage, the desired colour can be obtained. The process, which has produced results of artistic merit, is due to Cotton and Hayfield.[2] Anodizing is usually carried out in H_2SO_4 or H_3PO_4, but oxalic acid and various salts such as NaF or KF have been used. The relationship between voltage and colour in H_3PO_4/H_2SO_4, along with the thicknesses obtained by ellipsometry, is shown in Table IX.

TABLE IX

COLOURS PRODUCED ON TITANIUM (J. B. Cotton and P. C. S. Hayfield)

Formation voltage	Thickness Å	Colour
0 to 2	15 to 25	silver
4 to 6	105 to 132	gold just detectable
8 to 10	160 to 180	pale gold
12	220	rich gold
14	242	dark gold
16	254	gold with purple tint
20	272	purple with gold tint
24	364	dark blue
28	469	middle blue
32	715	pale blue

Tints on Stainless Steel. The process for colouring stainless steel, due to T. E. Evans and his colleagues,[3] requires no current in the first stage which leads to the production of the tints, although a cathodic treatment is used for the second (hardening) stage; here also attractive results have been obtained. Immersion in CrO_3 (250 g/l) and H_2SO_4 (490 g/l) at 70°C is followed by a hardening process based on cathodic treatment in CrO_3 (250 g/l) and H_2SO_4 (2·5 g/l) at 2·5 amp/dm^2 for 15 minutes. This greatly improves the mechanical properties, so that 200 rubs with a pencil eraser loaded at

1.   J. O'M. Bockris, M. A. Genshaw and V. Brusic, *Symp. Faraday Soc.* 1970, **4**, 177.
2.   J. B. Cotton and P. C. S. Hayfield, *Trans. Inst. Met. Finishing* 1967, **45**, 48.
3.   T. E. Evans, A. C. Hart, H. James and V. A. Smith, *Trans. Inst. Met. Finishing* 1972, **50**, 77.

500 g are needed to penetrate the film, in a case where the unhardened film can be removed by 2 to 4 rubs. The colour produced during immersion, being determined by the thickness, depends on the time of treatment. Blue is obtained after 15 minutes, gold after 18 minutes, mauve after 20 minutes and green after 22 minutes. Various precautions, such as careful degreasing both before and after the immersion process, are necessary.

A later paper[1] provides information regarding the nature of the coloured film, which is porous and strongly hydrated; apparently it possesses a spinel structure. The alloy suffers anodic attack at the base of the pores, where the potential lies in the trans-passive range. This provides Fe, Cr and Ni ions, whilst further Cr^{3+} is produced by cathodic reduction of $Cr_2O_7^{2-}$ on the outer surface of the film; the cathodic reduction involves the consumption of H^+ ions, so that the pH rises to a point where solid will be precipitated. The film contains Cr, Fe and Ni; the Cr content is higher than the Fe content, being clearly derived in part from the bath. One analysis showed 19·6% Cr, 11·7% Fe and 2·1% Ni by weight.

1. T. E. Evans, A. C. Hart and A. N. Skedgell, *Trans. Inst. Met. Finishing* 1973, **51**, 108.

Methods of Deposition

Plating Cell.   For plating with metals a rectangular glass cell of 5×3 cm area and 7 cm high was used, it was generally filled to a depth of 4 cm. The anode was a stainless steel gauze clipped vertically above the water-line to one of the 7×5 cm walls; the specimen to be plated was made the cathode and was held close to the opposite wall.

For the deposition of molybdenum dioxide a narrower cross section was required; a similar rectangular glass trough was used but rubber blocks were introduced to reduce the effective liquid space to $3 \cdot 0 \times 1 \cdot 1 \times 4 \cdot 0$ cm; the two vertical electrodes were 3 cm apart, and the depth of the liquid was 4 cm. A bath consisting of fresh liquid seemed to behave somewhat irregularly, and a few dummy specimens were generally coated and discarded before proceeding to the experiments proper; the necessity for this preliminary treatment of the solution was probably connected with the establishment of a steady state at which the rates of formation and destruction of various ions representing different valencies becomes equal.

Current was provided by a relatively high EMF applied through an extremely high resistance. In many of the experiments, a 9 volt battery applied through a resistance of the order of 10^4 ohms provided the required small current (1 or 2 milliamps); the main part of the resistance was provided by a horizontal strip of filter paper soaked in $M/100$ copper sulphate solution with two pieces of copper gauze pressed down at points sufficiently distant to obtain the required resistance. A variable wire resistance, connected in series, served to provide fine adjustment. The arrangement constituted a sufficient approach to a galvanostatic system, except in the coating of the magnesium alloy AZ31 where the current fluctuated wildly, as already mentioned. In other cases, small variations of current with time could be controlled by means of the wire resistance; but in cases where the application of a given number of millicoulombs was desired, this could be more conveniently obtained by varying the time than by altering the current.

Preparation and Treatment of Specimens.   The sheet was generally cut into rectangles 7×5 cm, which were degreased in xylene. In some cases, notably on mild steel, the surface was abraded with emery paper, and then again cleaned in xylene and wiped to remove powder particles. The copper rectangles were not abraded but were washed in the $H_2SO_4/CuSO_4$ mixture already mentioned.

Strips of filter paper, 1 cm or 2 cm broad were stretched across what was to form the 'front' side of the 7×5 cm rectangle, folded round the edges and secured at the back by means of adhesive zinc oxide surgical plaster (Fig. 47). The whole of the front, except the uppermost strip (which was left bare for contact with the terminal clip needed for current conveyance) and the 1 cm or 2 cm strip sheathed by the filter paper strip, was coated with polystyrene lacquer; after some hours, the lacquer was dry, and the second coat

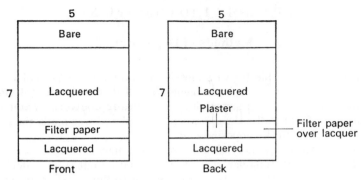

Fig. 47 Masking the greater part of the specimen surface with polystyrene lacquer, only a narrow strip (generally 5 × 1 cm) is left bare.

was applied. When this was dry, the filter paper stips were removed, and the whole of the back except the uppermost portion received two coats of polystyrene lacquer. If electroplating with nickel (or other metal) was called for, it was carried out at this stage; only the bare strips (5 × 1 or 5 × 2 cm) were plated.

The 5 × 7 cm rectangle was then cut into five strips each measuring 7 × 1 cm, and carrying bare areas 1 × 1 or 2 × 1 cm according to the breadth of the filter paper sheath; in some cases these areas would carry a coating of nickel or other metal.

Procedure for uniform colouring.   In cases where a uniform coating of molybdenum dioxide was required, the specimen was joined to the external circuit by flexible wire attached with a spring clip before being introduced into the liquid. At the chosen moment, noted on the watch, it was rapidly pushed down into position and held there until the watch indicated the correct moment for withdrawal. Thus the current flowed from the moment of immersion. The current was read at suitable moments on a milliammeter connected in series with the battery and resistance; in some cases two meters were used in series as a check.

Procedure for obtaining a sequence of colours.   When it was desired to obtain the whole sequence of colours on a single specimen, the *Emergence Method* was used. The resistance was adjusted so that the current flowing was about 20 mA when the cell was short-circuited. The specimen (often about 7 × 1 cm, but sometimes 7 × 0·5 cm), was plunged into the liquid to a depth of 4 cm, and immediately a slow withdrawal commenced; the specimen was first raised by about 1 mm per second, but later more slowly, so that emergence was completed in about 1 minute. After a few trials, it became easy to regulate the rate of withdrawal in such a way as to obtain a satisfactory spread of the colours representing the range shown in Table VII (p. 382); the 5th and 6th order colours were sometimes absent. The procedure could be varied if special information was desired about some part of the colour sequence. For instance, if there was special interest in the second order colours, the specimen would be plunged in and left fully immersed for such a time as would serve to reach the silvery hiatus over the whole surface. Then raising of the specimen would start, but more slowly than in the standard procedure, so that the second order colours would be distributed over the length of the specimen.

APPENDIX II

Relationship between Millicoulombs and Micrograms

Object. For three reasons it was desirable to establish the weight change corresponding to a film deposit at a definite number of millicoulombs.

(1) It was of interest to know which of the oxides of molybdenum was being deposited; exact agreement with a stoichiometric formula was not expected (the composition of oxide-films generally departs from the theoretical formula, owing to defect structure), but rough agreement was hoped for.

(2) It was right that readers should be enabled—should they desire it—to translate thickness expressed in millicoulombs per square centimetre (the units used in the main portion of this article) into micrograms per square centimetre.

(3) It was desirable to know whether the current efficiency varied greatly with conditions; variation would to some extent invalidate the arguments used.

Method. For the micro-balance experiments described below, masking of parts of the specimens (with lacquer) was found to be impossible, as there continued to be a loss of weight long after the lacquer coat appeared to be dry. Thus specimens were prepared with both sides bare; these were coated by introduction vertically into the bath; they were held at right angles to the anode surface whilst current flowed. Such an arrangement involved a non-uniform CD, so that the colours became gradually less advanced on passing from the front edge (nearest the anode) to the back edge; this produced a series of nearly vertical bands; for determining the relationship of millicoulombs to micrograms, uniformity of deposition is not needed. There was no evidence of a competing electrochemical reaction over the area near the water-line (for instance, cathodic reduction of O_2 competing with that of $MoO_4{}^{2-}$); such a competing reaction—had it occurred to any appreciable extent—would have rendered the method invalid.

Micro-balance Measurements. The weighings needed to establish the relationship between millicoulombs and micrograms were carried out by C. A. J. Taylor, using an Oertling balance. Specimens were prepared and kept in a desiccator over silica gel between weighings. After the first weighing they received cathodic treatments for different times and at different currents. After a period in the desiccator, they were reweighed. The average ratio of the weight-gain (expressed in micrograms) to the electricity applied (expressed in millicoulombs) was found to be 1·06.

If the film-substance is assumed to be an anhydrous oxide of fixed composition, the following equations deserve consideration:

Theoretical value of
micrograms/millicoulombs

$$2MoO_4^{2-} + 6H^+ + 2e = Mo_2O_5 + 3H_2O \qquad 1{\cdot}41$$

$$2MoO_4^{2-} + 8H^+ + 4e = MoO_2 + 4H_2O \qquad 0{\cdot}67$$

$$2MoO_4^{2-} + 10H^+ + 6e = Mo_2O_3 + 5H_2O \qquad 0{\cdot}42$$

The numbers suggest that the composition of the film does not correspond to a simple formula. Again, the fact that the films can be deposited in sufficient thickness to produce high-order colours suggests that the material possesses good electronic conductivity; such conductivity is usually associated with the co-existence of two ions of different valency (in magnetite, the high conductivity is generally ascribed to the co-existence of Fe^{2+} and Fe^{3+}). It seems probable that the film consists of MoO_2 carrying vacant cation sites. For convenience it has been called 'dioxide'. Definite knowledge of composition and structure would require an elaborate research. The question is not of vital importance. The validity of the argument does not depend on the composition of the film-substance.

Author Index

418

Young, J. F., 250
Yuno, T. K., 274

Zahavi, J., 162, 164
Zak, T., 311
Zamanov, B. A., 184
Zamin, M., 355
Zanker, L., 123
Zappfe, C. A., 219, 329

Zelenin, V. M., 93
Zelwer, A., 191
Zhigalova, K. A., 279
Zhukov, L. L., 54
Zimmerman, R. P., 115
Zirkl, A., 338
Zucchi, F., 92, 93, 115, 140, 154, 223, 274
Zucchini, G. L., 92, 93, 101, 164

Subject Index